T4-ABK-915

Passive Seismic Monitoring of Induced Seismicity
Fundamental Principles and Application to Energy Technologies

The past few decades have witnessed remarkable growth in the application of passive seismic monitoring to address a range of problems in geoscience and engineering from large-scale tectonic studies to environmental investigations. Passive seismic methods are increasingly being used for surveillance of massive, multi-stage hydraulic fracturing and development of enhanced geothermal systems. The theoretical framework and techniques used in this emerging area draw on various established fields, such as earthquake seismology, exploration geophysics and rock mechanics. Based on university and industry courses developed by the author, this book reviews all the relevant research and technology to provide an introduction to the principles and applications of passive seismic monitoring. It integrates up-to-date case studies and interactive online exercises, making it a comprehensive and accessible resource for advanced students and researchers in geophysics and engineering as well as for industry practitioners.

David W. Eaton is Professor of Geophysics in the University of Calgary's Department of Geoscience, where he served as Department Head from 2007 to 2012. He is presently co-director of the Microseismic Industry Consortium, a novel initiative dedicated to the advancement of research, education and technological innovations in microseismic methods and their practical applications for resource development. His current research is focused on microseismic monitoring and induced seismicity, intraplate earthquake swarms and the lithosphere–asthenosphere boundary beneath continents.

"It is now well established that human activities in the subsurface create induced seismicity. While large events can be extremely problematic from both a seismic hazard and operational safety perspective, smaller induced events, known as microseismic events, can tell us a great deal about changes in the subsurface. Eaton provides a clear and comprehensive description of such seismicity, starting from first principles and then progressively taking the reader through to real examples and case studies. A focus is on seismicity associated with oil and gas exploitation, but the book will also appeal to scientists interested in seismicity in a range of other settings (e.g., geothermal, mining, etc.). Eaton has crafted an excellent seismology text for students and earthquake seismologists in general."

– Professor Michael Kendall, University of Bristol

"*Passive Seismic Monitoring of Induced Seismicity* is a comprehensive textbook covering basic theoretical concepts of seismic and ancillary topics through to practical implementation in industrial settings. The book is an essential reference text on this topical technology."

– Dr Shawn Maxwell, IMaGE

"This comprehensive text is a much-needed and timely overview of topics related to seismic monitoring of induced earthquakes. It not only provides a thorough treatment of how microseismic monitoring is done and the data are analyzed, it provides a valuable overview of how and why injection-induced seismicity occurs."

– Professor Mark Zoback, Stanford University

Passive Seismic Monitoring of Induced Seismicity

Fundamental Principles and Application to Energy Technologies

DAVID W. EATON

University of Calgary

CAMBRIDGE UNIVERSITY PRESS

CAMBRIDGE
UNIVERSITY PRESS

University Printing House, Cambridge CB2 8BS, United Kingdom

One Liberty Plaza, 20th Floor, New York, NY 10006, USA

477 Williamstown Road, Port Melbourne, VIC 3207, Australia

314–321, 3rd Floor, Plot 3, Splendor Forum, Jasola District Centre, New Delhi – 110025, India

79 Anson Road, #06–04/06, Singapore 079906

Cambridge University Press is part of the University of Cambridge.

It furthers the University's mission by disseminating knowledge in the pursuit of education, learning and research at the highest international levels of excellence.

www.cambridge.org
Information on this title: www.cambridge.org/9781107145252
DOI: 10.1017/9781316535547

© David W. Eaton 2018

This publication is in copyright. Subject to statutory exception and to the provisions of relevant collective licensing agreements, no reproduction of any part may take place without the written permission of Cambridge University Press.

First published 2018

Printed in the United Kingdom by TJ International Ltd. Padstow Cornwall

A catalog record for this publication is available from the British Library.

Library of Congress Cataloging-in-Publication Data
Names: Eaton, David W., 1962– author.
Title: Passive seismic monitoring of induced seismicity :
fundamental principles and application to energy technologies /
David W. Eaton, Professor and NSERC-Chevron Industrial Research Chair,
Department of Geoscience, University of Calgary.
Description: Cambridge, United Kingdom ; New York, NY :
Cambridge University Press, 2018. | Includes bibliographical references and index.
Identifiers: LCCN 2017045854 | ISBN 9781107145252 (hardback) |
ISBN 1107145252 (hardback)
Subjects: LCSH: Induced seismicity. | Seismology.
Classification: LCC QE539.2.I46 E28 2018 | DDC 622/.1592–dc23
LC record available at https://lccn.loc.gov/2017045854

ISBN 978-1-107-14525-2 Hardback

Additional resources for this publication at www.cambridge.org\eaton

Cambridge University Press has no responsibility for the persistence or accuracy of URLs for external or third-party internet websites referred to in this publication and does not guarantee that any content on such websites is, or will remain, accurate or appropriate.

Contents

Preface	page ix
List of Symbols	xi

Part I Fundamentals of Passive Seismic Monitoring   1

1 Constitutive Relations and Elastic Deformation   3
   1.1 Stress and Strain   3
   1.2 Linear Elasticity   8
   1.3 Elastic Anisotropy   11
   1.4 Effective Media   16
       1.4.1 Voigt–Reuss–Hill Averaging   16
       1.4.2 Hashin–Shtrikman Extremal Bounds   17
       1.4.3 Backus Averaging   18
       1.4.4 Fractured Media   19
   1.5 Poroelasticity   21
       1.5.1 Fluid-Substitution Calculations   22
       1.5.2 Pore-Pressure Diffusion   23
       1.5.3 Fluid Flow   24
   1.6 Summary   25
   1.7 Suggestions for Further Reading   26
   1.8 Problems   26

2 Failure Criteria and Anelastic Deformation   29
   2.1 Brittle Structures in Rock   29
       2.1.1 Stress Field Near a Fracture   31
   2.2 Effective Stress and Brittle-Failure Criteria   33
       2.2.1 Mohr–Coulomb Criterion   36
       2.2.2 Griffith Criterion   37
       2.2.3 Other Brittle-Failure Criteria   38
   2.3 Frictional Sliding on a Fault   40
       2.3.1 Rate–State Friction   42
   2.4 Earthquake Cycle and Stress Drop   43
   2.5 Ductile Deformation   46
   2.6 Summary   48
   2.7 Suggestions for Further Reading   48
   2.8 Problems   49

3 Seismic Waves and Sources — 51
- 3.1 Equations of Motion — 51
- 3.2 Wave Solutions — 53
  - 3.2.1 Body Waves — 53
  - 3.2.2 Surface Waves — 56
  - 3.2.3 Green's Functions and Geometrical Spreading — 61
  - 3.2.4 Wave Amplitude Partitioning at Interfaces — 63
- 3.3 Effects of Anisotropy — 66
  - 3.3.1 Thomsen Parameters for TI Media — 68
- 3.4 Anelastic Attenuation — 69
- 3.5 Seismic Sources — 72
  - 3.5.1 Moment Tensors — 72
  - 3.5.2 Double-Couple Sources — 75
  - 3.5.3 Non-Double-Couple Sources — 79
- 3.6 Magnitude Scales — 82
- 3.7 Source Scaling and Spectral Models — 85
- 3.8 Magnitude Distributions — 88
- 3.9 Aftershocks — 90
- 3.10 Summary — 91
- 3.11 Suggestions for Further Reading — 93
- 3.12 Problems — 93

4 Stress Measurement and Hydraulic Fracturing — 95
- 4.1 How Subsurface Stress Is Determined — 95
  - 4.1.1 Focal-Mechanism Inversion — 96
  - 4.1.2 Crossed-Dipole Sonic Logs — 97
  - 4.1.3 Wellbore Failure Mechanisms — 97
  - 4.1.4 Diagnostic Fracture-Injection Tests — 100
  - 4.1.5 Overcoring — 102
  - 4.1.6 Simplified Mathematical Models — 103
- 4.2 Hydraulic Fracturing — 106
  - 4.2.1 Completion Methods and Treatment Strategies — 111
- 4.3 Numerical Models of Hydraulic Fractures — 114
- 4.4 Flowback and Flow Regimes — 121
- 4.5 Summary — 123
- 4.6 Suggestions for Further Reading — 124
- 4.7 Problems — 124

Part II Applications of Passive Seismic Monitoring — 127

5 Passive-Seismic Data Acquisition — 129
- 5.1 A Brief History of Microseismic Monitoring — 131
- 5.2 Sensor Configurations — 134
  - 5.2.1 Deep-Downhole Arrays — 134

		5.2.2 Surface and Shallow-Well Arrays	138
		5.2.3 Regional Seismograph Networks	141
	5.3	Background Model Construction and Calibration	143
		5.3.1 Calibration Sources	144
	5.4	Survey Design Considerations	146
		5.4.1 Sensor Types	146
		5.4.2 Noise	148
	5.5	Survey Design Optimization	149
	5.6	Summary	154
	5.7	Suggestions for Further Reading	156
	5.8	Problems	156

6 Downhole Microseismic Processing — 158

	6.1	Input Data	160
	6.2	Coordinate Systems and Transformations	161
	6.3	Event Detection and Arrival-Time Picking	165
		6.3.1 Single Receiver Methods	166
		6.3.2 Multi-Receiver Methods	169
		6.3.3 Wavefield Separation	172
		6.3.4 Matched Filtering and Subspace Detection	173
	6.4	Hypocentre Estimation	179
	6.5	Background Model Determination	182
	6.6	Source Characterization	185
	6.7	Summary	188
	6.8	Suggestions for Further Reading	188
	6.9	Problems	189

7 Surface and Shallow-Array Microseismic Processing — 190

	7.1	The Free-Surface Effect and Wave Amplification	192
	7.2	Beamforming and Vespagrams	194
	7.3	Basic Processing Workflow	195
	7.4	Elastic Imaging	197
		7.4.1 Imaging Condition	201
		7.4.2 Resolution and Uncertainty	202
	7.5	Case Examples	203
	7.6	Summary	205
	7.7	Suggestions for Further Reading	208
	7.8	Problems	208

8 Microseismic Interpretation — 209

	8.1	Interpretation Workflow	210
	8.2	Data Preconditioning	211
	8.3	Event Attributes	212
	8.4	Clustering Analysis	213
	8.5	Interpretation Methods	216

		8.5.1 Estimated Stimulated Volume	216
		8.5.2 Frequency–Magnitude Distributions and Fractal Dimension	221
		8.5.3 Microseismic Facies Analysis	224
	8.6	Source Mechanism Studies	226
		8.6.1 Waveform Modelling	230
	8.7	Applications	233
	8.8	Interpretation Pitfalls	237
	8.9	Summary	238
	8.10	Suggestions for Further Reading	238
	8.11	Problems	239
9	**Induced Seismicity**		**241**
	9.1	Background	241
		9.1.1 Pioneering Studies	242
		9.1.2 Distinguishing Between Natural and Induced Seismicity	243
		9.1.3 Activation Mechanisms	245
	9.2	Tools of the Trade	250
		9.2.1 Event Detection	252
		9.2.2 Hypocentre Estimation	254
		9.2.3 Moment–Tensor Inversion	257
	9.3	Case Studies	258
		9.3.1 Engineered Geothermal Systems (EGS)	259
		9.3.2 Saltwater Disposal (SWD)	264
		9.3.3 Hydraulic Fracturing (HF) Induced Seismicity	268
	9.4	Traffic Light Systems	272
	9.5	Probabilistic Seismic Hazard Assessment (PSHA)	274
	9.6	Natural Analogs of Injection-Induced Earthquakes	278
	9.7	Summary	278
	9.8	Suggestions for Further Reading	280
	9.9	Problems	280
Appendix A	**Glossary**		**281**
Appendix B	**Signal-Processing Essentials**		**292**
Appendix C	**Data Formats**		**299**

References 302
Index 340
Colour plates section can be found between pages 178 and 179

Preface

The past few decades have witnessed remarkable growth in the application of passive seismic monitoring to address a range of problems in geoscience and engineering, from large-scale tectonic studies to environmental investigations. Microseismic methods are a prime example of a passive-seismic approach applied to the study of brittle deformation in rocks. These methods are increasingly being used for in situ monitoring of fracture processes, including hydraulic-fracture stimulation of tight reservoirs, development of enhanced geothermal systems, assessment of caprock integrity for CO_2 sequestration, life-cycle reservoir monitoring for heavy-oil production and monitoring of mining operations. The theoretical framework and techniques used in this emerging discipline draw from various established fields, such as earthquake seismology, exploration geophysics and rock mechanics. The aim of this book is to synthesize research and technology within this topic, which at present is widely scattered across disparate scientific and engineering communities and published in discipline-specific journals and conference proceedings.

This book is grounded in seismology, but draws from related disciplines including reservoir engineering and rock-, fracture-, earthquake-, continuum- and geo-mechanics. It provides an introduction to the principles and applications of microseismic monitoring and is aimed at undergraduate and graduate students in geophysics or engineering, as well as working geoscience professionals. The applications of microseismic methods are myriad and include surveillance of hydraulic stimulation for unconventional hydrocarbon development and enhanced geothermal systems, monitoring and verification of long-term underground storage such as CO_2, and ensuring the safety of workers in deep underground mines. The theoretical underpinnings of passive-seismic monitoring include mathematical aspects of seismology, mechanics and signal processing, so it is assumed that readers will have a suitable background that includes mathematics and physics at the junior undergraduate level. A strong fundamental knowledge of these topics is key to achieving a quantitative understanding of the important industrial applications.

The interdisciplinary nature of passive-seismic monitoring means that mathematical expressions are rife with conflicting notation coming from established usage, across different disciplines, of identical symbols with utterly different meanings. This has necessitated a rather large dose of creativity in the use of subscripts and modifiers, in order to define a set of unique symbols for parameters and expressions that are used repeatedly. Nevertheless, some repetition of common symbols is virtually unavoidable, where the meaning is context sensitive. For example, it is contextually clear, thoughout this book, whether E is used to represent energy or Young's modulus. Similarly, specific terminology has evolved within different disciplines that can, at times, be virtually incomprehensible to those outside that field. In an effort to reduce this barrier to understanding, a Glossary is included here as

Appendix A. Throughout the text, new terms (many of which appear in the Glossary) are introduced using *italics*.

With a few exceptions, SI units are used in this book, although this usage may seem slightly foreign to those who are accustomed to the use of units of measure such as "barrels" and "psi." One exception to the use of SI units is permeability, for which the non-standard unit of Darcies is more convenient. The following list of Symbols specifies the units, as applicable, for repeatedly used quantities.

This book is largely based on notes and interactive online materials from a graduate course that I have developed, entitled *Introduction to Microseismic Methods*. While delivering this course, guest speakers, students and industry participants have broadened my knowledge horizons considerably. I am deeply indebted to all, especially guest speakers who have graciously contributed their expertise. In addition, this course has been energized by research and academic–industry interactions that occurred as part of the Microseismic Industry Consortium. A number of field datasets have been acquired since 2011, under the auspices of the Microseismic Industry Consortium; these field experiments provided data examples that are used throughout this book. To my knowledge, this level of university-led field data acquisition is unparalleled. Although it has presented daunting challenges, these field activities have provided important insights, hands-on experience and unique training opportunities for students and postdoctoral researchers.

There are numerous individuals whom I wish to thank for their help in preparing this book. Colleagues from academia and industry are sincerely thanked for providing reviews and critical feedback on sections of this book, including Ed Krebes, Jan Dettmer, Jeff Priest, Ron Wong, Shawn Maxwell, Peter Duncan, Gail Atkinson, Ryan Shultz, Yajing Liu, Hersh Gilbert, Chris Clarkson and Hadi Ghofrani. Current and former students and postdoctoral researchers also contributed many ideas and suggestions, including help with preparing figures and proofreading. I am particularly appreciative of contributions from Nadine Igonin, Jubran Akram, Hongliang Zhang, Suzie Jia, Megan Zecevic, Thomas Eyre, Kim Pike, Anton Biryukov and Ron Weir. In addition, I am very grateful to Sarah Reid, who provided tireless and skilled assistance with figures. Mirko van der Baan is sincerely thanked for his collegial partnership in the Microseismic Industry Consortium. Finally, my heartfelt thanks go to my wife, Pam, who endured months of my irrational work hours and distraction while this book was taking shape.

Symbols

$\widehat{M}_{max}$	Maximum observed magnitude in an event sequence
α	Biot's coefficient. Unitless
ϵ	Strain tensor. Unitless
$\hat{\gamma}$	Unit polarization vector
κ	Permeability tensor. Units: m^2 (1 Darcy $\approx 10^{-12}$ m^2)
σ	Stress tensor. Units: Pa
$\Delta\tau$	Co-seismic stress drop (scalar). Units: Pa
δ	Dirac delta
δ_T	Thomsen parameter for transverse isotropy. Unitless
δ_{ij}	Kronecker delta
ϵ_{ijk}	Alternating tensor
η	Dynamic viscosity. Units: Pa s (1 Poise = 0.1 Pa s)
γ	Spring damping factor
γ_T	Thomsen parameter for transverse isotropy. Unitless
κ	Permeability (scalar). Units: m^2 (1 Darcy $\approx 10^{-12}$ m^2)
λ	Lamé parameter. Units: Pa
λ_g	Geophone damping factor. Unitless
$\hat{\mathbf{n}}$	Unit normal vector
$\hat{\mathbf{S}}_{\mathbf{Hmax}}$	Unit vector in direction of maximum horizontal stress
$\hat{\mathbf{S}}_{\mathbf{Hmin}}$	Unit vector in direction of minimum horizontal stress
b	Intermediate axis for moment-tensor source
C	Covariance matrix
k	Wavevector. Units radians/m
m	Model vector
s	Slowness vector. Units: s/m
T	Traction vector. Units: Pa
$\hat{\mathbf{p}}$	Axis of compression for moment-tensor source
$\hat{\mathbf{t}}$	Axis of tension for moment-tensor source
u	Displacement vector. Units: m
$\mathcal{C}$	Cohesion. Units: Pa
$\mathcal{L}$	Characteristic slip distance. Units: m
$\mathcal{S}$	Unconfined compressive strength. Units: Pa
$\mathcal{T}$	Tensile strength. Units: Pa
$\mathcal{T}_0$	Iinitial tensile strength. Units: Pa
$\mathcal{T}_R$	Residual tensile strength. Units: Pa
μ	Shear modulus. Units: Pa

List of Symbols

μ_i	Coefficient of internal friction. Unitless
μ_r	Coefficient of residual friction. Unitless
μ_s	Coefficient of static friction. Unitless
ν	Poisson's ratio. Unitless
ω	Angular frequency. Units: radians/s
Φ	Scalar wave potential. Units: m^2
ϕ	Porosity. Unitless
ϕ_σ	Stress parameter representing the relative magnitude of stress components. Unitless
ϕ_i	Angle of internal fraction. Units: radians
ϕ_j^{ab}	jth element of the discrete cross-correlation between time series a and b
Ψ	Vector wave potential. Units: m^2
$\psi'(k_x, z, \omega)$	Time-reversed wavefield. Arguments are horizontal wavenumber, k_x, depth, z and angular frequency, ω
$\psi(\mathbf{x}, \omega)$	Zero-offset wavefield. Arguments are position $\mathbf{x}'$ and angular frequency, ω
$\psi_m(\mathbf{x}, \omega)$	Migrated image. Arguments are position $\mathbf{x}$ and angular frequency, ω
ρ	Density. Units: kg/m^3
ρ_F	Fluid density. Units: kg/m^3
ρ_M	Matrix density. Units: kg/m^3
Σ	Seismogenic index. Unitless
σ_1	Maximum principal stress. Units: Pa
σ_2	Intermediate principal stress (in three dimensions) or minimum principal stress (in two dimensions). Units: Pa
σ_3	Minimum principal stress (in three dimensions). Units: Pa
σ_n	Normal stress. Units: Pa
$\sigma_{\theta\theta}$	Hoop stress. Units: Pa
σ_{rr}	Radial stress (from Kirsch equations). Units: Pa
τ	Shear stress. Units: Pa
τ_d	Rupture time constant. Units: s
τ_r	Rupture rise time. Units: s
$\tau_{\theta r}$	Shear stress (from Kirsch equations). Units: Pa
Θ	Angular width of borehole breakout. Units: degrees or radians
$\tilde{\alpha}$	Angle between slip vector and fault plane. Units: radians
$\tilde{\delta}$	Fault dip angle. Units: degrees or radians
$\tilde{\epsilon}_i$	ith component of strain in Voigt notation. Unitless
$\tilde{\lambda}$	Rake angle. Units: degrees or radians
$\tilde{\mu}$	Coefficient of dynamic friction. Unitless
$\tilde{\mu}_0$	Steady-state sliding friction. Unitless
$\tilde{\phi}$	Fault strike angle. Units: degrees or radians
$\tilde{\sigma}_i$	ith component of stress in Voigt notation. Units: Pa
$\tilde{\theta}_i$	Picked backazimuth at the ith receiver. Units: degrees or radians
$\tilde{C}$	Consequence
$\tilde{C}_{ij}$	ijth component of stiffness matrix in Voigt notation. Units: Pa

List of Symbols

$\tilde{E}$	Exposure
$\tilde{H}$	Hazard
$\tilde{R}$	Risk
$\tilde{t}_i$	Picked arrival time at the ith receiver. Units: s
ε_T	Thomsen parameter for transverse isotropy. Unitless
φ	Complex phase. Units: radians
A	Wave amplitude. Units: m
a, b	Rate-state friction parameters. Unitless
a_j, p_j	Poles and zeros of the instrument response
b	Gutenberg–Richter b value
C_F	Correlation integral, used to determine the fractal dimension
c_{ijkl}	$ijkl$th component of the elastic stiffness tensor. Units: Pa
D	Fractal dimension
E	Young's modulus. Units: Pa
$F(\mathbf{x}, t)$	Image function. Arguments are position $\mathbf{x}$ and time t
g_a	Amplifier gain. Units: dB
G_{cr}	Energy release rate. Units: J/m^2
H_g	Geophone instrument response
H_s	Seismometer instrument response
I'_j	jth stress invariant. Units: Pa
K	Bulk modulus (various subscripts used for elements of poroelastic media). Units: Pa
K_I	Stress intensity factor. Units: Pa m$^{1/2}$
k_N	Nyquist wavenumber. Units: m^{-1}
k_x	Spatial wavenumber in the x-direction. Units: m^{-1}
K_{IC}	Fracture toughness. Units: Pa m$^{1/2}$
m	Mass. Units: kg
M_0	Seismic moment. Units: N m
m_b	Body-wave magnitude
M_c	Magnitude cut-off (threshold) for Gutenberg-Richter relation
M_L	Local magnitude
M_S	Surface-wave magnitude
M_W	Moment magnitude
M_{ij}	ijth component of the seismic moment tensor. Units: N
P	Pore pressure. Units: Pa
P_B	Formation breakdown pressure. Units: Pa
$P_E(M)$	Probability to observe one or more earthquakes above magnitude M within a fixed time interval (e.g. one year)
P_F	Fracture propagation presure. Units: Pa
P_{net}	Net pressure. Units: Pa
Q	Quality factor. Unitless
R_P^k, R_S^k	Characteristic function at the kth receiver for P and S waves
s	Slowness of a medium (reciprocal of velocity). Units: s/m
S_g	Geophone sensitivity scalar. Units: V/m/s

S_i	Sensor sensitivity factor. Units: (m/s)/V
S_s	Seismometer sensitivity scalar. Units: V/m/s
S_V	Vertical stress. Units: Pa
S_{Hmax}	Maximum horizontal stress. Units: Pa
S_{Hmin}	Minimum horizontal stress. Units: Pa
s_{ijkl}	*ijkl*th component of the compliance tensor. Units Pa^{-1}
$V(t,p)$	Vespagram. Arguments are time (t) and slowness (p)
v_g	Group velocity. Units: m/s
v_L	Love wave velocity. Units: m/s
$V_N(t,p)$	Nth root vespagram
v_P	P-wave velocity. Units: m/s
v_R	Rayleigh-wave velocity. Units: m/s
v_r	Rupture velocity. Units: m/s
v_S	S-wave velocity. Units: m/s
xi_c	Instrument scaling factor. Units: (m/s)/V
Z_N	Normal compliance. Units: Pa^{-1}
Z_T	Transverse compliance. Units: Pa^{-1}
FCP	Fracture closure pressure. Units: Pa
ISIP	Instantaneous shut-in pressure. Units: Pa

PART I

FUNDAMENTALS OF PASSIVE SEISMIC MONITORING

There is a crack in everything, that's how the light gets in.
Leonard Cohen (Anthem, 1992)

1 Constitutive Relations and Elastic Deformation

> In the beginning, God said let the four-dimensional divergence of an antisymmetric second-rank tensor equal zero ... and there was light.
> Michio Kaku (The Universe in a Nutshell, 2012)

Constitutive relations provide a foundation upon which to construct a theoretical framework for the response of a system to external stimuli. Formally, a constitutive relation defines the mathematical relationship between physical quantities that determine the response of a given material to applied forces (Macosko, 1994). In general, constitutive relations are based on experimental observation or mathematical reasoning other than a fundamental conservation equation (Pinder and Gray, 2008). This chapter deals primarily with a particular constitutive relationship that applies to *elastic media*; this relationship, known as the *generalized Hooke's Law*, describes a linear deformation regime in which the response to applied forces is fully recoverable and proportional to the magnitude of the net force. Countless experimental results confirm the applicability of this relationship to Earth materials when subject to small strains. As outlined in subsequent chapters, combining this constitutive relation with a few basic physical principles and boundary conditions leads to a remarkable wealth of wave-propagation phenomena.

As well as a description of the constitutive relations for an anisotropic elastic continuum, this chapter provides a brief introduction to various effective-medium theories that can be used to represent a complex medium with models that are more easily described and characterized. The types of media considered are of particular interest for investigations of reservoir processes and induced seismicity in sedimentary basins, including multiphase materials, vertically inhomogeneous (stratified) media and fractured elastic media. In addition, constitutive relationships for a poroelastic medium are introduced. This type of medium has two components: an elastic frame, plus a network of fluid-filled pores. A mathematical framework is also briefly introduced that governs the diffusion of pore-pressure in a poroelastic medium.

1.1 Stress and Strain

Forces that operate in Earth's interior drive a variety of deformation processes. The net internal force per unit area that acts at a point **x** on an arbitrary surface within a medium is called the *traction*, denoted by the vector **T**(**x**) (Figure 1.1). The surface on which this is defined may not necessarily correspond with a boundary, like a fracture or bedding plane.

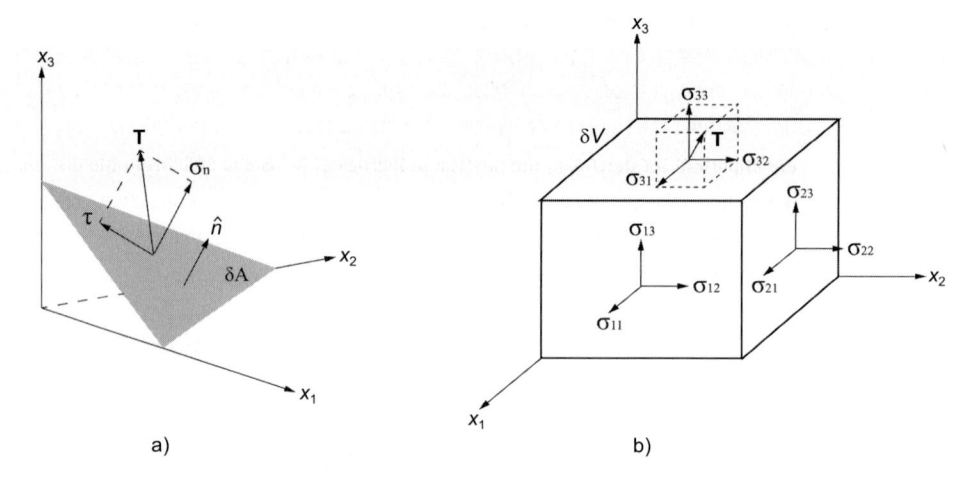

Fig. 1.1 Elements of the stress tensor. a) The traction acting on the shaded surface, denoted by **T**, can be decomposed into shear and normal components, denoted by τ and σ_n, respectively. b) Components of the stress tensor. δV represents an elementary volume.

To avoid the need to specify information about surface orientation, it is convenient to represent internal forces more generally using the *stress tensor*, which can be expressed with respect to an elementary volume as

$$\boldsymbol{\sigma} = \begin{bmatrix} \sigma_{11} & \sigma_{12} & \sigma_{13} \\ \sigma_{21} & \sigma_{22} & \sigma_{23} \\ \sigma_{31} & \sigma_{32} & \sigma_{33} \end{bmatrix}. \tag{1.1}$$

For each element of the stress tensor, the first index denotes the direction of the axis that is normal to the respective face for the elementary volume, while the second index denotes the direction in which the stress component acts. Both stress and traction have SI units of Pascal (Pa = N/m^2). For a surface with unit normal vector $\hat{\mathbf{n}}$, the stress tensor and jth element of the traction vector are related by

$$T_j = \boldsymbol{\sigma} \cdot \hat{\mathbf{n}} = \sum_{i=1}^{3} \sigma_{ij}\hat{n}_i \equiv \sigma_{ij}\hat{n}_i, \tag{1.2}$$

where the standard tensor summation convention for repeated indices is employed in the expression on the right side of Equation 1.2 (see Box 1.1). It follows that the net force **F** acting on a volume within a closed surface S may be written as

$$F_j = \int_S T_j dS = \int_S \sigma_{ij}\hat{n}_i dS. \tag{1.3}$$

In a state of equilibrium, force balance applies to the volume enclosed by S such that $|\mathbf{F}| = 0$.

The stress tensor has a number of salient characteristics. Maintaining continuity of a medium implies a condition of zero net torque on an elementary volume. This condition, in turn, implies that the stress tensor is symmetric (i.e. $\sigma_{ij} = \sigma_{ji}$). For a given surface

Box 1.1   Tensor Notation

Tensors are a generalization of vectors and provide a multidimensional representation of physical quantities that depend on spatial coordinates, including direction. Tensors are widely used in geophysics, as they are important for describing the physical properties of fields and systems in the disciplines of continuum mechanics and fluid mechanics. Tensor components are represented using index notation, where the *order* (also called *rank*) denotes the number of required indices. The stress tensor σ is a second-order tensor, sometimes referred to as a *dyadic*, and its ijth component is written as σ_{ij}. The *summation convention* for repeated indices, also known as *Einstein summation convention*, is used throughout this book. In this shorthand notation, repeated indices within products of tensors imply summation. Thus, the use of the summation convention means that

$$a_{ij}b_{jk} \equiv \sum_{j=1}^{N} a_{ij}b_{jk},$$

where N is the number of dimensions in the system (generally 2 or 3). The summation convention is sometimes applied to a single tensor quantity, such that

$$a_{ii} = a_{11} + a_{22} + a_{33}.$$

A tensor is invariant under a transformation of the coordinate system. For example, clockwise rotation of the coordinate system by θ about the x_3 axis can be expressed as

$$\sigma'_{mn} = R_{mi}R_{nj}\sigma_{ij}$$

where, in this case, the rotation operator $\mathbf{R}$ is given by

$$\mathbf{R} = \begin{bmatrix} \cos\theta & \sin\theta & 0 \\ -\sin\theta & \cos\theta & 0 \\ 0 & 0 & 1 \end{bmatrix}.$$

Spatial derivatives of tensors are represented using a subscripted comma. For example,

$$\sigma_{ij,j} \equiv \frac{\partial \sigma_{ij}}{\partial x_j}.$$

Finally, time derivatives are denoted with a dot, so that $\dot{u} \equiv \dfrac{\partial u}{\partial t}$ and $\ddot{u} \equiv \dfrac{\partial^2 u}{\partial t^2}$.

---

specified by a unit normal vector, **n**, the off-diagonal elements of the stress tensor ($i \neq j$) represent forces applied in the plane of the face and are called *shear* stresses, whereas the diagonal elements are called *normal* stresses. In general, any stress tensor is diagonalizable and may be written in the form

$$\boldsymbol{\sigma} = \boldsymbol{\Sigma}\boldsymbol{\Lambda}\boldsymbol{\Sigma}^{-1}, \tag{1.4}$$

where $\boldsymbol{\Sigma}$ is a matrix whose columns are unit eigenvectors of $\boldsymbol{\sigma}$, while $\boldsymbol{\Lambda}$ is a diagonal matrix whose elements are the corresponding eigenvalues. The eigenvectors are mutually perpendicular and are known as *principal stress* axes. These axes have particular physical significance, as they represent the normals to planes within which shear stresses vanish. The eigenvalues, called principal stresses, are denoted as σ_1, σ_2 and σ_3 and are ordered such that $\sigma_1 \geq \sigma_2 \geq \sigma_3$. For notational reference, in the case of Cartesian coordinates these principal stresses are sometimes, but not always, equivalent to the magnitude of stress (traction) acting in the vertical direction (S_V), the maximum stress magnitude acting in a horizontal direction (S_H) and the minimum stress magnitude in a horizontal direction (S_h).

Tensor quantities can be difficult to visualize. In the case of stress tensors, a *Mohr diagram*, named in honour of its inventor, Otto Mohr, is often used to depict the state of stress (Parry, 2004). As discussed in the next chapter, the Mohr diagram also provides a useful tool to represent fault/fracture stability with respect to various failure criteria. Consider an arbitrary plane defined by unit normal $\hat{\mathbf{n}}$; for any given stress state, the traction acting on this surface can be decomposed into normal and shear components, respectively denoted as the normal stress, $\sigma_n(\hat{\mathbf{n}})$, and the shear stress, $\tau(\hat{\mathbf{n}})$. As illustrated in Figure 1.2, these two stress components serve as the coordinate axes for constructing a Mohr circle. To understand how a Mohr diagram is produced, consider the stress tensor for a simplified two-dimensional scenario

$$\sigma = \begin{bmatrix} \sigma_1 & 0 \\ 0 & \sigma_2 \end{bmatrix}, \tag{1.5}$$

where σ_1 is the maximum principal stress and σ_2 is the minimum principal stress. With no loss of generality, a natural coordinate system is used here such that the x_1 and x_2 axes correspond to the principal stress axes, so that off-diagonal elements in the stress tensor (shear stresses) vanish. In general, the stress tensor can be expressed in a rotated coordinate system by applying a rotation transformation,

$$\begin{bmatrix} \sigma'_{11} & \sigma'_{12} \\ \sigma'_{21} & \sigma'_{22} \end{bmatrix} = \begin{bmatrix} \cos(\theta) & \sin(\theta) \\ -\sin(\theta) & \cos(\theta) \end{bmatrix} \begin{bmatrix} \sigma_1 & 0 \\ 0 & \sigma_2 \end{bmatrix} \begin{bmatrix} \cos(\theta) & -\sin(\theta) \\ \sin(\theta) & \cos(\theta) \end{bmatrix}, \tag{1.6}$$

where θ is the angle of rotation. Expanding the right side, the normal and shear stress values on the surface normal to the x'_1-axis, which makes an angle θ from the x_1 axis, are given by

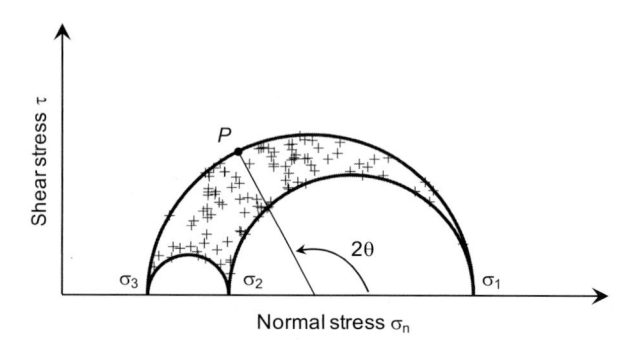

Fig. 1.2 3-D Mohr diagram, showing principal stresses, $\sigma_1 \geq \sigma_2 \geq \sigma_3$. Symbols represent the state of stress on 100 randomly oriented fractures. Point P defines the state of stress for a plane whose normal is co-planar with the maximum and minimum principal stress axes and that makes an angle θ with respect to the maximum principal stress axis.

$$\sigma_n = \sigma'_{11} = \sigma_1 \cos^2\theta + \sigma_2 \sin^2\theta = \frac{\sigma_1 + \sigma_2}{2} + \frac{\sigma_1 - \sigma_2}{2}\left(\cos^2\theta - \sin^2\theta\right), \tag{1.7}$$

$$\tau = \sigma'_{12} = (\sigma_2 - \sigma_1)\sin\theta\cos\theta = \frac{\sigma_1 - \sigma_2}{2} 2\sin\theta\cos\theta.$$

In the expression for σ_n, we have made use of the trigonometric identity $\sin^2\theta + \cos^2\theta = 1$. By further invoking the trigonometric identities $\cos 2\theta = \cos^2\theta - \sin^2\theta$ and $\sin 2\theta = 2\sin\theta\cos\theta$, we can write

$$\sigma_n = \frac{\sigma_1 + \sigma_2}{2} + \frac{\sigma_1 - \sigma_2}{2}\cos 2\theta,$$

$$\tau = \frac{\sigma_1 - \sigma_2}{2}\sin 2\theta, \tag{1.8}$$

which are parametric equations, with respect to the variable 2θ, for a circle with centre at $\frac{\sigma_1+\sigma_2}{2}$ and radius $\frac{\sigma_1-\sigma_2}{2}$. In two dimensions, the Mohr circle thus represents the stress state as a locus of points in (σ_n, τ)-space. Each point on the circle corresponds with a plane whose normal makes an angle θ with respect to the maximum principal stress axis. A 2-D Mohr diagram is commonly represented as a semicircle by plotting with respect to $|\tau|$ rather than τ.

In three dimensions, a similar approach can be applied. First, consider the subset of normal vectors **n** that are co-planar with the maximum (σ_1) and minimum (σ_3) principal stress axes. Based on the above arguments for the 2-D case, all of the possible stress states with respect to σ_1 and τ define a circle with centre $\frac{\sigma_1+\sigma_3}{2}$ and radius $\frac{\sigma_1-\sigma_3}{2}$. Similarly, planes defined by normal vectors that are co-planar with other pairs of principal stress axes define Mohr circles with smaller radii and different centres. This set of three semicircles creates a 3-D Mohr diagram (Figure 1.2). For a given stress state defined by principal stresses σ_1, σ_2 and σ_3, it can be shown that, for all possible normal vectors, including those that are not co-planar with pairs of principal stress axes, the large Mohr circle in a 3-D Mohr diagram forms an outer boundary in (σ_n, τ)-space, whereas the smaller two Mohr circles represent inner boundaries. Referring to Figure 1.2, planes with random orientations fall within the region between the three Mohr circles.

We now turn our attention to the concept of *strain*. As shown in Figure 1.3, strain is defined with respect to an elementary volume in terms of displacement, denoted as **u(x)**, as follows:

$$\epsilon_{ij} = \frac{1}{2}\left(u_{i,j} + u_{j,i}\right), \tag{1.9}$$

where the indicial comma notation for spatial derivatives has been employed on the right side of this expression. Displacement is specified here in a *Lagrangian* reference frame, which means that the coordinate system moves with a particle in the medium, consistent with seismological measurement systems (Aki and Richards, 2002). The velocity of a point in a medium is given by $\dot{\mathbf{u}}(\mathbf{x})$. Referring to Figure 1.3, it is evident that if the spatial derivatives of **u** are zero, the elementary volume is displaced with no change in shape or volume; consequently, the strain variable provides a measure of deformation in a medium. Because it is defined as a ratio of two quantities with units of length, strain is dimensionless. Strain

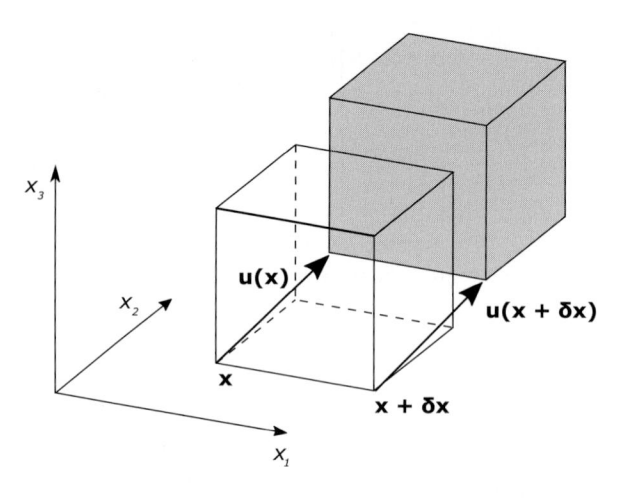

Fig. 1.3 Strain is defined by deformation of an elementary volume. The initial size and shape of the elementary volume (unshaded) are modified in response to an applied stress. The change in size and shape of the deformed volume (shaded) is described using the displacement field, $\mathbf{u}(\mathbf{x})$.

is a second-order tensor that, by definition, has the symmetry property $\epsilon_{ij} = \epsilon_{ji}$. Similar to the stress tensor, the diagonal elements are called normal-strain components and the off-diagonal elements are called shear-strain components.

1.2 Linear Elasticity

The fundamental constitutive relationship between stress and strain in a linear elastic medium is given by

$$\sigma_{ij} = c_{ijkl}\epsilon_{kl}, \tag{1.10}$$

where c_{ijkl} denotes the elastic *stiffness tensor* (recall the tensor summation convention that implies a double summation on the right side of this equation). This relationship is known as a generalized form of *Hooke's Law*. Since strain is unitless, the units of the stiffness tensor are Pa.

Since both stress and strain are second-order tensors, a fourth-order tensor (c_{ijkl}) is required to fully characterize all possible linear relationships between stress and strain components. In three dimensions, the stress tensor thus has 81 (3^4) components, where each individual scalar component is known as an *elastic modulus*. Due to various symmetries, for the general (triclinic) case the number of independent moduli can be reduced to 21. These symmetries arise from: 1) the inherent symmetry of the stress tensor ($\sigma_{ij} = \sigma_{ji}$); 2) the inherent symmetry of the strain tensor ($\epsilon_{kl} = \epsilon_{lk}$); and 3) the definition of strain-energy density, which implies that $c_{ijkl} = c_{klij}$ (Aki and Richards, 2002). A more compact notation for generalized Hooke's Law, known as *Voigt* notation, exploits these symmetries and reduces the stiffness tensor to a symmetric 6 × 6 stiffness matrix, **C**. Using Voigt notation, the elastic constitutive relation can be expressed as

$$\begin{bmatrix} \tilde{\sigma}_1 \\ \tilde{\sigma}_2 \\ \tilde{\sigma}_3 \\ \tilde{\sigma}_4 \\ \tilde{\sigma}_5 \\ \tilde{\sigma}_6 \end{bmatrix} = \begin{bmatrix} \tilde{C}_{11} & \tilde{C}_{12} & \tilde{C}_{13} & \tilde{C}_{14} & \tilde{C}_{15} & \tilde{C}_{16} \\ & \tilde{C}_{22} & \tilde{C}_{23} & \tilde{C}_{24} & \tilde{C}_{25} & \tilde{C}_{26} \\ & & \tilde{C}_{33} & \tilde{C}_{34} & \tilde{C}_{35} & \tilde{C}_{36} \\ & & & \tilde{C}_{44} & \tilde{C}_{45} & \tilde{C}_{46} \\ & & & & \tilde{C}_{55} & \tilde{C}_{56} \\ & & & & & \tilde{C}_{66} \end{bmatrix} \begin{bmatrix} \tilde{\epsilon}_1 \\ \tilde{\epsilon}_2 \\ \tilde{\epsilon}_3 \\ \tilde{\epsilon}_4 \\ \tilde{\epsilon}_5 \\ \tilde{\epsilon}_6 \end{bmatrix}, \quad (1.11)$$

in which pairs of indices are combined such that $()_{11} \to ()_1, ()_{22} \to ()_2, ()_{33} \to ()_3, ()_{23} \to ()_4, ()_{13} \to ()_5$ and $()_{12} \to ()_6$. For example, using this combination method $c_{1111} = \tilde{C}_{11}$ and $c_{1122} = (\tilde{C})_{12}$, where the tilde overbar notation is used here to distinguish Voigt parameters from the standard tensor representation. In addition, the Voigt strain parameters are assigned as follows: $\tilde{\epsilon}_1 = \epsilon_{11}, \tilde{\epsilon}_2 = \epsilon_{22}, \tilde{\epsilon}_3 = \epsilon_{33}, \tilde{\epsilon}_4 = 2\epsilon_{23}, \tilde{\epsilon}_5 = 2\epsilon_{13}$ and $\tilde{\epsilon}_6 = 2\epsilon_{12}$. Because $\tilde{\mathbf{C}}$ is a symmetric 6×6 matrix, this means that there is a maximum of 21 independent elastic moduli, as shown above. As a caution, it should be emphasized that the use of Voigt notation means that a more complex operator is required to transform from one coordinate system to another, known as the *Bond* transformation (see Winterstein, 1990 for details).

From an experimental perspective, the underlying mathematical model implied by the generalized form of Hooke's Law implies that 21 independent measurements of stress–strain response are required to fully characterize the elastic behaviour of a material – a daunting prospect that is seldom realized in practice. Fortunately, most rocks have inherent material symmetry properties that simplify the stress–strain relationship by reducing the number of independent coefficients needed to construct the stiffness tensor. These material symmetries arise from *rock fabric* elements that occur commonly in the subsurface, such as horizontal stratification, existence of parallel fracture sets and fabrics created by preferred alignment of minerals.

Consider the special case, albeit routinely invoked, of an *isotropic* medium. In such a medium, there is no directional dependence associated with the stress–strain relationship. Thus, for a given strain condition, a measurement of normal- or shear-stress components in a vertical orientation would yield the same result as a measurement in a horizontal orientation, or indeed at any angle of inclination. In simplistic terms, a subsurface rock mass could be considered as a fractured, fluid-saturated granular mineral aggregate. Isotropic elastic symmetry is often assumed to exist if these constituent elements, such as mineral grains or microfractures, are both small-scale and randomly oriented. Here "small scale" means small relative to the seismic wavelength, which is typically a few metres to a few hundred metres.

In the case of an isotropic medium, only two independent elastic moduli are required to fully characterize the stress–strain relationship. In this case, the elastic stiffness tensor may be written as

$$c_{ijkl} = \lambda \delta_{ij} \delta_{kl} + \mu \left(\delta_{ik} \delta_{jl} + \delta_{il} \delta_{jk} \right), \quad (1.12)$$

where λ and μ are independent constants known as the *Lamé* parameters and δ_{ij} is known as the *Kronecker delta*, which has the properties

$$\delta_{ij} = \begin{cases} 0 & \text{if } i \neq j, \\ 1 & \text{if } i = j. \end{cases} \quad (1.13)$$

For an isotropic material, Hooke's Law may be expressed in the form

$$\sigma_{ij} = \lambda \epsilon_{kk} \delta_{ij} + 2\mu \epsilon_{ij}, \tag{1.14}$$

where the term ϵ_{kk} (implied summation over index k) is called the *dilatation*, defined as $\Delta V/V$.

Although the Lamé parameters are useful for expressing the constitutive relationship, it is often more convenient to express stiffness characteristics of a material using alternative elastic moduli that are directly linked to experimental measurements. In addition to the shear modulus, μ, other commonly used elastic moduli include *bulk modulus*, K, *Young's modulus*, E, and *Poisson's ratio*, ν. For a sample of volume V, the bulk modulus is given by

$$K \equiv -V \frac{\partial P}{\partial V}, \tag{1.15}$$

where P is confining pressure and the stress tensor has *hydrostatic* form,

$$\sigma = \begin{bmatrix} -P & 0 & 0 \\ 0 & -P & 0 \\ 0 & 0 & -P \end{bmatrix}. \tag{1.16}$$

The negative sign used here reflects the convention that, whereas pressure is positive under compressional conditions, stress is positive under tensile conditions.

Young's modulus is measured under uniaxial stress conditions. For a sample of cross-sectional area A, Young's modulus can be expressed as

$$E \equiv \frac{F/A}{\Delta L/L_0} = \frac{\sigma_{axial}}{\epsilon_{axial}}, \tag{1.17}$$

where F is the force applied to the ends of the sample (positive for tensile and negative for compressional), L is the length of the sample measured along its axis in the direction of the applied force, ΔL is the change in sample length and L_0 is the sample length prior to application of the force. In addition, σ_{axial} denotes axial stress and ϵ_{axial} denotes axial strain. The shear modulus can be measured by applying a shear force to the sides of a sample and is defined as

$$\mu = \frac{\sigma_{ij}}{2\epsilon_{ij}}, \quad i \neq j \text{ (no summation)}. \tag{1.18}$$

Note that in many engineering texts the shear modulus is represented by G. The elastic moduli K, E and μ for Earth materials are typically expressed in units of GPa.

Another parameter that is commonly used to describe the properties of an elastic solid is Poisson's ratio. Like Young's modulus, this is measured under uniaxial stress conditions. Poisson's ratio is unitless and is given by

$$\nu = -\frac{\epsilon_{trans}}{\epsilon_{axial}}, \tag{1.19}$$

where ϵ_{trans} is the transverse strain.

Table 1.1 Relationships Between Pairs of Isotropic Elastic Moduli

a	K	E	λ	μ	ν
K, E	K	E	$\frac{2K(3K-E)}{9K-E}$	$\frac{KE}{9K-E}$	$\frac{3K-E}{6K}$
K, λ	K	$\frac{9K(K-\lambda)}{9K-\lambda}$	λ	$\frac{3(K-\lambda)}{2}$	$\frac{\lambda}{3K-\lambda}$
K, μ	K	$\frac{9K\mu}{3K+\mu}$	$K - \frac{2\mu}{3}$	μ	$\frac{3K-2\mu}{2(3K+\mu)}$
K, ν	K	$3K(1-2\nu)$	$\frac{3K\nu}{1+\nu}$	$\frac{3K(1-2\nu)}{2(1+\nu)}$	ν
E, λ	$\frac{E+3\lambda+R}{6}$ [b]	E	λ	$\frac{E-3\lambda+R}{4}$	$\frac{2\lambda}{E+\lambda+R}$
E, μ	$\frac{3\mu}{3(2\mu-E)}$	E	$\frac{\mu(E-2\mu)}{3\mu-E}$	μ	$\frac{E}{2\mu} - 1$
E, ν	$\frac{E}{3(1-2\nu)}$	E	$\frac{E\nu}{(1+\nu)(1-2\nu)}$	$\frac{E}{2(1+\nu)}$	ν
λ, μ	$\lambda + \frac{2\mu}{3}$	$\frac{\mu(3\lambda+2\mu)}{\lambda+\mu}$	λ	μ	$\frac{\lambda}{2(\lambda+\mu)}$
λ, ν	$\frac{\lambda(1+\nu)}{3\nu}$	$\frac{\lambda(1+\nu)(1-2\nu)}{\nu}$	λ	$\frac{\lambda(1-2\nu)}{2\nu}$	ν
μ, ν	$\frac{2\mu(1+\nu)}{3(1-2\nu)}$	$2\mu(1+\nu)$	$\frac{2\mu\nu}{1-2\nu}$	μ	ν

[a] K, E and ν denote bulk modulus, Young's modulus and Poisson's ratio, respectively, while λ and μ are the Lamé parameters. Table modified from Sheriff (1991).
[b] For brevity, the notation $R = \sqrt{E^2 + 9\lambda^2 + 2E\lambda}$ is used in some expressions in this table.

As outlined in Chapter 3, wave propagation in an isotropic elastic medium can be fully characterized using density, ρ, plus any pair of independent elastic moduli. For an isotropic medium, Table 1.1 summarizes the relationships between various pairs of elastic moduli. Given any two independent moduli, the values of other moduli are easily calculated.

1.3 Elastic Anisotropy

When a constitutive model is chosen such that properties of the Earth at a point are directionally dependent, it is said to be *anisotropic*. Virtually all of the constituent minerals in the Earth's crust have crystal lattice structures that produce anisotropic elastic properties (Musgrave, 2003). For example, kaolinite (Figure 1.4) is a clay mineral that is commonly found in shale, with a crystal lattice structure similar to mica (Gruner, 1932). The principal axes X_1, X_2 and X_3 shown in Figure 1.4 describe the orientation of the optical axes that determine optical properties that can be used to identify minerals (Putnis, 1992). The crystallographic axes a, c and c are defined by the *unit cell*, the basic building block of crystallographic lattice structure (Musgrave, 2003). For example, the measured elastic stiffness matrix for a single kaolinite crystal (Sato et al., 2005) is given in GPa by:

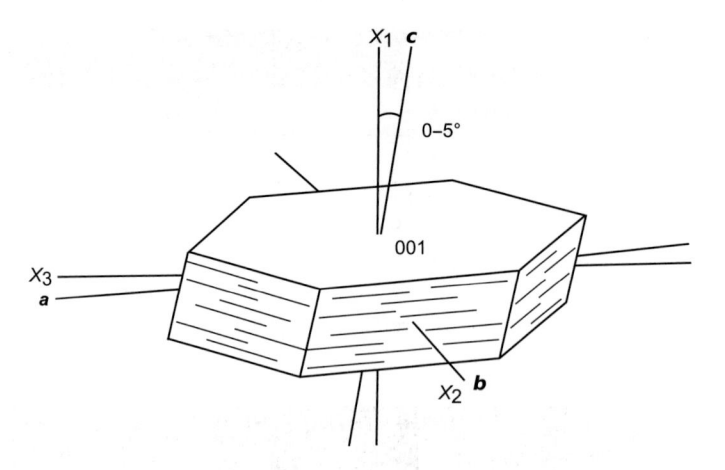

Fig. 1.4 Single crystal of kaolinite with crystallographic axes labelled (source: Imperial College Rock Library). The orientation of crystal axes, denoted as a, b and c, are controlled by the unit cell. Optical axes are denoted as x_1, x_2 and x_3. Used with permission of Imperial College.

$$\tilde{C} = \begin{bmatrix} 178 \pm 8.8 & 71.5 \pm 7.1 & 2.0 \pm 5.3 & -0.4 \pm 2.1 & 41.7 \pm 1.4 & -2.3 \pm 2.7 \\ & 200.9 \pm 12.8 & -2.9 \pm 5.7 & -2.8 \pm 2.7 & 19.8 \pm 0.6 & 1.9 \pm 1.5 \\ & & 32.1 \pm 2.0 & -0.2 \pm 1.4 & 1.7 \pm 1.8 & 3.4 \pm 2.2 \\ & & & 11.2 \pm 5.6 & -1.2 \pm 1.2 & 12.9 \pm 2.4 \\ & & & & 22.2 \pm 1.4 & 0.8 \pm 2.4 \\ & & & & & 60.1 \pm 3.2 \end{bmatrix}.$$

The coordinate system used for a single-crystal stiffness matrix is defined by the unit cell. On the basis of lattice-dependent symmetry properties such as axes of rotation, planes of symmetry and centres of inversion, crystals can be classified into seven distinct symmetry systems: triclinic, monoclinic, orthorhombic, tetragonal, trigonal, hexagonal and cubic (Musgrave, 2003). The symmetry system of the elastic stiffness tensor is generally equivalent to the crystal symmetry system for the corresponding single crystal, although in some cases higher-order symmetries exist (Winterstein, 1990). The crystal structure of kaolinite (Figure 1.4) is classified as triclinic (Sato et al., 2005), a symmetry system that requires either 18 or 21 independent elastic constants depending on the specific mineral (see Table 1.2). Elastic properties for many rock-forming minerals have been compiled by Simmons and Wang (1971).

Seismic wavelengths are generally orders of magnitude greater than the dimensions of single crystals. Wave propagation is therefore sensitive to average values of elastic moduli over a much larger region of the subsurface than that of a single crystal. The averaging region may also include fractures, which affect the directional dependance of the average properties and thus contribute to seismic anisotropy. In sedimentary rocks, microstructural factors that contribute to bulk elastic anisotropy in sedimentary rocks include: 1) lattice-preferred orientation of constituent minerals; 2) morphology of platy mineral grains leading to a shape-preferred orientation (SPO); and 3) parallel alignment of microfractures

Table 1.2 Summary of Anisotropic Symmetry Systems

Symmetry	Number of Independent Moduli
Isotropic	2
Transversely Isotropic (TI)[a]	5
Orthorhombic	8
Monoclinic	12 or 13
Triclinic	18 or 21

[a] Sometimes referred to as hexagonal symmetry

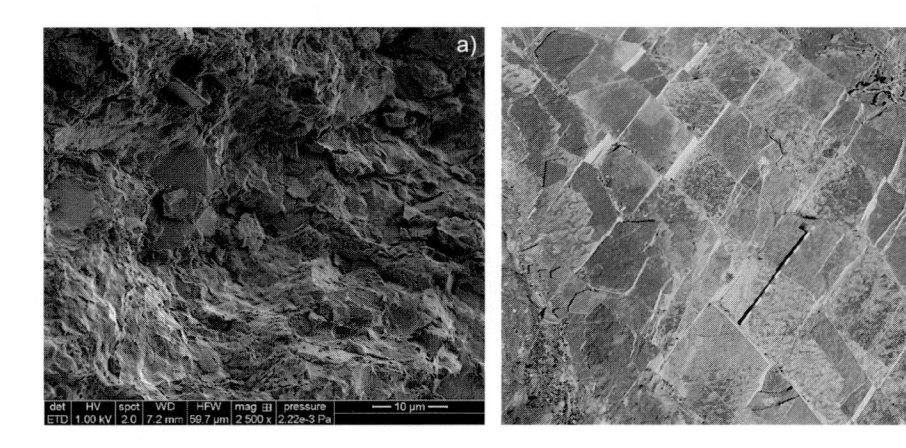

Fig. 1.5 Examples of anisotropy at different scales. a) Scanning electron microscope image of the Second White Specks (2WS) Formation in the deep basin, Alberta. Preferred orientation of platy minerals can impart intrinsic seismic anisotropy. b) Fracture set in horizontal bedding plane of the Jumping Pound Sandstone (2WS Formation) near the Highwood River, Alberta, with 1.5 m ruler for scale. Aligned fractures in otherwise isotropic rocks can also impart seismic anisotropy. Images courtesy of P. K. Pedersen, used with permission.

and pores (Valcke et al., 2006). Elastic anisotropy on a macroscopic scale can also be caused by periodic thin layering and stress-aligned fracturing (Crampin et al., 1984).

Figure 1.5 shows examples of factors that contribute to bulk anisotropy, at two widely different scales. The images were both obtained from samples of the Cretaceous Second White Speckled shale in Alberta, Canada. A scanning electron microscope (SEM) image (Figure 1.5a) reveals conspicuous platy mineral grains in the form of euhedral crystals at the micron scale. LPO- and SPO-causing mechanisms associated with detrital grains include gravitational deposition, mechanical compaction, diagenetic growth of minerals and currents that influence the orientation of elongate mineral grains (Valcke et al., 2006). An outcrop photograph (Figure 1.5b) shows intersecting metre-scale conjugate fracture sets exposed in a bedding plane. The geometrical characteristics of these fracture sets, which have different spacing and intersect at an angle other than 90°, are expected to generate monoclinic symmetry (Winterstein, 1990) when the properties are averaged at a much larger scale applicable to seismic wave propagation.

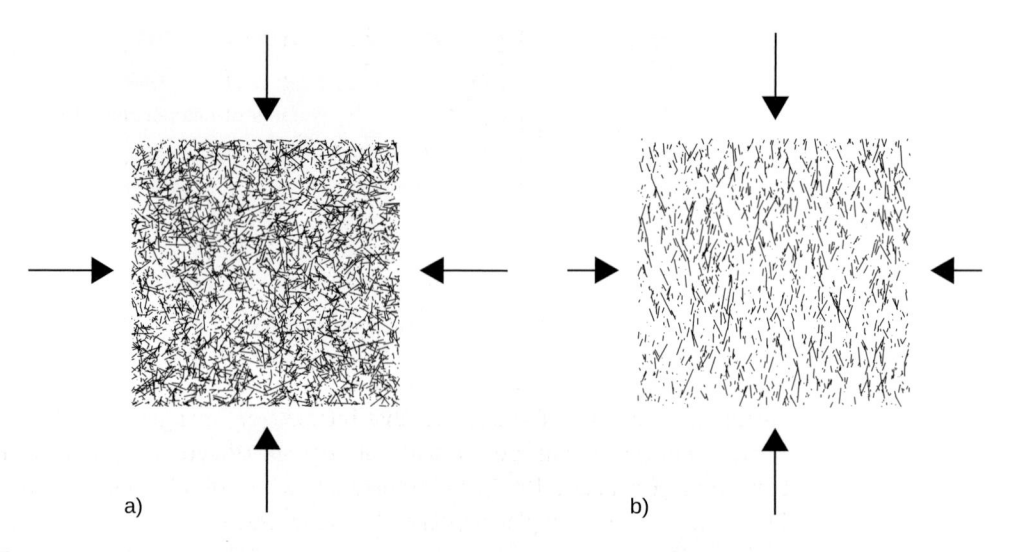

Fig. 1.6 Closure of randomly oriented fractures under isotropic and anisotropic confining stress. a) Randomly oriented fractures under isotropic, or *lithostatic*, confining stress. b) Preferential closure of fractures oriented perpendicular to the maximum confining stress, leading to an anisotropic fracture fabric. Arrows indicate principal stresses, with length proportional to stress magnitude. Modified from: Tectonophysics, Vol. 580, Douglas R. Schmitt, Claire A. Currie and Lei Zhang, Crustal stress determination from boreholes and rock cores: Fundamental principles, Pages 1–26, Copyright 2012, with permission from Elsevier.

Although constituent minerals are almost always anisotropic, if mineral grains are randomly oriented in a homogeneous unfractured rock mass, then the medium is usually assumed to be isotropic; similarly, if an otherwise homogeneous isotropic rock mass contains randomly oriented (micro)fractures, then under isotropic stress conditions ($\sigma_1 = \sigma_2 = \sigma_3$) the medium will tend to be isotropic with respect to its elastic properties (Figure 1.6). On the other hand, in the presence of stress anisotropy some fractures will close preferentially, leading to a medium with anisotropic bulk elastic properties (Crampin et al., 1984).

The characteristic symmetry properties for an otherwise homogeneous and isotropic medium with a single set of cracks are: 1) a principal, infinite-fold axis of rotation perpendicular to the cracks, and 2) an infinite number of twofold rotation axes perpendicular to the principal rotation axis (Winterstein, 1990). This symmetry system, called *transverse isotropy* (TI), has no full equivalence in crystallography[1], but it is of fundamental importance in seismology. The TI system is further classified based on the orientation of the principal axis of symmetry. Horizontal transverse isotropy (HTI) is characterized by a horizontal symmetry axis and commonly occurs in the presence of a single set of vertical cracks, whereas vertical transverse isotropy (VTI) is characterized by a vertical symmetry axis and commonly occurs in the presence of fine stratification. Referred to a coordinate system where the x_3-axis is aligned with the principal axis of symmetry, the stiffness matrix for a TI medium has six independent components and takes the form

---

[1] Crystals with hexagonal symmetry are characterized by elastic properties that are the same as a TI medium.

$$\tilde{\mathbf{C}} = \begin{bmatrix} \tilde{C}_{11} & \tilde{C}_{12} & \tilde{C}_{13} & 0 & 0 & 0 \\ \tilde{C}_{12} & \tilde{C}_{11} & \tilde{C}_{13} & 0 & 0 & 0 \\ \tilde{C}_{13} & \tilde{C}_{13} & \tilde{C}_{33} & 0 & 0 & 0 \\ 0 & 0 & 0 & \tilde{C}_{44} & 0 & 0 \\ 0 & 0 & 0 & 0 & \tilde{C}_{44} & 0 \\ 0 & 0 & 0 & 0 & 0 & \tilde{C}_{66} \end{bmatrix}, \quad (1.20)$$

where $\tilde{C}_{12} = \tilde{C}_{11} - 2\tilde{C}_{66}$. This form represents the specific case for a VTI medium; the equivalent form for a horizontal principal axis of symmetry is easily obtained by permutation of indices, or by applying a coordinate rotation.

Orthorhombic symmetry is also important in seismology, and can occur in the case of two identical equidistant fracture sets that intersect at an angle other than 90°, two non-identical fracture sets that are mutually orthogonal (Figure 1.7), or vertical cracks in a VTI medium (Tsvankin, 1997). For a medium with orthorhombic symmetry characterized by vertical symmetry planes perpendicular to the x_1- and x_2-axes, the stiffness matrix generally has nine independent components and takes the form (Musgrave, 2003)

$$\tilde{\mathbf{C}} = \begin{bmatrix} \tilde{C}_{11} & \tilde{C}_{12} & \tilde{C}_{13} & 0 & 0 & 0 \\ \tilde{C}_{12} & \tilde{C}_{22} & \tilde{C}_{23} & 0 & 0 & 0 \\ \tilde{C}_{13} & \tilde{C}_{23} & \tilde{C}_{33} & 0 & 0 & 0 \\ 0 & 0 & 0 & \tilde{C}_{44} & 0 & 0 \\ 0 & 0 & 0 & 0 & \tilde{C}_{55} & 0 \\ 0 & 0 & 0 & 0 & 0 & \tilde{C}_{66} \end{bmatrix}. \quad (1.21)$$

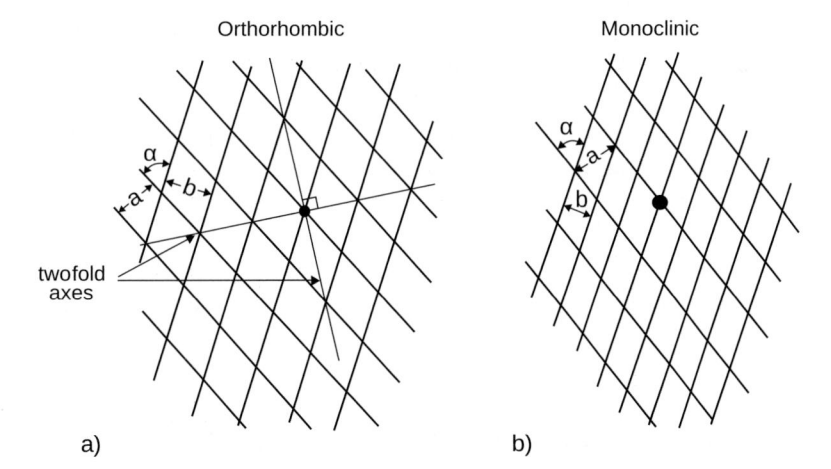

Fig. 1.7 Examples of idealized systems of cracks (plan view) to produce orthorhombic and monoclinic symmetry. Cracks are assumed to be smooth, planar and of infinite extent; compare with natural fractures in Fig. 1.5b. Orthorhombic symmetry can also be produced by two mutually perpendicular sets of cracks (shown by intersecting sets of lines) with spacing $a \neq b$, or two sets of cracks with $\alpha \neq 0$ or $90°$ and $a = b$. The solid dot shows a vertical twofold axis of rotation. Orthogonal lines passing through this point show edge-on view of mirror planes. For the monoclinic system, $a \neq b$ and $\alpha \neq 0$ or $90°$. Modified from Winterstein (1990). Used with permission from the Society of Exploration Geophysicists.

This form of $\tilde{\mathbf{C}}$ has the same non-zero components as the form for a TI medium, but there is a greater number of independent moduli. In some cases, such as a set of vertical fractures in a VTI medium, not all nine of the components in the stiffness matrix are independent (Schoenberg and Helbig, 1997). Fracture sets that are not identical and non-orthogonal (Figure 1.7) lead to monoclinic symmetry, with a twofold symmetry axis along the axis of intersection of the fracture systems (Winterstein, 1990). The stiffness matrix for a monoclinic medium has 12 or 13 independent moduli, while for a triclinic material, such as a single-crystal of kaolinite, there are 21 independent moduli.

1.4 Effective Media

An *effective medium* is a macroscopic model that mimics the bulk properties of a more complex medium, such as one with multiphase materials (Wang and Pan, 2008). In this context, it is important to recognize the distinction between anisotropy and *heterogeneity*; the latter applies to media where the physical properties vary with position but not with orientation. Effective-medium theory provides a set of mathematical tools with which to simulate the behaviour of a complex heterogeneous medium using a simpler, often anisotropic, medium. It can be used, for example, to determine the bulk elastic moduli for a rock mass that is represented mathematically as a polycrystalline aggregate, for a stratified medium with periodic thin layering, or for a medium containing fracture sets. Such tools can be very powerful for improving numerical simulation efficiency or for gaining general insights into complex physical systems.

1.4.1 Voigt–Reuss–Hill Averaging

Bulk elastic moduli for a N-phase composite material can be computed using the widely used *Voigt* or *Reuss* estimates (Li and Wang, 2005). These estimates may be considered as extreme rule-of-mixture models, where the Voigt estimate assumes equal strain in all phases under an applied load, whereas the Reuss estimate assumes equal stress. For the Voigt estimate, since the overall stress is the sum of stresses carried by each phase, for a given elastic modulus (C) the estimated bulk value is the average of the corresponding moduli of the individual phases,

$$C_V = \sum_{i=0}^{N-1} C^{(i)} \chi^{(i)}, \quad (1.22)$$

where $\chi^{(i)}$ is the volume fraction of the ith phase of the N-phase medium. Conversely, for the Reuss estimate, the total strain is the sum of the net strain carried by each phase, so the reciprocal of the estimated bulk average is given by

$$C_R^{-1} = \sum_{i=0}^{N-1} \frac{\chi^{(i)}}{C^{(i)}}. \quad (1.23)$$

As noted by Jones et al. (2009), these two estimates predict bounds that are widely separated. Hill (1963) showed that a better estimate is usually provided by the arithmetic average of the Voigt and Reuss estimates. The average of C_V and C_R is known as the Voigt–Reuss–Hill estimate (Hill, 1963).

1.4.2 Hashin–Shtrikman Extremal Bounds

In a series of papers during the early 1960s, Hashin and Shtrikman used a variational approach to develop a more rigorous theory to predict the physically realizable range for bulk values of a multiphase medium (Hashin and Shtrikman, 1962, 1963). The Hashin–Shtrikman (HS) extremal bounds have since been applied in many fields of science (Jones et al., 2009). Here we will focus only on one application of HS extremal bounds, for estimation of bulk elastic properties of composite materials. For isotropic elastic moduli, this theory gives exact results for assemblages of spheres (Watt et al., 1976). For purposes of computation, if the phases are ordered so that the 0th phase has the smallest modulus and the Nth phase has the largest, the HS lower bound for the bulk modulus, K, can be expressed as (Watt et al., 1976)

$$K_{HS}^- = K_0 + \frac{A_1}{1 + \alpha_1 A_1}, \tag{1.24}$$

where the "−" superscript denotes the lower bound,

$$\alpha_1 = \frac{-3}{3K_0 + 4\mu_0}, \tag{1.25}$$

and

$$A_1 = \sum_{i=1}^{N} \frac{\chi^{(i)}}{(K_i - K_0)^{-1} - \alpha_1}, \tag{1.26}$$

and where $\chi^{(i)}$ is the volume fraction of the ith phase in a medium with $N+1$ phases. The formulae for the bulk-modulus HS upper bound are similar:

$$K_{HS}^+ = K_N + \frac{A_2}{1 + \alpha_2 A_2}, \tag{1.27}$$

where

$$\alpha_2 = \frac{-3}{3K_N + 4\mu_N}, \tag{1.28}$$

and

$$A_2 = \sum_{i=0}^{N-1} \frac{\chi^{(i)}}{(K_i - K_N)^{-1} - \alpha_2}. \tag{1.29}$$

Similarly, the HS lower-bound formulae for the shear modulus are:

$$\mu_{HS}^- = \mu_0 + \frac{B_1}{1 + \beta_1 B_1}, \tag{1.30}$$

where
$$\beta_1 = \frac{-3(K_0 + 2\mu_0)}{5\mu_0(3K_0 + 4\mu_0)}, \qquad (1.31)$$
and
$$B_1 = \sum_{i=1}^{N} \frac{\chi^{(i)}}{2(\mu_i - \mu_0)^{-1} - \beta_1}. \qquad (1.32)$$

Finally, the HS upper-bound formulae for shear modulus are:
$$\mu_{HS}^{+} = \mu_N + \frac{B_2}{1 + \beta_2 B_2}, \qquad (1.33)$$
where
$$\beta_2 = \frac{-3(K_N + 2\mu_N)}{5\mu_N(3K_N + 4\mu_N)}, \qquad (1.34)$$
and
$$B_2 = \sum_{i=0}^{N-1} \frac{\chi^{(i)}}{2(\mu_i - \mu_N)^{-1} - \beta_2}. \qquad (1.35)$$

The HS bounds represent the largest possible physically realizable range of values for a particular parameter, yet they yield a range distribution that is nearly always considerably tighter than the difference between Voigt and Reuss estimates (Jones et al., 2009).

1.4.3 Backus Averaging

Within sedimentary basins, stratification and layering of rock units introduces a fundamental heterogeneity even in the absence of fracturing. *Periodic thin layering* (PTL) is a term that is applicable to stratified media with layers that are much thinner than the dominant seismic wavelength. Extensive theory concerning the nature of the equivalent homogeneous TI medium has been developed for such a scenario (Postma, 1955; Krey and Helbig, 1956). Backus (1962) gave formulae for computing the effective elastic constants for a TI-equivalent medium, for which each of the thin layers in the stratified model is transversely isotropic or isotropic. In terms of elastic moduli, these formulae may be written

$$\overline{C}_{11} = \left\langle \tilde{C}_{11} - \left(\tilde{C}_{13}\right)^2 \left(\tilde{C}_{33}\right)^{-1} \right\rangle + \left\langle \left(\tilde{C}_{33}\right)^{-1} \right\rangle^{-1} \left\langle \tilde{C}_{13} \left(\tilde{C}_{33}\right)^{-1} \right\rangle^2, \qquad (1.36)$$

$$\overline{C}_{33} = \left\langle \left(\tilde{C}_{33}\right)^{-1} \right\rangle^{-1}, \qquad (1.37)$$

$$\overline{C}_{13} = \left\langle \left(\tilde{C}_{33}\right)^{-1} \right\rangle^{-1} \left\langle \tilde{C}_{13} \left(\tilde{C}_{33}\right)^{-1} \right\rangle, \qquad (1.38)$$

$$\overline{C}_{44} = \left\langle \left(\tilde{C}_{44}\right)^{-1} \right\rangle^{-1}, \qquad (1.39)$$

and
$$\overline{C}_{66} = \left\langle \tilde{C}_{66} \right\rangle. \qquad (1.40)$$

In these expressions, the overbar signifies moduli that are associated with the equivalent medium. In addition, the $\langle\ \rangle$ brackets denote a weighted averaging, so that for a sequence of N layers,

$$\langle \tilde{C}_{66} \rangle = \frac{\sum_{i=1}^{N} h_i \tilde{C}_{66}^{(i)}}{\sum_{i=1}^{N} h_i}, \tag{1.41}$$

where h_i is the thickness of the ith layer. Specialized versions of these formulae are given by Levin (1979) for the case where each layer is isotropic. In practice, the use of Backus averaging has the advantage that when an equivalent TI medium is used in the place of a N-layered PTL medium, only five elastic moduli are required, rather than $2 \times N$ elastic moduli for isotropic layers or $5 \times N$ for TI layers.

1.4.4 Fractured Media

In addition to layering, fractures are pervasive in shallow crustal rocks. In sedimentary rocks, the orientation and distribution of fractures reflects the tectonic history of the rock. If fractures occur in sets with preferred orientation, this imposes a characteristic symmetry onto a medium as illustrated in Figure 1.7. Since fractures that are open at depth tend to be normal to the minimum principal stress (Figure 1.6), elastic anisotropy due to fractures has the potential to inform our understanding of the present-day stress field (Schoenberg and Sayers, 1995). These factors are particularly relevant to geothermal reservoir characterization, seismic monitoring of enhanced oil recovery operations and development of low-permeability hydrocarbon reservoirs. In some oil and gas fields, primary recovery is possibly only because of the presence of open fractures; elsewhere the migration of injected fluids is controlled by fracturing (Babcock, 1978).

As in PTL anisotropy, the calculation of effective elastic constants for a fractured elastic solid involves averaging over a scale length that is large compared to the fracture dimensions. Formulae for effective elastic moduli in a fractured solid medium have been developed by Garbin and Knopoff (1975) and Hudson (1981). Following the perturbation approach by Hudson (1981), consider an otherwise homogeneous and isotropic medium that is permeated with circular, infinitesimally thin, fluid-filled microfractures normal to the x_3-axis (fractures in other orientations can be described by performing the appropriate rotation to the stiffness tensor). In this case, the only elastic modulus that is affected by the fracturing is $\tilde{C}_{44}$ (Hudson, 1981). Denoting the crack radius as a and the crack density as ξ, the perturbation is given to the first order in ξa^3 by Hudson (1981) as

$$\Delta \tilde{C}_{44} = -\frac{32}{3} \xi a^3 \mu \left(\frac{\lambda + 2\mu}{3\lambda + 4\mu} \right), \tag{1.42}$$

where λ and μ are the Lamé parameters of the unfractured background medium.

Schoenberg and Sayers (1995) have developed a method for estimation of the elastic properties of a fractured medium that is based on the use of the *compliance* tensor, **s**, which can be used to recast the elastic constitutive relation as

$$\epsilon_{ij} = s_{ijkl}\sigma_{kl}. \tag{1.43}$$

This approach extends the range of applicability of the effective medium theory to higher crack density and also provides a tool for analysis of the effects of compressibility of fracture-filling fluids. Under the assumption of invariance with respect to rotation around an axis normal to the fractures, Schoenberg and Sayers (1995) showed that the overall compliance of the fractured rock mass can be expressed as the sum of the compliance of the intact, unfractured rock and a perturbation introduced by the presence of the fractures. Using Voigt notation, this relationship can be written as

$$\tilde{\mathbf{S}} = \tilde{\mathbf{S}}^r + \Delta\tilde{\mathbf{S}}, \quad (1.44)$$

where $\tilde{\mathbf{S}}, \tilde{\mathbf{S}}^r$ and $\Delta\tilde{\mathbf{S}}$ denote the overall compliance matrix, the compliance matrix for the unfractured rock and the perturbation due to fracturing, respectively. The compliance perturbation matrix is sparse and can be written in terms of the normal and transverse fracture compliance values, Z_N and Z_T, as

$$\Delta\tilde{\mathbf{S}} = \begin{bmatrix} Z_N & 0 & 0 & 0 & 0 & 0 \\ 0 & 0 & 0 & 0 & 0 & 0 \\ 0 & 0 & 0 & 0 & 0 & 0 \\ 0 & 0 & 0 & 0 & 0 & 0 \\ 0 & 0 & 0 & 0 & Z_T & 0 \\ 0 & 0 & 0 & 0 & 0 & Z_T \end{bmatrix}, \quad (1.45)$$

where this form is for fractures that are oriented perpendicular to the x_1-axis. The stiffness matrix for a TI medium can be determined by computing the matrix inverse of $\tilde{\mathbf{S}}$ after incorporation of the effects of fractures (Schoenberg and Sayers, 1995).

The normal compliance, Z_N, provides a measure of susceptibility to closure, whereas the transverse compliance, Z_T, provides a measure of susceptibility to slip. The fracture compliance ratio, Z_N/Z_T, generally falls within a range from 0 to 1 and has potential to be used as a way to characterize the internal architecture of a fracture network as well as properties of fracture-filling fluids (Verdon and Wüstefeld, 2013). For example, under the assumption that fractures are penny-shaped (circular) and drained, this ratio can be approximated by (Sayers and Kachanov, 1995)

$$Z_N/Z_T = 1 - \nu/2, \quad (1.46)$$

where ν is the Poisson's ratio of the unfractured rock. From a seismological perspective, the concept of drained fractures can be viewed as broadly representative of a scenario in which fluid can escape from fractures on a timescale that is shorter than the dominant period of seismic waves (Verdon and Wüstefeld, 2013). Assuming that fracture surfaces are smooth, Z_N is expected to be sensitive to the compressibility of fracture-filling fluids, whereas Z_T is not; hence, in the limiting case of an incompressible fluid, $Z_N/Z_T \to 0$. This limit may similarly be approached at high seismic frequencies, or in cases where fractures are hydraulically isolated, the fluid is highly viscous or the permeability of the host medium is too low (Hudson et al., 1996).

There are several observational parameters that may be used to constrain Z_N/Z_T. A theoretical model for the frequency dependence of seismic waves was developed by Chapman (2003). This model has been used to investigate fractured reservoirs by Xue et al.

(2017), based on amplitude-versus-offset and azimuth from 3-D seismic data. An interpretive approach to investigate Z_N/Z_T has been developed by Verdon and Wüstefeld (2013), using observations of shear-wave splitting from passive seismic monitoring of hydraulic fracturing.

1.5 Poroelasticity

A *poroelastic* medium is a two-phase medium that consists of a solid frame plus a network of fluid-filled pore spaces (Figure 1.8). Some types of fractured media, as outlined in the previous section, can be viewed as particular examples. A constitutive model for poroelastic media was initially developed by Biot (1962a) and has since been applied to a host of problems in engineering and seismology. The basic assumptions of this model are:

- the solid frame is composed of isotropic elastic material;
- the fluid phase is *Newtonian*, which means that the fluid constitutive relationship can be expressed as $\tau = \eta \dot{\epsilon}$, where τ is shear stress, $\dot{\epsilon}$ is shear strain rate and the scalar coefficient η is the *dynamic viscosity*;
- pores are connected, such that fluid can flow within the pore space;
- deformations are sufficiently weak that nonlinear mechanical effects can be neglected.

The dynamic viscosity of a fluid provides a measure of molecular interaction during flow. In the case of a fluid with high viscosity (e.g. honey), the molecules do not easily

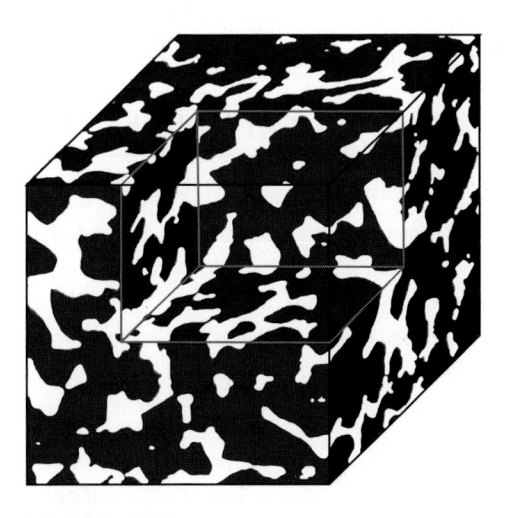

0.5 mm

Fig. 1.8 Characterization of a porous sandstone by X-ray computed tomography from microtomographic images with 5.7 μm resolution. Image is rendered by thresholding a grey level value of 140. Solid fraction is dark and pore space is white. Modified from Louis et al. (2007). Used with permission from the Geological Society of London.

slide past each other in response to an applied shear stress, whereas for a low viscosity fluid (e.g. acetone) the molecules exhibit limited interaction during flow (Pinder and Gray, 2008).

With these assumptions, Biot (1962b) developed the following constitutive relationships for a poroelastic medium:

$$\sigma_{ij} = \left[\lambda \delta_{ij}\delta_{kl} + \mu\left(\delta_{ik}\delta_{jl} + \delta_{il}\delta_{jk}\right)\right]\epsilon_{kl} - \alpha M \zeta \delta_{ij}, \qquad (1.47)$$

and

$$P = -\alpha M \epsilon_{kk} + M\zeta, \qquad (1.48)$$

where P denotes the pore pressure, λ and μ are the Lamé parameters of the solid matrix and the parameter α is called *Biot's coefficient*, defined as

$$\alpha = 1 - \frac{K_D}{K_M}. \qquad (1.49)$$

In this expression, K_D is the bulk modulus of the drained rock, while K_M is the bulk modulus of the nonporous solid material. In addition, the constant M is a measure of the coupling between the fluid and rock frame, defined as

$$M = \left(\frac{\phi}{K_F} + \frac{\alpha - \phi}{K_M}\right)^{-1}, \qquad (1.50)$$

where ϕ is the porosity of the medium and K_F is the bulk modulus of the fluid. Finally, the parameter ζ is given by

$$\zeta = -w_{k,k}, \qquad (1.51)$$

where w_k is the kth component of the displacement of the fluid relative to the solid frame. Equivalently, using the *divergence* operator, ζ can be expressed as $-\nabla \cdot \mathbf{w}$.

1.5.1 Fluid-Substitution Calculations

A low-frequency theory for porous media developed by Gassman (1951) provides an effective-medium framework for a widely used fluid-substitution method (Smith et al., 2003). This approach assumes that pore pressures are equalized over a length scale that is much greater than the pore dimension. This means that the medium must have a relatively high *permeability* (κ), a measure of the ease with which fluid can flow within a porous medium. Application of this approach to seismic wave propagation further assumes that pore pressure is equalized over a region with dimensions that are much smaller than the seismic wavelength. Gassmann defined the bulk modulus of the fluid-saturated medium as

$$K_S = K_D + \frac{\left(1 - \frac{K_D}{K_M}\right)^2}{\frac{\phi}{K_F} + \frac{1-\phi}{K_M} - \frac{K_D}{K_M^2}}, \qquad (1.52)$$

while the density of the fluid-saturated porous medium is given by

$$\rho = \phi \rho_F + (1-\phi)\rho_M, \qquad (1.53)$$

where ρ_F and ρ_M are the density of the fluid and solid (matrix) phases, respectively. Most of the parameters on the right side of the formula for K_S are generally well known, with the exception of the bulk modulus of the dry frame, K_D. The use of these equations for fluid-substitution modelling is thus a two-part process (Smith et al., 2003); the first step is to obtain an independent estimate of K_D, and the second is to calculate the bulk modulus for a medium that is saturated with any desired fluid, such as oil, or gas. For example, Moradi (2016) determined K_S using well log data and rearranged Equation 1.52 to solve for K_D using

$$K_D = \frac{K_{S0}\left(\frac{\phi K_M}{K_{F0}} + 1 - \phi\right) - K_M}{\frac{\phi K_M}{K_{F0}} + \frac{K_{S0}}{K_M} - 1 - \phi}, \tag{1.54}$$

where K_{S0} and K_{F0} are the saturated bulk modulus of the medium and the estimated bulk modulus of the pore fluid under in situ conditions. To determine K_M using the Reuss formula (Equation 1.23), Moradi (2016) estimated the volume fractions of constituent minerals and combined this with laboratory-derived bulk moduli. The calculated value of K_D from Equation 1.54 along with other parameters was then used to perform fluid-substitution calculations with different values of K_F, using Equation 1.52.

1.5.2 Pore-Pressure Diffusion

Pore-pressure *diffusion* is a physical mechanism that causes pore pressure to spread from areas of relatively high pressure to regions of relatively low pressure without bulk transport of pore fluid. This phenomenon is important for understanding fluid-induced seismicity (Shapiro, 2015). Based on Biot's theory, the diffusion of pore pressure in a poroelastic medium is described by the differential equation (Shapiro et al., 2003)

$$\dot{P} = \left(D_{ij} P_{,j}\right)_{,i}, \tag{1.55}$$

where D_{ij} is the *hydraulic diffusivity* tensor (defined below). In a homogeneous medium, this simplifies to

$$\dot{P} = D_{ij} P_{,ij}. \tag{1.56}$$

For an isotropic medium the differential equation simplifies further to the standard form of the *diffusion equation*, which may be written as

$$\frac{\partial P}{\partial t} = D \nabla^2 P. \tag{1.57}$$

Processes governed by the diffusion equation are common in many areas of biology, chemistry and physics. Conduction of heat from hot to cold regions is a familiar example of this process in everyday physics.

In a poroelastic medium with a single fluid phase, the hydraulic diffusivity tensor is given by (Dutta and Odé, 1979)

$$\mathbf{D} = \frac{N\kappa}{\eta}, \tag{1.58}$$

where κ is the *permeability tensor* (defined below) and N is a poroelastic parameter defined as

$$N = \frac{MP_D}{H}, \tag{1.59}$$

and where

$$M = \left(\frac{\phi}{K_F} + \frac{\alpha - \phi}{K_M}\right)^{-1}. \tag{1.60}$$

In addition to poroelastic parameters defined above, the following parameters are used in these expressions (Shapiro et al., 2003):

$$\alpha = 1 - \frac{K_D}{K_M}; \quad H = P_D + \alpha^2 M; \quad P_D = K_D + \frac{4}{3}\mu_D. \tag{1.61}$$

1.5.3 Fluid Flow

Fluid flow in porous media is of fundamental importance for a wide range of applications, including reservoir simulation, groundwater studies, contaminant transport and materials science (Pinder and Gray, 2008). In this section, several basic concepts are introduced.

A quantitative foundation for flow and transport processes in porous media was established in a series of filtration experiments during the nineteenth century conducted by Henry Darcy (King Hubbert, 1956). These experiments involved one-dimensional flow of water through a vertical column of sand, and showed that the volume of water q crossing a unit area in unit time can be written as

$$q = -\xi \frac{(h_2 - h_1)}{l}, \tag{1.62}$$

where h_1 and h_2 are measured water heights above a reference level, l is the length of the sand column and ξ is a constant of proportionality. This relationship can be generalized to three dimensions and expressed in terms of the pore-pressure gradient as (Shapiro, 2015)

$$q_i = -\frac{\kappa_{ij}}{\eta}\frac{\partial P}{\partial x_j}, \tag{1.63}$$

where κ and η are, respectively, the previously introduced permeability tensor and dynamic viscosity. Notably, the term $\frac{\kappa_{ij}}{\eta}$ is similar to the hydraulic diffusivity in Equation 1.55. The fluid-flux vector $\mathbf{q} \equiv \dot{\mathbf{w}}$ represents the average displacement of the fluid relative to the solid skeleton, and is also called the seepage velocity or the filtration velocity. This velocity differs from the velocity of fluid particles relative to the skeleton, which is related to seepage velocity $\Delta \dot{\mathbf{u}}$ as follows:

$$\Delta \dot{\mathbf{u}} = \frac{\mathbf{q}}{\phi}, \tag{1.64}$$

where ϕ is the porosity and $\Delta \mathbf{u}$ is the relative displacement of an average (or representative) fluid particle relative to the skeleton (Shapiro, 2015). Although it was initially developed experimentally, Equation 1.63 has now been rigorously derived using homogenization theory, as an asymptotic solution for laminar flow at low values of the local *Reynolds number* (Firdaouss et al., 1997). Indeed, it is recognized that this relationship, known as *Darcy's law*, plays an equivalent role in describing the transport of fluids through porous media as Ohm's law in the conduction of electricity, or of Fourier's law in the conduction of heat (King Hubbert, 1956). The local Reynolds number, given by

$$R_e = \frac{\Delta P a^3 \rho}{L \eta^2} \,, \qquad (1.65)$$

is a scalar parameter that measures the ratio of inertial forces to viscous forces acting on a fluid. Here, L is a characteristic macroscopic length scale, $\Delta P/L$ is a macroscopic pressure gradient that drives the flow, ρ is the fluid density and a is microseismic distance scale (e.g. grain size) (Firdaouss et al., 1997). Darcy's law is applicable under laminar flow conditions in which $R_e \ll 1$.

1.6 Summary

This chapter introduces basic concepts for stress, strain and moduli for an elastic medium. The stress tensor describes the force per unit area that operates on surfaces within a medium, in the limit as the surface area approaches zero. In some cases the surface may be a physical surface, such as a fault or fracture, but the concept of stress can equally be applied to an arbitrarily oriented hypothetical surface. For a given surface normal direction, the stress can be decomposed into a normal stress that acts perpendicularly to the surface, and a shear stress that acts in the plane of the surface. Strain is a unitless measure of deformation. Like stress, strain is a tensor quantity.

In an elastic medium, the constitutive relationship is straightforward: stress is proportional to strain. This constitutive relationship is known as generalized Hooke's Law. Since stress and strain are both second-order tensors, the elastic constitutive relationship requires a fourth-order tensor to express the linear relationship between stress and strain components, called the elastic stiffness tensor. In the most general case, the stiffness tensor contains 21 independent parameters known as elastic moduli. Under isotropic conditions in which the stress–strain relationship is independent of orientation, the number of independent moduli is reduced to two. Common subsurface scenarios, including preferred orientation of platy mineral grains, layering and sets of parallel fractures, can introduce a directional dependence that leads to elastic anisotropy.

Various effective medium theories have been developed to simplify constitutive relationships for materials that contain features that are small relative to wavelengths of seismic waves. These effective-medium theories include Voigt–Reuss–Hill averaging and Hashin–Shtrikman extremal bounds for polycrystalline aggregates. A different approach, known as Backus averaging, can be used to determine effective transversely isotropic (TI) moduli for a thinly bedded material. In the presence of fractures, various theories have been developed to compute effective anisotropic medium characteristics.

Sedimentary rocks contain pore spaces that are usually saturated with fluids. A poroelastic theory developed by Biot has been widely used to express the constitutive relationships for a porous, fluid-saturated rock mass. An approximate form of Biot theory, represented by the Gassmann equation, is commonly used for fluid-substitution modelling to determine effective bulk elastic parameters for a porous medium. A poroelastic model for pore-pressure diffusion has been extensively used to analyze the growth of expanding regions of microseismicity. Darcy's law is a widely used constitutive relationship that describes laminar fluid flow in a porous material.

1.7 Suggestions for Further Reading

This chapter provides a basic introduction to a variety of mathematical concepts and methods that are important for understanding and simulating induced seismicity. The following authoritative references provide a great deal more information on the topics covered here.

- Fundamental concepts of continuum mechanics in seismology: Aki and Richards (2002).
- *Physical Properties of Rocks* (Schön, 2015).
- Crystal acoustics and seismic anisotropy: Musgrave (2003).
- Effective-medium theory of sedimentary rocks (Sheng, 1990).
- Linear poroelasticity: Wang (2017).

1.8 Problems

1. Evaluate the following.

    a) What is Poisson's ratio for a Poisson solid (i.e. a material with $\lambda = \mu$)?
    b) Given Young's modulus $E = 8 \times 10^{10}$ Pa and Poisson's ratio $\nu = 0.28$, determine K, λ and μ.
    c) Assuming that these parameters correspond to an isotropic elastic solid, write the stiffness matrix in Voigt form (as in Equation 1.11 – see also Problem 3, below).
    d) How would the stiffness matrix change after applying an arbitrary rotation?
    e) What is the value of the elastic stiffness c_{1111}?

2. Consider a hypothetical granitic material that can be approximated as a 6-phase assemblage of minerals, as tabulated below. Determine the bulk modulus of the polycrystalline aggregate using the following methods.

Mineral	Volume Fraction (%)	K [GPa]	μ [GPa]
quartz	30	36	42
orthoclase	30	40	24
plagioclase	25	65	39
muscovite	5	45	27
biotite	5	40	24
amphibole	5	100	60

a) Voigt averaging.
b) Reuss averaging.
c) Voigt–Reuss–Hill (VRH) averaging.
d) Hashin–Shtrikman extremal bounds (calculate upper and lower bounds).

3. In Voigt notation, the stiffness matrix of an isotropic solid can be written in terms of Lamé parameters as

$$\tilde{\mathbf{C}} = \begin{bmatrix} \lambda + 2\mu & \lambda & \lambda & 0 & 0 & 0 \\ \lambda & \lambda + 2\mu & \lambda & 0 & 0 & 0 \\ \lambda & \lambda & \lambda + 2\mu & 0 & 0 & 0 \\ 0 & 0 & 0 & \mu & 0 & 0 \\ 0 & 0 & 0 & 0 & \mu & 0 \\ 0 & 0 & 0 & 0 & 0 & \mu \end{bmatrix}.$$

Suppose that the elastic properties of an unfractured granite can be approximated by the following stiffness matrix:

$$\tilde{\mathbf{C}} = \begin{bmatrix} 80 & 20 & 20 & 0 & 0 & 0 \\ 20 & 80 & 20 & 0 & 0 & 0 \\ 20 & 20 & 80 & 0 & 0 & 0 \\ 0 & 0 & 0 & 30 & 0 & 0 \\ 0 & 0 & 0 & 0 & 30 & 0 \\ 0 & 0 & 0 & 0 & 0 & 30 \end{bmatrix} \text{[GPa]}.$$

Now, suppose that the granite contains fractures with a transverse fracture compliance $Z_T = 3.0 \times 10^{-12}$ Pa^{-1}. Determine the stiffness matrix for the case of:

a) penny-shaped fractures that are well drained (Equation 1.46);
b) smooth fractures filled with an incompressible fluid.

4. Consider a binary sequence of shale and coal layers, where the shale is characterized by Lamé parameters $\lambda = 8.0$ GPa and $\mu = 9.0$ GPa, while the coal is characterized by Lamé parameters $\lambda = 4.0$ GPa and $\mu = 2.0$ GPa. Using Backus averaging, determine the Voigt matrix for the equivalent TI medium based on the following relative abundances:

a) equal-thickness layers of coal and shale;
b) coal layers that are 10% of the thickness of shale layers, on average.

5. Given the poroelastic parameters in the table below, use Equations 1.52–1.54 to perform fluid-substitution calculations, based on Gassmann's formula, to determine the saturated bulk modulus and bulk density of a porous rock that is initially saturated with brine and then becomes fully saturated (after fluid substitution) with supercritical CO_2.[2]

---

[2] A *supercritical* fluid has characteristics of both a liquid and a gas; its transport properties in a porous medium resemble a gas, but it is capable of dissolving materials like a liquid.

Property	Symbol	Value
Initial fluid density (brine)	ρ_{F0}	1230 kg/m^3
Substituted fluid density (CO_2)	ρ_{F1}	625 kg/m^3
Matrix density	ρ_M	2650 kg/m^3
Matrix bulk modulus	K_M	38.8 GPa
Initial fluid bulk modulus (brine)	K_{F0}	3.8 GPa
Substituted fluid bulk modulus (CO_2)	K_{F1}	0.25 GPa
Initial saturated bulk modulus	K_{S0}	22.0 GPa
Porosity	ϕ	18%

6. Given a pore-pressure gradient of 1.5×10^5 Pa/m, a medium permeability of 1×10^{-14} m^2 (0.01 Darcy) and viscosity values of 10^{-3} Pa-s for fluid 1 (brine) and 10^{-4} Pa-s for fluid 2 (a supercritical fluid), determine the following.

a) Estimate the fluid velocity using Darcy's Law (Equation 1.63).
b) For a porosity of 18%, what is the seepage velocity?
c) Determine Reynolds numbers for these parameters assuming a grain size of 1 mm and the fluid density values from question 5.
d) Do these values of Reynolds number meet the assumptions for Darcy's Law?

7. **Online exercise**: Recent studies have highlighted the potential significance of pore-fluid pressure as a factor that may control the onset, termination and distribution of seismicity induced by fluid injection into the subsurface. In a porous medium, the diffusion of pore-fluid pressure is described by Equation 1.56. A simple Matlab tool is provided to visualize pore-pressure diffusion and to investigate parameter sensitive, including:

- Permeability of the background medium.
- Viscosity of the injected fluid.
- Injection duration.
- Fracture orientation.
- Injection pressure.

2 Failure Criteria and Anelastic Deformation

> Although we often hear that data speak for themselves, their voices can be soft and sly.
> Frederick Mosteller (Beginning Statistics with Data Analysis, 1983)

The previous chapter emphasizes the elastic constitutive paradigm applicable to transient, recoverable deformation processes, which are accompanied by sufficiently small strain so that the stress–strain relationship is effectively linear. This chapter focuses on *anelastic* behaviour, which departs from the elastic constitutive model and leads to permanent deformation of a medium. Building upon concepts of the stress field and its tensor representation, this chapter deals with *brittle* and *ductile* failure processes that occur at levels of strain that are generally higher than for elastic behaviour. Brittle deformation is highly localized within the rock mass and accommodated by abrupt dislocation on new or pre-existing fractures and faults, whereas ductile deformation results in permanent strain that is distributed more pervasively and occurs without fracturing, in response to an applied stress. The framework for understanding these distinct but interrelated processes derives from the diverse disciplines of geomechanics, fracture mechanics and earthquake mechanics. This chapter provides a brief introduction to fundamental principles related to these disciplines that are important for development of a complete understanding of induced seismicity.

2.1 Brittle Structures in Rock

Fractures are quasi-planar discontinuities in a rock mass. In an idealized sense, fractures are often described as surfaces, but at small scale they can be viewed as narrow tabular features with finite aperture (Fossen, 2016). *Joints* are a type of fracture across which there is negligible apparent shear displacement; joints can, however, have *tensile* (opening) displacement and are therefore sometimes referred to as opening fractures or *dilatant* fractures (Aydin, 2000). The generic term *crack* is often used interchangeably with either joint or fracture, especially in the material-science and rock-mechanics literature. Fractures tend to occur in *sets* that are approximately mutually parallel and regularly spaced. Intersecting sets of fractures form a fracture *network*.

Fractures and joints occur at a broad range of scales. In crystalline rocks the size distribution of fractures is usually characterized by a fractal or power-law distribution (Bonnet et al., 2001); however, in sedimentary rocks, mechanical bedding caused by lithologic layering can constrain the height distribution of vertical fracture sets as a result of fracture

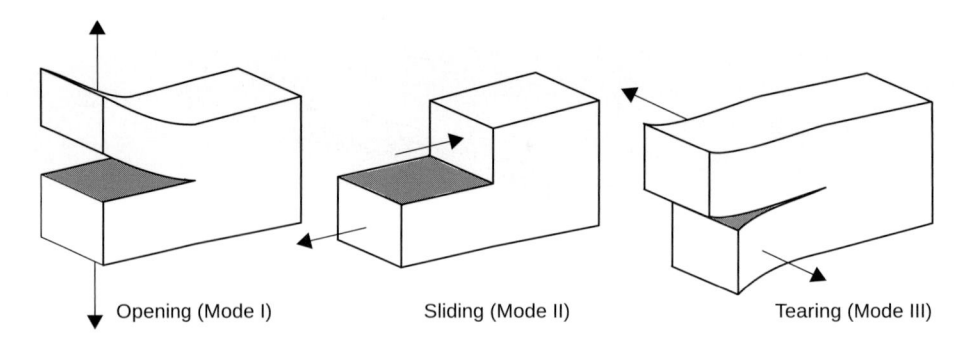

Fig. 2.1 Three basic modes of failure on a new or existing fracture.

terminations at bedding boundaries, thus leading to *stratabound* fracture networks (Odling et al., 1999; Eaton et al., 2014a). Rock fabric is defined as a configuration of planar or linear elements expressed within a penetratively deformed rock;[1] if the spacing between individual fractures is less than a few cm, they may be considered as a constituent *fabric element* of the rock mass (Jaeger et al., 2009).

As illustrated in Figure 2.1, fracture-formation mechanisms can be classified into three distinct modes on the basis of the associated displacement field. Mode I represents tensile failure, with the fracture opening perpendicular to the face of the fracture. Modes II and III both involve shear dislocation parallel to the fracture face. For mode II (sliding), the shear displacement is parallel to the direction of fracture growth; for mode III (tearing) the shear displacement is perpendicular to the direction of fracture growth. An additional type of failure (mode IV: closing) is sometimes used; this mode is represented in the present work as a negative polarity of mode I.

Fractures can be further classified as *closed* or *open*, depending upon the type of fracture-filling material (Jaeger et al., 2009). Closed fractures contain precipitated minerals such as calcite, dolomite, quartz or clay, whereas open fractures contain formation fluids, such as brine or hydrocarbon (Jaeger et al., 2009). Open fractures can serve as major conduits for subsurface fluid transport, while closed fractures may be orders of magnitude less permeable than the host rock and form a barrier to fluid flow within a porous medium (Aydin, 2000).

From a kinematic perspective, a fault is defined as a structural discontinuity across which there is net shear displacement of $\gtrsim$ 1 m (Fossen, 2016). As such, a fault can be viewed as a type of fracture (albeit at the large end of the size spectrum). Most faults in the geological record are not seismically active on historical timescales, so it is sometimes useful to distinguish *active* faults from *quiescent* faults. Seismologists are primarily concerned with active (or *seismogenic*) faults; thus, in the earthquake seismology literature the qualifier "active" is often omitted.

From a mechanistic perspective, a *fault zone* is an approximately tabular volume of rock with a central core and a surrounding, less deformed *damage zone* (Figure 2.2). Exhumed segments of the San Andreas fault system reveal a core zone composed of *ultracataclasite*

---

[1] Penetrative fabrics are lineations or planar structural elements that are recognizable throughout the rock mass.

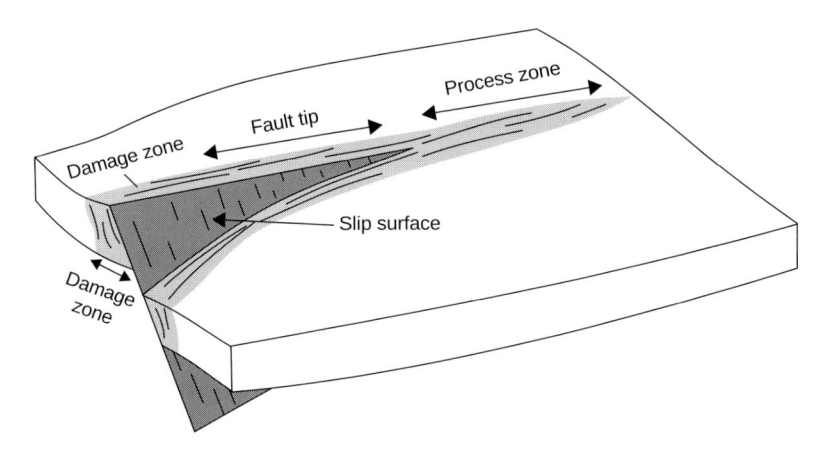

Fig. 2.2 Terminology used to describe a fault zone. During fault rupture and growth, complex small-scale deformation on linked fractures occurs within a *process zone* located near the tip of the fault, or at the leading edge of the rupture front during an earthquake. One or more slip surface(s) are located within the fault core. Around the fault core, an enclosing volume of rock forms a *damage zone*. Modified from Fossen et al. (2007). Used with permission from the Geological Society of London.

that is 10s-cm thick, with a surrounding damage zone that is 100s-m thick (Chester et al., 1993). Cataclastic material is a cohesive fine-grained rock and is classified as ultracataclasite if more than 90 percent of the rock mass is matrix (Fossen, 2016). The texture is created by mechanical grain-size reduction through a grinding process known as *comminution*. Displacement in the fault core is strongly localized along discrete *slip surfaces*. Other terms used to describe a fault zone are illustrated in Figure 2.2.

2.1.1 Stress Field Near a Fracture

Linear elastic fracture mechanics (LEFM) is a continuum mechanics approach that provides a basic constitutive framework to analyze the stress field of a fracture in an isotropic elastic material. LEFM theory has its roots in the pioneering work by Griffith (1921), who strove to explain large differences between measured tensile strength of materials and the predicted tensile strength obtained using *ab initio* calculations based on the energy required to break atomic bonds. He found that the low tensile strength of most materials, including rocks, can be explained by the presence of microdefects in the material.

Griffith's approach, subsequently modified by Irwin (1948), posed the problem at a fundamental level by defining the total energy, U, for a static crack as (Scholz, 2002)

$$U = (-W + U_e) + U_s . \tag{2.1}$$

In this expression, W is the work done by external forces, U_e denotes internal strain energy and U_s denotes the energy required to create new fracture surface area. Griffith postulated that a fracture evolves to a state that minimizes the system energy level. For example, in the case of a mode I fracture of length $2a$ in an infinite plate, the critical energy release rate can be written as (Anderson, 2005)

$$G_{cr} = \frac{K_I^2}{E}, \qquad (2.2)$$

where E is Young's modulus of the host medium, σ is the applied tensile stress and $K_I = \sigma\sqrt{\pi a}$ is called the *stress-intensity factor*. The energy release rate, G_{cr}, is thus proportional to the product of the fracture half-length and the square of the tensile stress. A fracture will propagate if K_I exceeds a critical value, K_{IC}, known as the *fracture toughness*. If the crack is assumed to be planar and perfectly sharp with no cohesion, the stress field produced by a mode I fracture is strongly concentrated at the fracture tips, providing a driving force for dynamic self-sustaining fracture growth. In two dimensions, the near-field stress tensor components can be approximated by (Anderson, 2005)

$$\sigma_{11} = \frac{K_I}{\sqrt{2\pi r}} \cos\left(\frac{\theta}{2}\right)\left[1 - \sin\left(\frac{\theta}{2}\right)\sin\left(\frac{3\theta}{2}\right)\right], \qquad (2.3)$$

$$\sigma_{22} = \frac{K_I}{\sqrt{2\pi r}} \cos\left(\frac{\theta}{2}\right)\left[1 + \sin\left(\frac{\theta}{2}\right)\sin\left(\frac{3\theta}{2}\right)\right], \qquad (2.4)$$

and

$$\sigma_{12} = \frac{K_I}{\sqrt{2\pi r}} \cos\left(\frac{\theta}{2}\right)\sin\left(\frac{\theta}{2}\right)\cos\left(\frac{3\theta}{2}\right), \qquad (2.5)$$

where r, θ and orientation of the x_1 and x_2 axes are defined in Figure 2.3. The near-field approximation in these expressions means that $r \ll a$.

Okada (1992) developed closed-form analytical solutions for internal stresses and displacements due to tensile or shear slip on a rectangular fault in an elastic half-space. These expressions are lengthy and not repeated here; the interested reader is referred to Okada (1992) for details. This formulation has a number of advantages over the approximate formulas given above, since:

1 it provides an exact solution, rather than a near-field approximation;
2 Okada's solution is more realistic, as it accounts for the finite dimensions of the fracture as well as a fundamental asymmetry arising from the presence of a free surface at the top of an elastic half-space.

Figure 2.4 shows normal and shear stress fields associated with a vertical mode I fracture at 3000 m depth, obtained using Okada's analytical solution.[2] The fracture is 100 m in strike

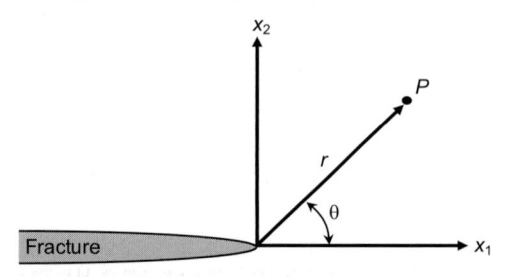

Fig. 2.3 Coordinate system and parameters r and θ used in approximate formulas for near-field fracture-tip stress at a point P.

[2] Note that this formulation does not depend on the specific driving mechanism for tensile failure; rather, it represents stresses in the medium that result from it.

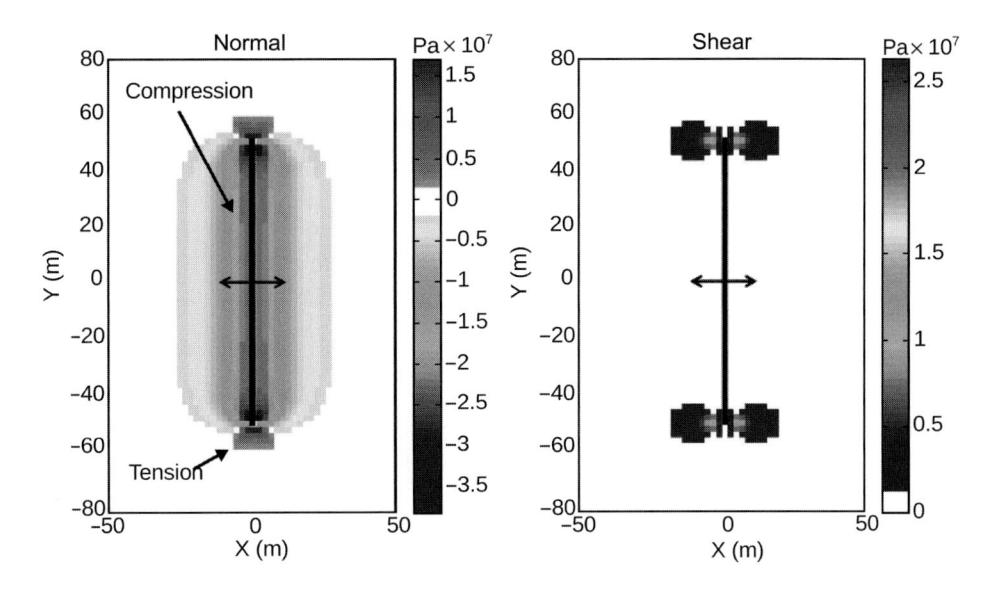

Fig. 2.4 Map view of normal stress and shear stresses around a vertical mode I crack. The crack is shown by a black line and is 100 m in length, 10 m in height, and located 3000 m below the surface of a homogeneous elastic half-space with $E = 2.25 \times 10^{10}$ Pa and $\nu = 0.28$. Fracture opening is 1.0 cm. Tensile stress is positive and localized near the crack tip. Stress calculations used the method of Okada (1992) and are resolved for the crack orientation. (A colour version of this figure can be found in the Plates section.)

length, 10 m in height, and has an aperture of 1 cm. Normal- and shear-stress components are calculated throughout the medium based on the orientation of the normal to the fracture plane. The normal stress is plotted using a polarity convention that is positive under tension, whereas the absolute shear stress is plotted. There is prominent stress concentration near the tips of the fracture, as expected based on the approximate formulas above. Moreover, the region alongside the fracture has negative normal stress, reflecting compression of the elastic material; the elastic host medium thus resists opening of the tensile fracture.

Figure 2.5 shows similar images for shear slip (mode II) on a fracture with the same configuration as in Figure 2.4. In this case the normal stress exhibits a characteristic polarity reversal across the fracture at the tip. A lobate pattern of shear stress is also apparent at the tip, as well as elastic shear resistance within the medium alongside the shear fracture. As with the stress field in Figure 2.4 for a tensile fracture, the stresses shown in Figure 2.5 are additive with respect to the background stress field. It should be noted that, in order to carry out the summation, both the fracture stress field and the background stress field must be expressed using the same coordinate system.

2.2 Effective Stress and Brittle-Failure Criteria

Various criteria have been established to determine stress conditions under which shear or tensile deformation of a rock mass are expected to occur. The criteria vary depending

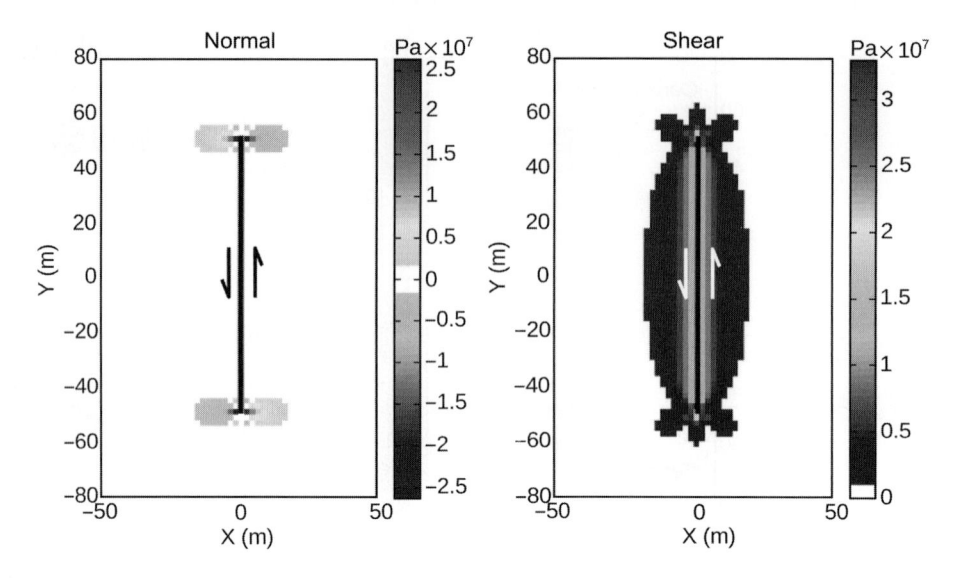

Fig. 2.5 Map view of normal stress and shear stress around a vertical mode II crack. The crack is shown by a black line and is 100 m in length, 10 m in height, and located 3000 m below the surface of a homogeneous elastic half-space with $E = 2.25 \times 10^{10}$ Pa and $\nu = 0.28$. Slip is 1.0 cm. Stress calculations used the method of Okada (1992) and are resolved for the crack orientation. (A colour version of this figure can be found in the Plates section.)

on the mode of failure, as well as whether the rock mass is intact and whether there are any pre-existing fractures or faults. Details of experimental procedures used to determine material parameters for these criteria are described by Zoback (2010). In contrast to LEFM and other theories of brittle-fracture processes, failure criteria are empirical and do not consider explicit physical processes for rock failure.

It is convenient to represent brittle-failure criteria using a Mohr diagram, described in Chapter 1. Using this representation, a particular failure criterion partitions the stress space into a stable region where failure is not expected, and an unstable region where the state of stress favours brittle failure. We begin with an empirical curvilinear criterion called the Mohr failure envelope. This can be determined using an experimental procedure, wherein cylindrical samples of a material are placed under a uniform confining pressure, followed by application of an increasing uniaxial stress until the sample fails. Although this type of test constitutes a special case where $\sigma_2 = \sigma_3$ it is customarily called a *triaxial* test. The magnitude of the uniaxial stress under which a material fails with no confining pressure is a material parameter called the *unconfined compressive strength*, which is denoted here by S. This parameter is sometimes called the uniaxial compressive strength. The Mohr envelope has a single point of tangency to each of the failure-state Mohr circles (Figure 2.6).

In the case of a poroelastic medium, both the state-of-stress and failure criterion are usually plotted with respect to the *effective* stress, σ'_{ij}. As a first approximation, the effective stress can be represented by

$$\sigma'_{ij} = \sigma_{ij} - P\delta_{ij}, \tag{2.6}$$

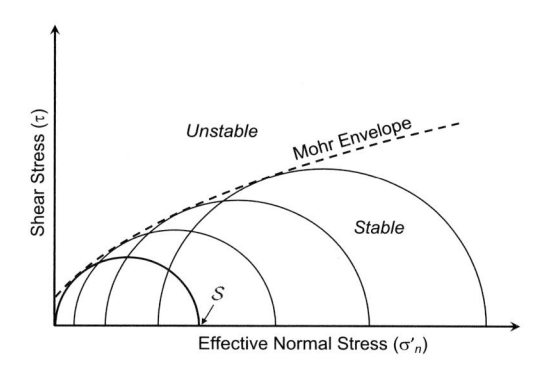

Fig. 2.6 Hypothetical set of 2-D Mohr circles depicting states of stress at the onset of failure, under varying confining conditions. The Mohr envelope is constructed as an empirically derived curve that is tangent to the failure-state Mohr circles, thus partitioning the stress space into stable and unstable regimes. S denotes the unconfined compressive strength.

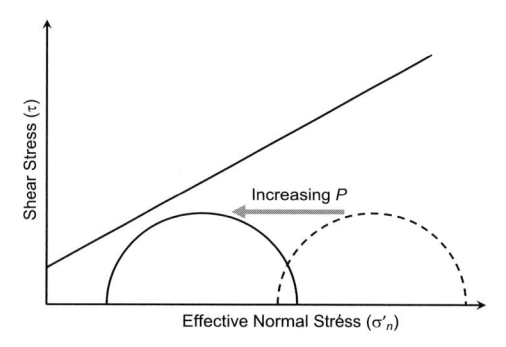

Fig. 2.7 2-D Mohr diagram illustrating a commonly invoked approximation to the effective stress arising from an increase in pore pressure. The dashed circle shows an initial (pre-injection) stress state; increasing pore pressure translates the Mohr circle with no change in deviatoric stress. Such a translation is consistent with full poroelastic coupling within a strike-slip regime (Lavrov, 2016). The black line shows a representative failure envelope.

where P is the pore pressure and δ_{ij} is the Kronecker delta. For any surface orientation, the use of this approximation for effective stress reduces the normal stress by P regardless of orientation, but does not alter the shear stress. This approximate form of effective stress is widely used and is often represented as a simple translation of the Mohr circle with no change in deviatoric stress (Figure 2.7).

A more general approximation for effective stress incorporates the Biot coefficient, α (see Chapter 1), into the pore-pressure term:

$$\sigma'_{ij} = \sigma_{ij} - \alpha P \delta_{ij}, 0 \leq \alpha \leq 1. \tag{2.7}$$

This form reduces to the same form as in Equation 2.6 in the special case where there is no poroelastic coupling between the fluid phase and the elastic framework ($\alpha = 1$).

This is still an approximation, however, and if poroelastic coupling is fully accounted for using Biot theory, changes in pore pressure are coupled to changes in deviatoric stress

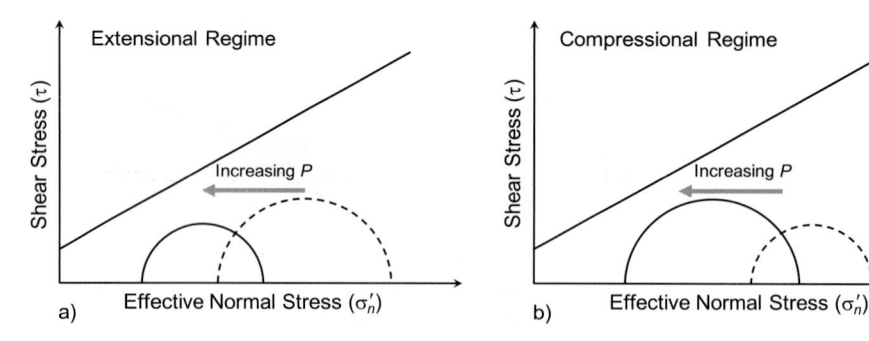

Fig. 2.8 Schematic illustration of the change in deviatoric stress within a porous reservoir (Lavrov, 2016) due to pore-pressure increase. Two cases are considered: a) an extensional stress regime and b) a compressional stress regime. The dashed Mohr circles show the initial stress state and the black line shows a representative failure envelope. Reprinted from *Energy Procedia*, Vol 86, Alexandre Lavrov, Dynamics of Stresses and Fractures in Reservoir and Cap Rock under Production and Injection, Pages 381–390, copyright 2016, with permission from Elsevier.

in a manner that depends upon stress regime (Lavrov, 2016). Figure 2.8 shows several examples of effective stress changes in a porous reservoir that incorporate full poroelastic coupling, for extensional and compressional stress regimes. Understanding these diagrams requires some basic knowledge of the classification of stress regimes. Anderson (1951) argued that in the Earth's crust one of the principal axes is likely to be perpendicular to the free surface and therefore close to vertical. The assumption that one of the principal stress axes is vertical leads to a natural classification scheme for stress regimes and faults, known as *Anderson's* fault classification, that is based on the relative magnitude of the *vertical stress* (S_V) as follows:

- Extensional (or normal) stress regime: $S_V = \sigma_1$;
- Strike-slip stress regime: $S_V = \sigma_2$;
- Compressional (or reverse) stress regime: $S_V = \sigma_3$.

As discussed in the following chapters, this classification scheme provides an important framework for understanding stress and fault behaviour as well as earthquake source mechanisms. The determination of stress magnitudes is covered in Chapter 4; in particular, S_V can usually be determined with a high degree of confidence, as shown explicitly in Equation 4.10. The vertical stress parameter therefore has a special significance for classifying stress regimes.

2.2.1 Mohr–Coulomb Criterion

Understanding the mechanical strength of rocks has been of considerable interest throughout civilization, due to the importance of underground mining as well as the use of rocks as building materials. In the late seventeenth century, the French physicist Charles Augustin de Coulomb proposed a failure criterion that bears his name. The Coulomb criterion for rock failure may be written as

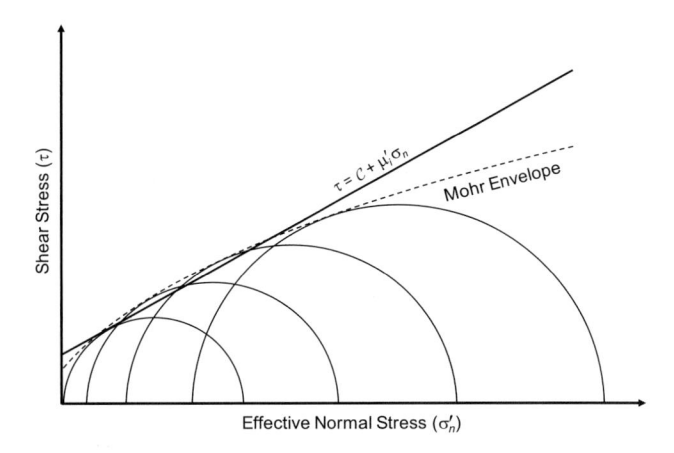

Fig. 2.9 The Mohr–Coulomb failure criterion is a linearized form of the Mohr envelope, with slope defined by the coefficient of internal friction (μ_i) and intercept defined by the cohesion, $\mathcal{C}$.

$$\tau = \mathcal{C} + \sigma'_n \tan \phi_i = \mathcal{C} + \mu_i \sigma'_n , \qquad (2.8)$$

where τ denotes the shear stress at which failure occurs under effective normal stress σ'_n. In this expression, $\mathcal{C}$ is called the *cohesion*, ϕ_i is the *angle of internal friction* and μ_i is the slope of the failure line, called the *coefficient of internal friction*. Although the cohesion exhibits a high degree of variation depending on the intrinsic material strength, the coefficient of internal friction shows less variability and generally falls within a range from 0.5 to 2.0, with a median value of 1.2 (Zoback, 2010). As described below, the internal friction angle has a close geometrical relationship to fault orientation for the common case where $\mathcal{C} = 0$.

The Coulomb criterion is a linearized form of the Mohr envelope (Figure 2.9) with slope μ_i and intercept $\mathcal{C}$. This relationship is also known as Mohr–Coulomb failure criterion. It should be noted that cohesion is not a physically measurable parameter, but it is related to the unconfined compressive strength ($\mathcal{S}$, see Figure 2.6) and μ_i as follows:

$$\mathcal{C} = \frac{\mathcal{S}}{2\left[\left(\mu_i^2 + 1\right)^{0.5} + \mu_i\right]} . \qquad (2.9)$$

Since both $\mathcal{S}$ and μ_i can be readily determined from laboratory strength tests (Zoback, 2010), the cohesion ($\mathcal{C}$) can be obtained from measured quantities using this formula.

2.2.2 Griffith Criterion

Rocks are weaker under tension than under compression. This is one of the reasons why the stresses at the fracture tips (Figure 2.4) can promote continued growth of a tensile fracture. The failure envelope under tension can be represented by the Griffith failure criterion,

$$\tau = \left(4T^2 - 4T\sigma'_n\right)^{1/2} , \qquad (2.10)$$

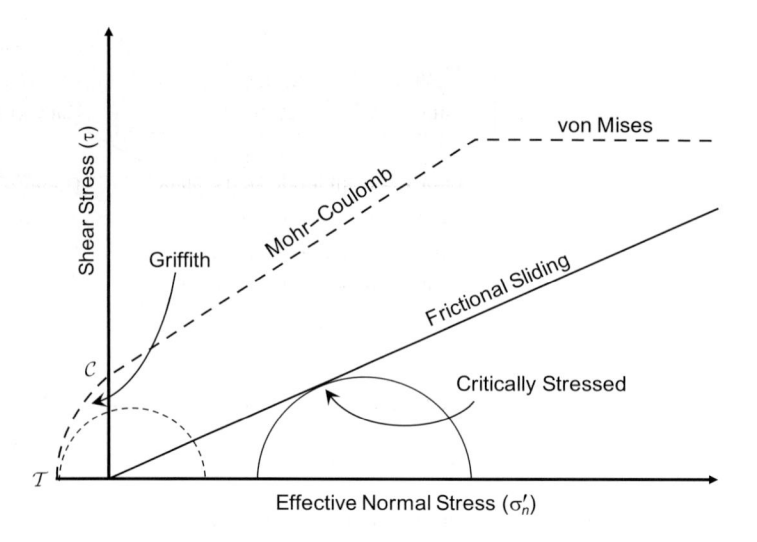

Fig. 2.10 Composite failure envelope (dashed) showing Griffith, Mohr–Coulomb and von Mises failure envelopes. The slope of the Mohr–Coulomb failure line is $\mu_i \simeq 1.2$. Continuity of the Griffith and Mohr–Coulomb criteria imposes a condition that $\mathcal{C} = 2\mathcal{T}$. A Mohr circle is shown (dashed) corresponding to a stress state that is on the verge of tensile failure. A cohesionless Mohr–Coulomb line is also shown to represent the condition for frictional sliding, together with a different Mohr circle that highlights a critically stressed state. The slope of the frictional sliding line is $\mu_s \simeq 0.6$.

where $\mathcal{T}$ is called the *tensile strength*. The Griffith criterion is parabolic in shape. As shown in Figure 2.10, the Griffith criterion can be combined with the Mohr–Coulomb criterion to give a composite failure envelope that encompasses both tensile and shear modes of failure. By combining Equations 2.8 and 2.10, the condition for continuity of these criteria at $\sigma'_n = 0$ is $\mathcal{C} = 2\mathcal{T}$. This relationship implies that only one of $\mathcal{C}$ and $\mathcal{T}$ is required for unique specification of a composite failure envelope; it is also consistent with the general observation that rocks are weak under tension.

Fracturing processes that combine shear and tensile failure have been observed for some cases of fluid-injection induced seismicity (Fischer and Guest, 2011; Eaton et al., 2014d). The source theory for this class of hybrid failure mechanism, referred to in this book as a *shear-tensile* source model, was developed by Vavryčuk (2011). Assuming a Griffith failure envelope, shear-tensile failure can occur for fractures whose strike orientation falls within $\approx 22.5°$ of the maximum principal stress axis. Fischer and Guest (2011) argued that shear-tensile fractures may be commonly triggered by fluid injection in the presence of a pre-existing fracture network.

2.2.3 Other Brittle-Failure Criteria

A number of other empirical failure criteria are used for rock-mechanics and geomechanics applications. The *Hoek–Brown* criterion was developed as a failure criterion for design of underground excavations, based on research into brittle fracture of intact rock (Hoek et al., 1968) and simulation of failure of fractured rock masses (Brown, 1970). Expressed in terms

Table 2.1 Hoek–Brown Material Parameter and Lithology

Lithologic Description	Range of m
Dolomite, limestone, marble	$5 < m < 8$
Mudstone, siltstone, shale, slate	$4 < m < 10$
Sandstone, quartzite	$15 < m < 24$
Fine-grained igneous rocks	$16 < m < 19$
Coarse-grained igneous rocks	$22 < m < 33$

of principal stresses, the original Hoek–Brown criterion is given by

$$\sigma_1 = \sigma_3 + S\left(m\frac{\sigma_3}{S} + s\right)^{0.5}, \tag{2.11}$$

where S is the unconfined compressive strength of the intact rock, m is a material constant and s is a scalar measure of the degree of fracturing, with $s = 1$ for an intact rock and $s = 0$ for unconsolidated material. Expressed in terms of σ'_n and τ, the Hoek–Brown formula yields a parabolic envelope in a Mohr diagram (Zoback, 2010). Table 2.1 summarizes the dependence of m on rock type (lithology) from (Hoek and Brown, 1997). A generalized form of the Hoek–Brown relationship (Hoek et al., 2002) has been developed, which links material parameters in the modified failure criterion to the *Geological Strength Index* (GSI). The GSI is a system of rock-mass characterization that is used in the field of rock mechanics (Marinos et al., 2005).

Both the Mohr–Coulomb and the Hoek–Brown criteria do not consider the influence of the intermediate principal stress, σ_2. Another rock failure criterion, known as the *modified Lade criterion*, includes the intermediate principal stress and has been applied for wellbore stability predictions (Ewy, 1999). If two of the principal stresses are known, this criterion can be assessed by solving for the unknown principal stress using

$$\frac{(I'_1)^3}{I'_3} = 27 + \tilde{\eta}, \tag{2.12}$$

where I'_1 and I'_3 are modified stress invariants given by

$$I'_1 = (\sigma_1 + C/\mu) + (\sigma_2 + C/\mu) + (\sigma_3 + C/\mu), \tag{2.13}$$

and

$$I'_3 = (\sigma_1 + C/\mu)(\sigma_2 + C/\mu)(\sigma_3 + C/\mu). \tag{2.14}$$

In addition, the parameter $\tilde{\eta}$ is given by

$$\tilde{\eta} = \frac{4(\tan \phi_i)^2 (9 - 7\sin \phi_i)}{1 - \sin \phi_i}, \tag{2.15}$$

where ϕ_i is the angle of internal friction. For given values of σ_1 and σ_3, the modified Lade criterion predicts that the rock strength is maximized for a value of σ_2 that is intermediate between the other principal stresses, that is no two principal stresses are equal (Zoback, 2010).

2.3 Frictional Sliding on a Fault

Earthquakes occur through stick-slip frictional instability on faults and seldom, if ever, occur through the development of a new shear crack (Scholz, 1998). Persistence of rock-deformation processes on a geological timescale can ultimately lead to development of a fault zone (Ben-Zion and Sammis, 2003), the inception of which is interpreted to take place within a damage zone through a progressive process that links up individual fractures and deformation bands (Fossen et al., 2007). These linkages form ahead of the fault tip, within a process zone (Figure 2.2). One feature of this model presents a physical conundrum: whereas dynamic growth of a mode I fracture can be explained by fracture-tip stresses, especially in view of the inherent weakness of rocks under tension, fracture-mechanics models imply that growth of mode II and mode III fractures is inherently self-limiting. Consequently, fault tips tend to propagate into intact rocks through more complex fracture linkages, such as wing faults (Fossen, 2016). These linkages may be energetically favoured by *strain weakening* processes that reduce frictional contact at *asperities* (an area of a fault that is stuck or locked) through cataclasis (grinding).

Once a mature fault has formed, subsequent deformation is governed by frictional processes, rather than fracturing. It has been understood since the time of Leonardo da Vinci that friction measures the tangential sliding resistance along a slip surface and scales with the normal stress acting on the surface. Based upon a compilation of experimental data, Byerlee (1978) showed that, under conditions of low normal stress ($\lesssim$ 5 MPa), the shear stress required to initiate sliding exhibits a high degree of variability; this was attributed to a corresponding variability in surface roughness related to lithology dependent characteristics of asperities. Under high normal stress, the lithologic differences diminish, leading to a simple constitutive relationship between the shear stress at failure (τ) and the effective normal stress acting on the slip surface (σ'_n),

$$\tau = \mu_s \sigma'_n, \qquad (2.16)$$

where μ_s is the *coefficient of static friction* and where $0.6 \lesssim \mu_s \lesssim 0.85$. Although it is equivalent to the Mohr–Coulomb relationship with $\mathcal{C} \to 0$, it should be emphasized that the friction coefficient here differs from the coefficient of internal friction that applies to fracturing processes.

Deep drilling and induced seismicity experiments indicate that, in most intraplate regions, Earth's crust is in a state of incipient frictional failure that is maintained by natural pore-pressure feedbacks (Townend and Zoback, 2000). Known as a *critical stress state*, if this scenario applies globally it implies that favourably oriented crustal faults are available, or equivalently that crustal faults are widely abundant and their distribution is characterized by normal vectors that exhibit a broad range of orientations. A critical stress state means that, for all depths, the 2-D Mohr circle is approximately tangent to the failure criterion defined by Equation 2.16. With respect to the three Andersonian stress regimes, criteria for occurrence of a critically stressed state can be expressed in terms of the ratio of the maximum and minimum effective stresses as follows (Zoback, 2010):

Extensional stress regime: $\dfrac{\sigma'_1}{\sigma'_3} = \dfrac{S_V - \alpha P}{S_{Hmin} - \alpha P} \leq \left[(\mu_s^2 + 1)^{1/2} + \mu_s\right]^2;$ (2.17)

Strike-slip stress regime: $\dfrac{\sigma'_1}{\sigma'_3} = \dfrac{S_{Hmax} - \alpha P}{S_{Hmin} - \alpha P} \leq \left[(\mu_s^2 + 1)^{1/2} + \mu_s\right]^2;$ (2.18)

and

Compressional stress regime: $\dfrac{\sigma'_1}{\sigma'_3} = \dfrac{S_{Hmax} - \alpha P}{S_V - \alpha P} \leq \left[(\mu_s^2 + 1)^{1/2} + \mu_s\right]^2,$ (2.19)

where, by definition, $\sigma_1 \geq \sigma_3$. Zoback et al. (1986) developed a diagram called a *stress polygon* that depicts the allowable range of horizontal principal stresses for specified values of S'_V, μ_s and α, assuming that the crust is in a critically stressed state (Figure 2.11). The horizontal principal stresses are denoted by S_{Hmin} and S_{Hmax}, respectively. For axes, this diagram uses the effective horizontal principal stresses, S'_{Hmin} and S'_{Hmax}, where S'_{Hmin} is the effective stress equal to $S_{Hmin} - \alpha P$, etc. The stress polygon is constructed from three contiguous triangles that share a common vertex at the point $S'_{Hmin} = S'_{Hmax} = S'_V$. The lower boundary of the stress polygon follows the line $S'_{Hmax} = S'_{Hmin}$. Other boundaries include the horizontal boundary between extensional and strike-slip regimes, where $S_{Hmax} = S_V$, and the vertical boundary between strike-slip and compressional regimes, where $S_{Hmin} = S_V$, both of which arise from the definition of the Andersonian stress regimes. Finally the outer perimeter of the stress polygon corresponds with critically stressed conditions defined by Equations 2.17–2.19.

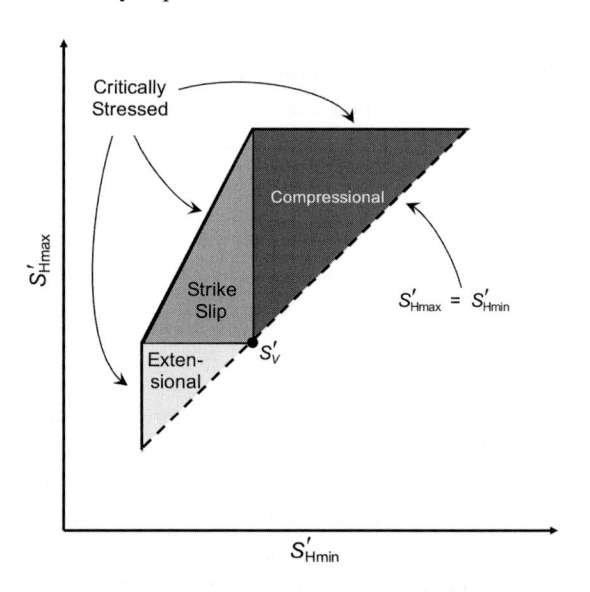

Fig. 2.11 Stress polygon showing the allowable range of $S'_{Hmax} - S'_{Hmin}$ effective stress values within extensional, strike-slip and compressional stress regimes, assuming that the crust is in a critically stressed state. Each regime forms a triangular region, determined based on specified values of S_V, pore pressure (P) and Biot parameter (α).

2.3.1 Rate–State Friction

Earthquake *nucleation* and propagation occur through runaway rupture processes that are driven by dynamic slip weakening on a fault (Garagash and Germanovich, 2012). Nucleation is a slip-acceleration process that occurs within a fault patch (Rubin and Ampuero, 2005). Earthquake nucleation requires unstable slip conditions on a fault, that is resistance to sliding must diminish faster than elastic unloading during fault slip. In the absence of any slip-weakening process, fault friction would retain its static value and dynamic slip would not occur. Early models based on the simple concept of a dynamic friction coefficient that is lower than the static coefficient have been replaced by a paradigm of *rate–state* friction, wherein the coefficient of friction depends on the slip velocity (rate) as well as the history of the slip surface (state). Rate–state models explain the following experimental observations of fault friction (Scholz, 1998, 2002):

1. the "static" friction coefficient, μ_s, increases logarithmically with time since the last slip event (Dieterich, 1972) due to fault-creep related gradual increase in surface contact area;
2. during steady-state sliding, the dynamic friction coefficient depends on the logarithm of the slip velocity (Scholz et al., 1972);
3. the tendency to produce a slip-induced increase (velocity strengthening) or decrease (velocity weakening) in dynamic friction depends upon rock type and temperature (Stesky et al., 1974);
4. if a fault is subjected to a sudden change in sliding velocity, the coefficient of friction approaches a new steady-state value over a characteristic slip distance, $\mathcal{L}$ (Dieterich, 1978).

Of various rate–state constitutive models, one that is in particularly good agreement with observations is called the *Dieterich–Ruina relationship*, or *slowness* law, in which the dynamic friction ($\tilde{\mu}$) can be expressed as (Scholz, 1998)

$$\tilde{\mu} = \tilde{\mu}_0 + a \ln\left(\frac{V}{V_0}\right) + b \ln\left(\frac{V_0 \theta}{\mathcal{L}}\right), \qquad (2.20)$$

where V denotes slip velocity, V_0 is a reference velocity, $\tilde{\mu}_0$ is the steady-state friction at $V = V_0$, a and b are constitutive properties and θ is a state parameter that satisfies:

$$\dot{\theta} = 1 - \theta V/\mathcal{L}. \qquad (2.21)$$

Figure 2.12 illustrates the relative changes in $\tilde{\mu}$ predicted by the Dieterich–Ruina rate–state constitutive model, in response to a sudden increase followed by a decrease in slip velocity. An initial instantaneous increase in frictional resistance is governed by the value of the a parameter. This is followed by a gradual decrease in friction that is asymptotic to a new steady-state value that differs from the initial friction by $a - b$. Similarly, the abrupt decrease in sliding velocity to its initial value is accompanied by an abrupt decrease in friction followed by a gradual recovery to the initial friction value. The frictional behaviour embodied in Figure 2.12 leads to a rich diversity of earthquake phenomena (Scholz, 1998).

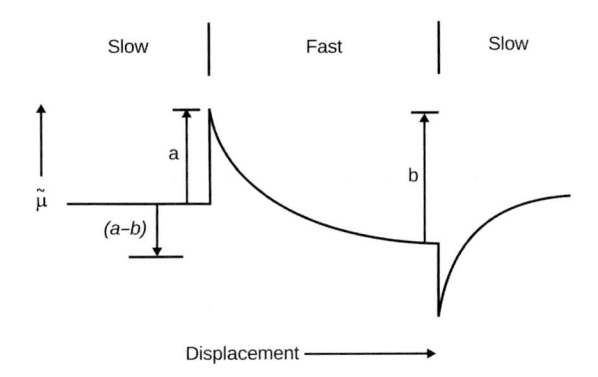

Fig. 2.12 Schematic diagram of frictional response to a suddenly imposed increase and then decrease in sliding velocity, based on the Dieterich–Ruina rate–state constitutive model. The parameter $\tilde{\mu}$ denotes dynamic friction, while a and b are constitutive parameters. Modified from Scholz (1998). Reprinted by permission from MacMillan Publishers Ltd.: *Nature*, 391, 37–42, copyright 1998.

The composite parameter $a-b$ strongly influences fault instability. A fault is intrinsically stable and is said to exhibit *velocity strengthening* behaviour if the criterion $a - b \geq 0$ is met. Conversely, a fault with $a - b < 0$ is said to exhibit *velocity weakening* behaviour. Indeed, earthquakes can only nucleate in regions where $a - b < 0$ and slip is arrested when propagation enters a region where $a - b > 0$ (Scholz, 2002). The instability behaviour of a velocity weakening fault in response to an abrupt increase in slip velocity depends on the magnitude of ΔV; under various conditions, a fault may be conditionally stable, unstable or it may exhibit oscillatory behaviour (Scholz, 1998).

Figure 2.13 shows experimental measurements of friction coefficient and the rate–state parameter $a - b$ for shale samples (Kohli and Zoback, 2013). These measurements show that shale samples with clay and organic content no greater than 30 percent are characterized by slip-weakening behaviour. This transition was interpreted by Kohli and Zoback (2013) to reflect a change in the shale grain-packing framework.

2.4 Earthquake Cycle and Stress Drop

The buildup and release of tectonic stress acting on a fault results in an earthquake cycle that is controlled by *stick–slip* behaviour. Stress on a fault increases gradually during the time interval when a fault is stuck. This stress is abruptly released during *co-seismic* rupture.[3] *Stress drop* is an earthquake source parameter that characterizes the spatially averaged co-seismic reduction in shear stress on a fault. In its most basic form, stress drop is defined as $\Delta \tau = \tau_1 - \tau_2$, where τ_1 is the initial stress (failure strength) on a fault and τ_2 is the residual stress (Wyss and Molnar, 1972). For a particular segment of a fault system, temporal variations in stress drop are closely linked to varying stress conditions,

---

[3] The term *co-seismic* refers to processes that take place during earthquake rupture.

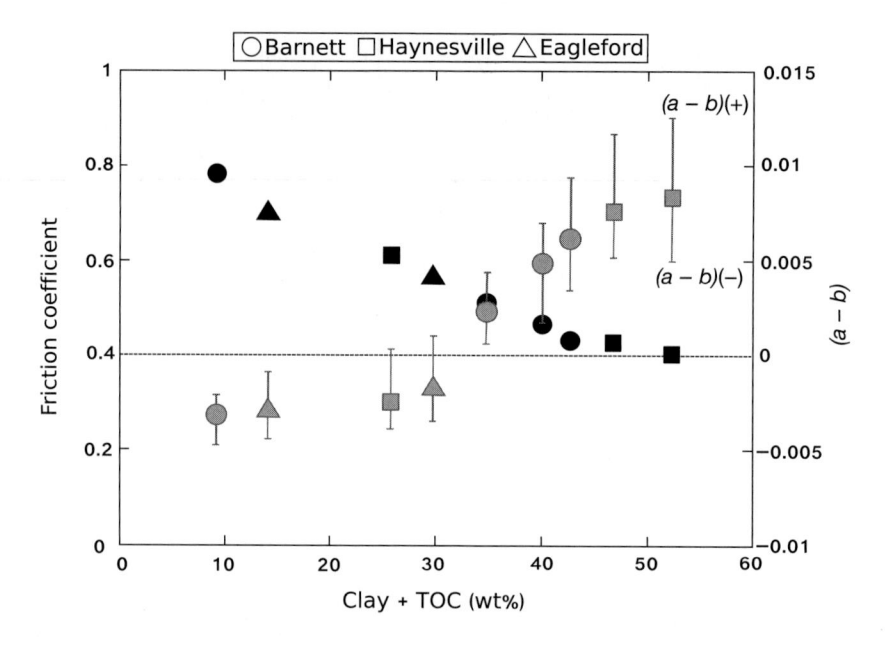

Fig. 2.13 Friction coefficient (black symbols) and rate–state friction parameter $(a - b)$ (grey symbols) for samples from three US shale plays (Kohli and Zoback, 2013). Samples with clay and organic content below about 30 percent exhibit velocity weakening behaviour. Reprinted with permission from Wiley.

fault friction and earthquake recurrence models (Shimazaki and Nakata, 1980). In a recent global compilation (Allmann and Shearer, 2009), stress drop and inferred fault strength have been observed to vary with tectonic regime, with average values ranging from 2.63 ± 0.5 MPa in continental collision boundaries to 6.03 ± 0.68 MPa along oceanic transform faults. For subduction zones, systematic changes in stress drop may be linked to depth-dependent variations in rigidity or friction along the plate-bounding fault (Bilek and Lay, 1999).

As illustrated in Figure 2.14, a strictly periodic process is implied for the simplest case of constant stress drop with time-independent initial and residual stresses. On the other hand, if a fault system is characterized by time-independent initial stress τ_1 and variable residual stress, then time-predictable behaviour is implied; conversely, if the system is characterized by time-independent residual stress τ_2 and variable initial stress, the behaviour is slip-predictable (Shimazaki and Nakata, 1980). In the general case in which both initial and residual stresses vary, then no regularity in fault slip behaviour exists (Kanamori and Brodsky, 2004). By linking magnitude scales or seismic moment with spatial characteristics of rupture processes, knowledge of stress drop can provide insights for fault strength and earthquake scaling relations (Kanamori and Anderson, 1975).

Faults that have experienced a prolonged period of quiescence are generally well healed and characterized by non-negligible cohesion, due to recrystallization and/or cementation processes during the interseismic period (Muhuri et al., 2003); however, as outlined above, most fault constitutive models derived from laboratory studies of active fault systems assume a cohesionless state. For well-healed faults, which are likely to be the

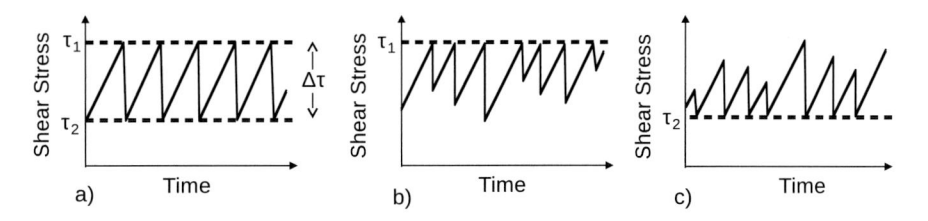

Fig. 2.14 Stress drop ($\Delta\tau$) and its relationship to simple earthquake recurrence models for constant loading rate, modified from Shimazaki and Nakata (1980). τ_1 denotes shear stress on the fault at the initiation of slip and reflects fault strength; τ_2 denotes shear stress at the termination of slip and reflects friction. a) Simple stick-slip recurrence model with constant stress drop, $\Delta\tau$. b) Time-predictable recurrence model, in which fault strength is assumed to be constant and slip initiates when stress reaches level τ_1, resulting in variable stress drop. c) Slip-predictable recurrence model, in which slip initiates randomly but terminates when shear stress reaches τ_2. Reprinted with permission from Wiley.

norm in previously quiescent regions where injection-induced seismicity is most abundant, earthquake-slip instability behaviour may therefore be strongly influenced by fault cohesive strength (Rutqvist et al., 2013). For a fault with non-negligible cohesion that is critically stressed, the relationship between the maximum (σ_1) and minimum (σ_3) principal stresses may be written as (Sattari, 2017)

$$\sigma_1 = \left[\sqrt{\mu_s^2 + 1} + \mu_s\right]^2 \sigma_3 + 2\mathcal{C}\left[\sqrt{\mu_s^2 + 1} + \mu_s\right]. \tag{2.22}$$

Beeler (2001) assumed that co-seismic stress drop can be expressed in terms of the difference between static friction (μ_s) and residual friction (μ_r):

$$\Delta\tau = \sigma_n'(\mu_s - \mu_r). \tag{2.23}$$

Within the rate–state frictional paradigm, the residual friction in this expression can be viewed as rejuvenation of the slip surface accompanied by establishment of a reduced friction coefficient (Dieterich, 1972). Underlying this expression is a further assumption that the normal stress is unchanged before and after slip on a fault, such that the stress drop represents a pure reduction in shear stress on a fault surface. Sattari (2017) tested this concept using a plane-strain finite-element method (Steffen et al., 2014) to simulate slip on a fault with cohesion. In practice, $\mathcal{C}$ is expected to drop to zero after slip initiates, as bonds between fault surfaces are broken (Muhuri et al., 2003). As illustrated in Figure 2.15, the finite-element simulation confirms that the shear stress on the fault is reduced by

$$\Delta\tau = \mathcal{C} + \sigma_n'(\mu_s - \mu_r), \tag{2.24}$$

with no change in the normal stress acting on the fault. This result provides a simple physical interpretation of coseismic stress drop, namely the establishment of a lower energy state after slip by reduction of average shear stress on a fault. According to this model, the reduction in shear stress reflects a change in friction and, in the case of slip on well healed faults, a loss of cohesion. A subtle yet significant aspect of this model is that, as a result of the change in frictional state, a fault that is initially optimally oriented is no

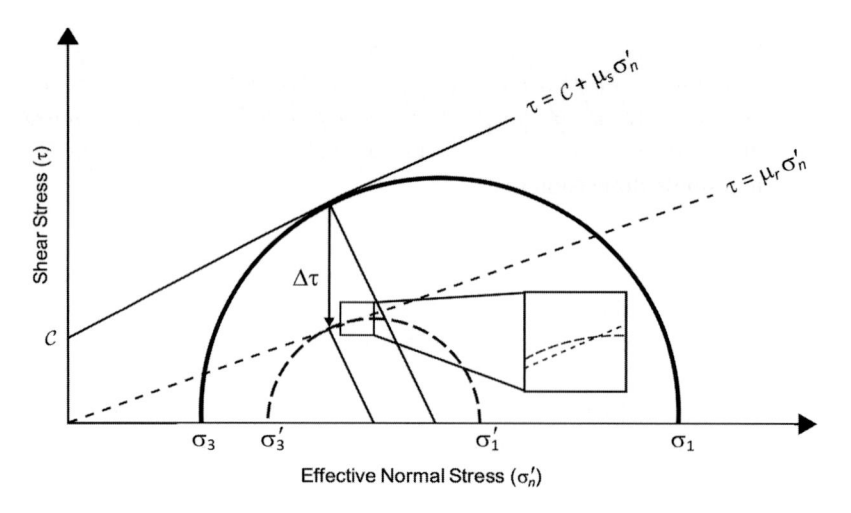

Fig. 2.15 Mohr circle representation of co-seismic stress drop based on finite-element modelling of fault slip for an optimally oriented fault (Sattari, 2017). The solid line shows the initial Mohr–Coulomb failure criterion, while the dashed line shows the fault strength after slip, characterized by reduced friction and cohesion. Similarly, the solid and dashed Mohr circles show the initial (critically stressed) and residual stress states for the fault, respectively. The downward-pointing arrow shows the computed co-seismic stress drop ($\Delta\tau$), which is a pure reduction in shear stress with no change in normal stress acting on the fault. Post-seismic principal stresses are indicated by σ'_3 and σ'_1. Inset shows that after slip, the fault is no longer precisely optimally oriented for slip. Modified with permission of the author.

longer precisely optimally oriented within the prevailing stress field. This is significant as it implies that nearby faults with slightly different orientations may then become critically stressed.

2.5 Ductile Deformation

Ductile deformation refers to the accumulation of strain without fracturing or stick–slip motion on faults; this type of deformation is commonly expressed in rocks by folding and bending of strata. Folding can occur in association with fault slip, such as *drag folds* located within the damage zone of a fault zone, or *fault-propagation folds* that occur in front of the tip of a propagating fault; indeed, when seismic profiles are used to identify faults, these associated ductile elements are sometimes the most clearly evident structures (Fossen, 2016).

The constitutive relationships associated with two types of ductile deformation are briefly considered here: *viscoelastic* and *plastic*. Viscoelastic behaviour describes materials that exhibit an instantaneous deformation in response to stress, followed by a gradual continuous deformation process known as *creep*. Although deformation of a viscoelastic material is recoverable in response to a transient applied stress, there is a time lag between the stress release and recovery. In general, the constitutive relationship for a viscoelastic medium can be split into two parts, a linear elastic component that satisfies generalized Hooke's Law, and a viscous component with a Newtonian rheology. For the sake of simplicity, the relationship between stress and strain for these components can be represented in a single dimension as

$$\sigma = E\epsilon , \qquad (2.25)$$

and

$$\sigma = \eta\dot{\epsilon} , \qquad (2.26)$$

where E and η represent scalar elastic stiffness and dynamic viscosity, respectively. Depending upon the choice of models, these can be combined together to produce different viscoelastic responses. For example, the constitutive relation for a 1-D Voigt material can be written as a simple combination of these components (Courtney, 2000),

$$\sigma = E\epsilon + \eta\dot{\epsilon} . \qquad (2.27)$$

This type of material can be illustrated schematically as a spring and a *dashpot* in parallel (Figure 2.16a), where the spring represents the elastic component of the medium and the dashpot represents the mechanical damping associated with the viscous component of the medium. It is sometimes useful to add additional elements in order to achieve a better representation of the temporal relaxation of a material. For example, a standard linear solid (Figure 2.16b) contains a Voigt material in series with an additional spring and dashpot (Courtney, 2000).

Plastic behaviour is used to describe a medium that undergoes continuous creep after reaching a stress or strain threshold. Many rocks deform in an elastic manner up until a critical stress value, known as the *yield point*. This point marks the onset of plastic (ductile) deformation, which can be accommodated at the microscale by *crystal-plastic* deformation processes that include *dislocation creep* (Fossen, 2016). After the onset of plastic deformation, a perfectly plastic material deforms continuously under uniform stress conditions. This is called the *von Mises failure criterion* (Figure 2.10). Many materials, on the other hand, exhibit strain *hardening* or *weakening* behaviour, in which continuing deformation is progressively harder or easier, respectively (Fossen, 2016).

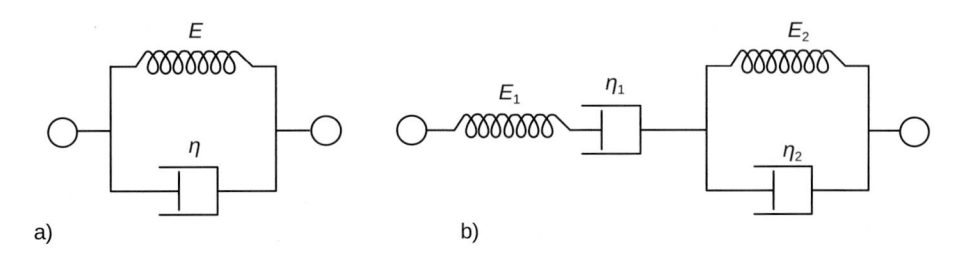

Fig. 2.16 Spring and dashpot model for a) a Voigt medium and b) a standard linear solid. The medium is parameterized using scalar elastic stiffness (E) and viscosity (η) parameters.

2.6 Summary

Failure criteria determine the state-of-stress that corresponds with the onset of failure. Various brittle failure criteria have arisen from experimental studies under different conditions, including Mohr–Coulomb, Hoek–Brown and Griffith. The Mohr–Coulomb critierion is a linearized failure envelope that is usually parameterized by a friction coefficient, which defines the slope of the failure line, and cohesion, which defines the y-intercept. The Hoek–Brown criterion is parameterized using the unconfined compressive strength, a lithology-dependent material constant and a scalar measure of the degree of fracturing of the rock mass. The Griffith criterion describes rock failure under tensile conditions and is parameterized by the tensile strength.

Failure is often localized along a fracture or fault surface. Within the discipline of linear-elastic fracture mechanics (LEFM), three fracture modes are recognized. A mode I fracture consists of tensile opening, whereas mode II (sliding) and mode III (tearing) represent shear dislocation mechanisms.

Anderson's theory of faulting assumes that one of the principal stress axes is vertical while the other two are horizontal, leading to three distinct stress regimes (normal, reverse and strike-slip) that are defined by a specific permutation of the three principal stresses. A diagram called a stress polygon depicts the allowable range of horizontal stress conditions within each of these stress regimes.

Laboratory friction experiments have led to the development of rate–state friction models that define instability conditions for dynamic rupture on a fault. According to these models, if a fault is subjected to a sudden change in sliding velocity, the coefficient of friction approaches a new steady-state value over a characteristic slip distance. The slowness law defines a rate–state constitutive model that provides a particularly good fit to experimental observations. Important parameters for this model are the a and b parameters, since earthquakes can only nucleate in regions where $a - b < 0$.

Active tectonic fault systems are characterized by earthquake cycles that are controlled by stick-slip fault behaviour. The scalar co-seismic stress drop is a measure of the relaxation of shear stress on a fault. In contrast with brittle failure processes during earthquakes, ductile processes lead to folding and bending of rock strata.

Viscoelastic behaviour describes materials that exhibit an instantaneous deformation in response to stress, followed by a gradual continuous deformation process known as creep, whereas plastic behaviour describes media that undergo continuous creep after reaching a stress or strain threshold.

2.7 Suggestions for Further Reading

- *The Mechanics of Earthquakes and Faulting* (Scholz, 2002).
- Rheology of Earth materials: Macosko (1994).
- Geomechanical principles: Zoback (2010).
- *Fundamentals of Rock Mechanics* (Jaeger et al., 2009).
- *Structural Geology* (Fossen, 2016).
- *Fracture Mechanics*: Fundamentals and Applications (Anderson, 2005).

2.8 Problems

1. Consider a tensile (mode I) crack as depicted in Figure 2.3.

   a) Given $K_I/\sqrt{2\pi r} = 1.0$ MPa, calculate the 2D stress tensor at $\theta = 0°, 30°$ and $90°$ from the tip of a mode I fracture (using Equations 2.3–2.5).
   b) Assuming $K_I/\sqrt{2\pi r} = 1.0$ MPa, what is the normal stress for each of the values of θ, for a planar surface that is oriented $30°$ from the fracture plane? What is the shear stress?
   c) Repeat the above calculations for a fracture orientation parallel to the main model I fracture.
   d) Which regions in the vicinity of the fracture tip are in tension? Which regions have elevated shear stress?

2. A set of laboratory measurements on a sample of dolomite indicates that the unconfined (or uniaxial) compressive strength (S) is 250 MPa.

   a) Determine the cohesion, for $\mu_i = 1.0$.
   b) Assuming $\sigma_1 = 15$ MPa, determine σ_3 at the point of failure, based on the Mohr–Coulomb criterion.
   c) Calculate σ_3 at the point of failure based on the Hoek–Brown criterion for an intact rock, using the mean value of the material constant m for dolomite given in Table 2.1.
   d) Calculate σ_3 at the point of failure using the modified Lade criterion, assuming that σ_2 is the mean of the maximum and minimum principal stresses.

3. Consider a point in the subsurface at a depth of 3000 m. Assuming a linear density gradient given by $\rho(z) = 2200 + 0.14z$, where ρ is in kg/m^3 and z is in m, calculate the following.

   a) Determine S_V.
   b) If the pore-pressure gradient is 15 kPa/m and the Biot coefficient (α) is 0.5, determine S'_V.
   c) For a critically stressed state in an extensional faulting regime with $\mu_s = 0.6$, what is S'_{Hmin}?
   d) For a critically stressed state in a strike-slip faulting regime with $\mu_s = 0.6$, what is S'_{Hmin} assuming that S'_V is the average of S'_{Hmax} and S'_{Hmin}?

4. Consider a rate-state friction relationship as defined by equation 2.20. How would you go about seeking a functional form of the state variable θ that satisfies equation 2.21?

5. Assuming that a slip surface is in a critically stressed state before and after rupture, as shown in Figure 2.15, calculate the stress drop if $\sigma_1 = 40$ MPa, $\sigma_3 = 15$ MPa, $\mu_s = 0.6$, $\mu_r = 0.4$ and $C = 5$ MPa.

6. Suppose that the Voigt model depicted in Figure 2.16a is subject to an abrupt increase in applied stress, σ. As outlined by Courtney (2000), the parallel configuration of the spring and dashpot in this model means that strain is equal for both elements, but the stress is

partitioned such that the total stress is $\sigma = \sigma_{sp} + \sigma_d$, where the subscripts sp and d denote the spring and dashpot, respectively. At the instant the stress is applied, the stress is carried entirely by the dashpot; over time, the stress is gradually transferred to the spring. Find analytical expressions for the temporal behaviour of $\sigma_{sp}, \sigma_d, \epsilon$ and $\dot{\epsilon}$.

7. **Online exercise**: Mohr diagrams are useful for visualizing the state of stress in addition to various forms of failure criteria. A Matlab tool is provided to depict 3-D Mohr diagrams under varying stress conditions, including the stress acting on fractures with random orientation.

3 Seismic Waves and Sources

> The footsteps of Nature are to be trac'd, not only in her ordinary course, but when she seems to be put to her shifts, to make many doublings and turnings, and to use some kind of art in endeavouring to avoid our discovery.
>
> Robert Hooke (Micrographia, 1665)

The point of departure for this chapter is development of a basic mathematical framework for waves that are generated by natural sources, such as earthquakes, or waves that are generated by artificial sources, such as buried explosive charges. Characteristics of the radiated wavefield depend on both the nature of the seismic source, including its geometry, size, kinematics and deformation mechanism, as well as the nature of the medium through which the waves propagate. In general, for an isotropic elastic medium the total wave energy is conserved during propagation. In such a non-attenuating medium, wavefield amplitude diminishes with distance due to the effects of geometrical spreading. Conversely, for an attenuating medium, wave amplitude decays more rapidly due to various dissipation mechanisms. If the medium is anisotropic, wave velocity depends on direction of propagation and polarization of the displacement vector. Moreover, waves in anisotropic media may exhibit certain diagnostic characteristics, such as shear-wave splitting.

The radiation pattern of P and S waves (or qP and qS waves in the case of anisotropic media) depends upon azimuth and inclination of the raypath from the source to the sensor, as well as the source type and orientation. Earthquakes typically produce mode II or mode III shear slip on a fault, but mode I (tensile) rupture processes have also been documented in response to fluid injection and/or volcanic processes. Seismic moment tensors provide a concise mathematical idealization to represent generalized deformation processes that act at a point in the subsurface, in the form of force couples. Unlike classical seismic-wave theory developed in most seismology texts, in this treatment the mathematical framework is developed assuming general anisotropy, from which isotropy emerges as a special case.

3.1 Equations of Motion

Consider an elementary volume, bounded by a surface S, within a continuous elastic medium (as depicted in Figure 1.1). Using Equation 1.3 and considering the symmetry properties of the stress tensor, the net force acting on this volume element may be written as

$$F_i = \int_S T_i dS = \int_S \sigma_{ij} \hat{n}_j dS. \tag{3.1}$$

For any vector field $\boldsymbol{\psi}(\mathbf{x})$, the *divergence theorem* states that

$$\int_V \nabla \cdot \boldsymbol{\psi} \, dV = \int_S \boldsymbol{\psi} \cdot \hat{\mathbf{n}} \, dS \, ; \tag{3.2}$$

or, using indicial notation,

$$\int_V \psi_{j,j} \, dV = \int_S \psi_j \hat{n}_j \, dS. \tag{3.3}$$

By combining Equations 3.1–3.3 and invoking Newton's second law of motion ($\mathbf{F} = m\mathbf{a}$), the net force per unit volume may be expressed as

$$\sigma_{ij,j} + f_i = \rho \ddot{u}_i, \quad i = 1, 2, 3 \tag{3.4}$$

where $\sigma_{ij,j}$ represents the net force due to the stress field acting within the medium, and $\mathbf{f}$ denotes internal forces such as a seismic source or gravitation. Since $\mathbf{u}$ is the displacement of a particle, $\ddot{u}_i$ represents the ith component of acceleration of a particle in the medium. Equation 3.4 is applicable to any type of medium and defines the *equations of motion*.[1] Next, by considering the definition of strain (Equation 1.9) and inherent symmetries in the tensor quantities, the constitutive relationship (Hooke's Law, Equation 1.10) may be written as

$$\sigma_{ij} = c_{ijkl} u_{k,l}. \tag{3.5}$$

By substituting Equation 3.5 into the equations of motion and rearranging, we obtain a form applicable to a general (anisotropic) elastic medium,

$$(c_{ijkl} u_{k,l})_{,j} - \rho \ddot{u}_i = -f_i. \tag{3.6}$$

In the case of a homogeneous medium where $c_{ijkl,j} = 0$, by applying the product rule to Equation 3.6 we obtain

$$c_{ijkl} u_{k,lj} - \rho \ddot{u}_i = -f_i. \tag{3.7}$$

Due to the inclusion of a source term on the right-hand side, in the context of the theory of partial differential equations this form is referred to as inhomogeneous (not to be confused with spatial inhomogeneity in properties of the medium). Finally, in the absence of body forces the equations of motion may be written in homogeneous form as

$$c_{ijkl} u_{k,lj} - \rho \ddot{u}_i = 0. \tag{3.8}$$

As elaborated below, wave propagation problems of interest can be investigated by solving the boundary-value problem corresponding to the applicable equations of motion (e.g. Equation 3.7 in the presence of internal forces, or Equation 3.8 otherwise), subject to a desired set of boundary conditions, elastic moduli, anisotropic symmetry system, etc.

---

[1] The plural is used here as a reminder that there are three distinct equations implied by the formulation in Equation 3.4.

3.2 Wave Solutions

Various types of waves represent solutions to boundary-value problems in seismology. For example, body-wave solutions represent waves that propagate within the interior of a continuum, whereas surface waves are guided by an external surface, often the Earth's free surface. Both of these types of waves can be investigated using solutions to the homogeneous form of the equations of motion. The inhomogeneous form of the equations of motion is used to obtain Green's functions, where the internal force is represented as an impulsive force acting at a point in a medium. Informally, Green's functions can be considered as an impulse response of the medium, and provide building blocks to develop solutions with more complex source-time functions. Finally, since much of the Earth's shallow crust is stratified, it is very important to consider scattering of waves at boundaries between two different types of materials. *Welded-contact* boundary conditions are typically used to represent an interface, which assume that both stress and displacement are continuous across the boundary.

3.2.1 Body Waves

Body waves propagate in the interior of a medium. In order to study the characteristics of body waves, we will consider the propagation of plane waves in an unbounded medium. To do so, a harmonic plane wave propagating in the $\hat{\mathbf{n}}$ direction is used as a trial solution or *ansatz*,

$$u_k(\mathbf{x}, t) = A\, \hat{\gamma}_k\, e^{i(\omega t - k_j x_j)} = A\, \hat{\gamma}_k\, e^{i\omega(t - s_j x_j)}. \tag{3.9}$$

In this expression, A is a wave amplitude coefficient, the unit vector $\hat{\gamma}$ denotes the *polarization* of the wave and ω is angular frequency. In addition, s_j is the jth component of the *slowness* vector,

$$s_j = \frac{\hat{n}_j}{v}, \tag{3.10}$$

where $v(\mathbf{k})$ is the *phase* velocity and $\mathbf{k} = \omega \mathbf{s}$ is the *wavevector*. The distinction between phase and *group* velocity is discussed below, for both anisotropic elastic media and attenuating media (e.g. poroelastic). Equation 3.9 uses *Euler's identity*,

$$e^{i\varphi} = \cos\varphi + i\sin\varphi, \tag{3.11}$$

where i is the imaginary unit and the argument φ is called the *phase* (see Appendix B). Since a *wavefront* can be defined as a surface of constant phase, then it is evident that at a given time t, for the kth component of the polarization vector, Equation 3.9 represents planar wavefronts with normal $\hat{\mathbf{n}}$. It should be emphasized that the use of this trial solution is quite general, since wavefronts of nearly arbitrary complexity can be constructed by superposition of harmonic plane waves.

Substitution of the plane-wave ansatz into Equation 3.8 yields

$$\left(c_{ijkl}n_j n_l - \rho v^2 \delta_{ik}\right)\hat{\gamma}_k = 0. \tag{3.12}$$

The equations of motion are now cast as an eigenvalue problem wherein, by specifying the propagation direction $\hat{\mathbf{n}}$ and the elastic stiffness tensor, c_{ijkl}, one can obtain the eigenvalues ρv^2 and eigenvectors, $\hat{\gamma}$. By introducing the expression

$$\Gamma_{ik} = c_{ijkl}n_j n_l, \tag{3.13}$$

this can be written as

$$\begin{bmatrix} \Gamma_{11} - \rho v^2 & \Gamma_{12} & \Gamma_{13} \\ \Gamma_{21} & \Gamma_{22} - \rho v^2 & \Gamma_{23} \\ \Gamma_{31} & \Gamma_{32} & \Gamma_{33} - \rho v^2 \end{bmatrix} \begin{bmatrix} \hat{\gamma}_1 \\ \hat{\gamma}_2 \\ \hat{\gamma}_3 \end{bmatrix} = 0. \tag{3.14}$$

This formula defines a system of equations known as the *Kelvin–Christoffel* equations (Musgrave, 2003). As a condition for existence of a nontrivial solution,

$$\begin{vmatrix} \Gamma_{11} - \rho v^2 & \Gamma_{12} & \Gamma_{13} \\ \Gamma_{21} & \Gamma_{22} - \rho v^2 & \Gamma_{23} \\ \Gamma_{31} & \Gamma_{32} & \Gamma_{33} - \rho v^2 \end{vmatrix} = 0, \tag{3.15}$$

where the matrix inside the determinant is called the *Christoffel* matrix. By solving the cubic polynomial equation for each of the three eigenvalues, ρv^2, the three mutually perpendicular polarization vectors, $\hat{\gamma}$ can then be determined by solving Equation 3.14. As outlined within the anisotropy section below, when the reciprocal of v is plotted $\forall\ \hat{\mathbf{n}}$, it defines a multifold surface in 3-D called the *slowness* surface.

When the isotropic form of the elastic stiffness tensor is used (Equation 1.12) with any wavefront normal to construct the Christoffel matrix, the largest eigenvalue is

$$\rho v_P^2 = \lambda + 2\mu, \tag{3.16}$$

which defines the P-wave velocity of an isotropic medium ($v_P = \sqrt{(\lambda + 2\mu)/\rho}$). The corresponding eigenvector is given by $\hat{\gamma}_P = \hat{\mathbf{n}}$; hence, the P-wave polarization vector is parallel to the propagation direction. Propagation of a P wave past a point within a medium results in a series of compression and dilations, as shown in Figure 3.1. Consequently, P waves are also known as compressional waves.

The isotropic case is degenerate, in that the two remaining eigenvalues are equal and are given by

$$\rho v_S^2 = \mu, \tag{3.17}$$

which defines the S-wave velocity of an isotropic medium ($v_S = \sqrt{\mu/\rho}$). The corresponding eigenvectors are perpendicular to the P-wave eigenvector and satisfy the relation $\hat{\gamma}_S \cdot \hat{\mathbf{n}} = 0$. In general, the polarization vector for an S wave in an isotropic medium is coplanar with the wavefront and thus produces a shearing type of motion during passage of the wave (Figure 3.1).

For isotropic media, it is customary to classify S waves into two categories; S_H waves are horizontally polarized perpendicular to $\hat{\mathbf{n}}$, while S_V waves are polarized perpendicular to both the P-wave polarization and the S_H-wave polarization. These two classes of S

3.2 Wave Solutions

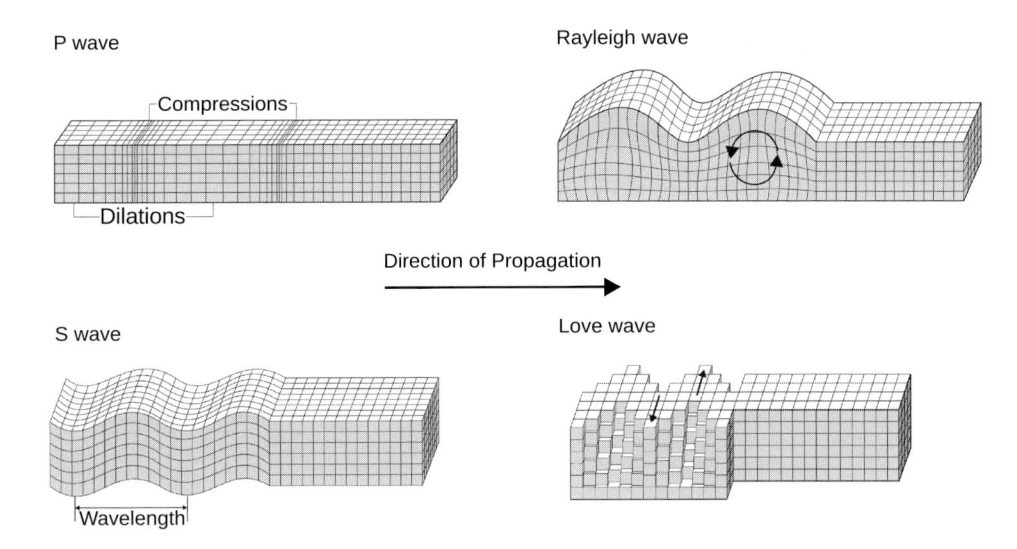

Fig. 3.1 Schematic illustration of wave propagation and particle motion associated with body waves (*P* and *S*) and surface waves (Rayleigh and Love waves). In each panel, the undisturbed region is on the right. The wavelength is the distance between successive peaks or troughs.

wave emerge naturally due to the decoupling of *P*-wave motion and S_H-wave motion at the free surface as well as at layer boundaries in stratified media, which in general are approximately horizontal. As discussed below, *P*-wave motion and S_V-wave motion is coupled at horizontal boundaries, meaning that the scattered wavefield includes mode conversions between different wave types. It is important to keep in mind that S_V waves are not necessarily vertically polarized; for propagation from a source to a receiver, the S_V polarization vector lies in the vertical plane that passes through both source and receiver, known as the *sagittal* plane. The representation of body waves in isotropic elastic media is often simplified through the use of wave potentials (see Box 3.1).

In a fluid, $\mu \to 0$ because fluids do not support shear stresses. Consequently, based on Equation 3.17, *S* waves do not propagate within a fluid. In a fluid, also called an *acoustic* medium, compressional waves are equivalent to familiar sound waves in air. In poroelastic media, the presence of pore fluids leads to viscous coupling between the fluid phase and the elastic frame, which generates slow waves that can be *dispersive*. Strictly speaking, the term *dispersion* means that the phase velocity depends on ω or $\mathbf{k}$.

In an anisotropic medium, for a given choice of $\hat{\mathbf{n}}$ there are generally three distinct eigenvalues that determine three different wave modes. For any Earth-like elastic material, one of the three waves has characteristics, including wave velocity and polarization, that are similar to a *P* wave in an isotropic medium. This mode is referred to as a *qP* (quasi-*P*) wave. The remaining two wave modes are *qS* waves. Unlike the isotropic case, the *qS* modes have well-defined polarizations that, in many cases, are not the same as the S_H and S_V polarizations for isotropic materials. For directions in which the *qS* wave velocities differ, observed arrivals are not simultaneous, leading to the well-known phenomenon of *shear-wave splitting*; this and other details of anisotropic wave propagation are discussed in §3.3.

> **Box 3.1**   **Wave Potentials and Vector Operators**
>
> Analysis of P and S motion in isotropic elastic media can be simplified through the use of *Helmholtz potentials*, which can be used to reduce the equations of motion to the form of the *wave equation* (Aki and Richards, 2002). Using wave-potential notation, the body-wave displacement field $\mathbf{u}(\mathbf{x}, t)$ may be written as
>
> $$\mathbf{u}(\mathbf{x}, t) = \nabla \Phi(\mathbf{x}, t) + \nabla \times \mathbf{\Psi}(\mathbf{x}, t),$$
>
> where $\Phi(\mathbf{x}, t)$ is called the *scalar potential* and $\mathbf{\Psi}(\mathbf{x}, t)$ is called the *vector potential*. The first part on the right side of the above equation represents P-wave motion, while the second part represents S-wave motion. The symbol ∇ is the *gradient operator*, which can be expressed using tensor notation as
>
> $$\nabla \Phi \equiv \Phi_{,i} \hat{\mathbf{x}}_i.$$
>
> It can be shown that Φ satisfies the scalar wave equation,
>
> $$\Phi_{,ii} = \frac{1}{v_P^2} \ddot{\Phi},$$
>
> where v_P is defined in Equation 3.16. Similarly, it can be shown that $\mathbf{\Psi}$ satisfies the vector wave equation,
>
> $$\Psi_{i,jj} = \frac{1}{v_S^2} \ddot{\Psi}_i, \quad i = 1, 2, 3,$$
>
> where v_S is defined in Equation 3.17 and the symbol $\nabla \times$ is the curl operator, which can be written using tensor notation as
>
> $$(\nabla \times \mathbf{\Psi})_i \equiv \epsilon_{ijk} \Psi_{k,j}.$$
>
> The symbol ϵ_{ijk} is called the *alternating* (or permutation) tensor and is given by
>
> $$\epsilon_{ijk} = \begin{cases} 0 \text{ for } i = j, j = k, \text{ or } i = k \\ +1 \text{ for } (i, j, k) \in \{(1, 2, 3), (2, 3, 1), (3, 1, 2)\} \\ -1 \text{ for } (i, j, k) \in \{(1, 3, 2), (3, 2, 1), (2, 1, 3)\}. \end{cases}$$

3.2.2 Surface Waves

Surface waves are guided waves that propagate along or near Earth's surface, with rapidly diminishing wave amplitude with increasing depth. Surface waves often produce the highest-amplitude recorded signals generated by shallow (< 30 km) earthquakes. They also represent a strong source of noise for passive-seismic monitoring and controlled-source seismic profiling. Indeed, surface waves generated by ocean waves at coastal regions produce ambient-noise signals within the microtremor period band (5–20 s) that can be detected far into the interior of continents (Pawlak et al., 2011). Mathematical expressions for surface waves are presented here without full derivation; the theory developed from first principles can be found in standard textbooks on seismology, such as Stein and Wysession (2009).

3.2 Wave Solutions

Surface waves arise as solutions to the elastic equations of motion subject to certain boundary conditions. A boundary condition that is common to all surface-wave derivations is the *free-surface* boundary condition. If we approximate Earth's surface as the boundary to an elastic half-space at $z = 0$, force-balance considerations imply that stresses on the free surface vanish; otherwise the shear and normal tractions acting on an infinitesimal volume just below the boundary would not be balanced by tractions from an infinitesimal volume above the boundary, resulting in unbounded acceleration. Another common boundary condition is called the *radiation condition*, which requires that the wave amplitude $\to 0$ as $z \to \infty$, where a coordinate system is used in which z increases with depth.

As a starting point to illustrate the characteristics of surface waves, we consider the simple scenario of a Rayleigh wave propagating along the top surface of a homogeneous isotropic elastic half-space. Rayleigh waves are named in honour of the pioneering scientist who first described them, Lord Rayleigh (1842–1919). Here, the trial solution consists of a coupled $P - S_V$ wave propagating in the x-direction; this may be written in terms of wave potentials as

$$\Phi = A e^{i(\omega t - k_x x - k_x r_P z)}$$
$$\Psi = B e^{i(\omega t - k_x x - k_x r_S z)}, \qquad (3.18)$$

where

$$r_P = \left(v_R^2/v_P^2 - 1\right)^{1/2}$$
$$r_S = \left(v_R^2/v_S^2 - 1\right)^{1/2}. \qquad (3.19)$$

In the above expressions, v_R denotes the phase velocity of the coupled (Rayleigh) wave. The radiation condition requires that $v_R < v_S$ in order to ensure that the dependence on z has a decaying exponential form, meaning that the negative, not positive, imaginary square root must be chosen for r_P and r_S. After applying the free-surface boundary condition, the existence of a nontrivial solution requires that (Stein and Wysession, 2009)

$$\left(2 - v_R^2/v_S^2\right)^2 + 4\left(v_R^2/v_S^2 - 1\right)^{1/2}\left(v_R^2/v_P^2 - 1\right)^{1/2} = 0. \qquad (3.20)$$

Of the two nonzero roots to this equation, only one satisfies the condition that $v_R < v_S$. For a *Poisson solid*, which is characterized by $\lambda = \mu$, the Rayleigh-wave velocity simplifies to $v_R = (2 - 2/\sqrt{3})v_S \simeq 0.92\, v_S$. For this case, the x and z displacements at the surface are given by

$$u_x = 0.42 A k_x \sin(\omega t - k_x x)$$
$$u_z = 0.62 A k_x \cos(\omega t - k_x x), \qquad (3.21)$$

which describes retrograde elliptical motion of the ground surface (Figure 3.1).[2] The major axis of the ellipse is vertical and approximately 1.5 times greater than the minor axis of the ellipse.

---

[2] It should be noted that $k_x = \omega/c$ suggests that the sinusoid coefficient is frequency-dependent. This is not true, however, because A is the potential amplitude. Since $A \sim U/\omega$, where U is the displacement amplitude, that eliminates the apparent frequency dependence.

In most cases of practical interest, the assumption of a uniform half-space is relaxed and the subsurface is approximated instead as a vertically inhomogeneous medium, of which a horizontally stratified medium is a special case. By applying the same approach used for a half-space in a stratified medium, it can be demonstrated that v_R exhibits frequency dispersion. In general, the *sensitivity kernel* for Rayleigh waves, a measure of the sensitivity of v_R to a change in v_S of the medium, is characterized by a peak that increases with wavelength. Since wavelength (λ) is related to frequency by $\lambda = v_R/f$, it transpires that the depth of peak sensitivity is inversely proportional to frequency. In most stratified sedimentary rocks, v_S increases with depth due to compaction; consequently, in layered media, Rayleigh waves exhibit frequency dispersion (i.e. v_R depends on frequency).

The dispersive nature of Rayleigh waves can be exploited to obtain information about the subsurface using the method of *Multichannel Analysis of Surface Waves* (MASW) (Park et al., 1999). This method is applied to *ground roll*, the name used in exploration seismology to describe Rayleigh waves in the 10–30 Hz band, in order to characterize the top $\approx$ 100 m of the subsurface. This approach is based upon a method for determination of phase velocity, which utilizes the multichannel redundancy of modern seismic acquisition systems. Once the dispersion curve is obtained, a model for v_S as a function of z can be obtained by fitting the observed dispersion trend using a least-squares inversion approach.

As a second illustration of surface-wave analysis, we will consider the case of *Love* wave propagation for a model with a single layer of thickness h overlying a half-space. Love waves, which are named in honour of their discoverer A. E. H. Love (1863–1940), do not exist on the surface of a half-space; hence this single-layer model is the simplest scenario that can be analyzed. Here the trial solution consists of coupled upgoing and downgoing S_H waves in the top layer, with wave amplitudes B_2 and B_1, respectively, together with a downgoing S_H wave in the half-space with wave amplitude B'. The S_H waves are propagating in the x direction and therefore produce ground motion in the y direction (Figure 3.1). Following Stein and Wysession (2009), the coupled S_H waves in the top layer may thus be written as

$$u_y^-(x,z,t) = B_1 e^{i(\omega t - k_x x - k_x r_1 z)} + B_2 e^{i(\omega t - k_x x + k_x r_1 z)}. \tag{3.22}$$

Similarly, the S_H wave in the half-space may be written as

$$u_y^+(x,z,t) = B' e^{i(\omega t - k_x x - k_x r_2 z)}, \tag{3.23}$$

where

$$r_j = \left(v_L^2/v_j^2 - 1\right)^{1/2} \quad j = 1, 2. \tag{3.24}$$

In these expressions, v_1 is the shear-wave velocity of the top layer and v_2 is the shear-wave velocity of the half-space, with $v_1 < v_2$. As in the previous example, the radiation condition requires that $v_L < v_2$ so that the dependence on z of the Love wave solution in the half-space has a decaying exponential form. After applying the free-surface boundary conditions and some algebra, the existence of nontrivial solutions of the boundary-value problem requires that

$$\tan\left[\omega(h/v_L)(v_L^2/v_1^2 - 1)^{1/2}\right] = \tan(\omega\xi) = \frac{\mu_2(1 - v_L^2/v_2^2)^{1/2}}{\mu_1(v_L^2/v_1^2 - 1)^{1/2}}, \qquad (3.25)$$

where μ_1 and μ_2 are the shear moduli for the top layer and the half-space, respectively. This can be considered as an implicit *dispersion relation*, that is an expression that defines velocity as a function of ω. A graphical solution to this equation, shown in Figure 3.2, illustrates how this equation can be used to determine v_L as a function of frequency. Using separate curves for the left side and right side of Equation 3.25, solutions exist where the curves intersect. It should be noted that a condition for a real argument of the tangent function is that $v_L > v_1$. A further condition to obtain real values on the right side is that $v_L < v_2$.

For a given frequency and within the range of allowable values of v_L, there are sometimes multi-valued solutions to Equation 3.25. In Figure 3.2, the intersections marked by $n = 0$ represent the *fundamental-mode* solution for a model with a 15 m thick layer with $v_S = 200$ m/s overlying a half-space with $v_S = 1400$ m/s. Other intersections for $n > 0$ represent higher-order modes. As shown by the dispersion curves in Figure 3.3, given the constraint that $v_1 < v_L < v_2$, the higher-order modes have a restricted frequency range for the solutions. In general, for each individual mode, at low frequencies that correspond to relatively long wavelengths, the Love wave velocity is close to the v_S for the half-space, whereas at high frequency the Love wave velocity approaches the v_S for the top layer. Dispersion that is related to velocity structure, rather than an intrinsic material property, is known as *geometrical dispersion*. For this model, it is evident that MASW recordings using 10 Hz geophones would require the use of higher-order modes ($n > 0$) in order to resolve the velocity of the half-space, since the fundamental mode exhibits negligible dispersion and maintains a velocity close to the asymptotic limit of the upper-layer velocity above 10 Hz.

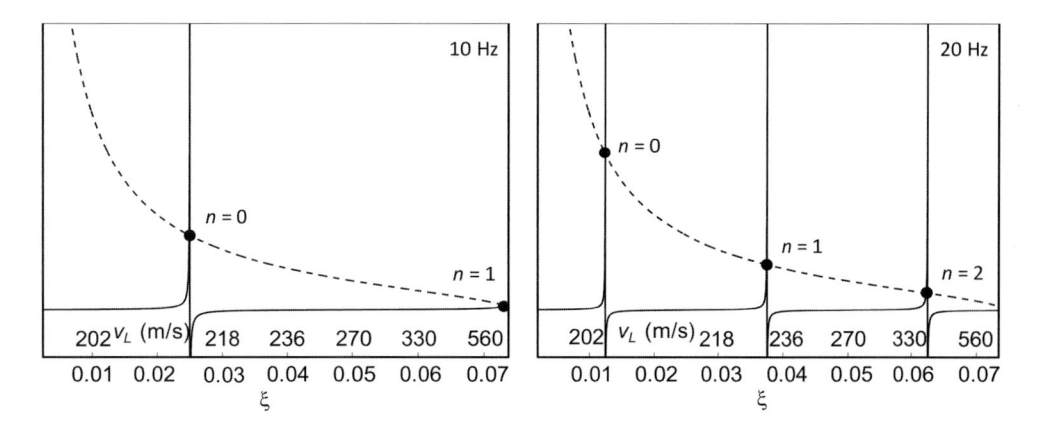

Fig. 3.2 Graphical method to determine Love wave phase velocity at 10 Hz and 20 Hz, for a model with a 15 m thick layer with $v_S = 200$ m/s overlying a half-space with $v_S = 1400$ m/s. For each graph, the solid curves show the function $\tan(\omega\xi)$ on the left side of Equation 3.25, while the dashed curves show the right side of the equation. The frequency-dependent solutions are indicated by the intersections of these curves. The first intersection ($n = 0$) is the fundamental mode, while the higher-order solutions represent overtones.

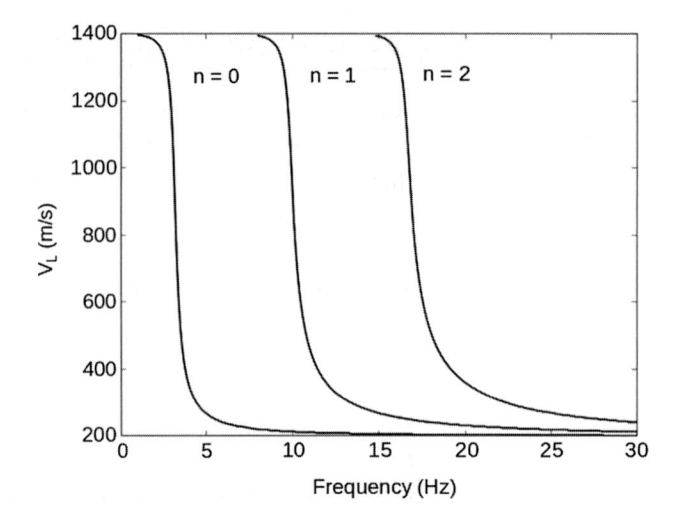

Fig. 3.3 Love wave dispersion curves for the fundamental mode ($n = 0$) and two higher-order modes, for a model with a 15 m thick layer with $v_S = 200$ m/s overlying a half-space with $v_S = 1400$ m/s.

The discussion thus far has emphasized phase velocity, which may be written as $v = \omega/k$. A harmonic plane wave propagates at the phase velocity; however, when a wave pulse is observed that contains a range of frequencies, in the presence of dispersion the pulse will generally change in shape as the wave propagates to greater distance. This phenomenon occurs due to the differing velocity of the constituent frequency components in the pulse. An important characteristic of dispersive wave propagation is that the *pulse envelope* travels at the *group* velocity, $v_g \equiv \frac{\partial \omega}{\partial k}$. If the phase velocity is known for a range of frequencies, the group velocity can be determined using

$$v_g = v + k\frac{\partial v}{\partial k}. \tag{3.26}$$

This section concludes with a brief look at the effects of seismic anisotropy on surface-wave propagation. Smith and Dahlen (1973) showed that for a stratified half-space, where each layer is weakly anisotropic with arbitrary symmetry (the meaning of "weak" anisotropy is discussed below), the azimuthal dependence of phase (or group) velocity of Rayleigh and Love waves can be expressed as

$$v(\omega, \theta) = A_1(\omega) + A_2(\omega)\cos 2\theta + A_3(\omega)\sin 2\theta + A_4(\omega)\cos 4\theta + A_5(\omega)\sin 4\theta, \tag{3.27}$$

where θ is the azimuth of the wavevector. This expression includes a component that has no azimuthal dependence (A_1), a component that has a 2θ dependence (measured by A_2 and A_3) and a component that has a 4θ dependence (measured by A_4 and A_5). It is becoming more common for teleseismic surface-wave inversion studies to make use of this formulation in order to investigate depth-dependent azimuthal dependence of anisotropy in the mantle. For example, Darbyshire et al. (2013) carried out a dispersion analysis and inversion for azimuthal parameters to interpret layered mantle structure including "fossil" anisotropy beneath Hudson Bay in northern Canada.

3.2.3 Green's Functions and Geometrical Spreading

The wave functions considered to this point are based upon the use of harmonic plane-wave trial solutions for the equations of motion. We now turn our attention to solutions that account for a source, as shown by the internal force term in Equations 3.6 and 3.7. In particular, we consider a *Green's function*, which, in general, is a solution for a boundary-value problem corresponding to a source that acts at a point in the medium. A Green's function is useful as it provides an impulse response for the system that provides insights into characteristic wave behaviour for the system of interest as well as a basis for construction of more complex solutions.

Aki and Richards (2002) developed a theoretical foundation for elastodynamic Green's functions for an unbounded isotropic homogeneous elastic medium. The source is located at the origin and is represented by $s(t)\delta(\mathbf{x})\hat{\mathbf{x}}_j$. The symbol $\delta(\mathbf{x} - \boldsymbol{\xi})$ used in this expression for the source is called the *Dirac* delta function; this is a concentrated distribution function that is zero for all values of the argument except $\mathbf{x} = \boldsymbol{\xi}$ (see Appendix B). The Dirac function has the further property that

$$\int_{-\infty}^{\infty} \delta(\mathbf{x} - \boldsymbol{\xi})d\mathbf{x} = 1. \tag{3.28}$$

With this source term, the ith component of displacement due to a body force having the source-time function $s(t)$ acting in the $\hat{\mathbf{x}}_j$ direction at the origin is given by

$$\begin{aligned} u_i(\mathbf{x},t) &= s * G_{ij} \\ &= \frac{1}{4\pi\rho}\left(3\gamma_i\gamma_j - \delta_{ij}\right)\frac{1}{r^3}\int_{r/v_P}^{r/v_S} \tau s(t-\tau)d\tau + \frac{1}{4\pi\rho v_P^2}\gamma_i\gamma_j\frac{1}{r}s\left(t - \frac{r}{v_P}\right) \\ &\quad - \frac{1}{4\pi\rho v_S^2}\left(\gamma_i\gamma_j - \delta_{ij}\right)\frac{1}{r}s\left(t - \frac{r}{v_S}\right). \end{aligned} \tag{3.29}$$

The Green's function notation G_{ij} denotes the ith component of displacement for a source acting in the $\hat{\mathbf{x}}_j$ direction, where $*$ denotes the convolution operator (discussed in Appendix B). In addition, $\gamma_i \equiv x_i/r$ is the ith component of the *direction cosine* and the variable r represents distance away from the source.

The first term on the right side of Equation 3.29 depends on r^{-3} and is called the *near-field* term. For a source of duration T, this term produces an emergent wave pulse that begins at $t = r/v_P$ and ends at $t = r/v_S + T$ (Aki and Richards, 2002). Since it decays much faster with distance than the other terms, called the *far-field* terms, it is usually negligible for most signals and therefore is rarely described. However, for downhole passive seismic monitoring it is possible that some recordings could capture the near-field expression of the source. An example of a recorded waveform for which the near-field term is important is shown in Figure 3.4. This recording was made using a broadband seismometer deployed at a deep (2km) underground observatory and is located within $\approx$ 400m of the source.

The far-field terms are split into a P-wave arrival with onset at $t = r/v_P$ and a S-wave arrival with onset at $t = r/v_S$. The wave-like character is evident from the terms with the form $X_0(t - r/v)$, which represents a pulse propagating away from the source at velocity

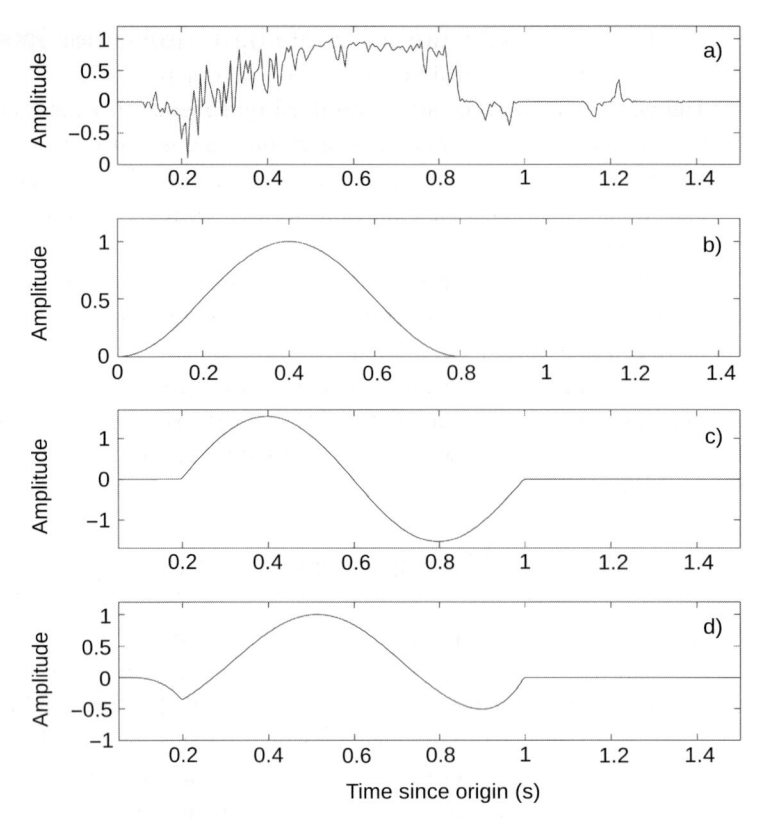

Fig. 3.4 a) Unfiltered vertical-component recording of a M 4.1 earthquake on 29 November 2006, recorded at a deep underground observatory located only 400m from the focus (Atkinson et al., 2008). b) Source time function used to compute synthetic seismograms, with period $T = 0.8$ s. c) Synthetic seismogram computed using the moment tensor determined by regional waveform modelling with only far-field terms. The radiation pattern is dominated by the S wave, which has a waveform that is the time derivative of the source time function in (b). d) Synthetic seismogram computed using the same moment tensor, but with near-source terms included. Interference of the near- and far-field terms reproduces some features of the observed waveform, such as the initial negative ground motion. Gail M. Atkinson, SanLinn I. Kaka, David W. Eaton, Allison Bent, Veronika Peci, Stephen Halchuk, A Very Close Look at a Moderate Earthquake near Sudbury, Ontario, *Seismological Research Letters*, 70, 119–131, 2008, copyright Seismological Society of America.

v. The far-field terms represent body waves with $1/r$ amplitude decay. This rate of decay is called *geometrical spreading*.[3] This rate of amplitude decay can also be deduced on the basis of conservation of energy, which stems from the equality of the *strain-energy* density and *kinetic-energy* density for seismic waves,

$$\frac{1}{2}\sigma_{ij}\epsilon_{ij} = \frac{1}{2}\rho \dot{u}_i \dot{u}_i. \tag{3.30}$$

---

[3] The simple r^{-1} form of geometrical spreading is only applicable to a homogeneous medium. It is much more complicated for a heterogeneous medium.

The kinetic energy density term on the right measures the flux rate of energy transmission per unit time normal to the wavefront. For a point source in a homogeneous isotropic medium the wavefronts are represented by spherical surfaces with area $4\pi r^2$. Conservation of energy requires that the energy flux rate is constant for wavefronts of any radius, which implies that the amplitude of the wave must decrease as $1/r$, since $\dot{u}$ is squared in Equation 3.30. In the case of a homogeneous isotropic medium, $1/r$ geometrical spreading is also called *spherical divergence*.

A similar argument can be applied to the amplitude decay for surface waves. Since surface waves propagate on a 2-D surface, the problem is suited to cylindrical coordinate system. For a homogeneous isotropic medium, conservation of energy requires a constant energy flux rate passing through circular wavefronts of any radius. This implies that the geometrical spreading term for surface waves has a $1/r^{1/2}$ dependence. Consequently, at large distances surface wave amplitudes tend to exceed body-wave amplitudes in wavefield observations.

3.2.4 Wave Amplitude Partitioning at Interfaces

Scattering refers to the interaction of waves with material heterogeneities and/or discontinuities within or at the boundary of a medium. The scattering of elastic waves at planar interfaces between layers in a stratified elastic medium is a particularly well studied scattering phenomenon, for which there is an extensive literature. The *Zoeppritz equations* are closed-formed expressions for the relative amplitudes of scattered phases, in the case of plane elastic waves that impinge onto a planar interface between two isotropic elastic half-spaces. Details of these equations can be found in sources, including Young and Braile (1976). The nomenclature for scattered waves, which includes reflected, transmitted and converted waves at an interface, is shown in Figure 3.5. The nomenclature combines the incident and scattered wave type (*P* or *S*) into a pair of letters, with subscript notation to indicate whether the scattered phase is reflected (*r*) or transmitted (*t*). If the incident and scattered wave types differ, this is referred to as a mode conversion.

In Figure 3.5, normals to wavefronts are shown as *rays*, a mathematical construct from the field of geometrical optics that shows the direction of wave-energy flux. More generally, rays are curves that are everywhere normal to a wavefront, including within heterogeneous media. Geometrical rays provide a powerful tool for visualization of wave scattering phenomena as well as for the computation of synthetic seismograms (discussed below). At an interface between two materials with velocities v_1 and v_2, the reflection and refraction (including mode conversion between *P* and *S* waves) satisfies *Snell's Law*, which is given by

$$\frac{\sin \theta_1}{v_1} = \frac{\sin \theta_2}{v_2} = p, \qquad (3.31)$$

where the angle θ is measured between the ray and the normal to the interface, as exemplified in Figure 3.5, and p is called the *ray parameter*. The term *refraction* refers to bending of rays due to a change in wave velocity. A noteworthy special case is a *critical angle*,

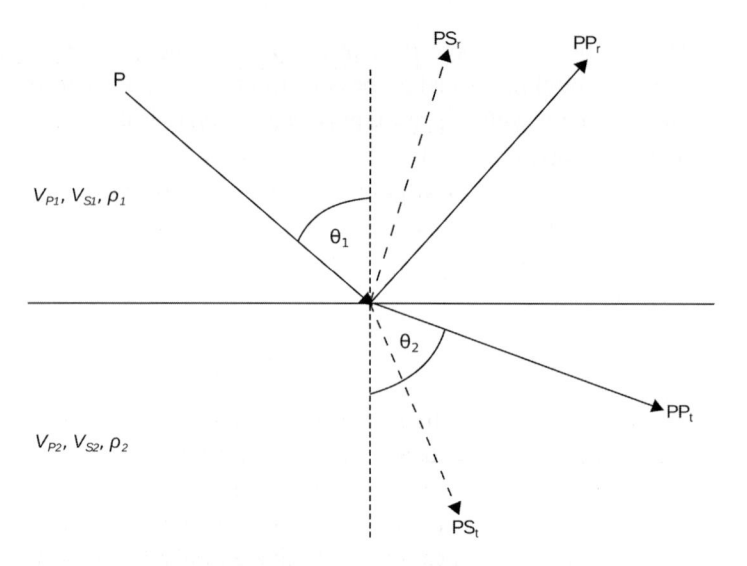

Fig. 3.5 Nomenclature for reflected, refracted and converted waves at an interface (see text). The incident P ray makes an angle of θ_1 with respect to the normal to the interface. The angle of the transmitted P wave is denoted by θ_2. All of the ray angles, including mode conversions, are governed by Snell's Law.

which is the incident angle that corresponds with a transmitted wave that propagates along the interface (i.e. $\theta_2 = \pi/2$). Since $\sin(\pi/2) = 1$, based on Snell's Law a critical angle is thus given by

$$\theta_c = \sin^{-1}\left(\frac{v_1}{v_2}\right). \tag{3.32}$$

Depending upon the velocities of the two media, when mode conversions are taken into account there may be multiple critical angles for a given incident wave.

The derivation of the Zoeppritz equations can found in textbooks on seismology, such as Aki and Richards (2002) or Stein and Wysession (2009). The equations are obtained by applying welded-contact boundary conditions at the interface. These boundary conditions stipulate that the three components of stress and the three components of displacement are continuous across the boundary. For a given type of incident wave, equating each of these six components on both sides of the boundary leads to a set of six equations in six unknowns, which are the relative amplitude of the four scattered phases in Figure 3.5 as well as the reflected and transmitted S_H waves. The amplitude coefficients determined using the Zoeppritz equations are normalized with respect to the amplitude of the incident wave; thus a reflection coefficient of 0.1 means that the reflected wave has an amplitude that is 10 percent of the amplitude of the incident wave. Different coefficients apply to the relative energy of the scattered phases.

The reflection and transmission coefficients are real-valued for pre-critical incidence, but take on complex values for post-critical incidence. The physical significance of the complex coefficients stems from Euler's identity, discussed above. A complex coefficient A can be expressed in terms of an amplitude and a phase as follows:

$$A = |A|e^{i\varphi} \text{ where } \varphi = \tan^{-1}\left(\frac{\text{Im}(A)}{\text{Re}(A)}\right), \tag{3.33}$$

where Im(A) and Re(A) denote the imaginary and real components, respectively, of the complex variable A.

A graph showing an example of a P-wave reflection coefficient is shown in Figure 3.6. The values shown were computed using a computer program to solve the Zoeppritz equations written by Young and Braile (1976). At normal incidence ($\theta = 0$), the reflection coefficient is given by

$$R = \frac{Z_2 - Z_1}{Z_1 + Z_2}, \tag{3.34}$$

where $Z = \rho v_P$ is called the *acoustic impedance*. For the medium parameters in Figure 3.6, the normal-incidence reflection coefficient works out to be 0.2871, which is useful for illustration but exceptionally high for most interfaces in sedimentary basins. Here, the critical angle works out to be 36.9°. Note that for pre-critical incidence, the phase angle is zero; physically, this means that the reflected pulse has the same shape as the incident pulse. For post-critical incidence, the phase angle is nonzero; this means that the reflected pulse is characterized by a change in shape due to a phase rotation relative to the incident pulse (see Appendix B). At pre-critical incidence, the amplitude of the reflection coefficient gradually decreases with θ. In practice, measurement of amplitude versus offset (the distance at the surface between the source and receiver) for reflected waves can be used to infer

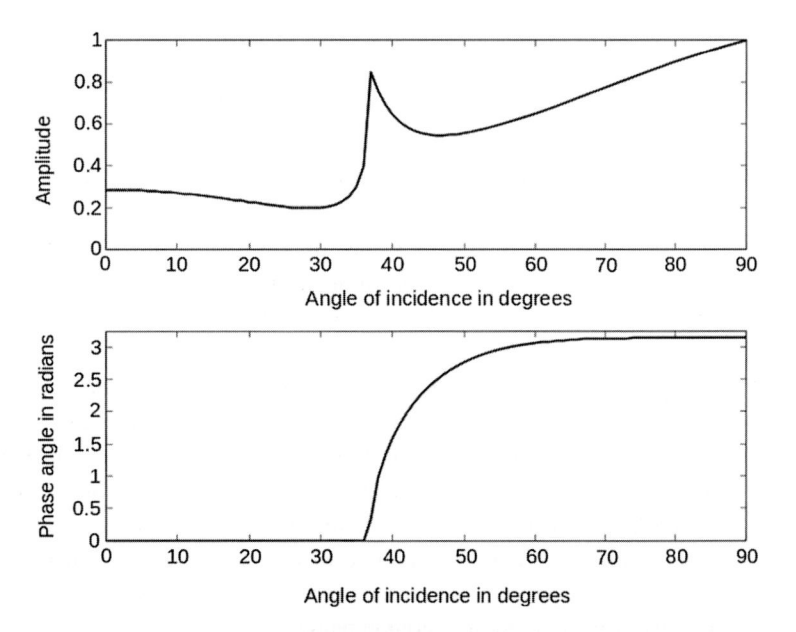

Fig. 3.6 Graphs of amplitude and phase lead for P-wave reflection coefficient (PP_r) at a planar boundary between an incident medium with $v_P = 3000$ m/s, $v_S = 1500$ m/s and $\rho = 2400$ kg/m^3 and a second medium with $v_P = 5000$ m/s, $v_S = 2941$ m/s and $\rho = 2600$ kg/m^3. The critical angle for this case is 36.9°. Pre-critical reflections have a phase angle of zero, whereas post-critical reflections are characterized by a phase rotation.

the change in properties of the medium across the boundary. At post-critical incidence, the amplitude of the reflected wave increases sharply, which indicates that most of the incident wave energy is reflected back. Understanding the behaviour of post-critical scattered phases is particularly significant for passive seismic monitoring using downhole arrays, since the rays from sources to the receivers often make a high angle with the normals to interfaces.

Chaisri and Krebes (2000) have solved the equivalent boundary value problem for a non-welded contact, in which tractions are continuous across a boundary between two different layers, but dislocation in displacement is permitted to occur. At such a contact, reflections can occur even in the case where there is no material discontinuity across the contact. Slip surfaces in active faults may be examples of such boundaries.

3.3 Effects of Anisotropy

In addition to the directional dependence of phase velocity that arises from anisotropy of the elastic stiffness tensor, the propagation of waves in an anisotropic elastic medium is characterized by a number of distinctive features. A few of the salient characteristics of seismic anisotropy are considered in this section.

Although the concept of dispersion is sometimes limited to cases where phase velocity varies with ω, a more general definition of dispersion relates to variation of phase velocity with the wavenumber vector, **k**. Using this more general definition, anisotropic elastic media can be seen to exhibit a directional dispersion. As such, group velocity differs from the phase velocity in a similar sense to the previous discussion for surface waves. In particular, for a homogeneous anisotropic medium, waves from a point source produce wavefronts with directionally varying radius that depends upon the group velocity. The geometrical surface obtained by plotting the group velocity $\forall \hat{n}$ is called the *wave surface*. A cross section through the wave surface of a strongly anisotropic material is given in Figure 3.7a. The strong anisotropy leads to a conspicuous departure from spherical wavefronts in isotropic media, including features that are discussed below, such as cusps and triplications.

The phase velocity surface for anisotropic media differs from the wave surface. It is customary, however, to plot *slowness* (the reciprocal of phase velocity), rather than phase velocity. Like the wave surface, the *slowness surface* is a geometrical surface that depicts the directional dependance of slowness. An example is given in Figure 3.7b. A geometrical procedure for construction of the wave surface using the slowness surface is described by Winterstein (1990), while the inverse construction is described by Vavryčuk (2006).

For Earth-like materials, it transpires that the qP wave has a group (or phase) velocity that is, in general, significantly greater than the group (or phase) velocity of the two qS waves. Relatively small differences in qS velocity lead to a *birefringent* phenomenon that is commonly known as *shear-wave splitting*. This phenomenon is equivalent to optical birefringence in some crystals. Shear-wave splitting occurs in anisotropic materials for most propagation vectors and manifested by a difference in group velocity for the two qS

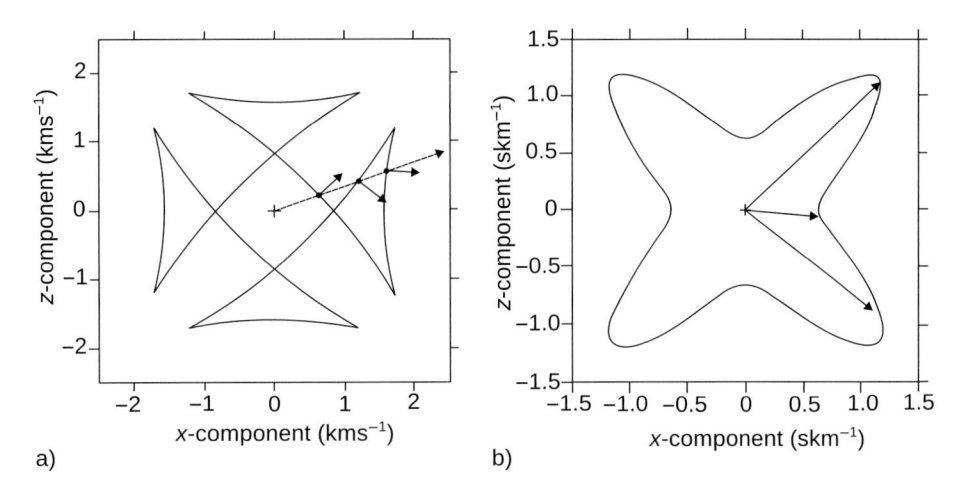

Fig. 3.7 Examples of S-wave wave surface (a) and slowness surface (b) for an anisotropic solid (Vavryčuk, 2006). Reprinted with permission of Royal Society Publishing.

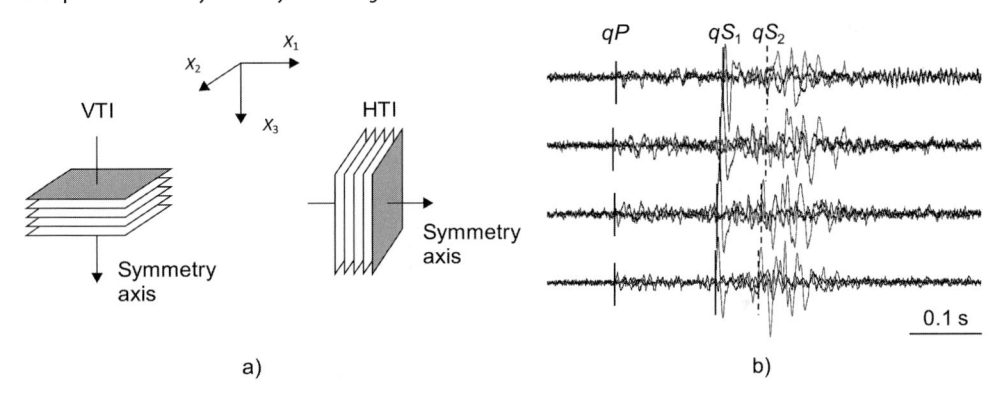

Fig. 3.8 a) Symmetry axes for media characterized by vertical transverse isotropy (VTI) and horizontal transverse isotropy (HTI), respectively (Rüger, 1997). Adapted with permission of the Society of Exploration Geophysicists. b) Shear-wave splitting example for a VTI medium. The microseismic event shown here was recorded using a downhole three-component geophone array. The traces show three mutually perpendicular directions of particle motion, where the fast S wave (qS_1) is horizontally polarized and the slow S wave (qS_2) is vertically polarized.

wave surfaces. For a given ray direction, there are two distinct qS arrivals with distinct and (nearly) mutually orthogonal polarizations.[4] These two arrivals are denoted here as qS_1 (fast shear wave) and qS_2 (slow shear wave). For a VTI medium, the slow qS_2-wave is typically polarized perpendicular to the layers, whereas the fast qS_1-wave is horizontally polarized (i.e. parallel to the layering). Similarly, in the presence of vertical fractures, the slow qS_2-wave is typically polarized perpendicular to the fractures, while the fast qS_1-wave polarized parallel to the fractures. An example of shear-wave splitting observed during passive seismic monitoring of hydraulic-fracturing operations is shown in Figure 3.8.

[4] For a given propagation vector, the polarizations defined by the corresponding eigenvectors are mutually perpendicular; however, for a point source, the polarization directions may not be exactly perpendicular.

A variety of interesting wave phenomena occur in the case of strongly anisotropic materials. For example, in some cases, a wave surface can fold back onto itself and thus produce three arrivals for a given type of wave. This scenario is called a *triplication* and is illustrated in Figure 3.7a. Similarly, wave surfaces with triplications contain cusps, which map to inflections in the corresponding slowness surface (Vavryčuk, 2006). Full-wave theories show that cusps in the wave surface generate *diffractions*, which are waves that are scattered from a spatially localized region. Finally, the intersection of two qS slowness surfaces is called a *shear-wave singularity*. Waves that propagate along a singularity are sometimes accompanied by anomalous amplitudes. Sub-categories of singularities are distinguished on the basis of the geometry of the intersection (e.g. point, curve, tangent, etc.).

3.3.1 Thomsen Parameters for TI Media

Thomsen (1986) introduced convenient notation and derived expressions for phase velocity and other quantities that are applicable to transversely isotropic (TI) media, including linearized forms of velocity formulas that are tailored for *weak anisotropy*. Although Thomsen's formulas are specified for VTI media, where the symmetry axis is vertical, they are no less applicable (subject, of course, to an appropriate change of coordinates) for horizontal transversely isotropy (HTI) and tilted transversely isotropy (TTI) media. Expressed in terms of Voigt stiffness parameters, the velocity for quasi-P wave and quasi-S wave propagation in a homogeneous VTI medium along the direction of the symmetry axis (x_3) are given, respectively, by

$$\alpha_0 = \sqrt{\tilde{C}_{33}/\rho} \qquad (3.35)$$

and

$$\beta_0 = \sqrt{\tilde{C}_{44}/\rho} \qquad (3.36)$$

where ρ denotes density and where both of the qS phases have the velocity β_0 along the symmetry axis. For the remainder of this section, the quasi designation will be omitted, and we will refer instead to S_H and S_V waves, which are natural wave modes for the TI symmetry system.

Thomsen (1986) showed that the propagation velocities with respect to the angle θ between the ray direction and the symmetry can be expressed as

$$v_P^2(\theta) = \alpha_0^2 \left[1 + \varepsilon_T \sin^2\theta + D^*(\theta) \right]; \qquad (3.37)$$

$$v_{SV}^2(\theta) = \beta_0^2 \left[1 + \frac{\alpha_0^2}{\beta_0^2} \varepsilon_T \sin^2\theta - \frac{\alpha_0^2}{\beta_0^2} D^*(\theta) \right]; \qquad (3.38)$$

and

$$v_{SH}^2(\theta) = \beta_0^2 \left[1 + 2\gamma_T \sin^2\theta \right]. \qquad (3.39)$$

These expressions are exact, and make use of two of Thomsen's well-known parameters:

$$\varepsilon_T \equiv \frac{\tilde{C}_{11} - \tilde{C}_{33}}{2\tilde{C}_{33}}; \qquad (3.40)$$

and
$$\gamma_T \equiv \frac{\tilde{C}_{66} - \tilde{C}_{44}}{2\tilde{C}_{44}}. \tag{3.41}$$

The formula for the parameter D^* is lengthy and can be found in Thomsen (1986). In the case of weak anisotropy (i.e. $\varepsilon_T \ll 1$ and $\gamma_T \ll 1$), these angle-dependent velocities can be approximated as

$$v_P(\theta) \simeq \alpha_0 (1 + \delta_T \sin^2\theta \cos^2\theta + \varepsilon_T \sin^4\theta), \tag{3.42}$$

$$v_{SV}(\theta) \simeq \beta_0 \left[1 + \frac{\alpha_0^2}{\beta_0^2}(\varepsilon_T - \delta_T)\sin^2\theta \cos^2\theta \right] \tag{3.43}$$

and

$$v_{SH}(\theta) \simeq \beta_0 (1 + \gamma_T \sin^2\theta). \tag{3.44}$$

The velocities $v_P(\theta)$ and $v_{SV}(\theta)$ both make use of a third parameter,

$$\delta_T \equiv \frac{(\tilde{C}_{13} + \tilde{C}_{44})^2 - (\tilde{C}_{33} - \tilde{C}_{44})^2}{2\tilde{C}_{33}(\tilde{C}_{33} - \tilde{C}_{44})}. \tag{3.45}$$

The role of the Thomsen parameters becomes more clear in these approximate forms. We see, for example, that for a horizontally propagating SH wave ($\theta = \pi/2$),

$$v_{SH}(\pi/2) = \beta_0 (1 + \gamma_T). \tag{3.46}$$

Similarly, for a horizontally propagating P-wave velocity,

$$v_P(\pi/2) = \alpha_0 (1 + \varepsilon_T) ; \tag{3.47}$$

thus, the parameters γ_T and ε_T are scalar measures of the relative velocity difference between vertically and horizontally propagating SH and P waves, respectively. The SV wave, by contrast, has the same velocity for horizontal and vertical directions of propagation. For oblique propagation angles, the velocity for P and SV waves is partly controlled by the δ_T parameter, while the SH propagation is simpler as it depends only on γ_T. Similar notation for orthorhombic media has been developed by Tsvankin (1997).

3.4 Anelastic Attenuation

In a perfectly elastic medium, a plane wave propagates through the medium with no energy loss. In realistic Earth materials, however, waves can experience energy loss due to processes such as heat generated by intergranular friction or viscous coupling in poroelastic media between the elastic frame and viscous pore fluid. Materials where these processes occur are examples of *anelastic* media, or media for which the constitutive relationship between stress and strain differs from the generalized Hooke's Law. Here we adopt a phenomenological approach by considering heuristic models for anelastic amplitude decay, without direct consideration of the underlying small-scale physical processes.

Anelastic attenuation can be incorporated into calculations of wave amplitude with the use of an exponential decay factor. Using this approach, the wave amplitude, A, in the harmonic plane-wave ansatz (Equation 3.9) is replaced with

$$A = A_0 e^{-\alpha x} = A_0 e^{-\frac{\pi x}{\lambda Q}} = A_0 e^{\frac{-\omega x}{2vQ}}, \tag{3.48}$$

where α denotes the attenuation coefficient, x is propagation distance, $\lambda = 2\pi v/\omega$ is the wavelength and the parameter Q is called the seismic *quality factor*. This formulation derives from the *damped harmonic oscillator*,

$$u(t) = A_0 e^{\frac{-\omega_0 t}{2Q}} \cos(\omega t). \tag{3.49}$$

This is a solution to the ordinary differential equation for a mass-spring system (Figure 3.9a) with spring constant k and damping factor γ,

$$m\frac{d^2 u}{dt^2} + \gamma m \frac{du}{dt} + ku = 0, \tag{3.50}$$

where $\omega_0 = (k/m)^{1/2}$ is the resonant frequency of the mass-spring system and $Q \equiv \omega_0/\gamma$. Once set into motion, the damped harmonic oscillator exhibits sinusoidal oscillatory behaviour within an exponentially diminishing amplitude envelope. The notion of quality arises because, as Q increases, damping is reduced and the oscillatory motion persists for longer.

Returning to Equation 3.35, it is evident that for a harmonic plane wave propagating in a constant-Q medium (i.e. a medium for which Q does not depend on frequency), the decay in amplitude over one wavelength is

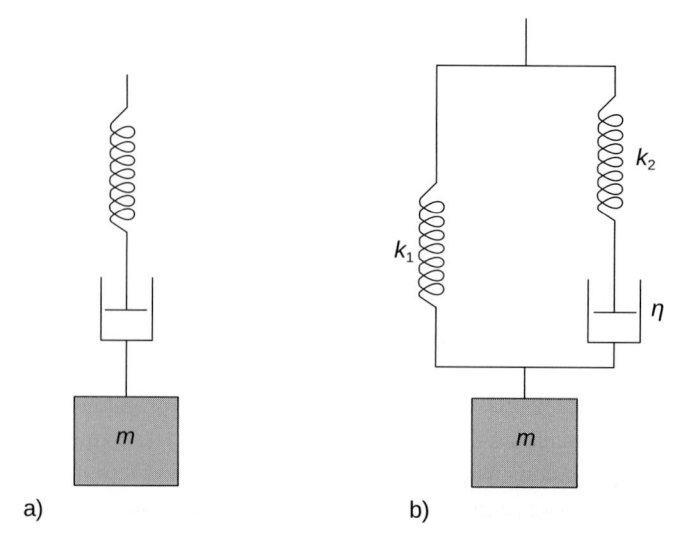

Fig. 3.9 Models for harmonic oscillators based on mass-spring systems. a) Simple harmonic oscillator. Here, the spring constant is k and the damping factor is given by γ. b) Linear solid model with spring constants (k) and viscous damping (η) as indicated.

$$\frac{A}{A_0} = e^{-\frac{\pi}{Q}}. \tag{3.51}$$

Alternatively, after propagation over a distance of Q wavelengths, the amplitude loss for a harmonic plane wave is $e^{-\pi}$. From Equation 3.30, kinetic energy density of a wave scales with $\dot{u}$, and at a given frequency $|\dot{u}| = \omega|u|$; thus, Q is related to the energy loss in one cycle as follows:

$$Q^{-1} = \frac{\Delta E}{2\pi E}. \tag{3.52}$$

Thus, it is evident that a rock with a high quality factor experiences less anelastic dissipation of wave energy per wavelength than a rock with a low quality factor.

It can be shown that the pulse associated with a wave propagating in a constant-Q medium exhibits a change in shape with distance, which is characteristic of dispersive wave propagation. For example, in the case of a pulse that is in the form of a Dirac delta function at $x = 0$ and $t = 0$, the displacement at time t and distance x is given by (Stein and Wysession, 2009)

$$u(x,t) = \pi^{-1}\left[\frac{x/2vQ}{(x/2vQ)^2 + (x/v - t)^2}\right]. \tag{3.53}$$

This formula represents a symmetric pulse with a peak at the geometrical arrival time, $t = x/v$ and a width that broadens with increasing time. The pulse symmetry is problematic, since it implies that wave energy arrives before the geometrical arrival time. Such a scenario is called *acausal*, and contrasts with relatively sharp pulse onset that is characteristic of most observed seismicity pulses in practice. Indeed, this acausality constitutes an unphysical attribute of a simple constant-Q model. A more physically realistic alternative model for wave attenuation is represented by the linear solid depicted in Figure 3.9b. In this case, the Q parameter for a damped harmonic oscillator varies with frequency and can be expressed as

$$Q(\omega) = \frac{k_1}{k_2}\left(\frac{1 + (\omega\tau)^2}{\omega\tau}\right), \tag{3.54}$$

where $\tau = \eta/k_2$ is the relaxation time of this system. This frequency-dependent expression for Q yields the following dispersion relation (Stein and Wysession, 2009)

$$v(\omega) = v_0\left[1 + \frac{k_2}{2k_1}\frac{(\omega\tau)^2}{1 + (\omega\tau)^2}\right], \tag{3.55}$$

where $v_0 = (k_1/\rho)^{1/2}$. This dispersion model results in increasing phase velocity with increasing frequency, which mitigates the problem of acausality for a constant Q material. This is a relatively simple case, and more elaborate frequency-dependent dispersion models can be created in order to improve the fit to observations, through the addition of more dashpots and springs in the model (Figure 3.9).

3.5 Seismic Sources

Building upon the foregoing discussion of Green's functions, we return now to the topic of transient internal body forces. Such forces can generate a *seismic source*, which is understood here simply as a spatially localized emitter of any type of seismic wave arising from a force or dislocation in the interior of a medium, or along one of its boundaries. The source may be artificial, as in buried explosive charges used in controlled-source seismology, or naturally occurring, as in earthquakes. A few mathematical models that are used to describe such sources are briefly introduced below.

3.5.1 Moment Tensors

A *moment tensor* is an idealized representation of a seismic source, given by an equivalent set of force couples that act at a point in a medium (Figure 3.10). It is expressed as a symmetric second-order tensor,

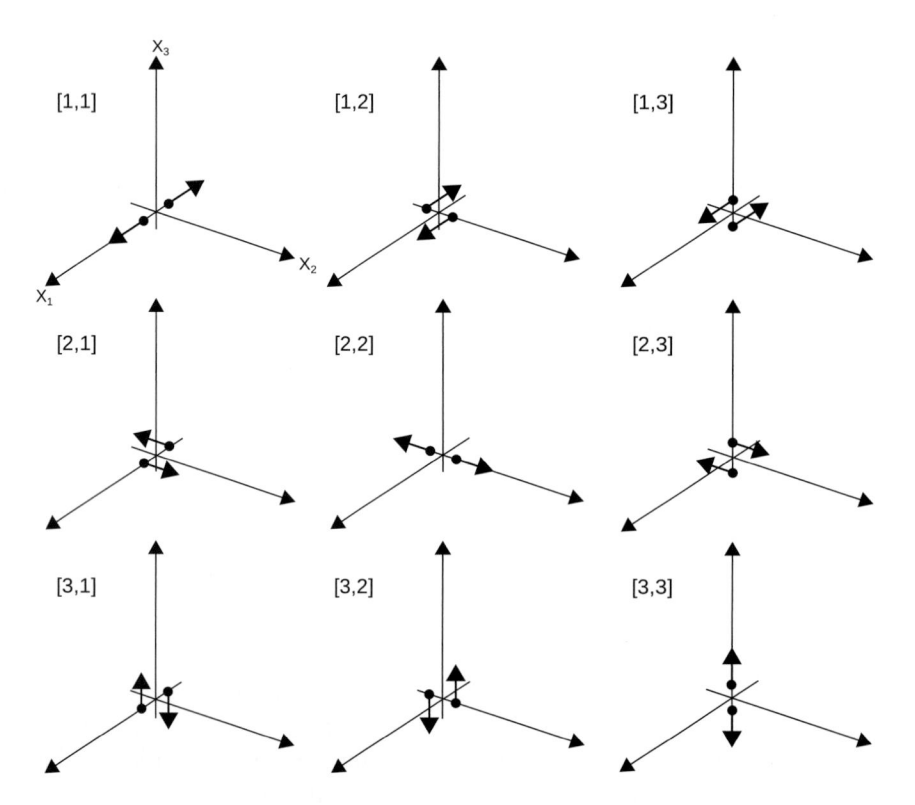

Fig. 3.10 Seismic moment-tensor components (force couples), M_{ij}. The tensor indices are shown inside square brackets. Diagonal elements are normal couples and off-diagonal components are shear couples.

$$\mathbf{M} = \begin{bmatrix} M_{11} & M_{12} & M_{13} \\ M_{21} & M_{22} & M_{23} \\ M_{11} & M_{12} & M_{13} \end{bmatrix}, \tag{3.56}$$

where the *ij*th element of **M** is a force couple, composed of opposing forces pointed in the $\pm\hat{x}_i$ direction and mutually offset in the $\hat{x}_j$ direction. It should be noted that this form of the moment tensor is the zeroth-order term in a series expansion for the displacement field. This approximation is valid if the seismic signal used in the analysis has wavelength much greater than the source dimensions (Jost and Herrmann, 1989).

Diagonal components of the moment tensor are called *normal force couples*. The sign convention in this case is that outward pointing force couples, which favour a volumetric increase, have a positive sign. A seismic moment tensor for which the trace (sum of the diagonal elements) is zero will produce no net change in volume. Off-diagonal elements are *shear-force couples*; the sign convention for these is that right-lateral shear is positive. The seismic moment tensor has the symmetry $M_{ij} = M_{ji}$. This symmetry ensures that the resulting point source does not produce a net torque and it means that only six independent elements are required to fully characterize a general moment tensor.

The moment tensor concept was introduced by Gilbert (1971), building upon theory for seismic body-force equivalence for dislocation models (Burridge and Knopoff, 1964). It has since become a standard model for characterizing earthquake sources. This includes the rapid generation of *centroid* moment tensor solutions for significant earthquakes (Dziewonski et al., 1981; Ekström et al., 2012), where the centroid can be viewed as an optimal source location based on certain metrics (Julian et al., 1998). Combined with the use of Green's functions (Equation 3.29), the moment-tensor formalism can be used to compute the displacement wavefield based on the expression

$$u_i(\mathbf{x}, t) = M_{jk} \left[G_{ij,k} * s(t) \right], \tag{3.57}$$

where $s(t)$ is the source-time function. For a moment-tensor source in a homogeneous isotropic elastic medium, the far-field radiated body waves at a distance r from the source may be written as (Madariaga, 2007)

$$\frac{1}{4\pi r \rho v_P^3} \left[\gamma_i \gamma_j \gamma_k M_{jk} \right] \dot{s}\left(t - \frac{r}{v_P}\right), \tag{3.58}$$

for *P* waves, and

$$\frac{1}{4\pi r \rho v_S^3} \left[(\delta_{ij} - \gamma_i \gamma_j) \gamma_k M_{jk} \right] \dot{s}\left(t - \frac{r}{v_S}\right), \tag{3.59}$$

for *S* waves. As in previous expressions, γ_i is the *i*th direction cosine and $s(t)$ is the source-time function. It should be noted that the spatial derivative of the Green's function in Equation 3.44 has been replaced with the time derivative of the source-time function. In addition, the terms inside square brackets [ ] in Equations 3.45 and 3.46 are known as the isotropic *radiation patterns* for *P* and *S* waves, respectively (Figure 3.11).

Like the stress tensor, the seismic moment tensor can be decomposed into eigenvalues, along with an orthonormal basis of eigenvectors. Various schemes have been proposed for

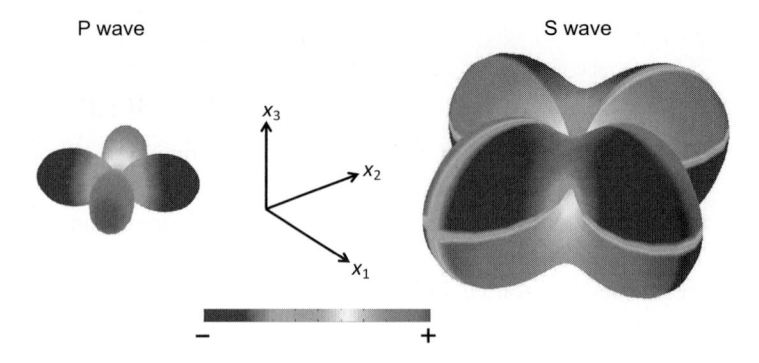

Fig. 3.11 Perspective view of the P-wave and S-wave radiation pattern for a double-couple source with $M_{12} = M_{21} = 1$. Colours denote amplitude, with positive outwards for the P wave and positive upwards for the x_3-component of the S wave. The P wave has two mutually orthogonal nodal planes, while the S wave has mutually perpendicular nodal axes. (A colour version of this figure can be found in the Plates section.)

decomposition of moment tensors into constituent components (Jost and Herrmann, 1989). These decomposition approaches generally make use of the eigenvalues of the moment tensor, which are ordered as

$$M_1 \geq M_2 \geq M_3. \tag{3.60}$$

The corresponding eigenvectors, $\mathbf{e}_1, \mathbf{e}_2$ and $\mathbf{e}_3$, define the $\mathbf{t}$ (tension), $\mathbf{b}$ (intermediate) and $\mathbf{p}$ (compression) *principal strain axes*, respectively (Vavryčuk, 2015).

Insights into the nature of an earthquake source can be gained by decomposition of the moment tensor into constituent elements that correspond to specific rupture mechanisms. In general, three basis functions are required to represent a moment tensor completely in decomposed form. Although there is no unique set of basis functions (Jost and Herrmann, 1989), the most common decomposition uses elementary moment tensors ($\mathbf{E}$) given by isotropic (ISO), double couple (DC) and compensated linear-vector dipole (CLVD) components. These elementary moment tensors are given by

$$\mathbf{E}_{\mathrm{ISO}} = \begin{bmatrix} 1 & 0 & 0 \\ 0 & 1 & 0 \\ 0 & 0 & 1 \end{bmatrix}, \tag{3.61}$$

$$\mathbf{E}_{\mathrm{DC}} = \begin{bmatrix} 1 & 0 & 0 \\ 0 & 0 & 0 \\ 0 & 0 & -1 \end{bmatrix}, \tag{3.62}$$

$$\mathbf{E}_{\mathrm{CLVD}}^{+} = \frac{1}{2}\begin{bmatrix} 2 & 0 & 0 \\ 0 & -1 & 0 \\ 0 & 0 & -1 \end{bmatrix}, \tag{3.63}$$

and

$$\mathbf{E}_{\mathrm{CLVD}}^{-} = \frac{1}{2}\begin{bmatrix} 1 & 0 & 0 \\ 0 & 1 & 0 \\ 0 & 0 & -2 \end{bmatrix}, \tag{3.64}$$

where the elementary form $\mathbf{E}_{\text{CLVD}}^{+}$ is used when $M_1 + M_3 - 2M_2 \geq 0$, whereas $\mathbf{E}_{\text{CLVD}}^{-}$ is used otherwise (Vavryčuk, 2015).

Each of these elementary moment tensors has a distinct physical meaning. The ISO elementary moment tensor contains the volumetric component of the rupture process (since the trace of the other elementary moment tensors is zero, it is evident that they produce no volumetric component). An underground explosion, for example, can be represented by a positive ISO moment tensor, whereas an implosion is given by an ISO moment tensor with negative polarity. The radiation pattern for an ISO moment tensor is composed of P waves radiated in every direction with equal amplitude, with no radiation of an S wave. As discussed below, the DC elementary moment tensor is represented by two orthogonal force couples. It can easily be shown that the diagonal form of the double-couple elementary moment tensor in Equation 3.49, which contains two normal force couples of opposite polarity, is equivalent (after rotation) to a moment tensor with two orthogonal shear couples. The source mechanism for the vast majority of earthquakes is dominated by a double-couple component. The CLVD force system was initially conceived as a mechanism for deep-focus earthquakes caused by phase transformation (Knopoff and Randall, 1970) and can be considered as the superposition of two double-couple ruptures with different polarity.

The scalar *seismic moment* (M_0) for a general moment tensor can be expressed in several ways (Vavryčuk, 2015). An optimal representation (Silver and Jordan, 1982) uses the Euclidean norm of the eigenvalues,

$$M_0 = \sqrt{\frac{1}{2}\left(M_1^2 + M_2^2 + M_3^2\right)}. \tag{3.65}$$

For a pure double-couple source, this representation yields the same results as the standard definition of seismic moment,

$$M_0 = \mu A \bar{d}, \tag{3.66}$$

where μ is a representative shear modulus of the fault, A is the slip area and $\bar{d}$ denotes the average coseismic slip (net displacement) over the area A.

3.5.2 Double-Couple Sources

As shown by Burridge and Knopoff (1964), mode II or mode III slip on a fault can be represented in the form of a *double-couple* mechanism. For a pure double-couple mechanism, the moment tensor can be written as

$$M_{ij} = \mu A \left(\hat{d}_i \hat{n}_j + \hat{d}_j \hat{n}_i\right), \tag{3.67}$$

where $\hat{\mathbf{n}}$ is the unit normal to the fault plane and $\hat{\mathbf{d}}$ is the unit slip vector. For example, for a vertical fault with $\hat{\mathbf{n}} = (1, 0, 0)$ and slip vector $\hat{\mathbf{d}} = (0, 1, 0)$ the moment tensor may be written as

$$\mathbf{M} = \mu A \begin{bmatrix} 0 & 1 & 0 \\ 1 & 0 & 0 \\ 0 & 0 & 0 \end{bmatrix}, \tag{3.68}$$

which makes the double-couple nature of the source mechanism more clear.

The P and S wave radiation patterns for this mechanism are shown in Figure 3.11 as perspective diagrams. This representation of the amplitude radiation was computed using the terms inside [ ] in Equations 3.45 and 3.46. Although there is a strong directional dependence, the overall amplitude of the radiated S wave is larger than the P wave by a factor of $\sim \left(\frac{v_P}{v_S}\right)^3$. The radiation patterns for both waves have a distinct four-lobe shape. For the P wave, pairs of lobes are bisected by *nodal planes*, which are planar surfaces on which the P wave amplitude is zero and across which there is a polarity reversal. The normals to the nodal planes are given by $\hat{\mathbf{n}}$ and $\hat{\mathbf{d}}$. Conversely, the S wave amplitude achieves a maximum within the nodal planes, while there are two *nodal axes* in the directions of $\hat{\mathbf{n}}$ and $\hat{\mathbf{d}}$.

A double-couple source is typically parameterized using three angular values: *strike* ($\tilde{\phi}$), *dip* ($\tilde{\delta}$) and *rake* ($\tilde{\lambda}$). The strike is the clockwise angle of the intersection of the fault plane with the surface, measured from geographic north. The *right-hand rule* is applied, which means that $\tilde{\phi}$ is measured when facing along the fault as it is dipping toward the right. The dip ($\tilde{\delta}$) is the angle between the surface and the fault plane, measured perpendicular to the strike. The rake is measured in the fault plane, and defines the angle between the slip vector and the strike direction. To be more specific, the slip vector shows the motion of the *hanging wall* relative to the *footwall* of the fault.[5] For a dipping fault, the hanging wall lies above the fault, as is marked by **H** in Figure 3.12. For a vertical fault (the case for most strike slip faults), the hanging wall is not defined so the slip vector measures the motion of the right fault block relative to the left fault block, when viewed in the strike direction. The rake has a positive sign if the slip vector is pointed upwards.

The normalized displacement vector can be expressed in terms of these fault parameters as

$$\hat{\mathbf{d}} = \left(\cos\tilde{\lambda}\cos\tilde{\phi} + \cos\tilde{\delta}\sin\tilde{\lambda}\sin\tilde{\phi}\right)\hat{\mathbf{x}}_1 \\ + \left(\cos\tilde{\lambda}\sin\tilde{\phi} - \cos\tilde{\delta}\sin\tilde{\lambda}\cos\tilde{\phi}\right)\hat{\mathbf{x}}_2 \\ - \sin\tilde{\delta}\sin\tilde{\lambda}\,\hat{\mathbf{x}}_3. \quad (3.69)$$

Similarly, the fault normal can be expressed in terms of strike, dip and rake as

$$\hat{\mathbf{n}} = -\sin\tilde{\delta}\sin\tilde{\phi}\,\hat{\mathbf{x}}_1 + \sin\tilde{\delta}\cos\tilde{\phi}\,\hat{\mathbf{x}}_2 - \cos\tilde{\delta}\,\hat{\mathbf{x}}_3. \quad (3.70)$$

Recall from §2.2 that Anderson developed a classification scheme based upon the assumption that one of the principal stresses is vertical. Based upon the Mohr–Coulomb failure criterion, the normal to an optimally oriented fault plane makes an angle of $\tan^{-1}\mu_s$ with respect to the minimum principal stress axis and an angle of $\pi/2 - \tan^{-1}\mu_s$ with respect to the maximum principal stress axis, where μ_s is the coefficient of static friction. As illustrated in Figure 3.12, by combining these concepts for we find that for each of the Andersonian stress regimes, there is a characteristic type of fault:

- Extensional stress regime $\longleftrightarrow$ normal faulting
- Strike-slip stress regime $\longleftrightarrow$ strike-slip faulting
- Compressional stress regime $\longleftrightarrow$ reverse (or thrust) faulting

---

[5] The terms hanging wall and footwall come from mining, since many ore deposits form along faults and the footwall forms the floor of the mine gallery.

3.5 Seismic Sources

Normal Fault

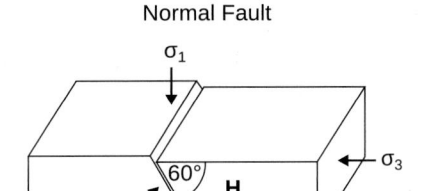

Extensional Stress Regime

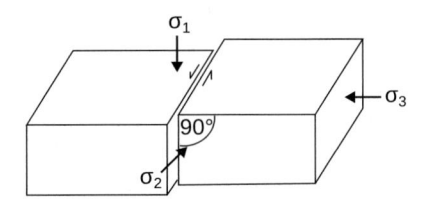

Strike-Slip Fault

Strike-Slip Stress Regime

Reverse Fault

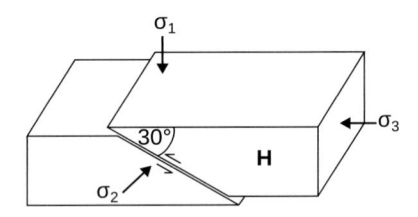

Compressional Stress Regime

Fig. 3.12 Relationship of normal, strike-slip and reverse faults to stress regimes based upon Anderson's classification scheme. For dipping faults, **H** denotes the hanging wall.

The attributes for each of these fault types are summarized in Table 3.1. Note that: a) depending upon the rake, strike-slip faults are further classified as *right-lateral* and *left-lateral*; b) low-angle reverse faults ($\tilde{\delta} < 30°$) are known as *thrust* faults. In the case of a right-lateral strike-slip fault, an observer standing on one fault block and looking across the fault would observe a displacement to the right. The opposite sense of displacement applies to a left-lateral strike-slip fault.

Double-couple events can have hybrid source mechanisms comprised of a superposition of two characteristic fault types. On the basis of the rake, the classification for a hybrid mechanism combines the term *oblique* with the name of the primary fault type; for example, a double-couple source with a rake of 20° is classified as a left-lateral oblique reverse fault, or more simply as an oblique-slip reverse fault. Other important descriptive terms for earthquakes include:

Table 3.1 Fault Classification

Fault Type	Stress Regime	S_V	Dip[a] ($\tilde{\delta}$)	Rake ($\tilde{\lambda}$)
Normal	Extensional	σ_1	60°	−90°
Strike Slip	Strike Slip	σ_2	90°	0°[b] or 180°
Compressional	Reverse[c]	σ_3	30°	90°

[a] Approximate dip for an optimally oriented fault assuming $\mu_s \approx 0.6$
[b] A left-lateral strike slip fault has a characteristic rake of $\approx 0°$, whereas a right-lateral strike-slip fault has a rake of $\approx 180°$
[c] Low-angle reverse faults ($\tilde{\delta} < 30°$) are known as *thrust faults*

- *Focus*: the point in the subsurface where slip on a fault initiates.
- *Hypocentre*: the estimated location of the earthquake focus.
- *Epicentre*: the point at the surface that lies vertically above the hypocentre.
- *Focal depth*: depth of the hypocentre below the surface.

The mechanism for a double-couple event is commonly shown using a graphical representation called a *focal-mechanism* diagram or a *beachball* diagram (or sometimes as a fault-plane solution). Focal mechanism diagrams are based on the concept of *focal sphere*, which is a hypothetical sphere of unit radius that is centred on the earthquake focus. A *lower-hemisphere projection* is used, which is a projection of the lower half of the focal sphere onto a horizontal plane that passes through the focus. For a double-couple source, the aforementioned P-wave nodal planes subdivide the focal sphere into four regions. In general, the intersection of the nodal planes with the focal sphere is manifested as an arc, once a lower-hemisphere projection is formed. Some regions defined by the nodal planes are filled (often in black), such that areas within which the P-wave polarity points away from the focus (as determined using Equation 3.45) are filled in. In practice, beachball diagrams can be produced by careful projection of the observed P-wave first motion onto the focal sphere; a good description of this process is given by Fowler (2004).

Beachball diagrams are effective because they provide, at a glance, a number of useful attributes, including a basic representation of the type of fault mechanism as well as two possible orientations for the fault plane (Figure 3.13). Recall that the P-wave amplitude vanishes on the mutually perpendicular nodal planes, whose normals are given by $\hat{\mathbf{n}}$ and $\hat{\mathbf{d}}$. The nodal plane with normal $\hat{\mathbf{d}}$ is called the *auxiliary plane*. In general, these two planes cannot be distinguished from P-wave first motion data alone – it is necessary to have some other source of information, such as a priori geological knowledge of the likely strike direction for a fault, in order to determine which of the nodal planes is the fault plane. Once nodal planes have been established, the principal strain axes (eigenvectors of the moment tensor) can be determined as follows:

$$\hat{\mathbf{t}} = \frac{1}{\sqrt{2}}\left(\hat{\mathbf{n}} + \hat{\mathbf{d}}\right) \quad \text{Axis of Tension}$$
$$\hat{\mathbf{b}} = \hat{\mathbf{n}} \times \hat{\mathbf{d}} \quad \text{Intermediate Axis} \quad (3.71)$$
$$\hat{\mathbf{p}} = \frac{1}{\sqrt{2}}\left(\hat{\mathbf{n}} - \hat{\mathbf{d}}\right) \quad \text{Axis of Compression.}$$

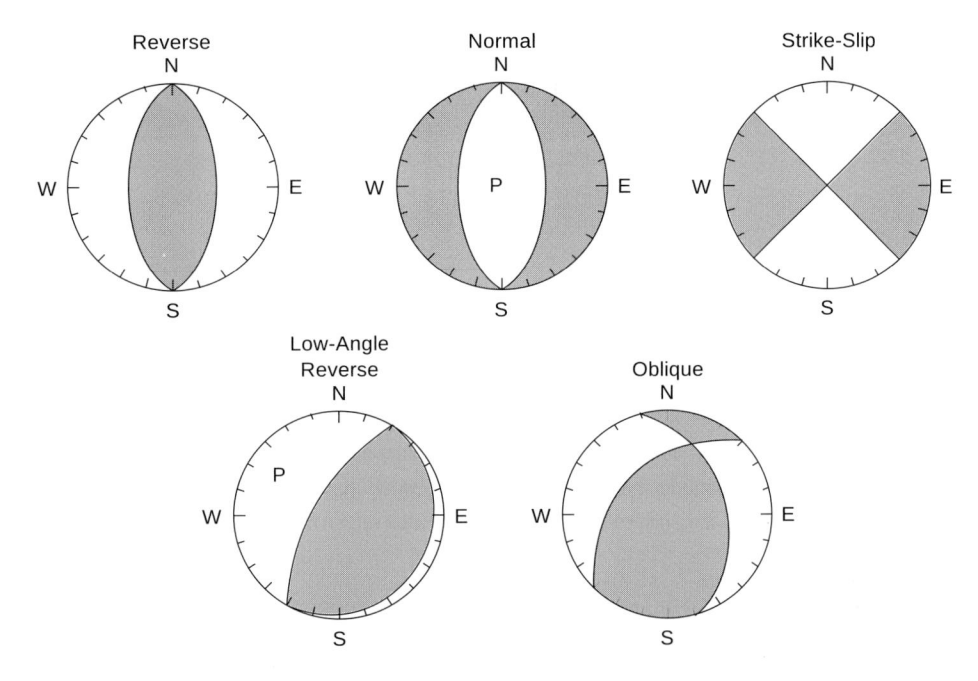

Fig. 3.13 Focal-mechanism (beachball) diagrams for various types of double-couple sources. Adapted from USGS Earthquake Glossary (https://earthquake.usgs.gov/learn/glossary).

The intersection of these axes with the lower hemisphere can be plotted as points in a focal-mechanism diagram.

While useful and well established, beachball diagrams have their limitations. This is especially true for non-double-couple sources, such as volumetric (implosion/explosion) sources and tensile (mode I) failure. A more general way to show the source mechanism is to display the radiation pattern for P and S waves, as exemplified in Figure 3.11 for a double-couple source.

The foregoing discussion assumes that the earthquake source is located in an isotropic material. Vavryčuk (2005) showed that the seismic wave radiation is more complex in the case of an earthquake hosted by an anisotropic medium. Pure double-couple mechanisms occur only in cases where the fault plane coincides with a plane of symmetry for media with orthorhombic or higher-order symmetry. In other cases, the focal mechanism generally includes non-double-couple components, which are discussed in the next section. Determination of the fault-plane solution and slip vector requires knowledge of the full elastic stiffness tensor at the source. Failure to account for the effects of anisotropy at the source can lead to errors of 5–10° in terms of fault orientation (Vavryčuk, 2005).

3.5.3 Non-Double-Couple Sources

Although the waves radiated from most earthquakes can be well represented as a double-couple source, non-double-couple sources are more common in volcanic and geothermal

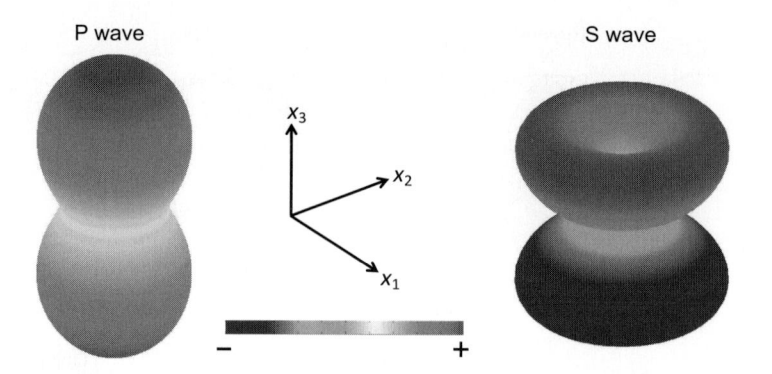

Fig. 3.14 Perspective view of the P-wave and S-wave radiation pattern for a tensile crack opening source with $M_{11} = M_{22} = \lambda$ and $M_{33} = \lambda + 2\mu$. Colours denote amplitude, with positive outwards for the P wave and positive upwards for the x_3-component of the S wave. The P wave has no nodal planes, while the S wave has a nodal plane and a perpendicular nodal axis. (A colour version of this figure can be found in the Plates section.)

areas (Shimizu et al., 1987; Miller et al., 1998; Foulger et al., 2004). Non-double-couple mechanisms have also been reported for microseismic events induced by hydraulic fracturing (Baig and Urbancic, 2010; Fischer and Guest, 2011; Eaton et al., 2014d). For example, a mode I crack generated by tensile opening perpendicular to the x_3 axis can be represented using the following elementary moment tensor

$$\mathbf{E_T} = \begin{bmatrix} \frac{1}{\lambda} & 0 & 0 \\ 0 & \frac{1}{\lambda} & 0 \\ 0 & 0 & \frac{1}{\lambda+2\mu} \end{bmatrix}, \qquad (3.72)$$

where λ and μ represent the Lamé parameters of the isotropic elastic host medium. The radiation pattern for this type of source mechanism is shown in Figure 3.14.

A theory for moment tensors of tensile earthquakes, herein described as a *shear-tensile* source model to avoid confusion with a tensile crack opening, has been developed by Vavryčuk (2011). In terms of failure criteria, shear-tensile rupture processes occur when the Mohr circle is tangent to the Griffith criterion (Fischer and Guest, 2011). The shear-tensile model is similar to the double-couple model, but it has an additional degree of freedom since the displacement vector is not constrained to lie within the fault plane. Thus, in addition to the geometrical parameters ϕ, δ and λ used for a double-couple source, a parameter $\tilde{\alpha}$ is introduced to measure the angle subtended between the displacement vector and the fault plane. Using this parameterization, the fault normal is still given by Equation 3.57, whereas the displacement vector differs from a double-couple mechanism and may be written as

$$\hat{\mathbf{d}} = \bar{d}\left[\left(\cos\tilde{\lambda}\cos\tilde{\phi} + \cos\tilde{\delta}\sin\tilde{\lambda}\sin\tilde{\phi}\right) - \sin\tilde{\delta}\sin\tilde{\phi}\sin\tilde{\alpha}\right]\hat{\mathbf{x}}_1$$
$$+ \bar{d}\left[\left(\cos\tilde{\lambda}\sin\tilde{\phi} - \cos\tilde{\delta}\sin\tilde{\lambda}\cos\tilde{\phi}\right) + \sin\tilde{\delta}\cos\tilde{\phi}\sin\tilde{\alpha}\right]\hat{\mathbf{x}}_2 \qquad (3.73)$$
$$- \bar{d}\left[\sin\tilde{\delta}\sin\tilde{\lambda} - \cos\tilde{\delta}\sin\tilde{\alpha}\right]\hat{\mathbf{x}}_3,$$

where $\tilde{\alpha}$ is positive if there is a tensile-opening component of displacement and negative if there is a tensile-closing component. A shear-tensile source is equivalent to a double-couple source when $\tilde{\alpha} = 0$ and it is equivalent to a pure tensile crack opening when $\tilde{\alpha} = 90°$. The moment tensor for a shear-tensile source is given by

$$M_{ij} = \lambda D_{kk}\delta_{ij} + 2\mu D_{ij}, \tag{3.74}$$

where λ (not to be confused with the fault rake angle) and μ are Lamé parameters. The symmetric tensor $\mathbf{D}$ is called the *potency* and is given by

$$D_{ij} = \frac{dA}{2}\left(\hat{n}_i \hat{d}_j + \hat{d}_i \hat{n}_j\right). \tag{3.75}$$

Focal-mechanism diagrams can also be constructed for non-double-couple events. Like double-couple sources, these are lower-hemispheric projections of *P*-wave first motion. In general, for non-double-couple events the *P*-wave nodal surfaces are not planar, resulting in patterns that do not necessarily look like a beachball. An example of a focal-mechanism diagram for a non-double-couple event is shown in Figure 3.15. As previously noted, a radiation-pattern diagram (e.g. Figure 3.14) provides a more complete representation of the source mechanism as it also conveys the radiated *S*-wave amplitudes and polarities.

Hudson et al. (1989) introduced a type of diagram that provides a convenient visual representation of the source type for a general moment tensor. A source-type diagram (or *Hudson diagram*) is constructed using a projection of the eigenvalues (M_i) of a moment tensor onto the parameters u and v, defined as follows:

$$\begin{aligned} u &= -\frac{2}{3}\left(\bar{M}_1 + \bar{M}_3 - 2\bar{M}_2\right), \\ v &= \frac{1}{3}\left(\bar{M}_1 + \bar{M}_2 + \bar{M}_3\right), \end{aligned} \tag{3.76}$$

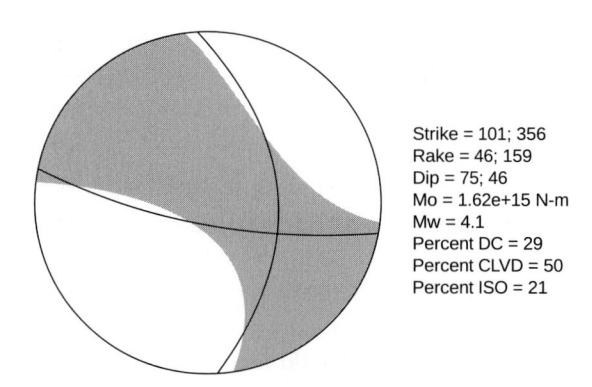

Strike = 101; 356
Rake = 46; 159
Dip = 75; 46
Mo = 1.62e+15 N-m
Mw = 4.1
Percent DC = 29
Percent CLVD = 50
Percent ISO = 21

Fig. 3.15 Focal-mechanism diagram for a non-double-couple earthquake source, showing the percentage of double couple (DC), compensated linear-vector dipole (CLVD) and isotropic (ISO) components and the strike, dip and rake of best-fitting DC nodal planes (Zhang et al., 2016). The shaded region shows the part of the local focal hemisphere in which the *P*-wave first motion is positive (away from the source). Reprinted with permission from Wiley.

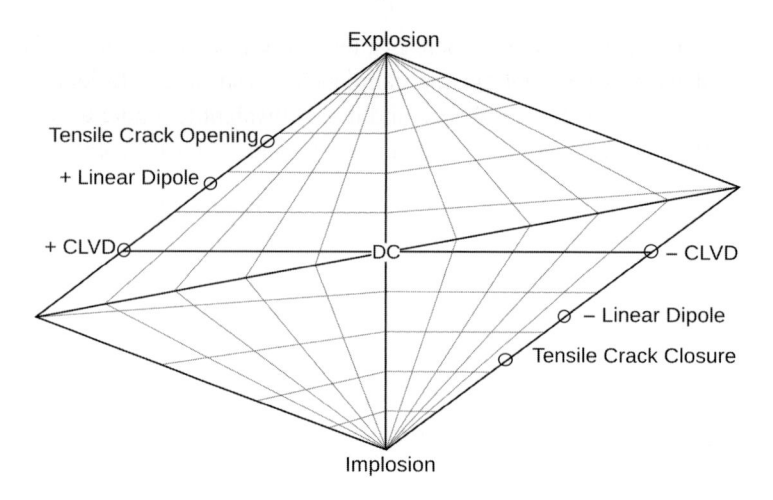

Fig. 3.16 Source-type diagram developed by Hudson et al. (1989). DC denotes double couple and CLVD denotes compensated linear-vector dipole. Adapted with permission from Wiley.

where

$$\bar{M}_i = \frac{M_i}{\max(|M_1|,|M_2|,|M_3|)}, \quad i = 1, 2, 3. \tag{3.77}$$

Hudson's (1989) parameterization leads to a source-type plot with a skewed diamond shape (Figure 3.16); this parameterization was chosen such that the a priori probability density of the moment ratios is uniform throughout the plot, which is useful for contour diagrams produced using a distribution of different moment tensors (Baig and Urbancic, 2010; Eaton and Forouhideh, 2011). This graphical representation of a moment tensor emphasizes the type of source mechanism; however, since the number of independent parameters is reduced from six components of the symmetric moment tensor to two parameters (u and v), some information is necessarily lost. For example, neither the magnitude nor the orientation of a source mechanism can be inferred from a Hudson plot. The coordinates of various elementary moment tensors with positive and negative polarity are shown in Figure 3.16, for reference.

3.6 Magnitude Scales

The *magnitude* of an earthquake is of fundamental importance, as a first-order estimate of source size based on observations of radiated seismic waves. Although universal use of a single magnitude scale would be desirable, a number of different magnitude scales are in common use, reflecting among other things the measurement of different components of the wavefield, observation at various distance ranges, variable conditions at different geographic locations and the use of diverse instrumentation (Kanamori, 1983). Challenges of unifying magnitude scales ultimately stem from difficulties to represent adequately such a complex spatiotemporal process using a scalar value.

C. F. Richter developed the first instrumental magnitude scale (Richter, 1935), based on the observation that earthquakes in California all tend to exhibit similar fall-off in amplitude with distance. Richter postulated that a range-independent scale could be developed based on the logarithmic offset, with respect to a reference event, of the amplitude versus distance curve for an earthquake. His proposed scale, now known as the *local magnitude* (M_L), was based on the peak ground-motion amplitude recorded on a standard Wood-Anderson (WA) seismometer, located at a distance of 100 km from an event. The formula can be approximated as (Shearer, 2009)

$$M_L = \log_{10} A + 2.56 \log_{10} \Delta - 1.67, \qquad (3.78)$$

where the peak amplitude A is measured in μm and Δ denotes the *epicentral distance* (i.e. the distance of an observation point from the earthquake epicentre), in km. This formula is valid for the epicentral distance range $10\,\text{km} \leq \Delta \leq 600\,\text{km}$. Subsequently, a *surface-wave* magnitude was introduced and calibrated to M_L (Gutenberg, 1945b). This scale is suitable for determination of the magnitude of shallow earthquakes at teleseismic distances ($\Delta \geq 1000\,\text{km}$) by various types of instruments, and is given by

$$M_S = \log_{10}(A/T) + 1.66 \log_{10} \Delta + 3.3, \qquad (3.79)$$

where A is the peak surface-wave amplitude in μm, T is the measured period of the wave in seconds and Δ is epicentral distance in degrees (this is the angle subtended at the centre of the Earth by radii to the source and observation point). Shortly thereafter, a teleseismic *body-wave* magnitude scale was developed (Gutenberg, 1945a); this is given by

$$m_b = \log_{10}(A/T) + Q(h, \Delta), \qquad (3.80)$$

where h denotes the focal depth in km and Q is an empirical function (Shearer, 2009). The magnitude notation uses a naming convention that if body waves are used the magnitude symbol is shown in lower case, whereas if surface waves are used, or if the measurement is phase-independent, then it is shown as an upper case M.

Local magnitude scales are convenient for rapid magnitude determination and remain widely used by operators of regional seismograph networks (Yenier, 2017). Richter's local magnitude scale is calibrated for southern California using then-standard Wood-Anderson seismometers, and a regional calibration is required to account for the attenuation attributes in other geographic locations. Examples of such calibrated local magnitude scales include southern (Hutton and Boore, 1987), central (Bakun and Joyner, 1984) and northern California (Eaton, 1992), Switzerland (Bethmann et al., 2011), Italy (Di Bona, 2016) and western Canada (Yenier, 2017).

The *seismic moment* is arguably the most definitive scalar representation of earthquake size. The standard representation of seismic moment is given in Equation 3.53, but this is limited to double-couple source mechanisms and a more general definition based on moment-tensor eigenvalues is given in Equation 3.52. SI units for seismic moment are N-m, but older studies make use of units dyn-cm (1000 dyn-cm = 1 n-m). The moment-magnitude scale (Kanamori, 1977; Hanks and Kanamori, 1979) is defined on the basis of seismic moment, and is given in SI units by

Table 3.2 Event Classification

M_W	Class	Length	Displacement	Frequency	Seismic Moment
8–10	Great Earthquake	100–1000 km	4–40 m	0.001–0.1 Hz	1 kAk–1 MAk[a]
6–8	Large Eq.	10–100 km	0.4–4.0 m	0.01–1.0 Hz	1 Ak–1 kAk
4–6	Moderate Eq.	1–10 km	4–40 cm	0.1–10 Hz	1 mAk–1 Ak
2–4	Small Eq.	0.1–1 km	4–40 mm	1–100 Hz	1 μAk–1 mAk
0–2	microearthquake	10–100 m	0.4–4.0 mm	10–1000 Hz	1 nAk–1 μAk
−2–0	microseism	1–10 m	40–400 μm	0.1–10 kHz	1 pAk–1 nAk
−4 to −2	nanoseism	0.1–1 m	4–40 μm	1–100 kHz	1 fAk–1 pAk
−6 to −4	picoseism	1–10 cm	0.4–4.0 μm	10–1000 kHz	1 aAk–1 fAk
−8 to −6	femtoseism	1–10 mm	0.04–0.4 μm	1–100 MHz	1 zAk–1 aAk

[a] The Aki (Ak), named for Keiiti Aki, is the recommended standard unit for seismic moment by the International Association of Seismology and Physics of the Earths Interior. 1 Ak = 10^{18} N-m. Table modified from Bohnhoff et al. (2010).

$$M_W = \frac{2}{3} \log_{10} M_0 - 6. \qquad (3.81)$$

This scale was designed to be more or less equivalent to M_L up to $M \approx 6$. At varying levels above that limit, other magnitude scales tend to saturate, whereas the M_W does not.

Bohnhoff et al. (2010) proposed a naming scheme for earthquakes that is based on M_W, or equivalently, seismic moment. Their classification scheme is modified in Table 3.2. This table also provides a summary of other characteristics, including approximate scales for rupture length, displacement and dominant frequency. The proposed naming convention adapts prefixes micro, nano, pico, femto, atto, zepto, etc. from the SI system, for application to low seismic events of relatively low magnitude, whereas for larger events standard terminology from earthquake seismology is adopted. In a departure from the SI convention, since moment magnitude is proportional to $\frac{2}{3} \log_{10} M_0$, the prefixes increment by 10^2 rather than the standard increment of 10^3.

The prefix "micro" has been used in the literature in reference to a wide range of different phenomena and scale sizes. Observing that the term "microearthquake" traditionally refers to earthquakes of $M < 3$, in order to achieve compatibility with even-ordered magnitude increments Bohnhoff et al. (2010) proposed that this prefix should be restricted to a magnitude range from 0 to 2. Within the engineering and applied geophysics literature on passive seismic methods, however, the term microseismic has been used for decades in reference to seismic events with magnitude below magnitude 0 (Albright and Pearson, 1982; Maxwell et al., 2010). In order to resolve this inconsistency in established terminology in different user communities, in this book the root word "seism" is applied to events with $M < 0$, to distinguish them from earthquakes which in this system have $M \geq 0$. According to this scheme, magnitude ranges associated for seismis with the prefixes micro, nano, pico and femto are each reduced by two magnitude units from the scheme proposed by Bohnhoff et al. (2010).

The scheme shown in Table 3.2 is used throughout this book; it is adopted from Eaton et al. (2016b) and modified from Bohnhoff et al. (2010). The naming scheme indicates a phenomenological subdivision as follows:

- the term "earthquake" is applied to events for which $M_W > 2$, which is sometimes regarded as the minimum threshold for felt natural events;
- the term "seismic events" applies to the magnitude range of $M_W < 2$, corresponding to events that are unlikely to be felt at the surface.

Independent determination of seismic moment using Equation 3.53 requires an estimate of the rupture area and average fault displacement. *Aftershocks* are events that occur following an earthquake, and their spatial distribution can be used to estimate the rupture zone. Care is required in the use of aftershocks for rupture-zone determination, however, as some aftershocks occur as result of slip triggered on proximal faults by static stress change (King et al., 1994) from the *mainshock* (the largest event in an earthquake sequence). Rupture area and associated uncertainties can also be determined by finite-fault inversion, by fitting radiated waveforms (Trifunac, 1974; Beresnev, 2003; Dettmer et al., 2014) as well as through the inversion of geodetic data (Zhang et al., 2011a).

3.7 Source Scaling and Spectral Models

Various empirical *scaling relations* have been developed that relate seismic moment, or M_W, to other earthquake parameters. The theoretical basis for a number of these relations is reviewed by Kanamori and Anderson (1975). In order to understand earthquake scaling relations, some knowledge of source spectra is needed. From Equations 3.45 and 3.46, we know that the far-field displacement field is proportional to the product of the seismic moment with the time derivative of the source-time function, $M_0 \dot{s}(t)$. Haskell (1964) proposed a source-time function whose time derivative is represented by a trapezoidal function of time. As shown in Figure 3.17, this shape is obtained by the *convolution* of two box-car functions of width τ_r and τ_d, which denotes the *rise time* and *rupture time constant*, respectively. Convolution is a linear filtering process that is described in Appendix B. The rupture time constant incorporates a correction for directivity (Doppler effect) and is given by

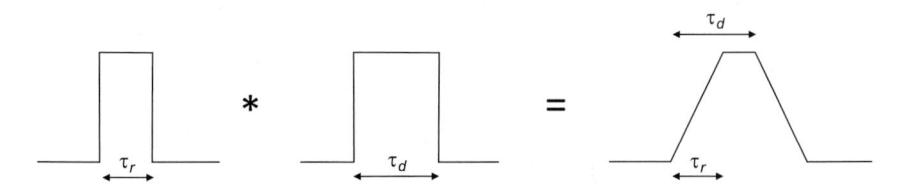

Fig. 3.17 Haskell fault model, formed as the convolution (denoted by *) of two boxcar functions of duration τ_r and τ_d, respectively. Modified from P. M. Shearer, *Introduction to Seismology*, 2009.

$$\tau_d = L_r/v_r - (L_r/v_r)\cos\theta_r, \qquad (3.82)$$

where L_r is the rupture propagation distance, v_r is the rupture velocity and θ_r is the azimuth of the station relative to the rupture direction. For a given azimuth, the source spectrum may be written as

$$|A(\omega)| = \tau_r \tau_d \, \text{sinc}(\omega\tau_r/2) \, \text{sinc}(\omega\tau_d/2), \qquad (3.83)$$

where the *sinc* function is defined as

$$\text{sinc}(x) = \frac{\sin x}{x}. \qquad (3.84)$$

By using the asymptotic properties of the sinc function, the Haskell source spectrum can be approximated as (Shearer, 2009)

$$\log_{10}|A(\omega)| - G = \begin{cases} \log_{10} M_0, & \omega < 2/\tau_d, \\ \log_{10} M_0 - \log_{10}\frac{\tau_d}{2} - \log_{10}\omega, & 2/\tau_d \leq \tau < 2/\tau_r, \\ \log_{10} M_0 - \log_{10}\frac{\tau_d \tau_r}{4} - 2\log_{10}\omega, & 2/\tau_r \leq \omega, \end{cases} \qquad (3.85)$$

where it is assumed that $\tau_d > \tau_r$ and G denotes the amplitude decay from geometrical spreading and anelastic attenuation. On a log–log plot of $A(\omega)$ versus ω, this defines a trilinear spectrum in which the first segment has a slope of zero, the second segment has a slope of -1 and the third segment has a slope of -2. The *corner frequencies* are points where there is a change in slope and occur at $\omega_1 = 2/\tau_d$ and $\omega_2 = 2/\tau_r$. In practice, it is often not possible to recognize both corner frequencies, so typically a single corner frequency is determined based on the intersection of the low-frequency asymptote and the high-frequency asymptote, as shown in Figure 3.18.

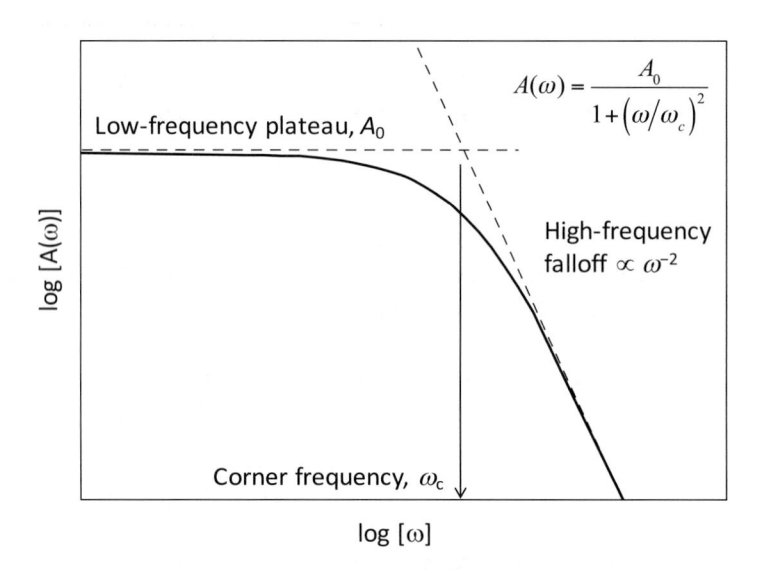

Fig. 3.18 Relationship of Brune model parameters to idealized far-field displacement spectrum for slip on a penny-shaped crack. Model spectrum is shown by the thick curve and does not include the effects of anelastic attenuation.

A more general model for the source spectrum in an attenuation medium with quality factor Q is given by (Abercrombie, 1995)

$$A(\omega) = \frac{A_0 \, e^{-(\omega t/2Q)}}{\left[1 + (\omega/\omega_c)^{n\gamma}\right]^{1/\gamma}}. \tag{3.86}$$

A widely used form of this source model, developed by Brune (1970) (see also Brune, 1971) and modified by Madariaga (1976), represents shear slip on a small circular crack. The corresponding source spectrum has the parameters $Q \to \infty$ (i.e. attenuation is ignored), $n = 2$ and $\gamma = 1$. These parameters can be used with Equation 3.73 to describe the shape of the source spectrum for tensile opening of a mode I crack (Walter and Brune, 1993). An alternative source spectrum with a sharper corner was proposed by Boatwright (1980). Boatwright's model was found by Abercrombie (1995) to provide a better fit to observed source spectra. This model has the same parameters as the Brune model, except that $\gamma = 2$.

The total radiated energy from an earthquake can be estimated using

$$\log_{10} E = 4.8 + 1.5 M_W, \tag{3.87}$$

where E is in Joules. This equation implies that each unit increment in moment magnitude corresponds to a factor of 30 increase in energy release. For both the Brune and Boatwright source models, the high-frequency falloff scales as ω^{-2}. Based on Equation 3.74, this is greater than the minimum falloff of $\omega^{-3/2}$ that is required to assure that the radiated energy is finite (Walter and Brune, 1993).

For the Brune source model, the seismic moment is related to the low-frequency plateau for the far-field S-wave displacement spectrum (A_0) by

$$M_0 = \frac{4\pi \rho v_S^3 A_0 r}{R_S}, \quad \text{for } S - \text{waves}, \tag{3.88}$$

where $R_S \simeq 0.63$ is the average S-wave radiation pattern (Boore and Boatwright, 1984). Similarly, the low-frequency plateau for the far-field P-wave spectrum is given by

$$M_0 = \frac{4\pi \rho v_P^3 A_0 r}{R_P}, \quad \text{for } P - \text{waves}, \tag{3.89}$$

where $R_P \simeq 0.52$ is the average P-wave radiation pattern (Boore and Boatwright, 1984). In an ideal scenario of a well-recorded event with pure double-couple rupture, both of these equations should yield the same value of seismic moment; in practice, this is rarely (if ever) achieved.

Assuming that the stress drop can be approximated by the co-seismic shear strain multiplied by 2μ, the stress drop implied by the Brune model for slip on a circular fault of radius a is given by

$$\Delta \tau = \frac{7 M_0}{16 a^3}. \tag{3.90}$$

Subject to a rather lengthy set of assumptions (Beresnev, 2001), according to the Brune model the source radius can be estimated using the corner frequency as follows:

$$a = 2.34 v_S / \omega_c. \tag{3.91}$$

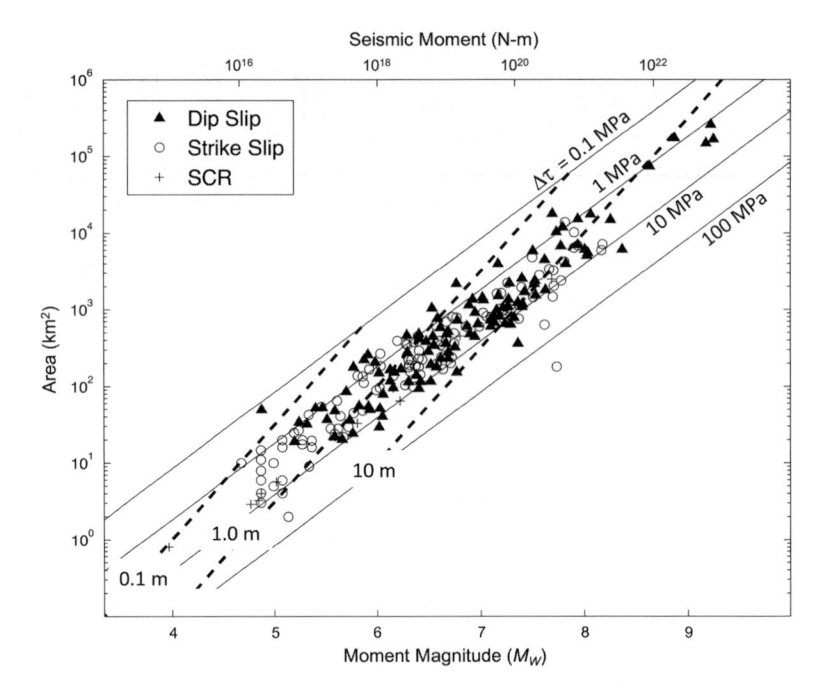

Fig. 3.19 Rupture area versus moment magnitude (M_W) or seismic moment, based on earthquake data compiled by Leonard (2010) and Leonard (2014). Lines of constant stress drop, $\Delta\tau$ (solid) and average slip (dashed) are based on scaling relationships for a circular fault in Kanamori and Anderson (1975). Most earthquakes fall within a stress-drop range of 1–10 MPa, including dip-slip, strike-slip and those located in stable continental regions (SCRs).

Figure 3.19 illustrates some salient scaling characteristics for earthquakes, using data compiled by Leonard (2010, 2014). This log–log scatter plot shows rupture area versus seismic moment, for events where the rupture area was estimated using data not derived from source spectra, such as from aftershock distributions. Although there is some scatter, the distribution is roughly linear in log–log space. The trend is consistent with $\Delta\tau$ in the range of 1–10 MPa for various different types of earthquake, including stable-continental regions (SCRs). This relatively narrow range of stress drop provides an indication of *self-similarity* of earthquakes over a large range of size. This graph also shows the average displacement based on the Brune model, $\bar{d}$, which is given by (Kanamori and Anderson, 1975)

$$\bar{d} = \frac{16\,a\,\Delta\tau}{7\pi\mu}. \tag{3.92}$$

3.8 Magnitude Distributions

It has long been recognized that, for a given region, the frequency-magnitude distribution of a set of earthquakes is usually well approximated by the *Gutenberg–Richter* relationship (Ishimoto and Iida, 1939; Gutenberg and Richter, 1944),

$$\log_{10} N = a - bM, \tag{3.93}$$

where $N(M)$ reflects a cumulative distribution and is the total number of earthquakes of magnitude $\geq M$. The parameters a and b describe, respectively, the earthquake productivity and the relative size distribution of events. The b-parameter typically falls within the range of 0.8–1.2 and can provide insights into the stability of a fault system. For example, the b-value can exhibit spatial and temporal changes that reflect variations in stress, focal depth, crustal heterogeneity and pore pressure (El-Isa and Eaton, 2014).

Various methods are used to estimate the b value for a set of earthquakes. Among the most robust is the maximum-likelihood method devised by Aki (1965), in which the estimated b value is given by

$$\hat{b} = \frac{\log_{10} e}{\overline{M} - M_c}, \tag{3.94}$$

where $\overline{M}$ denotes the average magnitude for $M \geq M_c$ and M_c is the *magnitude of completeness* – that is the magnitude level above which it is assumed that, for the region under study, the catalog is complete and there are no missing events. M_c can be approximated by the peak of the non-cumulative magnitude distribution (Wiemer and Wyss, 2000). For N events, the confidence limits for $\hat{b}$ are

$$\hat{b}\left(1 - d_\epsilon/\sqrt{N}\right) \leq \hat{b} \leq \hat{b}\left(1 + d_\epsilon/\sqrt{N}\right), \tag{3.95}$$

where $d_\epsilon = 1.96$ at the 95 percent confidence limit (Aki, 1965).

The probability density function $f(M)$ can be cast as a *power-law* relation with respect to the seismic moment Kagan (2002),

$$f(M_0) = \beta M_c^\beta M^{-1-\beta}, \quad \text{for } M \geq M_c, \tag{3.96}$$

where $\beta = 2b/3$. Based on the foregoing discussion regarding scaling relations, it is reasonable to suppose that any fault system has a maximum magnitude that is determined by the finite fault dimensions. Indeed, hazard models commonly invoke a tapered distribution such as (Kagan, 2010)

$$f_t(M) = \left[\beta + \frac{M}{M_t}\right] M_c^\beta M^{-1-\beta} \exp\left(\frac{M_c - M}{M_t}\right), \quad \text{for } M_c \leq M < \infty, \tag{3.97}$$

where the taper parameter M_t is used to adjust the upper-limit seismic moment.

Figure 3.20 exemplifies the analysis that can be used to determine parameters of the magnitude distribution for microseismic monitoring. Figure 3.20a shows a scatter plot of magnitude versus distance based upon a catalog from a microseismic experiment in central Alberta, Canada (Eaton et al., 2014a). The detection limit and completeness limit vary with distance, due to the effects of attenuation on the seismic signals. As explained in Chapter 5, it is important to take this distance-varying detection limit into account in the design of monitoring programs. Figure 3.20b shows the frequency-magnitude distribution for this catalog, as well as the best-fitting curves obtained using the maximum likelihood approach of Aki (1965) (MLGR) and the best-fitting tapered distribution. A tapered distribution appears justified for this dataset, based on the pronounced curvature of the frequency-magnitude distribution as the high-frequency limit of the catalog is approached. The magnitude of completeness for the catalog was determined using the

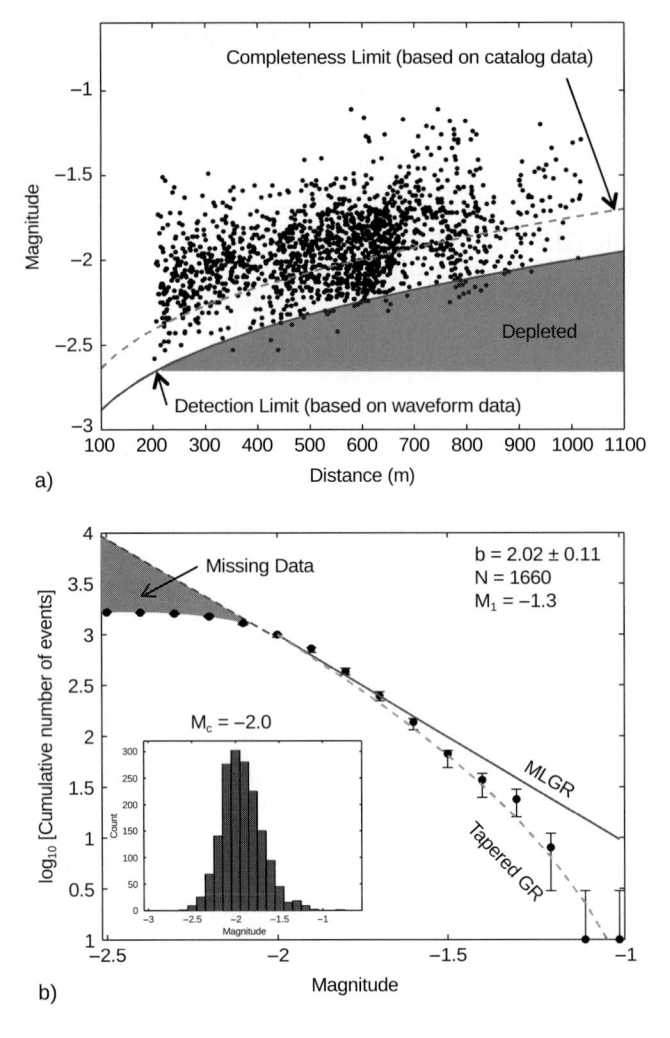

Fig. 3.20 a) Magnitude-distance scatter plot for a microseismic dataset from central Alberta (Eaton et al., 2014a). b) Data fit using the maximum-likelihood Gutenberg–Richter (MLGR) and tapered Gutenberg–Richter formulas, where the magnitude of completeness (M_c) parameter is measured from the peak of the magnitude distribution (inset). From Eaton and Maghsoudi (2015). Reprinted with permission of the EAGE.

magnitude histogram (inset) based on the method of Wiemer and Wyss (2000). This analysis yields $\hat{b} = 2.0$ and a magnitude of $M_1 = -1.3$ associated with the tapered distribution (which falls within the microseismic range identified in Table 3.2). Eaton and Maghsoudi (2015) discuss further details, including the potential bias introduced by catalog depletion below the detection threshold.

3.9 Aftershocks

Aftershocks are seismic events that occur after the largest event in a sequence (the mainshock). The occurrence of aftershocks is a manifestation of localized stress relaxation of a fault following an earthquake. The temporal rate of decay of an aftershock sequence, $r_a(t)$, is usually consistent with the empirical *Omori's Law* (Utsu, 1961)

$$r_a(t) = \frac{k}{(c+t)^p}, \tag{3.98}$$

where parameters c and k vary from one fault system to another and the exponent p generally falls within the range 0.7–1.5. A generalized form of Omori's Law that considers other scaling characteristics of aftershock sequences has been proposed by Shcherbakov et al. (2004). Conversely, *foreshocks* are events that occur prior to the mainshock. Foreshock sequences, albeit rare, can record accelerating seismicity leading up to a mainshock, via the inverse Omori Law (Jones and Molnar, 1979; Helmstetter et al., 2003)

$$r_f(t) = \frac{k}{(t_{MS} - t)^{p'}}, \tag{3.99}$$

where t_{MS} is the time of the mainshock. Foreshock sequences occur on a shorter timescale than aftershocks, on the order of weeks or months in advance of the mainshock (Helmstetter et al., 2003).

The *Epidemic Type Aftershock Sequence* (ETAS) model is a cascading point process derived from Omori's Law that can be used to simulate the temporal patterns of earthquake sequences in a given region (Ogata et al., 1993). According to the ETAS model, the rate of aftershocks associated with the ith event, occurring at time t_i with magnitude M_i, can be written as

$$r_i(t) = \frac{k}{(c+t-t_i)^p} e^{\alpha(M_i - M_0)}, \tag{3.100}$$

where M_0 is the minimum magnitude that is considered. The parameters k, α, c and p are global parameters that do not depend on i. The history of event occurrence, given by $H_t = \{(t_i, M_i); t_i < t\}$, is used to construct the conditional intensity function,

$$\lambda(t|H_t) = \nu + \sum_{\{i: t_i < t\}} r_i(t). \tag{3.101}$$

Ogata et al. (1993) developed a maximum-likelihood approach to estimate ETAS parameters based on a set of earthquake observations. Once ETAS parameters have been determined using past seismicity, the ETAS model can then be used to investigate dynamics of the fault system in a given area.

3.10 Summary

This chapter provides an overview of key concepts from seismology that are used to describe seismic sources and radiated waves in different types of media. Waves functions represent solutions to the equations of motion and can be classified as body waves, which propagate in the interior of a medium, or surface waves, which propagating on or near an external surface. Scalar and vector wave potentials can be used to simplify the derivation of wave solutions. Some wave functions are characterized by variation in velocity with wavevector (or frequency), known as dispersion. This leads to a distinction between phase and group velocity.

The subsurface contains a multitude of interfaces (or boundaries) between rock units with differing elastic properties and density. When a wave impinges onto such an interface, the wave energy is partitioned into various scattered modes that typically include reflected, transmitted and mode-converted (e.g. $P \rightarrow S$ mode conversion). For a plane wave incident onto a planar interface, the solutions to this boundary-value problem are described by the Zoeppritz equations. The propagation directions of scattered modes are governed by Snell's Law.

The velocity of waves propagating within an anisotropic elastic medium depends upon the polarization and direction of wave propagation. The directional dependence is customarily represented using wave and slowness surfaces to depict the group velocity and phase slowness relationships, respectively. In general, quasi-S waves (qS) exhibit birefringence (splitting), which is a diagnostic feature of seismic anisotropy.

Anelastic attenuation arises from energy losses due to processes such as heat generated by intergranular friction or viscous coupling in poroelastic media. The seismic quality factor (Q) is proportional to the frequency-dependent energy loss per cycle. A frequency-independent Q model is widely used to approximate anelastic attenuation. More general frequency dependence can be modelled using analogue mass-spring systems.

Seismic moment tensors have six independent parameters and provide an idealized representation of seismic sources as force couples acting at a point in a medium. When used in combination with Green's functions, seismic moment tensors can be used to model the far-field radiation patterns from a general seismic source. Most natural earthquakes can be well approximated as double-couple sources, which reduces the number of independent parameters for a geometrical source description to three parameters (typically these are specified as strike, dip and rake). Non-double-couple source types include shear-tensile earthquakes, which have an additional parameter ($\bar{\alpha}$) that denotes the angle between the slip vector and the fault plane, isotropic sources that represent a volumetric increase or decrease (e.g. an underground explosion), tensile-crack opening, and compensated linear-vector dipole (CLVD) sources. A general moment tensor can be decomposed into a combination of three different types of seismic sources.

Source size can be quantified using the seismic moment, $M_0 = \mu \bar{A} \bar{d}$. Various logarithmic magnitude scales are used to characterize earthquakes, of which the most general scale is the moment magnitude, M_W. Earthquake scaling relationships based on seismic moment and rupture area indicate that most events are characterized by a stress drop between 1 and 10 MPa. Spectral models, such as those developed by Brune, Boatwright and others, predict the shape of the source spectrum. In general, modelled displacement spectra approach an asymptotic value at low frequency that is proportional to seismic moment, and exhibit a high-frequency falloff above a corner frequency.

The frequency-magnitude relationship for natural earthquake systems can be characterized statistically by the Guternberg–Richter relation, which expresses a classic power-law dependence between the seismic moment and cumulative number of events. On a log–log plot, the slope of the cumulative frequency-magnitude distribution is called the b value, which is typically close to unity for active fault systems. The rate of aftershocks, which are events that occur after an earthquake, follows an empirical decay curve given by a generalized form of Omori's Law. The Epidemic Type Aftershock Sequence (ETAS) model is a cascading point process derived from Omori's Law.

3.11 Suggestions for Further Reading

- Fundamental concepts of theoretical seismology: Aki and Richards (2002).
- Earthquake and global seismology: Stein and Wysession (2009).

3.12 Problems

1. Find a rotation operator that transforms the diagonal form of the double-couple moment tensor in Equation 3.62 into a tensor form with two shear couples.

2. Within Anderson's classification scheme, what type of event is represented by the double-couple (DC) solution in Figure 3.15?

3. By solving Equation 3.25, determine the fundamental-mode Love-wave velocity at frequencies of 10 Hz and 20 Hz, for a simple model with a 12 m thick layer with $v_S = 250$ m/s and $\rho = 1800$ kg/m^3 overlying a half space with $v_S = 1250$ m/s and $\rho = 2250$ kg/m^3.

4. Suppose that phase velocity within the frequency band $0 < f < \pi$ is given by $v = v_1 + A\cos(f/B)$. Use Equation 3.26 to find an expression for the group velocity within this band.

5. Consider a planar interface between two isotropic elastic half-spaces characterized by $v_{P1} = 2100$ m/s, $v_{S1} = 1000$ m/s and $\rho_1 = 2200$ kg/m^3 in the upper half-space, with $v_{P2} = 4600$ m/s, $v_{S2} = 2600$ m/s and $\rho_2 = 2200$ kg/m^3 in the lower half-space. Determine all of the critical angles for incident plane waves (P, S_V and S_H) in the upper medium. Discuss how the post-critical reflections might impact wide-angle recordings using a downhole microseismic array.

6. Using $Q_P = 100, 500$, $Q_S = 50, 250$, plot Brune and a Boatwright spectra at $t = 5$ s for all Q values, for a source with a corner frequency of $f = 5.0$ Hz.

7. Consider a fault with strike of 30°, dip of 70° and rake of 10°.

   a) Calculate the vectors corresponding to the $\hat{d}$, $\hat{n}$, $\hat{t}$, $\hat{b}$ and $\hat{p}$ axes.

   b) Write the full moment tensor using geographic coordinates.

   c) Write the full moment tensor in diagonalized form.

8. The surface projection of the rupture zone for a 30°-dipping reverse fault is 3 km long and 0.25 km in width. The slip function on the fault can be approximated a three fault segments that occupy 50%, 30% and 20% of the fault surface area, with uniform slip of 40 cm, 30 cm and 20 cm (respectively) within each of these ruptures zones. Assuming that

the shear modulus is $\mu = 10$ GPa, what is the seismic moment? What is the corresponding moment magnitude?

9. After a large earthquake, the number of recorded aftershocks per day after 1, 4 and 20 days is 80, 18 and 3 per day. Using Equation 3.98 (generalized Omori's Law), estimate the parameters k, c and p.

10. **Online exercises**:

a) The Zoeppritz equations can be solved numerically to visualize the amplitude and phase of reflected and transmitted plane waves at a planar interface between two elastic half-spaces that are in welded contact. A Matlab tool is provided that facilitates graphical exploration of energy partitioning at an interface.

b) Source-type diagrams developed by Hudson et al. (1989) provide a convenient way to classify the type of rupture (double couple, tensile crack opening, compensated linear-vector dipole, etc.), irrespective of geometry of the source. A visualization tool is provided to project an arbitrary moment tensor into this diamond-shaped graph.

4 Stress Measurement and Hydraulic Fracturing

> But though he is confined to its crust, he may penetrate into all its secrets.
> Jules Verne (The Steamhouse, Part II, 1881)

Hydraulic fracturing is a reservoir-stimulation method used to create permeable pathways deep into a rock formation, thereby enhancing fluid transport within the reservoir medium (Smith and Shlyapobersky, 2000). Extensive deployment of this technology for development of low-permeability unconventional hydrocarbon reservoirs has had a "game-changing" effect on world markets (Montgomery and Smith, 2010). Simply put, a hydraulic fracture is initiated when engineered fluids are pumped into a wellbore at a rate that is faster than the fluids can escape into the formation, resulting in a pressure buildup that ultimately surpasses the tensile strength of the rockmass. Initiation and growth of a hydraulically stimulated fracture network is usually accompanied by brittle rock-failure processes that can be monitored using passive-seismic (microseismic) methods (Maxwell, 2014). Passive-seismic monitoring is also the primary surveillance technology for induced seismicity (Shapiro, 2015), which in rare instances can be caused by hydraulic fracturing (Bao and Eaton, 2016).

Numerical simulation of hydraulic fracturing is widely used for hydraulic fracturing and completion design (Adachi et al., 2007). The in situ stress field exerts a first-order control on evolution of a hydraulic-fracture network; hence, a robust and accurate model of the stress state is required in order to achieve meaningful results. The first part of this chapter provides a concise overview of methods that are used to determine components of the stress tensor in the subsurface. Stress is difficult to measure directly, so most methods are based upon indirect inference of stress from induced strain observations, coupled with knowledge of the constitutive properties of the medium. Next, this chapter summarizes salient aspects of technology that is used for hydraulic fracturing of low-permeability rock formations. Taken together, these concepts are foundational for understanding practical applications of passive-seismic monitoring that are covered in the following chapters.

4.1 How Subsurface Stress Is Determined

The in-situ state of stress in the subsurface represents a complex superposition of interacting stresses due to overburden and tectonic forces, as well as gradients in temperature and pore pressure arising from natural and anthropogenic factors (Amadei and Stephansson,

1997). Our knowledge of the contemporary stress field in the *lithosphere*, Earth's relatively rigid outermost shell, derives from a diverse set of tools that include inversion of earthquake focal mechanisms, geophysical well logging and in situ measurements such as borehole breakout, overcoring and small-volume injections (Zoback, 1992; Heidbach et al., 2010; Schmitt et al., 2012). Although regional stress models are well established for some areas such as the Western Canada Sedimentary Basin (Bell and Babcock, 1986), it is nevertheless useful to obtain detailed local constraints in order to capture adequately any localized heterogeneity in the stress field.

4.1.1 Focal-Mechanism Inversion

If a suitable set of earthquakes is available from a given region, a partial estimate of the stress tensor can be obtained by *focal-mechanism inversion* (Gephart and Forsyth, 1984; Lund and Slunga, 1999). The basic requirements for implementation of this procedure are:

1. fault orientations need to sample a sufficiently diverse range, as marked by variations in one or more principal strain axes of $\approx 30° - 45°$ (Hardebeck and Hauksson, 2001);
2. earthquakes all occur within a region characterized by a uniform stress field that is invariant in space and time (Barth et al., 2016);
3. focal mechanisms are mutually independent and thus lack any kinematic interaction resulting from static stress changes (King et al., 1994);
4. for each event, ambiguity with respect to identification of the fault plane based on the measured orientation of two possible nodal planes can be resolved using independent constraints (Barth et al., 2016).

Another key underlying assumption, called the *Wallace–Bott* hypothesis, holds that for the set of analyzed events, each dislocation vector is parallel to the maximum resolved shear stress on the corresponding fault plane (Wallace, 1951; Bott, 1959). Based on this set of conditions and assumptions, the inverse problem is typically formulated as a least-squares optimization, in which stress parameters are determined that minimize the misfit between the observed and calculated displacement vectors.

In general, estimation of the complete stress tensor using focal-mechanism inversion is an *ill-posed* problem, meaning that the solution is either under-constrained (the case for focal-mechanism inversion), or it is over-determined and inconsistent. Although the stress magnitude cannot be uniquely determined using earthquake focal mechanisms, it is possible to obtain robust estimates of the orientation of the three principal stress axes, the stress regime (normal, strike-slip or reverse), and ratios of the principal-stress components (Michael, 1984). Due to the wide availability of focal-mechanism solutions, especially near plate boundaries, stress estimates using earthquake focal mechanisms comprise the majority of data records in the database of the World Stress Map project (Barth et al., 2016). Based on tests using synthetic data, Vavryčuk (2014) found that, under realistic noise conditions, roughly 20 focal mechanisms are needed in order to reduce uncertainty in inferred orientations of principal axes to less than $5°$.

4.1.2 Crossed-Dipole Sonic Logs

Crossed-dipole sonic logging (Close et al., 2009) is an acoustic well-logging method that is used to infer anisotropic shear-wave velocity structure in the vicinity of a wellbore. The raw data consist of waveforms that are acquired using a wireline toolstring equipped with sonic dipoles. Each dipole consists of paired piezoelectric transducers of opposing polarity, which sum to generate a directional force (Close et al., 2009). The directional forces excite *flexural* wave modes, a class of guided wave that propagates along the wellbore (Schmitt et al., 2012). Like surface waves, borehole flexural waves exhibit geometrical dispersion; consequently, low-frequency components of these signals are sensitive to velocity structure that is farther from the wellbore than high-frequency components. Furthermore, in an anisotropic medium the flexural waves exhibit birefringence that, like qS waves, is characterized by mutually orthogonal fast and slow polarization directions whose orientations depend on the anisotropic symmetry of the elastic stiffness tensor (Schmitt et al., 2012). "Crossed" dipoles means that both source and receivers have mutually orthogonal dipoles that are designed for simultaneous excitation of fast and slow flexural modes.

The use of crossed-dipole logs for stress determination is a two-step procedure. In the first step, the recorded waveforms are processed to yield fast and slow qS-wave slowness values (reciprocal of velocity) and the fast qS polarization angle (Esmersoy et al., 1994). In addition, wavelength-dependent dispersion curves are measured for fast and slow flexural wave modes, which have the same polarization directions as the fast and slow qS waves (Schmitt et al., 2012). In the second step, stress parameters are estimated based upon the pattern of seismic anisotropy and dispersion for all of the observed wave models. This step is complicated by superposition of intrinsic seismic anisotropy due to rock fabrics together with seismic anisotropy induced by the present-day deviatoric stress environment (Figure 1.6). The stresses are further complicated by drilling-related perturbations to the near-wellbore stress field, which are discussed in the next section. Fortunately, there are diagnostic patterns of qS and flexural-mode dispersion that may be used to disentangle these effects (Boness and Zoback, 2006).

Crossed-dipole sonic logging is thus an in situ technique for measurement of seismic anisotropy, which in turn can be used to constrain stress parameters. Under favourable circumstances, where anisotropy effects of rock fabric are negligible or otherwise accounted for, fast polarization directions measured in, for example, a vertical wellbore could be used to infer the orientations of $\hat{S}_{Hmax}$ and $\hat{S}_{Hmin}$. The intensity of seismic anisotropy, moreover, provides semi-quantitative information about relative stress magnitudes (Schmitt et al., 2012).

4.1.3 Wellbore Failure Mechanisms

The presence of a wellbore strongly perturbs the ambient stress environment, due to the redistribution of stresses in order to maintain force equilibrium. Expressed in cylindrical coordinates (r, θ), the perturbed 2-D stress field for a circular hole of radius a, within an isotropic elastic plate characterized by an ambient uniaxial stress field σ_{xx}, is given by the *Kirsch* equations:

$$\sigma_{\theta\theta} = \frac{\sigma_{xx}}{2}\left(1 + \frac{a^2}{r^2}\right) - \frac{\sigma_{xx}}{2}\left(1 + \frac{3a^4}{r^4}\right)\cos 2\theta,$$

$$\sigma_{rr} = \frac{\sigma_{xx}}{2}\left(1 - \frac{a^2}{r^2}\right) + \frac{\sigma_{xx}}{2}\left(1 + \frac{3a^4}{r^4} - \frac{4a^2}{r^2}\right)\cos 2\theta, \qquad (4.1)$$

$$\tau_{\theta r} = -\frac{\sigma_{xx}}{2}\left(1 - \frac{3a^4}{r^4} + \frac{2a^2}{r^2}\right)\sin 2\theta.$$

The parameters $\sigma_{\theta\theta}$ and σ_{rr} denote the *hoop* stress and radial stress, respectively, while the parameter $\tau_{\theta r}$ denotes the shear stress arising from the ambient uniaxial stress (Schmitt et al., 2012). This method thus constrains components of the stress tensor in the plane orthogonal to the axis of the wellbore.

Figure 4.1 shows the normalized stress distribution in the vicinity of an empty hole, calculated using the Kirsch equations. Of particular note, under compressive ambient stress conditions the hoop stress is tensile and equal in magnitude to σ_{xx} in the direction parallel to the uniaxial stress, whereas in the orthogonal direction the hoop stress is compressive and has a magnitude of $3\sigma_{xx}$. As elaborated below, stress anisotropy arising from the hoop stress largely controls the main mechanisms of wellbore failure, namely *borehole breakout*

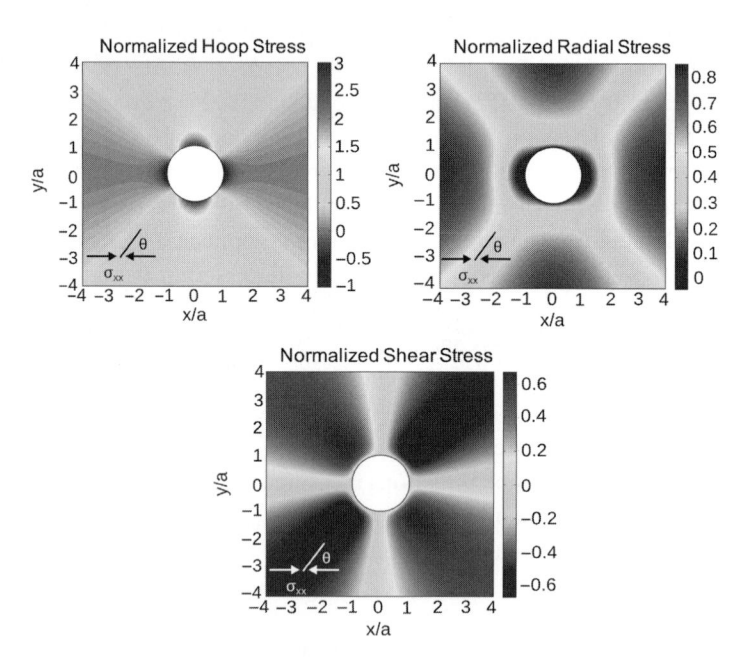

Fig. 4.1 Stress distribution in an elastic plate around an empty circular cavity of radius a, showing hoop stress, $\sigma_{\theta\theta}$, radial stress, σ_{rr} and shear stress, $\tau_{\theta r}$ computed using the Kirsch equations (Equation 4.1) and normalized based on the magnitude of the ambient uniaxial stress, σ_{xx}. The origin of the cylindrical coordinate system (r, θ) is at the centre of the cavity, with θ measured counterclockwise from the direction of the uniaxial stress. Stress amplitudes are plotted using the sign convention that tensile stress is negative. (A colour version of this figure can be found in the Plates section.)

and *drilling-induced tensile fractures* (DITFs). The latter can be used to constrain S_{Hmin} using the drilling mud weight (Zoback et al., 1993).

Not surprisingly, the formulas for stress in 3-D are more complicated, particularly if the axis of the wellbore is not aligned with a principal stress direction. The interested reader is referred to Schmitt et al. (2012), where the 3-D stress equations are provided under several scenarios. In addition, if the hole is filled with an incompressible fluid with pressure P, this produces an additive hoop stress component given by

$$\sigma_{\theta\theta}^{F} = -P\frac{a^2}{r^2}, \tag{4.2}$$

the effect of which increases the tensile stress component in the direction of the uniaxial stress (Figure 4.1). If the medium is poroelastic, an instantaneous change in fluid pressure, ΔP, produces an *infiltration* hoop stress at $r = a$ that is given by

$$\Delta\sigma_{\theta\theta}^{I}(r = a) = -\Delta P(1 - \xi), \tag{4.3}$$

where

$$\xi = \frac{1}{2}\alpha\frac{1 - 2\nu}{1 - \nu}. \tag{4.4}$$

In this expression, α is the Biot coefficient and ν denotes Poisson's ratio.

In addition to the effects of fluid pressure, changes in temperature can have a very large effect on the state of stress in the wellbore, denoted as $\Delta\sigma_{\theta\theta}^{T}$. Taking all of these effects into account, a generic generated hoop stress $\Delta\sigma_{\theta\theta}$ at $r = a$ can be expressed as (Schmitt et al., 2012)

$$\Delta\sigma_{\theta\theta} = \Delta\sigma_{\theta\theta}^{F} + \Delta\sigma_{\theta\theta}^{T} + \Delta\sigma_{\theta\theta}^{I}. \tag{4.5}$$

The aforementioned wellbore failure mechanisms associated with the generated hoop stress are schematically illustrated in Figure 4.2a. In a vertical wellbore, drilling induced tensile failure occurs in the direction of $\hat{S}_{Hmax}$ when the stress state exceeds the Griffith failure criterion. Borehole breakout is more common and occurs as a result of delamination mechanisms and shear failure, leading to elongation of the (originally circular) well cross section in the direction of $\hat{S}_{Hmin}$. The magnitude of the maximum horizontal stress (S_{Hmax}) can be estimated based on the angular breakout width, Θ (Figure 4.2), using the formula (Barton et al., 1988)

$$S_{Hmax} = \frac{S + \Delta\sigma_{\theta\theta} - S_{Hmin}(1 - 2\cos\Theta)}{1 + 2\cos\Theta}, \tag{4.6}$$

where S is the unconfined compressive strength. Assumptions that underlie this formula are (Zoback et al., 2003):

1. the borehole width does not change as the breakout progresses;
2. the edge of the breakout zone corresponds to the azimuth from $\hat{S}_{Hmin}$ at which the stress state is tangent to the Mohr–Coulomb failure envelope.

Borehole breakout observations can be performed using a caliper log measured using a four-arm caliper tool. As the caliper tool is pulled upwards, articulated arms press up against the walls of the wellbore to measure the borehole dimensions and orientation of

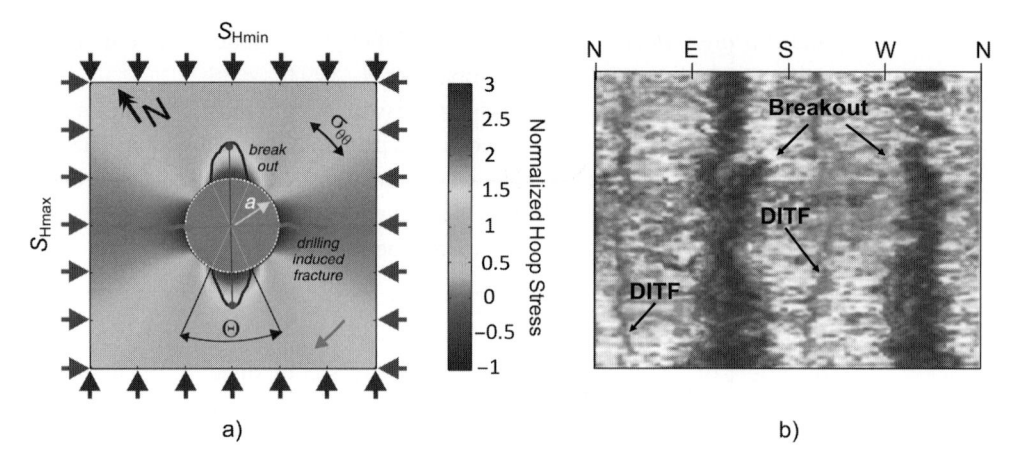

Fig. 4.2 a) Schematic illustration of regions of borehole breakout and drilling-induced tensile fracture (DITF) in relation to the generated hoop stress, $\sigma_{\theta\theta}$ around a wellbore of radius a. Angular breakout width is shown by Θ. Reprinted from *Tectonophysics*, Vol 580, Douglas R. Schmitt, Claire A. Currie and Lei Zhang, Crustal stress determination from boreholes and rock cores: Fundamental principles, Pages 1–26, copyright 2012, with permission from Elsevier. b) Borehole image data showing spatially oriented observations of borehole breakout and DITF. Geographic directions of north (N), east (E), south (S) and west (W) are indicated. Reprinted from *International Journal of Rock Mechanics & Mining Sciences*, Vol 40, M. D. Zoback et al., Determination of stress orientation and magnitude in deep wells, Pages 1049–1076, copyright 2003, with permission from Elsevier. (A colour version of this figure can be found in the Plates section.)

the elongation direction. The use of borehole breakout data to determine stress parameters was pioneered by Bell and Gough (1979), based upon extensive caliper measurements in western Canada. Caliper logs, however, have limited capability to detect drilling induced tensile failure; nor can they be used to measure the breakout width Θ. These measurements, in addition to in situ observations of fractures, are facilitated using borehole image logs, which provide high-resolution digital imaging capabilities within a wellbore. Various types of image logs exist, based on electrical, ultrasonic or optical imaging principles. Images derived from a cylindrical wellbore are generally flattened for viewing using an unwrapping procedure, as illustrated in Figure 4.2b.

4.1.4 Diagnostic Fracture-Injection Tests

Diagnostic fracture-injection tests (DFITs) are small-volume injections (Figure 4.3) that are performed prior to hydraulic fracturing, in order to gain important in-situ stress and permeability information. The same (over very similar) procedure is sometimes referred to as a mini-frac or extended-leakoff test. Following Nguyen and Cramer (2013), the basic DFIT procedure is:

- A surface pump begins to inject fluid at a constant rate, leading to increasing *bottom-hole* pressure. The bottom-hole pressure at depth z is calculated from the measured

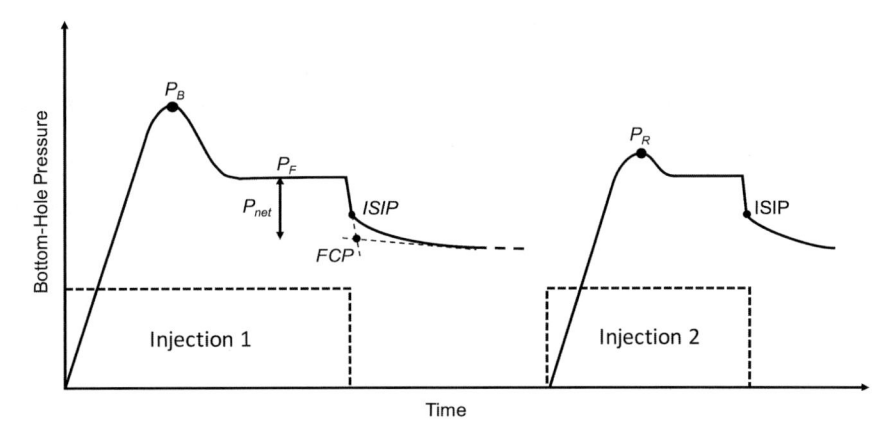

Fig. 4.3 Idealized pressure curves for two injection stages of an extended leakoff test. The first stage is equivalent to a DFIT. Dashed boxes show injection windows. The following abbreviations are used: P_B, formation-breakdown pressure; P_F, fracture-propagation pressure; ISIP, instantaneous shut-in pressure; FCP, fracture-closure pressure; P_{net}, net pressure; P_R, residual pressure.

surface injection pressure, by accounting for perforation friction, fluid-slurry friction, near-wellbore friction and hydrostatic pressure (Brown et al., 2000).

- The hydraulic fracture initiates once *formation-breakdown pressure* (P_B) is reached. This can usually be recognized at the surface by an abrupt drop in wellhead pressure.
- Fluid injection continues at a constant rate until a steady-state pressure condition is achieved. The bottomhole pressure under steady-state conditions represents the *fracture-propagation pressure* (P_F).
- Fluid injection is stopped and the well is *shut in*, resulting in a rapid drop in bottomhole pressure to the *instantaneous shut-in pressure* (ISIP) value.
- The shut-in pressure is monitored as it declines through the *fracture-closure pressure* (FCP). There are various methods to estimate FCP using the pressure-decline curve; White et al. (2002) recommended the use of a tangent-intersection approach (Figure 4.3). The *net pressure* P_{net} is the difference between P_F and FCP and provides a measure of frictional pressure drop and the fracture-tip resistance to propagation (Smith and Shlyapobersky, 2000). Potocki (2012) used an alternate definition for net pressure as the difference between ISIP and FCP, which they argued reflects the far-field fracture stimulation complexity.
- The pressure is monitored for a period of time after closure within a reservoir-dominated pressure regime.

The DFIT procedure is identical to the first injection stage in an *extended-leakoff test* (XLOT) used in scientific drilling programs; however, XLOTs usually include repeated injection stages that are designed to reduce uncertainty in the derived parameters (White et al., 2002). The subsequent XLOT stages also enable estimation of the *residual pressure* (P_R) and the *residual tensile strength* (T_R), as shown in Figure 4.3.

Various stress and reservoir parameters can be determined using DFIT analysis. The FCP provides the best available proxy for the magnitude of the minimum principal stress, σ_3. Based on Anderson's fault classification, σ_3 is equivalent to S_{Hmin} in strike-slip and normal stress regimes and S_V in a reverse faulting regime. Once the FCP is reached during the DFIT process, the fractures are closed and continued pressure decline during shut-in is related to fluid leakoff into the reservoir (Nguyen and Cramer, 2013). The post-closure pressure decline curve can be used to evaluate several reservoir parameters such as pore pressure and *reservoir conductivity* (the product of permeability and thickness), as well as fracture-closure dynamics related to progressive failure of small asperities on the fracture surface (Martin et al., 2012).

Under normal and strike-slip stress regimes, it is possible to estimate S_{Hmax} based on XLOT observations, albeit with much greater uncertainty than S_{Hmin} (Schmitt et al., 2012). This calculation makes use of the *hydraulic fracturing breakdown equation* (Schmitt and Zoback, 1993), which is given by

$$P_B \simeq 3S_{Hmin} - S_{Hmax} + T_0 - P, \qquad (4.7)$$

where T_0 is the *initial tensile strength* of the rock and P is the pore pressure. The breakdown pressure, P_B and the minimum horizontal stress, $S_{Hmin} \simeq$ FCP are measured from the pressure curves. If the residual pressure P_R can be determined using XLOT analysis, as shown in Figure 4.3, then the tensile rock strength may be obtained from (Schmitt et al., 2012)

$$T_0 \simeq P_B - P_R. \qquad (4.8)$$

This means that the maximum horizontal stress can be expressed in terms of parameters that are all measurable, at least in principle, using XLOT analysis:

$$S_{Hmax} \simeq 3S_{Hmin} - P_R - P. \qquad (4.9)$$

4.1.5 Overcoring

Overcoring stress analysis (Leeman, 1968) is a method for determination of the complete, in situ stress tensor in the subsurface. This method is carried out by installing a strain measurement cell into a small-diameter pilot hole that is drilled concentrically at the base of a larger diameter wellbore (Sjöberg et al., 2003). The annulus containing the strain cell is isolated from the stress field of the host medium by overcoring (re-drilling with a larger-diameter drill bit) and its deformation is monitored before and during re-equilibration. The ambient stress field is then back-calculated, based on the measured values of the strain tensor after re-equilibration. Overcoring is mainly used in mining and civil engineering for the design of underground openings, primarily for the purpose of gaining information about the local stress state, rather than to measure tectonic stresses (Reinecker et al., 2016).

The theoretical basis for overcoring derives from closed-form solutions for the stress state around a wellbore (Leeman, 1968). The determination of stress assumes that the sampled region is a homogeneous, linear elastic solid and thus makes use of generalized Hooke's Law. Analytical methods have been developed for isotropic, transversely isotropic

(TI) and orthorhombic media (Hakala et al., 2003). For an isotropic material, the two required elastic parameters (Young's modulus and Poisson's ratio) can be determined using core extracted from the pilot hole for the strain cell. Obtaining the state of stress requires additional elastic parameter measurements for TI (five independent measurements) and orthorhombic (eight independent measurements) media that include extraction of core in inclined wellbores. Measurement errors are mitigated by averaging stress values obtained for a closely spaced set of measurements (Sjöberg et al., 2003). Due to the erratic nature of stresses at shallow depths, the World Stress Map Project does not include any data from overcoring measurements for depths of less than 100 m (Reinecker et al., 2016).

4.1.6 Simplified Mathematical Models

In many cases, stress measurements are insufficient or unavailable in a given area, in which case it may be necessary to estimate stress components using a mathematical model. In this section, a number of commonly used stress models are discussed along with underlying assumptions and range of validity.

A simplified model is generally used to calculate the vertical principal stress, S_V. At depth z, S_V can be calculated using measured density value, such as from well logs, by integrating the weight of the overburden:

$$S_V(z) = \int_0^z g\rho(z')dz', \qquad (4.10)$$

where $\rho(z)$ is density and $g \simeq 9.8 m/s^2$ is the acceleration of gravity at Earth's surface. Although this approach is often treated as a direct calculation rather than a model per se, implicit in this expression is an assumption that there is no long-term flexural strength that can support the overburden load. Averaged over a large area, this is a reasonable assumption that is borne out by the observation that Earth's lithosphere is in a regional state of isostatic equilibrium. At smaller scales, however, heterogeneities exist with respect to vertical stress; for example, in an underground mine the weight of the overburden is locally supported by transferring the load to the surrounding rockmass.

Abrupt changes in the state of stress offer clues about the spatial and temporal scales of stress heterogeneity that are sustainable within Earth's crust. Co-seismic stress drop is one example. From Figure 3.19, we see that stress drops of up to 10 MPa are not uncommon; in the case of a shallow earthquake, a stress drop of this magnitude could represent a significant fraction of S_V. Similarly, the length scale of large earthquakes (Table 3.2) gives a sense for spatial scales over which a quasi-static departure from Equation 4.10 may be sustainable. In the absence of direct measurements, this formula nevertheless provides the best available model for vertical stress at depth.

Next, models for horizontal stress need to be considered. Zoback et al. (2003) argued that a critical stress state, as expressed by Equations 2.17–2.19 and represented by the stress polygon (Figure 2.11), provides an upper bound on the ratio of the minimum to maximum effective stress, σ'_1/σ'_3. Application of this condition requires a priori knowledge of the pore pressure, Biot coefficient, friction coefficient and stress regime. Given S_V from Equation 4.10, for a normal faulting regime this approach places a lower limit on S_{Hmin}

based on Equation 2.17, or an upper limit on S_{Hmax} based on Equation 2.19. In the case of a strike-slip regime, S_V is the intermediate principal stress and does not enter into the critical stress condition (Equation 2.18). Hence, for a strike-slip regime, some other constraint is required.

As an example of this approach to model stress, Roche and Van der Baan (2015) developed a layered stress model within a strike-slip stress regime. In order to determine the depth-dependent stress field, they calculated the effective vertical stress (see §2.2) using

$$S'_V(z) = \bar{\rho} g z - P(z), \tag{4.11}$$

where $\bar{\rho}$ is an average density value and pore pressure is defined by a linear depth gradient k,

$$P(z) = zk. \tag{4.12}$$

Roche and Van der Baan (2015) initialized the horizontal stresses by applying conditions for a cohesionless, critically stressed crust within a strike-slip regime,

$$\left[\frac{S'_{Hmax}(z) - S'_{Hmin}(z)}{2} \right] \sqrt{1 + \mu_s^2} - \mu \left[\frac{S_{Hmax}(z) + S_{Hmin}(z)}{2} \right] = 0. \tag{4.13}$$

They assumed a friction coefficient $\mu_s = 0.7$ and made the further assumption that S'_V is the average of the two horizontal stresses,

$$\frac{S'_{Hmax}(z) + S'_{Hmin}(z)}{2} = S'_V(z). \tag{4.14}$$

Well-log data provided depth curves of V_P, V_S and ρ, which were used to compute Young's modulus (E) and Poisson's ratio (μ) based on the following equations:

$$\mu = \rho V_S^2 \,;\ \lambda = \rho - 2\mu \,;\ E = \frac{\mu(3\lambda + 2\mu)}{\mu + \lambda} \,;\ \nu = \frac{\lambda}{2(\mu + \lambda)}. \tag{4.15}$$

Roche and Van der Baan (2015) constructed a layered elastic model using ρ, E and ν from well logs. Layered stress values (Figure 4.4) were then obtained numerically using a discrete-element method, in which the stress solutions to the system of Equations in 4.13 and 4.14 were applied as boundary conditions at the sides of the model domain and welded boundary conditions were applied at the interfaces between layers. This process was repeated for various trial values of pore-pressure gradient. As shown in Figure 4.4, a measured value of S_{Hmin} at this location (Bell and Bachu, 2003) agrees reasonably well with calculations for a hydrostatic pore-pressure gradient, which is given in general by $k = 10$ kPa/m.

The resulting model (Figure 4.4) contains layers with high deviatoric stress that likely exerted a strong localizing influence on failure processes during fluid injection (Roche and Van der Baan, 2015). For a stratified medium, the model predicts that mechanically stronger layers, characterized by relatively high Young's modulus, bear a disproportionate share of the overall stress load.

As an aside, it should be noted that, while it is common to use well-log data in order to compute elastic moduli for stress calculations, there are several uncertain factors. For

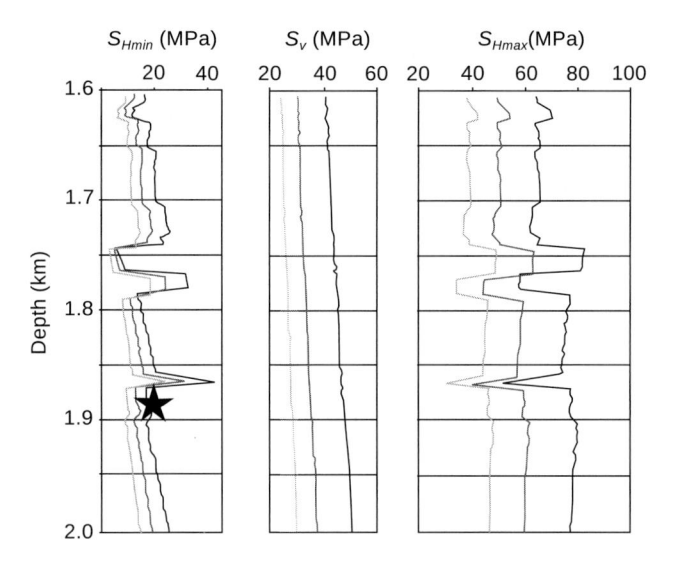

Fig. 4.4 Stratified model for stress based on a hydraulic fracturing study in western Canada. The model assumes a critically stressed state and produces layered variations in horizontal stress due to lithologic strength variations. Gray curves represent a range of pore pressures, from no fluid (light gray) to hydrostatic (dark gray). The star shows a calibration point where S_{Hmin} was independently measured. Modified from Roche and Van der Baan (2015). Reprinted with permission from Wiley.

example, at the frequencies used for sonic logs ($\approx$ 1–50 kHz), measured values in fluid-saturated formations are representative of undrained moduli (Kalahara, 1996). Moreover, at such frequencies log-derived moduli may not give accurate representations of static moduli (Ong et al., 2016). Finally, well logging measures velocities of wave propagation along the well (often vertical), which provides an incomplete representation of the elastic constitutive relation if the medium is anisotropic.

Another widely used model for effective horizontal stresses, known as the *uniaxial-elastic-strain* (UES) (Kalahara, 1996), is given by

$$S'_{Hmin} = \frac{v}{1-v} S'_V = K_0 S'_V, \tag{4.16}$$

where $S'_V = S_V - \alpha P$ is the effective vertical stress, α is the Biot coefficient (often assumed to be ≈ 1) and K_0 is called the *coefficient of Earth pressure at rest* (Thiercelin and Roegiers, 2000). The model is valid only within a normal-faulting stress regime where $S_{Hmax} \simeq S_{Hmin}$. The name given to this model describes an assumed hypothetical constitutive model as follows: elastic deformation resulting from the retrieval of an isotropic rock sample from its in situ stress state in the subsurface to an unconfined state, wherein stress and pore pressure are both reduced to zero, would lead to a uniaxial vertical strain ϵ_V but negligible horizontal strain (Figure 4.5a). A less restrictive model, called the *generalized-stain model* is similar to the UES approximation but includes an additional constant tectonic-stress term, S_T,

$$S'_{Hmin} = \frac{v}{1-v} S'_V + S_T, \tag{4.17}$$

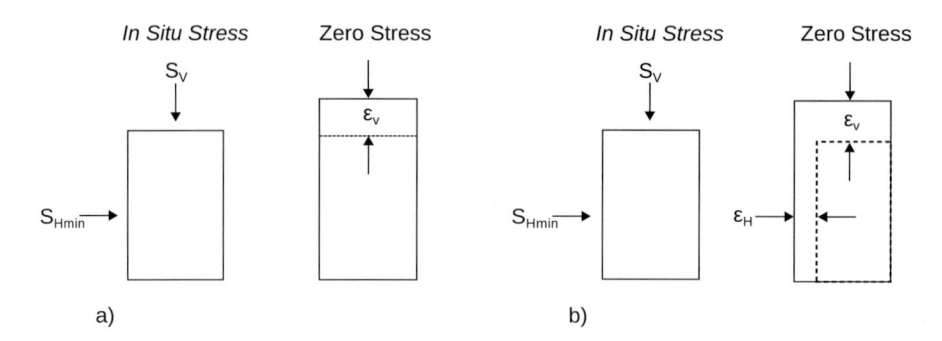

Fig. 4.5 Strain characteristics for several approximations for in situ stress.
a) Uniaxial-elastic-strain (UES) model. b) Generalized-stain model, which includes a constant tectonic stress term. Modified from Kalahara (1996). Used with permission of the Society of Petrophysicists and Well-log Analysts.

where

$$S_T = \frac{2\mu}{1-\nu}(\epsilon_H + \nu\epsilon_h). \tag{4.18}$$

For this model, elastic deformation resulting from unstressing an isotropic rock sample, as described above, would produce horizontal strain ϵ_H in the direction of the maximum horizontal stress and ϵ_h in the direction of minimum horizontal stress (Figure 4.5b), where $\epsilon_H \geq \epsilon_h$. According to this model, the horizontal stresses are given by

$$S'_{Hmax} = \frac{\nu}{1-\nu}S'_V + \frac{2\mu}{1-\nu}(\epsilon_H + \nu\epsilon_h), \tag{4.19}$$

and

$$S'_{Hmin} = \frac{\nu}{1-\nu}S'_V + \frac{2\mu}{1-\nu}(\nu\epsilon_H + \epsilon_h). \tag{4.20}$$

More elaborate stress models exist along similar lines. For example, Warpinski (1989) developed a viscoelastic stress model that can be used to match the strain history of a basin, and Kalahara (1996) discussed models that include inelastic and thermal effects.

4.2 Hydraulic Fracturing

Hydraulic fracturing is a process for injecting engineered fluids into a rock formation under pressure in order to enhance the connection between the wellbore and the reservoir, by creating a permeable pathway for fluid flow. Hydraulic fracturing is key enabling technology that has helped to unlock vast reserves of unconventional oil and gas resources within low-permeability hydrocarbon-bearing rocks. Although the method has been in commercial use since 1947, hydraulic fracturing for unconventional oil and gas development has proliferated during the last few decades. This has led to spectacular increases in oil and gas reserves and production, especially with the advent of massive *multi-stage hydraulic fracturing* (MSHF) in horizontally drilled wells (Dusseault and McLennan, 2011). Indeed, without the use of hydraulic fracturing, resources such as tight sands, coalbed methane and shale gas would remain largely undeveloped (Speight, 2016). Hydraulic fracturing is thus a prime example of a "disruptive technology" that has reshaped the dynamics of global oil and gas markets (King, 2012).

In its benchmark scenario, which accounts for policy changes to keep global warming to less than 2°C, the International Energy Agency forecasts a 30 percent growth in worldwide energy use by 2040, largely supplied by fossil fuels (EIA, 2015). In this context, the development of shale-gas resources has been cited as a potential bridge fuel that could be used to replace coal, which has higher greenhouse gas (GHG) emissions. The public discourse about regulation and use of hydraulic fracturing has nevertheless triggered intense debate. For example, Vengosh et al. (2014) summarized four potential risks for water resources arising from shale-gas development. Although these are not directly related to hydraulic fracturing processes, they warrant mention here within a broader context. These risks include:

1 contamination of shallow aquifers by fugitive gas;
2 contamination of surface water and shallow groundwater by spills or leaks;
3 accumulation of toxic and radioactive elements in soil or stream sediments near disposal or spill sites;
4 overextraction of water for high-volume hydraulic fracturing, which could produce water shortages or conflicts with other water use.

Concerns have also been raised about the level of greenhouse gas emissions from shale gas development, in general (Allen et al., 2013). The most relevant environmental concern that is addressed in this book, however, is the risk of activating fault slip during hydraulic fracturing (Bao and Eaton, 2016) or through other types of fluid injection such as brine disposal (Ellsworth, 2013). This aspect is discussed in Chapter 9.

Figure 4.6 is a schematic illustration of the key features of a horizontal well with multiple hydraulic-fracturing stages. The well is drilled from the surface in an approximately vertical orientation until a depth is reached close to the target formation. The heel of the well, also known as the build section, follows a curved well path from vertical into a roughly horizontal inclination. The stratigraphic interval within which the lateral well segment is drilled is called the *landing zone*. The lateral segment of the well is typically drilled for distances in excess of 2 km along an azimuth that is usually roughly along S_{Hmin} in order to produce transverse fractures that maximize the reservoir contact area (Dusseault and McLennan, 2011). The completion process is performed in multiple intervals called *stages*. After drilling, metal casing is cemented into the wellbore to maintain stability, to provide fluid isolation and for control of well pressure. Cemented casing is always used above the reservoir for containment; it is sometimes used in the reservoir, depending upon the completion style. The diameter of the casing decreases with depth, with the largest diameter near the surface for protection against contamination of potable water.

Hydraulic fracturing is considered to be a well-completion method, which is entirely distinct from the drilling process in terms of equipment, personnel and timing. In general terms, the equipment used for hydraulic fracturing (Brown et al., 2000) includes (Figure 4.7):

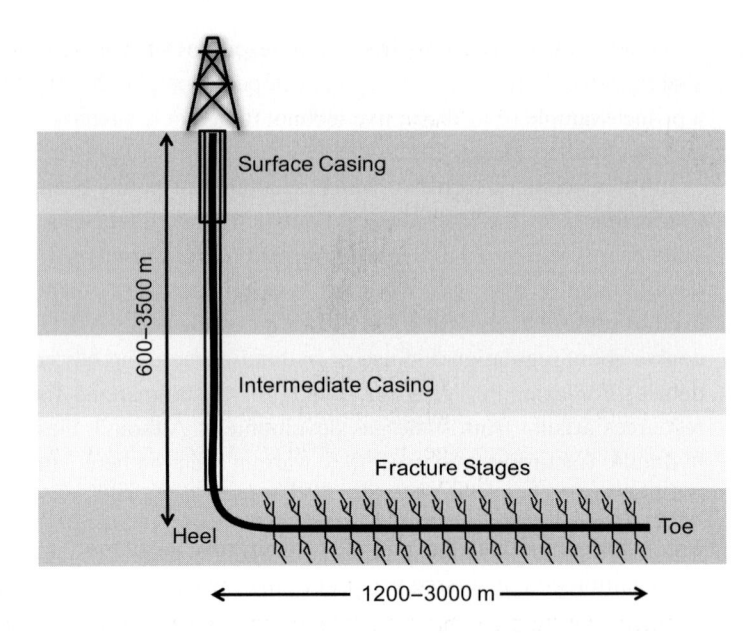

Fig. 4.6 Schematic diagram showing main elements of a multi-stage hydraulic fracturing completion of a horizontal wellbore (not to scale).

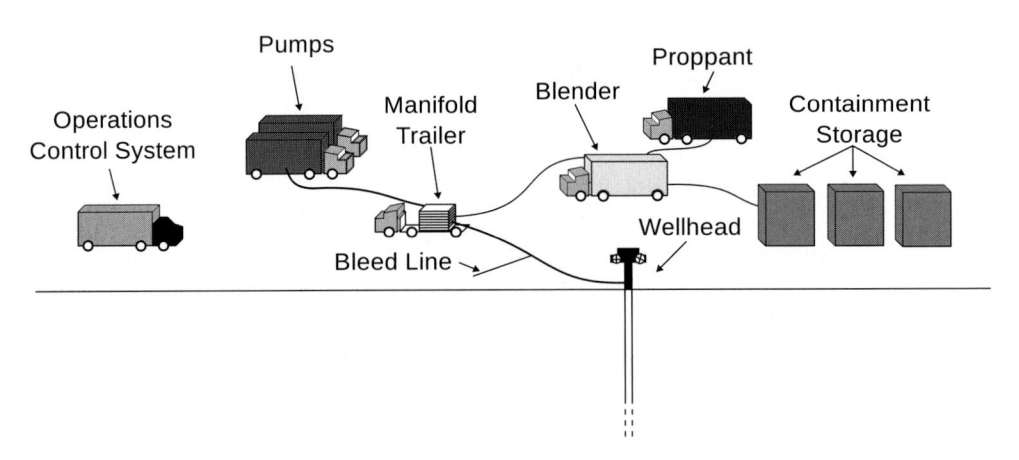

Fig. 4.7 Schematic diagram showing types of equipment used for hydraulic fracturing.

- containment and/or delivery systems for proppant, additives and the primary fracturing fluid;
- blending equipment for continuous mixing of additives and proppant with the fracturing fluid in order to create a slurry with the desired composition;
- high-pressure pumps, which operate in parallel to supply the requisite pressure and flow rate;

- an operations control system including a job monitoring unit and various types of sensors;
- a system for management of flowback fluids.

Many hydraulic-fracturing treatments for development of low-permeability hydrocarbon resources (including shale gas, tight oil and liquids-rich plays) use *slickwater* as the primary fracturing fluid. This type of fluid is predominantly composed of fresh water, with surfactants and friction reducers used as additives in order to reduce surface tension and pumping resistance. Other common additives are biocides, which prevent microbial growth in the reservoir in response to the injected fluid, and scale inhibitor, which reduces the development of deposits that can increase flow resistance within the system (King, 2012). Alternative types of fracturing fluids include energized foam, which contains gas such as nitrogen or CO_2, or gel-based fluids that use a gelling agent such as guar (Barati and Liang, 2014). Non-slickwater fracturing fluids are generally more costly but they may be better suited to sensitive formations with swelling clays, and/or provide greater control on dynamic viscosity of the injected fluid. In some cases, a hybrid approach is used, wherein gelling agents (Constien et al., 2000) are introduced near the end of the stage. The type of fracturing fluid can have a profound influence on the character and distribution of the induced microseismic events (Duhault, 2012).

Proppant materials, such as sand or ceramic beads, are mixed with the fracturing fluid at varying concentration to create a slurry. The purpose of injecting proppant is to ensure that fractures remain open after the pumping pressure is released. Without the use of proppant, fractures would close, leading to little if any residual fracture permeability except in the case of *self-propping* of fractures (Brown and Bruhn, 1998; Dusseault and McLennan, 2011).[1] Ideal properties of proppant are high strength in order to resist crushing, corrosion resistance and low density in order to reduce gravitational settling of proppant material to the bottom of fractures (EPA, 2004). Availability and cost economy are also important practical considerations; materials that best meet these characteristics include silica sand, resin-coated sand and manufactured ceramic beads. Mixing of proppant into the slurry substantially increases the dynamic viscosity of the fracturing fluid (Constien et al., 2000).

Proppants are characterized by a range of grain sizes. In practice, the designation of proppant grain size is referenced to the American Petroleum Institute (API) sieve-size designation (Table 4.1). For example, a proppant that is designated as 20/40 mesh represents a nominal grain-size distribution in which approximately 90 percent of the proppant mix would fall through the 20-mesh sieve (850 μm), but would remain on the 40-mesh sieve (425 μm). Other common API proppant designations include 30/50 mesh, 16/30 mesh, 12/18 mesh and 16/20 mesh, although the median particle diameter (MPD) may vary greatly for a given mesh designation (Schubarth and Milton-Tayler, 2004). In general, proppants with a larger diameter produce a proppant pack with higher permeability, but are less well suited to "dirty" formations that are subject to *fines migration*[2] into the fracture (Constien et al., 2000).

---

[1] Self-propping can occur due to juxtaposition of asperities on a fracture surface.
[2] Transport of fine grains within the reservoir formation due to drag forces during flowback or production.

Table 4.1 API Sieve Designation

US Sieve Number	Opening Size
12	1.7 mm
16	1.18 mm
18	1.0 mm
20	850 μm
30	1600 μm
40	425 μm
50	300 μm
70	212 μm
140	106 μm

Source: API (2014)

Parameters that are measured and used for monitoring and control of hydraulic fracturing include pressure, injection rate, fluid density, temperature, pH and viscosity (Brown et al., 2000). Fluid pressure sensors measure the deformation of strain gauges, which deform in response to pressure in the line. Downhole pressure is the primary diagnostic parameter and can be either measured or calculated using observed surface pressure. Injection rate can be measured in various ways, including simple pump-stroke counters that measure the rotational speed of pumps; these have the advantage over other types of rate sensors that they are not affected by the presence of proppant in the slurry. Density measurements are based on absorption, using a radioactive source and a gamma-ray counter on the other side of the pipe. All of these sensors provide critical data in real time to the control system so that problems can be detected immediately and appropriate action taken.

Well completion can be performed right after drilling, but it often occurs following a substantial delay (weeks or more). During hydraulic fracturing of a horizontal well, stages are completed in sequence starting at the toe (distal end of the lateral) and progressing back toward the heel of the well (Figure 4.6). Examples of representative pumping curves (surface-treatment pressure, injection rate and proppant concentration) for a single hydraulic-fracturing stage are shown in Figure 4.8. Although the pressure curve in this diagram is seemingly similar to the DFIT curve in Figure 4.3, pressure curves measured during hydraulic fracturing are generally more complicated for a variety of reasons, included varying rheology of proppant-laden slurry (Gulrajani and Nolte, 2000) and because hydraulic fracturing stages are pumped at high injection rates that lead to complex turbulent processes at the injection point (Tary et al., 2014a). Nevertheless, like the idealized DFIT pressure curve, the surface-treatment pressure curve in Figure 4.8 shows a characteristic peak pressure at the formation breakdown, which subsequently drops to a roughly steady-state value that marks the fracture-propagation pressure. The injection rate in this example is maintained close to 4 m^3/min after breakdown, whereas the proppant concentration is increased in a stepwise manner. The initial volume of fracture fluid that is injected without any proppant is called the *pad*, which is used to initiate the fracture and create open fracture volume for later proppant placement. Care is required to avoid increasing proppant concentration too quickly, as this can lead to a blockage that prevents continued fluid injection into the

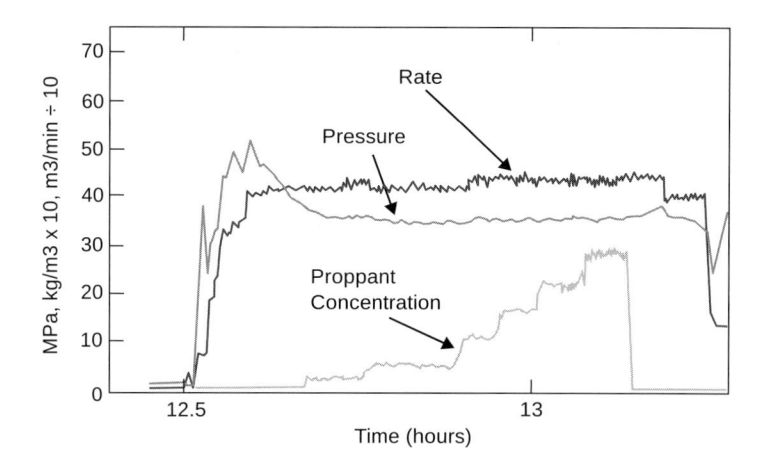

Fig. 4.8 Representative pumping curves for a single hydraulic fracturing stage, modified from Caffagni et al. (2016). Note pressure buildup until the breakdown pressure is achieved, after which the injection pressure stabilizes to a roughly steady-state value that defines the fracture-propagation pressure. Once the fracture is initiated, the injection rate is held nearly constant. Proppant concentration is increased during the stage in a step-wise manner. Note that the pressure increase at $\sim$ 13.5 hours is associated with initiation of the next stage of the open-hole completion. Used with permission from Oxford University Press.

fracture and produces a sudden pressure increase. This is known as *scree nout* and may be caused by a number of processes such as a proppant slurry bridge or proppant dehydration due to fluid loss (Gulrajani and Nolte, 2000).

Details of the hydraulic-fracturing method used in a particular well completion depends upon the well design, such as whether the completed interval is cased. Different completion designs have important implications for the acquisition, processing and interpretation of microseismic monitoring programs. Several types of completions are described in the next section.

4.2.1 Completion Methods and Treatment Strategies

The use of well casing in the reservoir interval depends on the choice of completion method. For wellbores that are cased and cemented throughout the target zone, a "perf and plug" method is employed, which requires the creation of perforations in the casing in order to allow egress of fracturing fluids into the formation (Figure 4.9). To accomplish this task, a plug is set to establish pressure isolation from completed sections of the wellbore. A *perforation gun* is then deployed with shaped explosive charges and fired to create ballistic channels in the well casing, cement and rock. The initial directions of the perforation channels are determined by the angular pattern of perforations around the circumference of the wellbore, known as peforation *phasing*. For example, 180° phasing means that perforations are made in diametrically opposing directions, which is useful if the stress state is known a priori (Smith and Montgomery, 2015). In order to promote the development of hydraulic fractures from multiple initiation points, perforations are typically grouped within two to

Fig. 4.9 Schematic illustration of cased and open-hole completion methods. For a cased well, it is common to perforate the casing using shaped charges, providing points of egress for pressurized fluids to access the formation. For an open wellbore, a sliding sleeve assembly can be used, in which an engineered sphere is pumped inside a liner. After the sphere seats onto the sliding sleeve assembly, fluid pressure builds inside the isolated section of the well between the packers, until the frac port opens, allowing pressurized fluid to enter the formation.

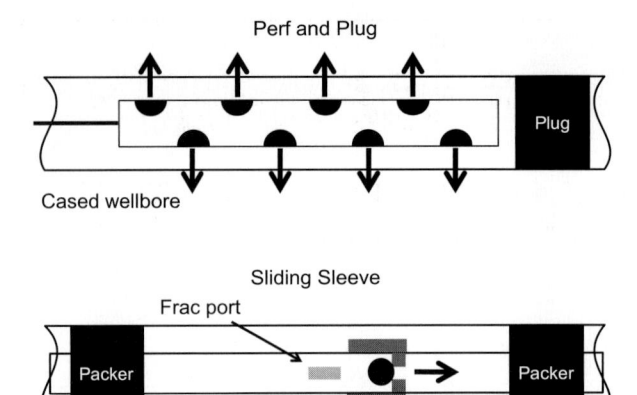

eight perforation clusters that are distributed uniformly along the stage length (Ajaya et al., 2013).

In order to reduce near-wellbore *tortuosity*[3] due to the effects of near-wellbore stress perturbations, it is desirable for perforation channels to extend into the formation for at least 2.5 well diameters (Smith and Montgomery, 2015). The penetration depth for perforation shots is a function of charge size and characteristics, fluid pressure in the wellbore and the mechanical properties of the target formation (Halleck, 2000). Perforation shots provide a method for velocity-model calibration during microseismic processing, since they have a known time and location.

Hydraulic fracturing that is performed in a wellbore without casing in the target interval is called an *open-hole* completion. In this case, open-hole *packers* are used for pressure isolation between sections of a wellbore. Sleeves equipped with ports that can be opened are used, thus enabling fracturing within specific intervals (King, 2012). Compared with a perf-and-plug completion that constrains the fracture initiative to occur at the perforations, this approach enables the fracture fluid to exploit pre-existing joints and weaknesses in the formation (Reynolds et al., 2012). A particular open-hole completion method is illustrated in Figure 4.9 that uses a metal sphere in order to activate the port to open the sliding sleeve. This "ball drop" method uses engineered spheres of different size, such that the smallest sphere is used for the most distal stage at the toe of the well, and progressively larger spheres are used thereafter. The sphere is pumped through a liner until it seats in a sliding-sleeve assembly (Figure 4.10). This allows pressure to build up within the pressure-isolated segment of the wellbore. After a design pressure is reached, this opens the port

---

[3] Tortuosity can be defined as the ratio of the length of the twisting and turning actual path to the net distance traversed by a fracture.

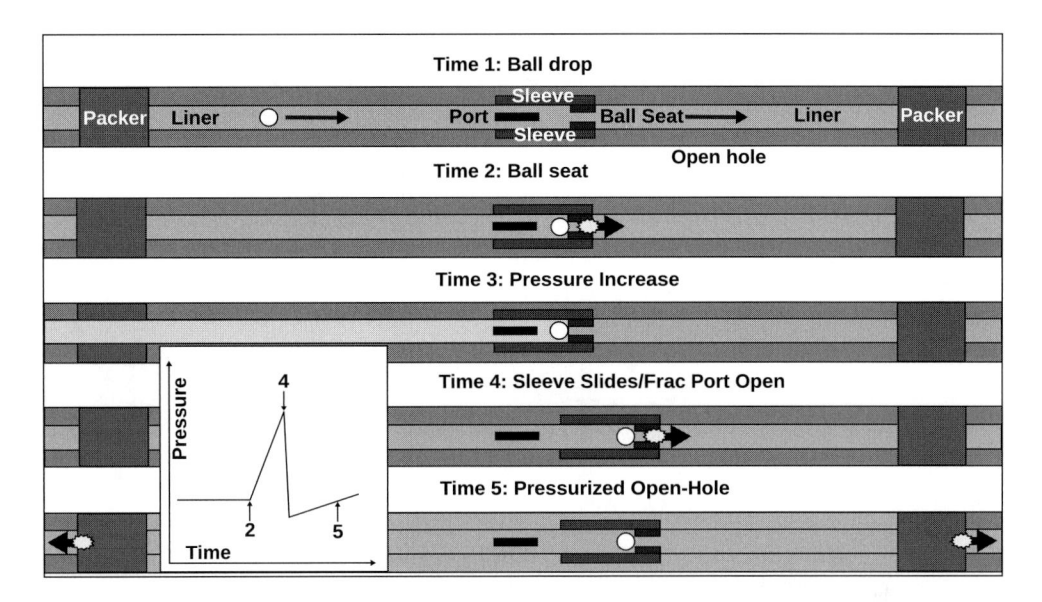

Fig. 4.10 Components of the sliding-sleeve mechanical process, modified from Maxwell and Parker (2012). At time 1, the sphere is pumped within the liner into the sleeve assembly. When the ball seats at time 2, pressure starts to increase. When the pressure reaches a design threshold, the fracture port opens, allowing fracture fluids access into the formation within the packed-off interval of the open wellbore. Inset shows the expected shape of the pressure pulse at the source.

and begins the fracture stage. In general, open-hole completions can be performed more rapidly than perf-and-plug completions, since it is not necessary to deploy a perforation gun between stages. Although perforation shots are not available for velocity-model calibration, a diagnostic sleeve-opening signal has been documented (Maxwell and Parker, 2012) that provides another type of signal for velocity-model calibration.

A variety of fracturing fluids, proppant types and treatment schedules have been evaluated in order to optimize the effectiveness of hydraulic fracturing (King, 2012). One method, known as *short-interval refracturing* (SIR) treatments, uses a modified injection schedule that is split into three phases (Kent et al., 2017). For each stage of a SIR treatment, the schedule consists of: 1) a typical hydraulic-fracturing stage; 2) a shut-in period, on the order of hours, known as the soak phase; and 3) a second hydraulic-fracturing stage is injected within the same zone. The soak phase is accompanied by fluid leak-off into the formation (see below). The third (re-injection) phase is often accompanied by increased rates of microseismic (MS) activity (Inamdar et al., 2010), which may be due to loss of cohesion on natural fractures (Kent et al., 2017).

Drilling and completion of horizontal wells for unconventional resource development is commonly undertaken from a *well pad*, from which multiple wells are drilled. The multi-well pad approach leads to a reduced overall environmental footprint compared to equivalent development of the resource using individual wells, in addition to substantial efficiencies in terms of construction of access roads, pipelines and other infrastructure. There is a current trend to place lateral wellbores closer together, both in the horizontal plane as well as vertically for so-called "stacked" unconventional plays such as the Midland

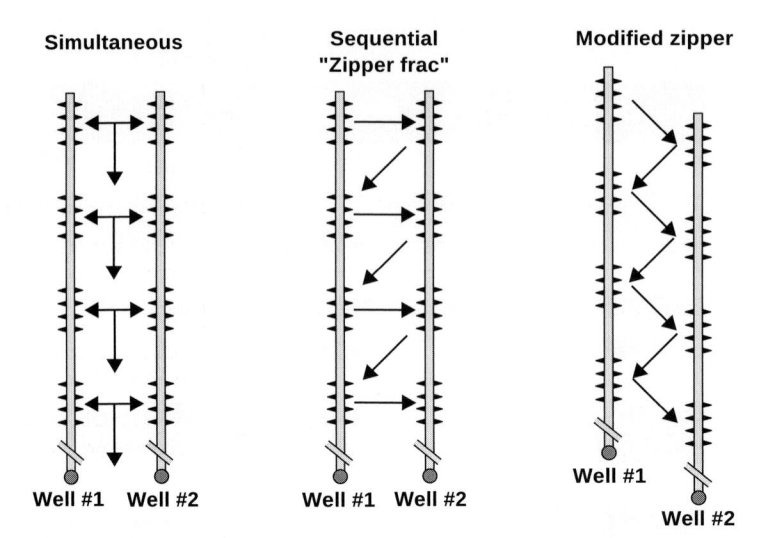

Fig. 4.11 Examples of multi-well completion schemes. Reprinted from Nagel et al. (2014) with permission of the Society of Petroleum Engineers.

Basin Wolfcamp shale (Flumerfelt, 2015). Stress inference between nearby hydraulic fractures, however, can inhibit the growth of some fractures, even at the scale of individual perforations, as shown by Warpinski et al. (2014).

Various completion schedules have been developed for use with pairs of parallel lateral wellbores. Some of these methods are summarized in Figure 4.11. In order to maximize the reservoir drainage, the most desirable scenario is for the fractures to extend together to a point where they are almost (but not quite) in contact. The simultaneous hydraulic fracturing approach progresses from the toe to the heel of both wells, using a concurrent injection schedule in the two wells. During completion, the growth of fractures in each well may be enhanced by the fracture-tip stresses produced by fractures in the adjacent well. An alternative approach, known as *zipper frac*, uses a sequential treatment schedule. A modified zipper method was proposed by Nagel et al. (2014). Using a numerical simulation approach that considers in situ stress and well configuration, Nagel *et al.* argued that this approach yields improved results compared with the simultaneous or standard zipper completion methods.

The next section outlines various analytical and numerical approaches used for hydraulic-fracture simulation. This builds upon the introductory discussion of linear elastic fracture mechanics (LEFM) in §2.1.1.

4.3 Numerical Models of Hydraulic Fractures

The primary design objective for a hydraulic-fracturing completion program is to optimize the surface area of the propped fracture network that is in contact with the reservoir.

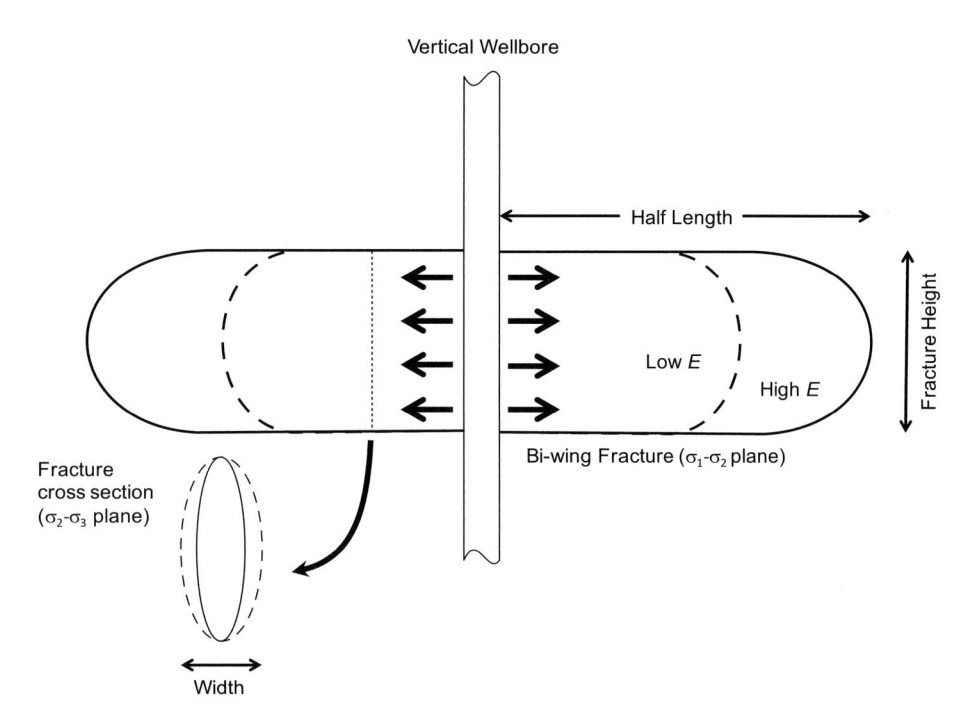

Fig. 4.12 Generic model for growth of a bi-wing hydraulic fracturing, showing effect of Young's modulus (E). For a given injected fluid volume, a hydraulic fracture in a medium with high Young's modulus will have a greater length but lower in width.

Another motivation of fracture models is to understand fracture height-growth characteristics based on the stress profile (Figure 4.4). Model calibration can be performed by matching shut-in pressure, or through the use of fracture-diagnostics tools including microseismic monitoring (Maxwell, 2014).

Neglecting, for the moment, complications such as tortuosity within the near-wellbore stress environment, classical fracture-mechanics considerations predict that hydraulic fractures within a homogeneous medium should have a symmetrical *bi-wing* geometry, as illustrated in Figure 4.12. In order to minimize the fracture energy, the fracture opens perpendicular to the minimum principal stress axis. In cases where the minimum principal stress is horizontal (corresponding to normal or strike-slip faulting regimes) hydraulic fractures are thus expected to be vertical. In map view, hydraulic fractures will be elongate parallel to the maximum horizontal stress direction. In cases where the minimum stress direction is vertical, corresponding to a reverse fault environment, fractures may be horizontal. This is sometimes known as *pancake* and may occur at very shallow depths.

Insights can be gained by considering a few of the classical asymptotic solutions for hydraulic fractures. For example, in the *plane-strain* approximation, the maximum width (w) of a hydraulic fracture of fixed height h_f is given by (Sneddon and Elliot, 1946)

$$w = \frac{2P_{net}\, h_f}{E'}, \tag{4.21}$$

where E' is the plane strain modulus, given by (Mack and Warpinski, 2000)

$$E' = \frac{E}{1-\nu^2}. \tag{4.22}$$

In this expression, E and ν are Young's modulus and Poisson's ratio of the medium, respectively. The plane-strain approximation means that strains are zero that are normal to a given axis (in this case, the minimum principal stress direction), such that planes perpendicular to this axis that are mutually parallel before deformation will remain parallel after deformation. This condition is well approximated if the fracture length is much greater than its height.

Within this approximation, Perkins and Kern (1961) developed analytical expressions applicable to hydraulic fractures whose height growth is confined by higher stresses in the overlying and underlying layers of a stratified sequence (*cf.* Figure 4.4). By making the additional assumptions that fracture toughness and leakoff can be neglected, for a Newtonian fluid of viscosity η the net pressure in a fracture of length L is given by (Mack and Warpinski, 2000)

$$P_{net} = \left[\frac{16\eta q_i E'^3}{\pi h_f^4} L\right]^{1/4}, \tag{4.23}$$

where q_i is the injection rate. Replacing P_{net} leads to the formulation

$$w(x) = 3\left[\frac{\eta q_i (L-x)}{E'}\right]^{1/4}. \tag{4.24}$$

We see that increasing the plane strain modulus (E') by increasing Young's modulus or decreasing Poisson's ratio of a material will tend to decrease the fracture width, as shown in Figure 4.12. For a fixed fluid volume, it follows that fracture length therefore increases with increasing E'. Conversely, a decrease in E' will increase the fracture width and decrease its length.

To this model, Nordgren (1972) added fluid leakoff and volume change by fluid storage in the fracture. The combined model is called the *Perkins–Kern–Nordgren* (PKN) 2-D fracture model. At a position **x** on a fracture surface, the fluid leakoff velocity u_L is given by the basic model

$$u_L(\mathbf{x}) = \frac{C_L}{\sqrt{t - t'(\mathbf{x})}}, \tag{4.25}$$

where C_L is called the *leakoff coefficient* and t' is the time at which the fracture opened at **x**. By applying the mass balance condition,

$$q_i = q_f + q_L, \tag{4.26}$$

where q_f and q_L are fluid fluxes, respectively, into the fracture and leaking off into the formation, the fracture face area for the PKN model can be approximated as a function of time as (Mack and Warpinski, 2000)

$$A_f = \frac{q_i t}{\bar{w} + 2C_L\sqrt{2t}}, \tag{4.27}$$

where $\bar{w}$ is the average fracture width.

The *Khristianovich–Geertsma–de Klerk* fracture model provides a 2-D asymptotic solution based upon a different set of assumptions, which provide a good approximation for a fracture with height that is much greater than the fracture length. In particular, the KGD model assumes that the fracture width is constant over the height of the fracture, implying slip at the upper and lower boundaries. Pressure is assumed to be constant except for a dry region near the fracture tip. Moreover, the LEFM approach with fracture toughness is applied at the fracture tip. For the KGD model, the approximate fracture width can be expressed in terms of P_{net} as

$$w = \frac{4LP_{net}}{E'}. \tag{4.28}$$

An approximate KGD solution for fracture length L is given by (Geertsma and De Klerk, 1969)

$$L \approx \frac{1}{2\pi} \frac{q_i \sqrt{t}}{h_f C_L}, \tag{4.29}$$

for the limit where

$$\alpha = \frac{8C_L \sqrt{\pi t}}{\pi w + 8S_p} \gg 1, \tag{4.30}$$

and where S_p denotes the spurt fluid loss. The PKN and KGD fracture models are two-dimensional models, in the sense that the fracture height parameter is fixed. Both of these classical models are thus height-confined, plane-strain approximations that make different assumptions about the variation in fracture width.

A more complete nonlinear model for fluid leakoff from a hydraulic fracture has been developed based on multiphase flow in a reservoir as a filtration and separation process (Settari, 1985). This model indicates that properties of the reservoir fluids that are displaced, including compressibility, play a significant role. Figure 4.13 illustrates, schematically but faithfully, stress and leakoff effects for a classical hydraulic fracture model in two different types of reservoirs (Cipolla et al., 2011). In both cases, the fracture-tip related patterns of shear, compressional and tensile stresses generated by the hydraulic fracture are equivalent to the previously discussed model for a mode I crack in a homogeneous elastic medium (Figure 2.4). In the case of a moderately permeable ($\kappa \geq 10^{-17} m^2 \sim 0.01$ mD) poroelastic medium with an incompressible fluid (i.e. oil, water), however, leakoff causes pressure changes that extend farther away from the hydraulic fracture than in the case of a gas-filled reservoir. As elaborated in Chapter 8, these differences have important implications for microseismic interpretation in these various types of reservoir.

Further complications arise from the presence of natural fractures in the reservoir. For the purpose of numerical simulations, natural fractures are sometimes represented using *discrete fracture networks*. In the subsurface, a hydraulic fracture system will develop following a path of least resistance; consequently, mechanical heterogeneities due to natural fractures are expected to produce hydraulic fracture systems with greater complexity than in the simple models discussed above. Dusseault and McLennan (2011), for example, emphasized the role of *incipient* fractures, which are pervasive features in in low-permeability rocks, with respect to permeability enhancement within a zone of dilation

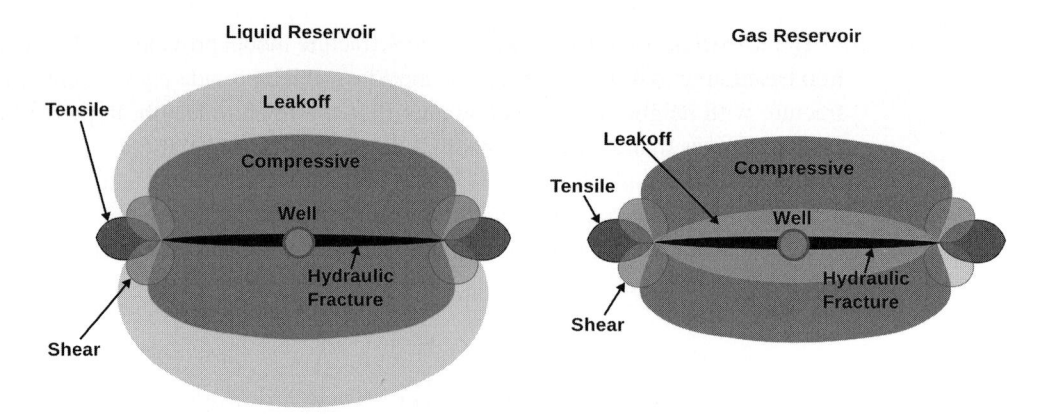

Fig. 4.13 Schematic but faithful representation of the effects of reservoir fluid compressibility on leakoff and stress. Stress patterns around the fracture include regions of compression beside the fracture, as well as regions of concentrated tensile and shear stress near the tip. Fluid leakoff from the hydraulic fracture is a nonlinear filtration process that is influenced by compressibility of the fluid in a poroelastic medium. A reservoir with a relatively incompressible liquid phase (oil, water) and moderate permeability ($\kappa \geq 0.01$ mD) is expected to have a much larger leakoff-influenced region than a low-permeability gas reservoir. Modified from Cipolla et al. (2011). Used with permission of the Society of Petroleum Engineers.

around the hydraulic fracture and placement of proppant into the reservoir. As illustrated in Figure 4.14, geomechanical considerations for simulating the development of hydraulic fracture systems in realistic reservoir media include (Dusseault and McLennan, 2011):

- a strong influence of the natural fracture fabrics, including incipient fractures, on the dilated zone that develops around a hydraulic fracture;
- permeability enhancement within the dilated zone arising from processes such as block displacement, aperture wedging and shear displacement across fractures and bedding planes that lead to self-propping;
- confinement of proppant to a smaller region than the dilated zone, the extent of which is determined by proppant grain size and characteristics of the carrier (fracture) fluid.

In order to extend classical semi-analytical methods for homogeneous elastic media and thus capture such medium complexities, numerical simulation methods are used. These methods can be classified into several different categories, as summarized in Figure 4.15. *Lumped-parameter* models provide a simple time-stepped approach that preserves the general features of classical models, by parameterizing the fracture front using a few control points. At each time step, these control points are connected to produce a continuous fracture surface using a geometrical shape such as concentric ellipses. At every time step, the growth of the fracture is driven by numerically fluid-pressure gradients. *Pseudo-3D* models use a cell-based numerical-simulation scheme that incorporates pressure gradients by dividing the fracture into segments (Mack and Warpinski, 2000). Unlike height-confined models such as PKN and KGD, the width and equilibrium height is computed independently for each segment.

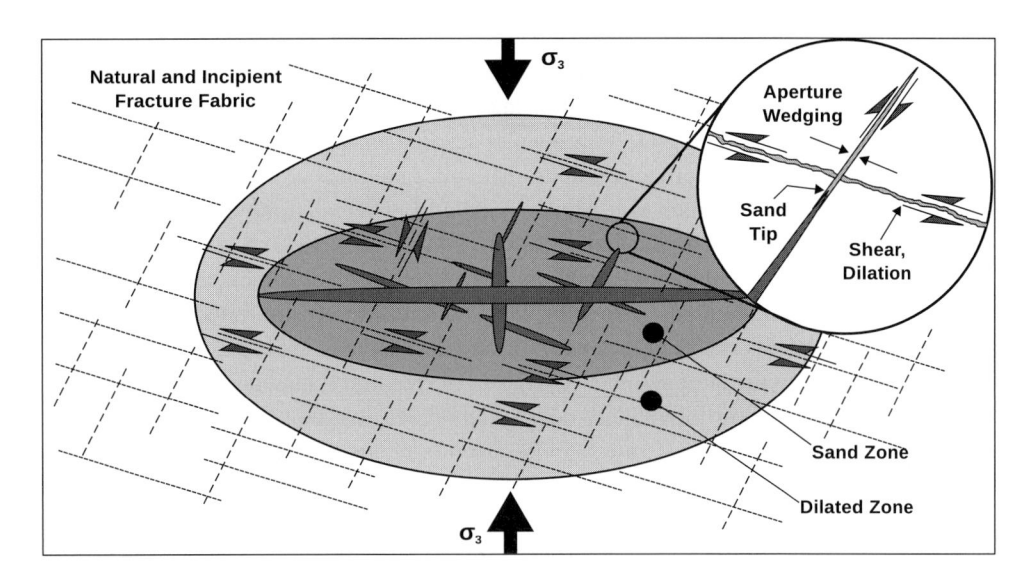

Fig. 4.14 Interaction of a hydraulic fracture with pre-existing fracture fabric. Permeability is enhanced in a dilated zone around a reservoir through processes such as block rotation, aperture wedging and self-propping due to shear dislocation. Proppant is confined to a smaller region within the dilated zone, providing a hydraulically conductive connection from the hydraulic fracture to surrounding parts of the reservoir. Modified from Dusseault and McLennan (2011). Reprinted with permission of the authors.

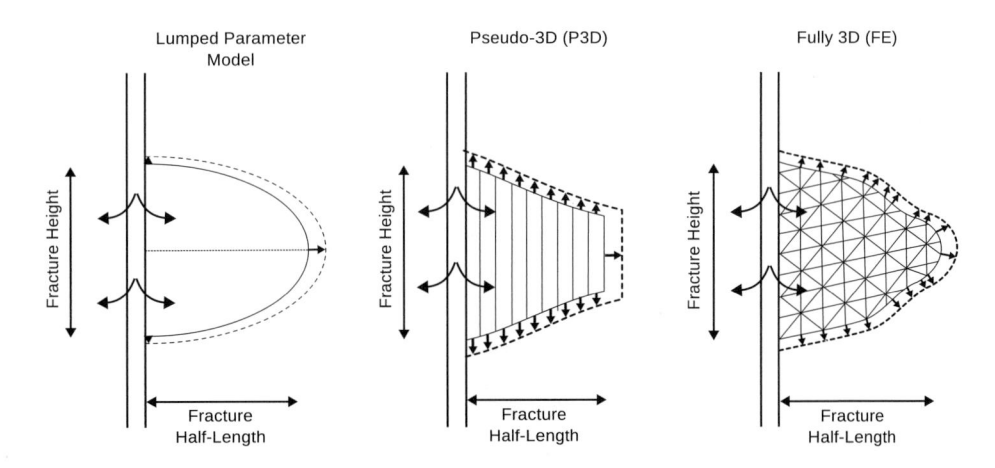

Fig. 4.15 Summary of various numerical schemes for simulating hydraulic-fracture growth, in increasing order of complexity and computational requirements. Lumped-parameter models use an efficient time-stepping approach based on interpolation of the fracture front using a small number of control points. Pseudo-3D models split the fracture into segments or cells, within which equilibrium height and fracture width are independently determined. Fully three-dimensional models discretize the underlying differential equations to obtain a numerical solution and can accommodate more general heterogeneity of the medium, but at higher computational expense. Curved arrows show fluid injection, while small arrows show fracture growth increment.

Fully three-dimensional numerical schemes are based on various approaches to discretize the underlying system of differential equations. Finite-difference, finite-element and distinct-element approaches each impose different constraints on the nature of the computational grid, with distinct advantages and disadvantages. In the case of continuum mechanics approaches, finite-difference methods are straightforward to implement but are generally limited to a regular fixed grid, whereas finite-element methods have the significant advantage of enabling the use of more flexible and general elements, which comes at a cost of additional computational overhead to keep track of individual elements. Distinct-element methods enable the definition of virtually unlimited types of interactions between elements such as blocks or particles, but this can lead to exceedingly long execution times due to the large number of degrees of freedom (Dusseault and McLennan, 2011).

A hybrid iteratively-coupled approach that combines finite-element and distinct element approaches into a finite-discrete element method (FDEM) was first suggested by Munjiza et al. (1995). This method splits up the computational tasks into those that apply to the continuum, where a finite-element solution is obtained, and those that apply to fractures, for which a distinct-element method is used. Thus, while the medium is undergoing elastic deformation the behaviour of intact material is explicitly modelled using finite-elements, whereas when the strength of the material is exceeded, fractures are initiated that produce discontinuous blocks whose interaction is represented using a distinct-element approach. The principles of nonlinear elastic fracture mechanics (Barenblatt, 1962) are applied in the development of new fractures. Figure 4.16 illustrates how the medium can be discretized within a FDEM framework (AbuAisha et al., 2017).

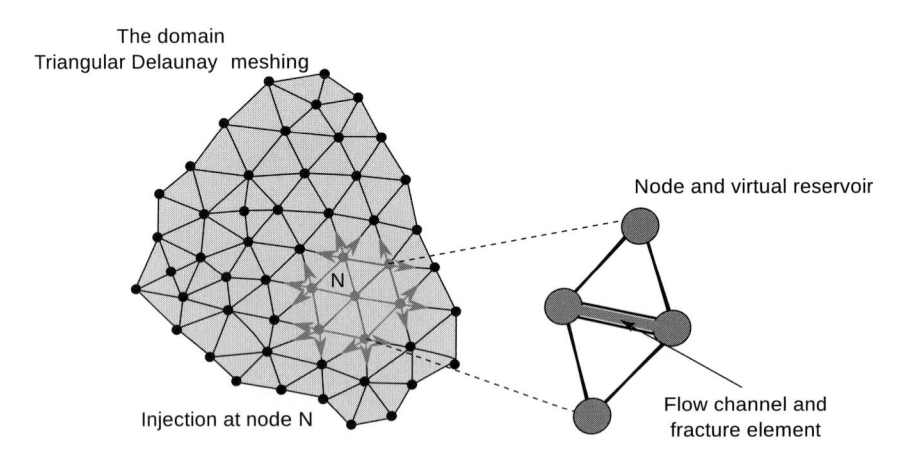

Fig. 4.16 Example of medium parameterization used for a hybrid finite-discrete element method (FDEM) where the numerical simulation is split into distinct tasks. Continuum modelling of the intact medium is performed using a finite-element approach that uses a triangular Delaunay mesh. Once a fracture forms, the flow channel and fracture elements are represented by distinct elements. Reprinted with minor modifications from *Journal of Petroleum Science and Engineering*, Vol 154, AbuAisha, M., Eaton, D. W., Priest, J., and Wong, R., Hydro-mechanically coupled FDEM framework to investigate near-wellbore hydraulic fracturing in homogeneous and fractured rock formations, Pages 100–113, copyright 2017, with permission from Elsevier.

Most current numerical simulation methods for hydraulic fracturing incorporate numerical models for proppant transport, such as Unwin and Hammond (1995). Proppant-transport processes include settling of heavy proppant particles and/or convective proppant transport within the fracture channel. Kern et al. (1959) were among the first to recognize that these processes lead to formation and evolution of a bed of settled proppant material at the base of a vertical fracture. Early work showed that these processes are complicated by *proppant bridging*, which occurs if the fracture width decreases to less than twice the size of the diameter of the proppant (Daneshy, 1978). In addition, proppant settling is retarded by fluid leakoff, which acts to increase the proppant concentration within the fracture (Daneshy, 1978).

Classical numerical models for proppant transport assume that proppant grains settle in the fluid-filled fracture at the terminal velocity given by Stoke's Law (Gadde et al., 2004),

$$v_S = \frac{g \Delta \rho d^2}{18 \eta}, \tag{4.31}$$

where v_S denotes the Stoke's setting velocity of a single particle, $\Delta \rho$ is the density difference between the proppant and the suspending fluid, d is the proppant grain diameter, $g = 9.8$ m/s^2 is the acceleration of gravity and η is the dynamic viscosity of the fluid. Since this expression is only valid for low Reynold's number ($R_e < 2$), Gadde et al. (2004) developed a method for correcting this for fluid turbulence effects in the case that $R_e \geq 2$. Advances in experimental modelling, including the use of 3-D printing to fabricate complex models (Ray et al., 2017), is helping to overcome known limitations (Warpinski et al., 2009) in existing methods for simulating proppant transport.

4.4 Flowback and Flow Regimes

Flowback is performed following a hydraulic-fracture stimulation in order to recover the maximum amount of injected fluids to avoid fracturing fluids suppressing production, while removing a minimum amount of proppant from the fracture (Brown et al., 2000). Design variables for flowback such as timing and rate can have a significant impact on the conductivity of the induced hydraulic-fracture network. As noted above for the SIR method, a fluid soaking period may reduce cohesion on natural fractures in a time-dependent manner. Furthermore, fracture closure is required in order to prevent movement of proppant during flowback (Brown et al., 2000).

The effectiveness of multi-stage hydraulic fracturing (MSHF) can be evaluated using a variety of methods that are based on measurement and analysis of flow rates and/or pressures during flowback as well as production phases (Clarkson et al., 2014). These approaches enable the characterization of *flow regimes* as well as estimation of hydraulic-fracture parameters such as fracture half-length and fracture conductivity. Flow-regimes refer to characteristic patterns of flow to the well through hydraulic fractures and the reservoir over time. Identification is necessary to determine what model is appropriate for

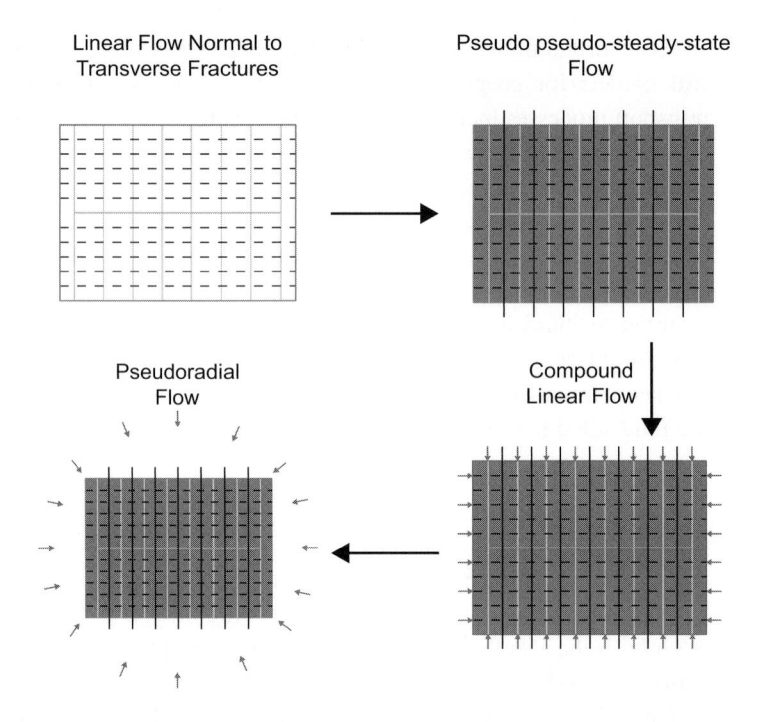

Fig. 4.17 Postulated sequence of flow regimes for a multi-fractured horizontal well completed in shale. Modified from Clarkson and Williams-Kovacs (2013a), with permission of the Society of Petroleum Engineers.

analysis. Flow regimes are distinguished by streamline geometries and can be recognized on the basis of distinctive pressure signatures (Ehlig-Economides and Economides, 2000).

Figure 4.17 illustrates the theoretical sequence of flow regimes occurring over time after MSHF. After a very short intra-fracture flow period, the dominant flow-regime is linear flow to the hydraulic fractures, followed by a pseudo-depletion period followed, in turn, by flow external to the fractured region (Clarkson et al., 2014). These are post-flowback flow regimes. For the flowback period, a sequence of intra-fracture flow-regimes (transient radial, then fracture linear then fracture depletion) occur before formation fluids break through to the fracture. The last flow-regime typically observed on flowback is the linear flow identified in Figure 4.17.

Linear flow is characterized by parallel streamlines within the reservoir. Given a simple hydraulic fracturing model that produces a symmetrical set of conductive fractures in a low-permeability reservoir, a transient flow regime is established at a relatively early time, as represented by linear flow normal to fractures (Figure 4.17). A linear flow regime can be analyzed using rate-transient analysis by plotting pressure against the square root of time (Ehlig-Economides and Economides, 2000). During production, this flow regime transitions through pseudo-steady-state flow into a compound linear flow regime, where roughly orthogonal drainage patterns develop. The appearance time and duration of these flow regimes depends on fracture spacing, fracture length, well spacing and formation permeability. For example, if well spacing is too close, you would not expect to see compound linear flow or pseudoradial. The last stage in the evolution of flow regimes is pseudo-radial flow; in terms of reservoir modelling, this is defined by convergence of streamlines toward a region rather than a single wellbore. Analysis of radial flow can be used to quantify the permeability within the plane of convergent flow (Ehlig-Economides and Economides, 2000).

Flowback analysis is complicated by multi-phase flow, which mixes together both fracture and reservoir fluids. New methods are under development that are based on rate-transient analysis during early flowback (Clarkson and Williams-Kovacs, 2013b). Such approaches may provide rapid estimates of fracture half-length, a key parameter for reservoir simulation, that are independent from and complementary to those obtained using other methods such as passive seismic monitoring.

4.5 Summary

Hydraulic fracturing is a transformative technology that has unlocked vast unconventional oil and gas reserves. Monitoring methods, including passive seismic monitoring, are important for design optimization as well as to address environmental concerns.

Knowledge of the subsurface stress field is important for simulating hydraulic fracturing and for understanding induced seismicity. The presence of a fluid-filled wellbore in an otherwise continuous medium strongly perturbs the near-field stress environment. These stress perturbations can be modelled using the Kirsch equations and are manifested by borehole breakout parallel to the direction of $\hat{S}_{Hmin}$ and drilling-induced tensile failure (DITF) in the direction of $\hat{S}_{Hmax}$. Borehole breakout has been extensively measured and provides a significant set of observational data for regional stress models. The Kirsch equations provide the basis for the overcoring method of in situ stress determination.

Interpretation of data from small-volume injections, including diagnostic fracture injection tests (DFITs) or extended leakoff tests (XLOTs), is arguably the most robust approach to determine the magnitude of S_{Hmin}, which is assumed to be equivalent to the fracture closure pressure (FCP). These types of analysis also provide a wealth of other information about the stress environment and permeability of the medium.

In areas where in situ stress data are incomplete, a number of simple mathematical models exist that can be used to approximate the state of stress. A depth profile of the vertical stress (S_V) can be calculated using measured density values from well logs. The horizontal principal stresses can then be approximated using a number of simplifying assumptions. The uniaxial-elastic-strain (UES) approach leads to a simple relationship between S_{Hmin} and S_V that is widely used, but this method is strictly valid only in a normal-faulting stress regime where $S_{Hmax} \simeq S_{Hmin}$. A more general method, called the generalized-strain model, incorporates assumed tectonic stresses into the calculation. Another approach assumes that the crust is in a critically stressed state and combines this with a priori information, such as the stress regime and stress ratio, in order to estimate the complete stress field.

Hydraulic fracturing is well-completion method used to enhance the permeability of a reservoir, by injecting large fluid volumes at pressures above the fracture pressure. Most hydraulic-fracturing treatments are performed in multiple stages along a horizontal wellbore. Various treatment schedules and strategies have been developed for proximal wells, such as the zipper frac method, which alternates between two parallel wells. Fracture fluids are injected in the form of a slurry containing proppant and other additives, where the use of proppant ensures that fractures remain open after flowing back the treatment well. Several different completion methods are in widespread use. For example, the perf-and-plug method is carried out within a cased wellbore, while a sliding-sleeve approach is carried out in a wellbore that is open to the reservoir. These differences are important for the calibration of velocity models that are used to estimate microseismic event locations. The fracture pressure, rate, proppant concentration and other parameters are monitored during the treatment; these pumping data curves are invaluable for the interpretation of microseismicity.

Our basic understanding of hydraulic-fracture growth comes from linear elastic fracture mechanics theory. Numerical simulation of hydraulic-fracture growth is a key part of the design process. Classical methods using a semi-analytical approach, such as the PKN model (suitable for fractures that are long relative to their height) or the KGD model (suitable for fractures with height that is much greater than the length), predict simple bi-wing hydraulic fractures. However, it is well established that hydraulic fractures interact with natural fractures, as well as mechanical and stress heterogeneities, to produce patterns that are often complex. This heterogeneity can be handled, to varying degrees of approximation, using lumped-parameter models, pseudo-3-D models and fully 3-D approaches. Some hybrid numerical simulation methods combine continuum approaches with discrete methods.

Injected fluids are flowed back to the surface after completion of the hydraulic-fracturing program. Posterior analysis of the effectiveness of the hydraulic-fracturing program can be undertaken by analyzing flow rates and/or pressures during flowback and early production phases. These observations are complementary to passive-seismic monitoring and can be used to infer *flow regimes* as well hydraulic-fracture parameters such as fracture half-length and fracture conductivity.

4.6 Suggestions for Further Reading

- Concise review of methods for determining crustal stress: Schmitt et al. (2012).
- Authoritative compilation on reservoir stimulation: Economides and Nolte (2000).
- White-paper summary of hydraulic-fracturing methods (in appendix): EPA (2004).
- Review of methods for computer simulation of hydraulic fractures: Adachi et al. (2007)

4.7 Problems

1. S_{Hmin} = 44.0 MPa has been estimated based on observation of the fracture closure pressure (FCP) from an extended leakoff test (XLOT).

a) Use Equation 4.6 to estimate S_{Hmax}, neglecting the generated hoop stress $\Delta\sigma_{\theta\theta}$ and assuming $\Delta\Theta$ (angular breakout width) = 20° and $\mathcal{S}$ (unconfined compressive strength) = 150 MPa.

b) If P_R = 49 MPa, use this estimate for S_{Hmax} to estimate the pore pressure, P.

c) Why is it not possible to infer S_{Hmax} in a reverse faulting environment?

2. Using the same model for density and pore-pressure gradient as in question 3 from Chapter 2, and assuming the generalized strain model, calculate the effective principal stresses S'_V, S'_{Hmax} and S'_{Hmin} given a Biot parameter of $\alpha = 0.5$, tectonic strains of $\epsilon_H = 10^{-3}$ and $\epsilon_h = 10^{-4}$, and a shear modulus of $\mu = 10$ GPa. Assume a Poisson solid ($\nu = 0.25$). This approach is sometimes called a mechanical Earth model.

3. Consider a hydraulic fracture of fixed height $h_f = 30$ m. Assuming a Perkins–Kern–Nordgren (PKN) model, calculate the following at time $t = 30$ minutes, given a constant injection rate of $q_i = 8$ m^3/min, dynamic viscosity $\eta = 10^{-3}$ Pa-s and an average fracture width (or aperture) of 2 mm. Assume that the medium is characterized by a Young's modulus of $E = 20$ GPa, Poisson's ratio of $\nu = 0.3$ and a leakoff coefficient of $C_L = 0.002$ m s$^{-1/2}$.

a) The fracture area as a function of time, $A_f(t)$.
b) The fracture length, for a simple case of a fracture with a rectangular cross section.
c) The net pressure, P_{net} (Equation 4.23).
d) The fracture width (also known as fracture aperture) at the $x = 0.5L$, using Equation 4.24. Compare this with the assumed average width given above.

4. Use the Khristianovich–Geertsma–de Klerk (KGD) model with the parameters given in the previous question to compute the following.

a) The fracture length (L).
b) The net pressure (using Equation 4.28).

5. Calculate the terminal proppant settling velocity based on Stoke's Law (Equation 4.31) using the dynamic viscosity given in Problem 3. Assume the maximum grain diameter at 90 percent probability for a 20/40 mesh, and a density contrast of 1500 kg/m^3 between the proppant grains and the fluid.

6. The hydraulic injected energy is given by $E = \int_{t_1}^{t_2} PR\, dt$, where P is the injection pressure and R is the injection rate. For a constant bottomhole injection pressure of $P = 50$ MPa and a uniform injection rate of $R = 8$ m^3/min, calculate the total hydraulic injected energy from $t_1 = 0$ to $t_2 = 30$ minutes. Use Equation 3.87 to calculate the magnitude of an earthquake with the equivalent radiated seismic energy.

PART II

APPLICATIONS OF PASSIVE SEISMIC MONITORING

The most exciting phrase to hear in science, the one that heralds new discoveries, is not "Eureka!" but "That's funny..."

Isaac Asimov (Usenet discussion forum, 1987)

5 Passive-Seismic Data Acquisition

> You can observe a lot by watching.
>
> Yogi Berra (John Wiley & Sons, 2008)

Passive-seismic monitoring makes use of observations of either naturally occurring seismic signals, or anthropogenic "sources of opportunity,"[1] in order to characterize the architecture, properties and dynamic processes that occur within a targeted region in the subsurface. Similar information can be gleaned from *active seismic* methods, wherein controlled sources, such as buried explosive charges, are used to generate wavefields with precisely known origin time and location. Passive monitoring has the advantage that the effort and expense to deploy seismic sources are avoided, but the disadvantage that precise timing and location of sources are not known a priori.

There are number of practical applications of passive-seismic monitoring. *Microseismic monitoring* is a key technology in the petroleum industry for surveillance and diagnostic characterization of hydraulic fracturing (Warpinski, 2009; Maxwell et al., 2010; Van der Baan et al., 2013), enhanced-oil recovery (McGillivray, 2005), reservoir characterization (Maxwell and Urbancic, 2001) and casing integrity (Smith et al., 2002). Similar methods are used at a more regional scale for monitoring induced seismicity, with applications in various sectors including the petroleum (Raleigh et al., 1976; Suckale, 2010) and geothermal industries (Majer et al., 2007). Microseismic arrays are also critical for safety protocols in underground mines, where they are widely used to monitor fault activity and rockbursts (Gibowicz and Kijko, 1994). Furthermore, passive-seismic methods based on ambient-noise interferometry have been developed for subsurface imaging (Artman, 2006; Wapenaar and Fokkema, 2006).

The capabilities of an acquisition system to capture the desired components of the seismic wavefield vary with configuration, sensitivity and bandwidth of the installed sensors (Zimmer, 2011; Maxwell, 2014). Seismic events generate signals with different frequency content, depending upon the magnitude of an event; hence, the type of passive-seismic monitoring system that is used depends upon the industrial application and associated seismic signal characteristics. Systems are tailored to record a particular frequency range, covering application-relevant dominant frequencies as shown in Figure 5.1. These systems may be permanent or temporary, and can be classified as follows:

- laboratory acoustic-emission systems;
- microseismic monitoring systems;

---

[1] Seismic sources that occur independently of the monitoring program, such as from hydraulic fracturing (Duncan, 2005).

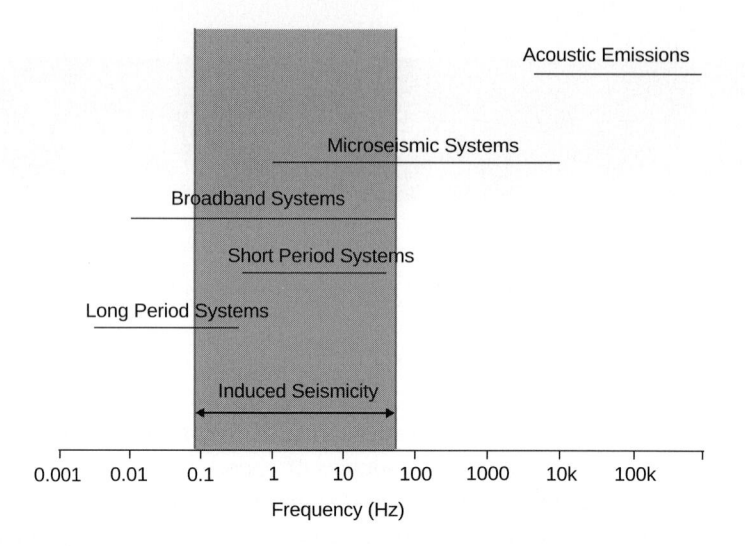

Fig. 5.1 Frequency ranges for passive-seismic monitoring systems. Modified from Urbancic and Wuestefeld (2013).

- regional short-period seismograph networks;
- broadband seismograph networks;
- long-period global networks.

Microseismic monitoring systems are generally designed for observation and analysis of events within a distance range of up to a few km, with magnitudes of less than zero. These systems typically make use of geophones, which are low-cost devices that passively record ground motion, typically at frequencies of 5 Hz or higher. Regional seismograph networks are used to monitor earthquake activity within a specific geographic area, generally at distances of less than $\sim$ 1400 km, within which seismograms are dominated by interaction with the crust-mantle boundary (Lay and Wallace, 1995). These networks are usually tailored to record small to moderate earthquakes, which typically produce strong signals in the short-period bandwidth (frequencies $\geq$ 1 Hz). Long-period seismograph networks are well suited to record *teleseismic*[2] surface waves and normal modes of the Earth. Historically, the distinction between short-period and long-period instrumentation reflects high signal-to-noise (S/N) recording bands on either side of a global noise peak at about 0.15 Hz (Webb, 2002). Broadband seismometers are sophisticated instruments that overlap the frequency range of both short-period and long-period instruments, providing coverage of the full frequency bandwidth for induced seismicity above magnitude zero (Figure 5.1).

This chapter begins with a review of the history of development of microseismic monitoring systems, with a particular emphasis on systems developed for hydraulic-fracture monitoring (HFM). Next, the characteristics of various types of sensors are discussed, along with attributes of different network/array configurations. Following this, various sources of random and coherent noise are considered, together with approaches that can

---

[2] The term teleseismic refers to seismograms recorded at distances greater than about 3000 km (Lay and Wallace, 1995).

be used to mitigate or attenuate noise. Next, the use of calibration sources such as perforation shots or string shots is considered. These signals are important for refinement and validation of velocity models required for estimation of accurate source locations. Finally, various approaches for survey-design optimization are reviewed.

5.1 A Brief History of Microseismic Monitoring

The widespread use of microseismic methods for HFM by the petroleum industry, starting in about the year 2000, was preceded by developments that took place over several decades (Maxwell et al., 2010). A few of the pioneering projects are highlighted here. Perhaps the earliest tests of microseismic monitoring of hydraulic fracturing were undertaken by the US Oak Ridge National Laboratory, at a site developed for disposal of radioactive waste into shale at a depth of around 300 m (McClain, 1971). A dyed and tracer-doped cement slurry was used, which hardened permanently within horizontal fractures.[3] The fracture extent was confirmed by delineation drilling, but the cost of drilling motivated the development of remote-sensing methods, such as microseismic monitoring, that could provide reliable fracture-mapping capabilities. Seismic events from the first fracture experiment were recorded by a newly installed (at the time) seismograph located 1.5 km from the injection site. Subsequent tests from 1967 to 1970 provided encouragement to continue to develop microseismic monitoring methods (McClain, 1971).

The first petroleum-related tests of HFM were performed by the El Paso Natural Gas Company in the San Juan Basin, New Mexico in 1973 and the Green River Basin, Wyoming in 1974 (Power et al., 1976). Using a combination of downhole and surface sensors, fracture-related signals were observed during and after treatment. It was unclear from the results of these tests, however, whether a new hydraulic fracture system was being formed, or fluids were being pumped into existing natural fractures (Power et al., 1976).

Los Alamos National Lab in the USA developed a downhole tool with 10 Hz multicomponent geophones (Figure 5.2). This downhole tool served as a prototype for continued technology development and was deployed from 1976 to 1979 in a series of tests, as part of the Hot Dry Rock (HDR) experiment at Fenton Hill, New Mexico (Albright and Pearson, 1982). In total, three hydraulic fracture tests and one circulation test were monitored. The observations showed that large numbers of microseismic signals were generated once a critical fluid volume had been injected. During the HF tests microseismic event locations outlined a vertical planar fracture geometry, whereas during the circulation test the observed microseismicity was more diffuse in the jointed Precambrian bedrock (Albright and Pearson, 1982). This work included a cautionary note for application of these results in sedimentary rocks, as it was presumed that seismic attenuation would generally be much greater than in the crystalline rocks at Fenton Hill. Over the next few years, passive-seismic methods were tested and developed for monitoring geothermal operations at the Geysers

---

[3] As discussed in §4.3, hydraulic fractures are perpendicular to the least principal stress. Fractures are horizontal in this case because S_V is the least principal stress at a shallow depth of 300 m.

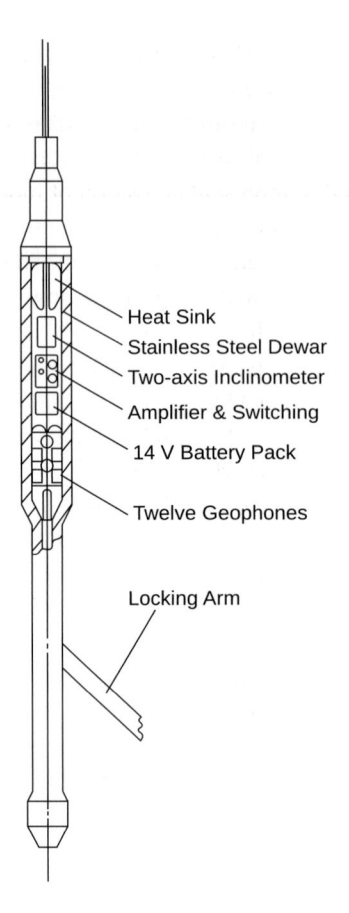

Fig. 5.2 Schematic drawing of the downhole tool used by Los Alamos National Laboratory at the Fenton Hill HDR Project from 1976 to 1979. This tool served as a prototype for continued technology development for deep-downhole microseismic monitoring. From Albright and Pearson (1982), with permission of the Society of Petroleum Engineers.

in California (Majer and McEvilly, 1979; Denlinger and Bufe, 1982; Eberhart-Phillips and Oppenheimer, 1984).

The Multi-Site Hydraulic Fracture Diagnostic (M-Site) Project from 1983 to 1996 provided a critical set of fracture-diagnostic tests in low-permeability sandstone. The M-Site Project was carried out by the Gas Research Institute and US Department of Energy in the Piceance Basin, Colorado (Warpinski et al., 1998). Microseismic observations using two downhole accelerometer arrays were complemented by tiltmeters and drilling to intersect the hydraulic fractures. The microseismic data revealed fracture containment, large discrepancies in inferred fracture area compared with modelling results and evidence of fracture complexity, including secondary and T-shaped fracturing. Complementary observations revealed multiple fracture strands and large pressure drops along the fracture length, along with a wealth of other information (Warpinski et al., 1998).

In 1997, a consortium of operators and service providers monitored six hydraulic fracture treatments in a vertical well at the Carthage Cotton Valley gas field in eastern

Texas (Walker, 1997). This gas field produces from low-permeability sand layers within an interbedded siliciclastic sequence. Rutledge and Phillips (2003) obtained precise hypocentre locations by upsampling the recorded traces using an interpolation function and re-picking phase arrivals using cross-correlation. The sharpened images of microseismicity were interpreted to represent activation of mode-II (strike-slip) failure on natural fractures that are isolated within individual sands, thus limiting the height grown. Natural fractures in this reservoir, inferred to control the distribution of microseismicity, have a strike direction that is subparallel to the expected hydraulic fracture orientation (Rutledge and Phillips, 2003). The Cotton Valley dataset has since been used in many studies as a benchmark for development and testing of microseismic methods.

Lessons learned from the projects at Fenton Hill, the Piceance Basin and Cotton Valley provided a foundation for the initial application of microseismic monitoring in the Barnett play in Texas (Maxwell et al., 2002). The Barnett shale is a naturally fractured reservoir, in which microseismic monitoring provided conclusive evidence for activation of complex fracture networks – contrary to expectations from theoretical models of hydraulic-fracture growth, as discussed in §4.3. The success of microseismic methods in the Barnett shale, initially using downhole systems but subsequently using other acquisition configurations, prompted considerable growth and widespread use of the technique in virtually every other active shale play (Maxwell, 2010). The advent over the next few years of surface and near-surface microseismic acquisition demonstrated the capabilities of large-aperture geophone arrays (Duncan, 2005; Robein et al., 2009). This opened the door to a number of different sensor configurations for microseismic surveys (Figure 5.3), as discussed in the next section.

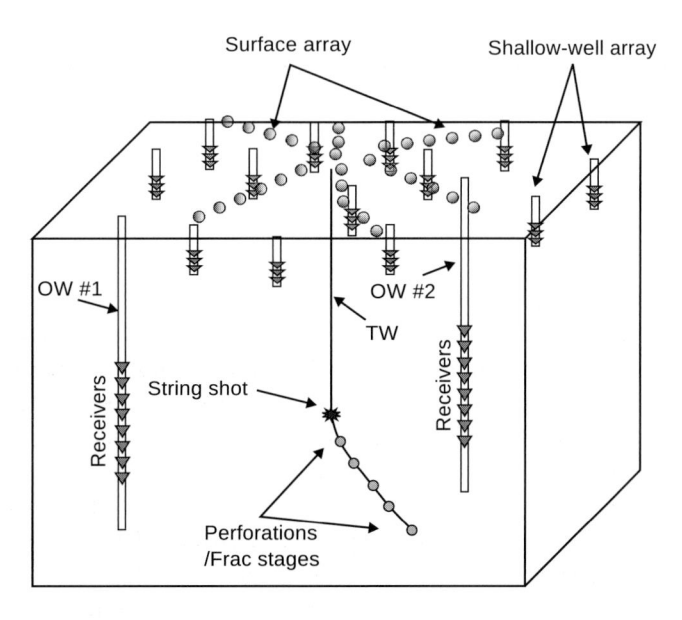

Fig. 5.3 Microseismic acquisition geometries used for hydraulic-fracture monitoring. TW denotes treatment well and OW denotes observation well. Modified from Akram (2014), with permission of the author.

5.2 Sensor Configurations

Since the initial development of microseismic monitoring methods, a range of different sensor configurations has evolved (Figure 5.3). The development of different acquisition approaches has occurred within a landscape of vigorous technical debate about the fundamental sensitivity of surface versus downhole recording systems (Maxwell, 2014). For example, sensors deployed in deep wellbores are removed from strong surface-noise sources and have shorter raypaths, thus reducing the effects of anelastic attenuation; individual waveforms are thus characterized by higher S/N than waveforms recorded at the surface, generally enabling the detection of weaker events. On the other hand, the positioning of sensors at the surface is far less constrained than for deep wellbores, thus facilitating the use of a greater number of sensors. This flexibility enables a higher degree of S/N enhancement by waveform stacking, along with more favourable aperture for event detection and source characterization. If the program objective is only to monitor anomalous induced earthquakes, the ability to detect weak events may be a secondary consideration, but sensor bandwidth is paramount. Finally, in some settings such as boggy terrain, seismic attenuation in the near surface is extraordinarily high and debilitating for high-fidelity recording; this has led to a hybrid approach that uses shallow wellbores. Ultimately, the choice of sensor configuration depends upon various factors such as availability of observation well(s), surface conditions and access. As elaborated in Chapters 6, 7 and 9, methods used to process raw microseismic data depend on the choice of sensor configuration used for data acquisition.

5.2.1 Deep-Downhole Arrays

Deep-downhole arrays refer to sensors, typically multi-component geophones installed in a pod similar to the prototype example in Figure 5.2, that are deployed in a wellbore within relatively close proximity (100s m) to the injection zone. Various installation scenarios can be used, as depicted in Figure 5.4. In some cases, the sensor array is located within the near-field for seismic wave propagation, which may introduce unusual complications in the analysis of signals, as discussed in §3.23.

Temporary sensor arrays are usually installed on a wireline into a steel-cased wellbore. In this case, a mechanism for clamping the sensor pods to the casing is required, such as a mechanical arm or magnets. Figure 5.5 shows an example of an downhole array deployment using a type of wireline system. Here, the array is being installed using a crane into a vertical observation well in western Canada. Alternatively, sensor arrays can be installed using coiled tubing-conveyed systems.[4] For any temporary deployment scenario, the clamping force should be at least a factor of 15 times the tool weight, in order to ensure adequate coupling (Maxwell, 2014). Coupling is important to reduce noise and to ensure signal *fidelity*.[5] In practice, the number of sensors used in an array varies from 8 to

---

[4] Coiled tubing refers to flexible metal piping, typically 2.5 to 8.3 cm in diameter, which is spooled into the wellbore from a large reel.
[5] In this context, signal fidelity means consistency and lack of signal distortion across the sensor array.

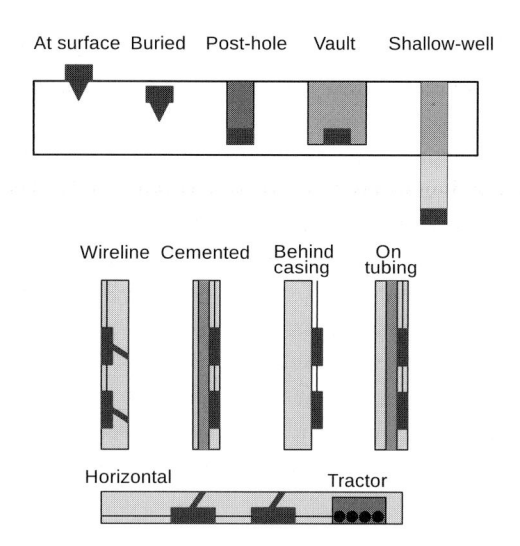

Fig. 5.4 Scenarios for installing sensors at the surface and in a deep wellbore. Modified from Maxwell (2014). Republished with permission of Society of Exploration Geophysicists, from *Microseismic Imaging of Hydraulic Fracture: Improved Engineering of Unconventional Shale Reservoirs*, S. Maxwell, copyright 2014; permission conveyed through Copyright Clearance.

Fig. 5.5 Installation of the University of Calgary's downhole geophone array, which uses a type of wireline system designed by Engineering Seismology Group. The work environment during installation is demanding and safe practices are of paramount importance. Photo credit Drew Chorney, used with permission.

36, with a typical spacing of 10s m between each sensor depending upon the design of the *toolstring*.

Simultaneous observation from multiple wells, as depicted in Figure 5.3, is effective for reducing location uncertainty, but only for a subset of events that are well recorded with more than one sensor array. The use of multiple observation wells also confers significant advantages for moment-tensor inversion (MTI) using downhole microseismic observations, since the array aperture of a single vertical observation well is insufficient to retrieve the full moment tensor (Vavryčuk, 2007). If data are available from only a single observation well, it is possible to estimate a composite source mechanism for a set of events, under the assumption that all of the events share a common source mechanism (Rutledge and Phillips, 2003). For the case of either single-event MTI from multiple observation wells, or composite-event MTI from a single observation well, a necessary condition is that the source-receiver geometry samples a sufficiently large solid angle for the inverse problem to be well-posed (Eaton and Forouhideh, 2011).

Downhole sensor arrays can also be installed permanently, behind casing and/or cemented into the wellbore. For example, Bell et al. (2000) describe a permanent geophone array installed behind production casing within the Athel formation in south Oman, where massive hydraulic-fracture stimulation is required in order to realize economic production rates. Permanent installation provides a high assurance of adequate coupling with the host medium and facilitates time-lapse data acquisition to investigate temporal changes in the subsurface. A disadvantage of permanent downhole arrays, however, is that it is impractical to repair or replace sensor elements that malfunction. Failure of sensor elements is not uncommon within the in situ environment of most reservoirs, which are typically unfavourable for manufactured components due to the effects of corrosion and prolonged exposure to elevated temperatures.

During wireline deployment of a permanent or temporary downhole microseismic array, the tools undergo twisting due to unavoidable torque in the wireline. Except in rare circumstances when the tool assembly includes a gyroscopic device, the net rotation of each element in the toolstring is unknown. Thus, for a standard triaxial geophone installed in a vertical wellbore, the azimuth of the horizontal geophones is not known, although the vertical geophone has a known orientation after deployment and the horizontal geophones have known horizontal inclination. Consequently, processing of downhole microseismic data includes a step to determine the tool orientation. In a wellbore with a deviated trajectory, a well survey is necessary to map wireline depth into geographic coordinates; the transformation of geophone orientations in a deviated wellbore is correspondingly more complicated, since several rotation operations are required.

In many cases, downhole arrays are installed into existing wells, due to the considerable expense of drilling a new well specifically for use as a microseismic observation well. One scenario is to deploy a downhole array into a nearby offset well; this scenario is not uncommon in mature oil or gas fields that are under redevelopment, where a first generation of vertical wells undergoes replacement by infill horizontal wells that are completed with hydraulic fracture stimulation (Reynolds et al., 2012). If an existing producing well is used as an observation well, it may be necessary to install a bridge plug in order to isolate the producing zone from the sensor array. This often means that the sensor array is located in a sub-optimal vertical position above the treatment depth. Alternatively, the cost of drilling a new well can be avoided by installing microseismic sensors into a treatment well, although exceptionally high noise levels during hydraulic fracturing operations mean that only post-injection seismicity can be recorded (Maxwell, 2014).

Due to the nature of pad drilling with multiple horizontal wells, it is becoming more common for multiple adjacent horizontal wells to be available, but with no vertical well sufficiently close to the treatment zone in order to achieve successful microseismic monitoring. In this case, a downhole array can be installed into a horizontal well using a tractor device (Figure 5.4). Maxwell and Le Calvez (2010) discuss the pros and cons of installing microseismic arrays into horizontal wells versus vertical wells. In such cases, the horizontal array may be expected to lead to improved sensitivity due to proximity to the treatment zone, whereas vertical arrays generally provide better accuracy for event locations, especially for out-of-zone seismicity. Pre-survey modelling can be undertaken to evaluate these different options.

Transfer-function characteristics of the recording system, including sampling rate, gain and sensitivity settings as well as poles and zeros of the system (see Appendix B), determine the *dynamic range* of the system[6] used in order to convert raw recorded data into units of m/s. Once this conversion has been applied, event magnitudes can be estimated based on preliminary hypocentre locations (methods for determining hypocentre locations are discussed in the next chapter). *Magnitude–distance scatter plots* provide a valuable quality-control (QC) tool for downhole microseismic monitoring (Cipolla et al., 2011; Zimmer, 2011) and serve here to illustrate a well-known detection-distance bias for a single downhole array. An example of a magnitude–distance scatter plot is given in Figure 5.6.

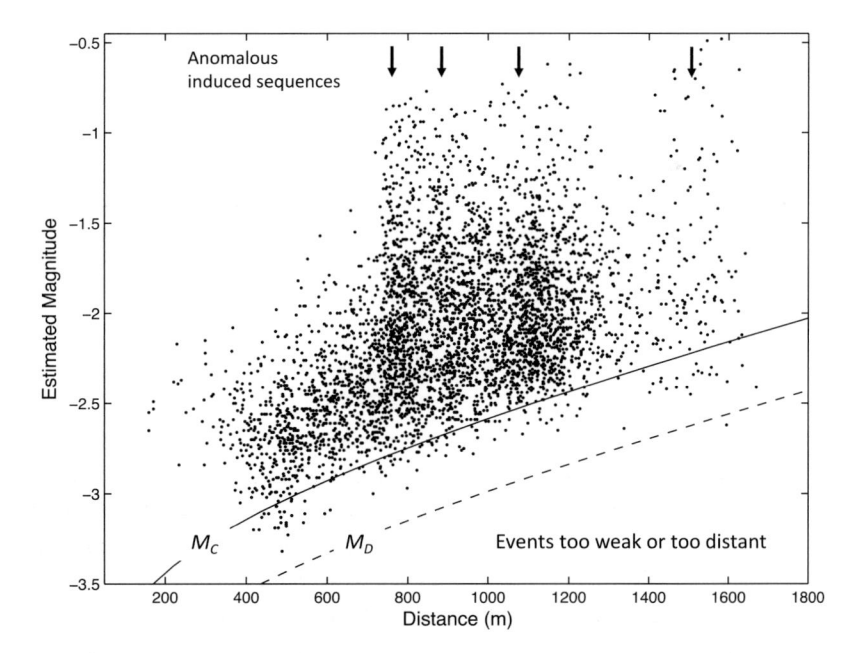

Fig. 5.6 Magnitude–distance scatter plot for hydraulic fracturing well completion in the Horn River basin, using data from Eaton et al. (2014a). Curves showing estimated magnitude of completeness (M_C) and detection (M_D) are calculated using $Q = 200$. Vertical bands indicate anomalous induced seismicity sequences associated with fault activation.

---

[6] Dynamic range is the ratio of the largest to the smallest recoverable signal (Sheriff, 1991).

The amplitude loss of seismic waves due to geometrical spreading and anelastic attenuation means that the weakest detected events are observed at the closest recording distances. For reference, theoretical curves are indicated on the plot. The curve labelled as M_C approximates the variation with distance of the magnitude of completeness (§3.8). The first step for computing this curve is to determine the average spectral amplitude level for which the probability of signal detection is 90 percent, denoted as A_{90}. This detection level depends on the S/N of the raw data as well as the detection method; for a discussion, see Schorlemmer and Woessner (2008). Next, by combining Equation 3.73 with 3.76, at a hypocentral distance r the seismic moment calculated using the P wave can be expressed as

$$M_0^P = A_{90} r \frac{4\pi \rho V_P^3}{R_P} e^{\frac{\pi f_0 r}{V_P Q_P}}, \tag{5.1}$$

where $R_P \simeq 0.52$ is the average P-wave radiation pattern (Boore and Boatwright, 1984), f_0 is the dominant frequency of the signal, and V_P and Q_P are the average P-wave velocity and quality factor. Similarly, the seismic moment calculated using the S wave can be written as

$$M_0^S = A_{90} r \frac{4\pi \rho V_S^3}{R_S} e^{\frac{\pi f_0 r}{V_S Q_S}}, \tag{5.2}$$

where $R_P \simeq 0.63$ is the average S-wave radiation pattern (Boore and Boatwright, 1984), and V_S and Q_S are the average S-wave velocity and quality factor. For a given distance, the M_C value is determined by solving Equations 5.1 and 5.2 and then computing

$$M_C = \frac{2}{3} \log_{10} \left(\max \left\{ M_0^P, M_0^S \right\} \right) - 6. \tag{5.3}$$

Similarly, the M_D curve in Figure 5.6 shows a *detection limit* that is determined using the same approach but with an average spectral amplitude level that has a 10 percent probability of detection. This distance-varying variable detection limit is an important consideration for the interpretation of microseismic events recorded using downhole arrays (Cipolla et al., 2011) as well as for analysis of magnitude distributions (Eaton and Maghsoudi, 2015).

5.2.2 Surface and Shallow-Well Arrays

Surface microseismic arrays are not affected by the type of distance-detection bias discussed above. Surface arrays, moreover, have other practical advantages in geographic areas where there are no suitable observation wells available, or where temperatures within the reservoir exceed the thermal tolerance of downhole toolstrings (Maxwell, 2014). One design for surface arrays makes use of sensors deployed at the surface along receiver lines with a radial configuration centred on the well pad (Figure 5.7), similar to the spokes of a wheel (Duncan and Eisner, 2010; Chambers et al., 2010). Array aperture considerations mean that the overall diameter of the surface array should be approximately twice the depth of the target (Duncan and Eisner, 2010). For a given number of sensors, this configuration achieves the highest possible spatial sampling along lines radial to the well pad, which is optimal for suppression of undesirable near-surface noise generated by pumps, blenders, etc.

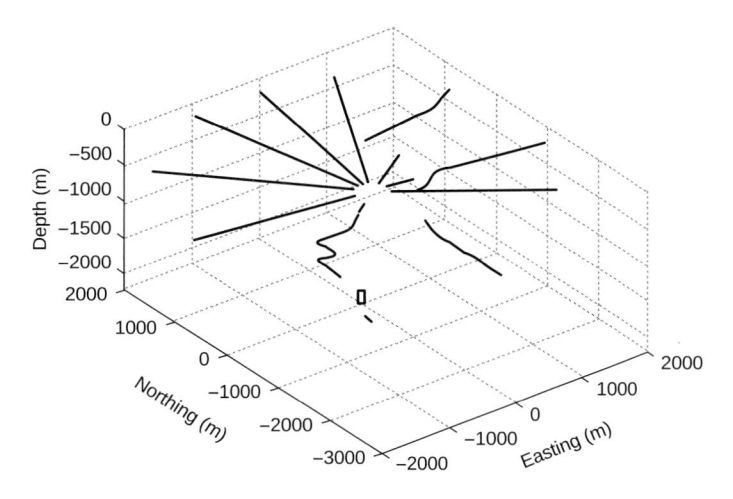

Fig. 5.7 Example of a star-shaped surface array, using receiver lines that are mainly configured in a radial orientation centred on the well pad located at the origin. Adapted from Eisner et al. (2010). Copyright 1999–2017 John Wiley & Sons, Inc. All Rights Reserved.

In general, sensor arrays have the following objectives (Pap, 1983):

1 Suppression of coherent noise, especially ground roll, a Rayleigh wave within the low-velocity overburden that creates a strong coherent linear noise pattern in seismic records on land (Xia et al., 1999).
2 Enhancement of S/N by *stacking* (summing) recorded traces; for n traces this leads to a S/N improvement of $\sqrt{n}$ with respect to random background noise, assuming that the signal is identical on every trace.
3 Prevention of spatial *aliasing*, a spectral ambiguity that arises when continuous data are discretized by sampling at a uniform spatial interval Δx that is more than half of the shortest wavelength in the original data. Put a different way, aliasing occurs if the continuous data has spectral components with *wavenumber* (the reciprocal of wavelength) that is higher than the *Nyquist* wavenumber, given by $k_N = \frac{1}{2\Delta x}$ (see Figure 5.8).

A *linear array* consists of a set of equally space, uniformly sensitive elements. Based on antenna theory, for n sensors at an interval Δx, the amplitude response for a linear array can be written as (Pap, 1983)

$$A(k_x) = \frac{1}{n} \left| \frac{\sin(\pi k_x n \Delta x)}{\sin(\pi k_x \Delta x)} \right|, \qquad (5.4)$$

where $k_x = \frac{1}{\lambda_A}$ denotes spatial wavenumber and λ_A is apparent wavelength for a propagating wave, measured in the direction of positive x. This filter response is plotted in Figure 5.8 for a linear array with $n = 10$ elements. The response is plotted in *decibels* (dB), a logarithmic measure of the ratio between two numbers, which in this case is given by

$$A(k_x)[\text{dB}] = 20 \log_{10} \left(\frac{A}{A_{max}} \right). \qquad (5.5)$$

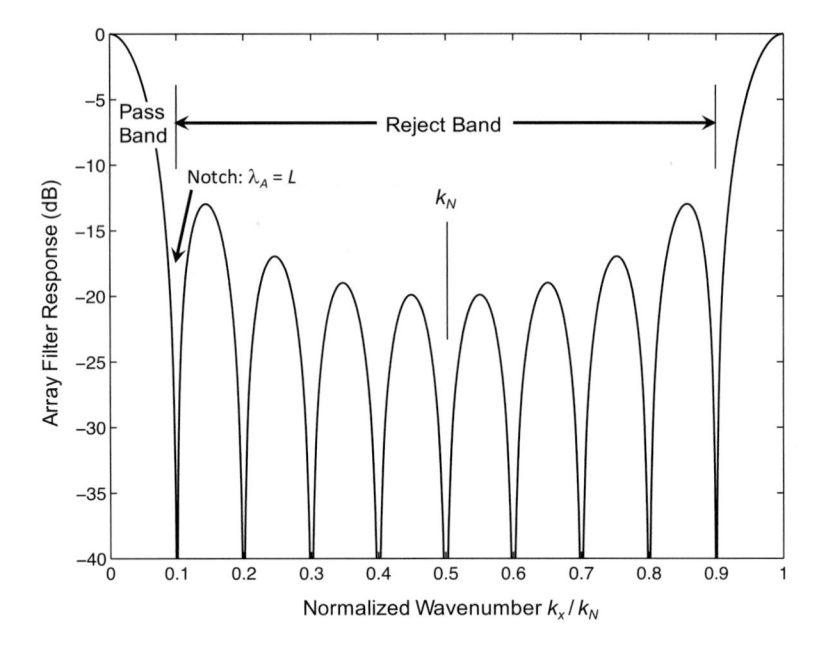

Fig. 5.8 Graph of the spatial filter response of a 10-element linear array with respect to normalized wavenumber, k_x/k_N. Arrays can be used to attenuate undesirable surface noise. See text for details.

For further context, based on this formula an increase of 6 dB corresponds with a doubling of the amplitude, whereas an increase of 20 dB represents a factor of 10 amplitude increase and an increase of 60 dB corresponds to a factor of 1000.

As shown in Figure 5.8, this filter contains a *notch* at $\lambda_A = L$, where $L = n\Delta x$ is called the *effective array length*. This spectral notch marks the transition from the filter pass band and reject band. This means that the stacking of waveforms recorded over a linear array with an effective length L will attenuate waves with a wavelength of λ_A. Suppression of ground roll, which typically has wavelengths in the range from 10–160 m (Xia et al., 1999) can thus be achieved using a linear geophone array with an effective length approximately equal to the ground-roll wavelength.

The array concept has been generalized to two dimensions using a patch array design for surface microseismic acquisition (Pandolfi et al., 2013; Roux et al., 2014). This design uses densely sampled geophone arrays that are deployed within small areas (patches). Stacked traces are produced within each patch, similarly realizing a $\sqrt{n}$ improvement in S/N with respect to random noise, while also providing omni-directional array filtering capabilities. This approach also has a practical advantage that can be used for cases where surface access is not available throughout the desired aperture of the recording system.

Another strategy for attenuating near-surface noise is to deploy sensors in shallow wells that penetrate below the low-velocity overburden. The overburden layer can have highly variable thickness, leading to wave scattering effects, and is often characterized by exceptionally high attenuation (Snelling and Taylor, 2013). Consequently, attenuation and scattering in the overburden may be the dominant contribution to signal losses along the

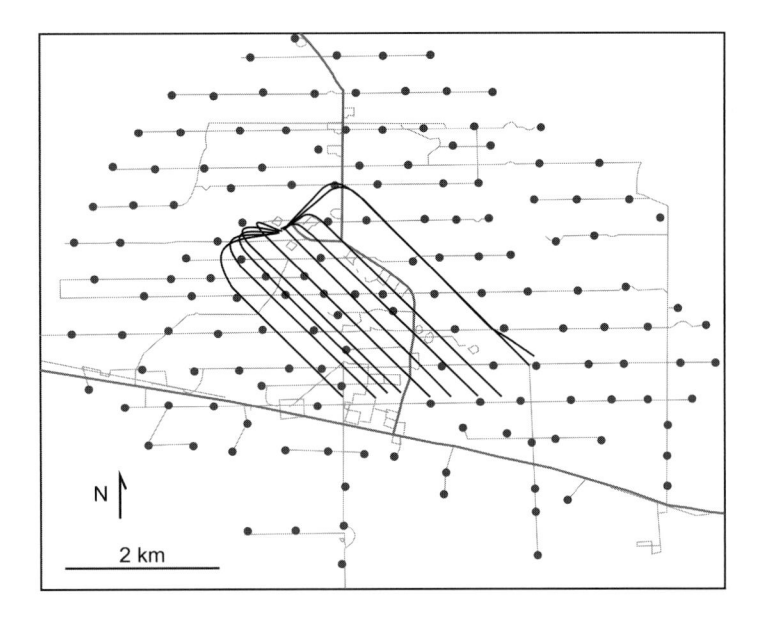

Fig. 5.9 Example of a shallow-well array from the Horn River basin in northwestern Canada. The monitored well pad has 10 horizontally drilled wells (black lines) that were completed in three formations within the Horn River group. The 151-element microseismic array, shown as dots, covers an area of 40 km^2 and was installed using seismic cut lines (fine gray lines). Thicker gray lines show roads. Modified from Snelling and Taylor (2013).

entire path from the microseismic source at depth up to the surface. An example of the station distribution for a shallow well array is shown in Figure 5.9. The array contains 151 shallow wells that were drilled to the base of the overburden; 10 Hz three-component geophones were cemented at the base of each well. This array was used for microseismic monitoring at a large well pad in the Horn River basin in northwestern Canada (Snelling and Taylor, 2013). Since unconventional reservoirs are typically developed over an extended time period, installation of a permanent shallow wellbore array may provide a good return on investment for long-term monitoring (Zhang et al., 2011b).

5.2.3 Regional Seismograph Networks

In areas that have experienced induced earthquakes, regulatory agencies, industry and the academic community have responded by installing or expanding regional seismograph networks. This is particularly true for the onset of seismic activity in areas of formerly sparse seismograph network coverage, which is likely to be the case within previously quiescent stable-continental regions (SCRs). For example, the Regional Alberta Observatory for Earthquake Studies Network (RAVEN) in Alberta, Canada (Schultz and Stern, 2015), which incorporates the earlier Alberta Telemetered Seismograph Network (ATSN) (Eaton, 2014), was initiated in 2013 by the Alberta Geological Survey (AGS) in response to concerns about induced seismicity. Similarly, the British Columbia Seismic Research Consortium was established in 2012 to improve understanding of induced seismicity in areas of unconventional resource development in northeastern British Columbia (NEBC), Canada. As a coordinated effort involving both public and private sectors to address critical knowledge gaps, this initiative led to installation of a regional seismic network in NEBC to complement sparse coverage by the Canadian National Seismograph Network (CNSN) in order to provide more uniform sensitivity and location accuracy (Mahani et al., 2016).

Figure 5.10 exemplifies recent growth of regional seismographic network coverage for monitoring induced seismicity in western Canada. New stations in this region have been installed for several reasons: 1) to improve monitoring capabilities in areas of active development by industry, and 2) to acquire baseline seismicity data in a region that is prospective for future unconventional resource development. At the onset of seismicity in this region in about 2009, the only seismograph station in this area was a CNSN broadband station

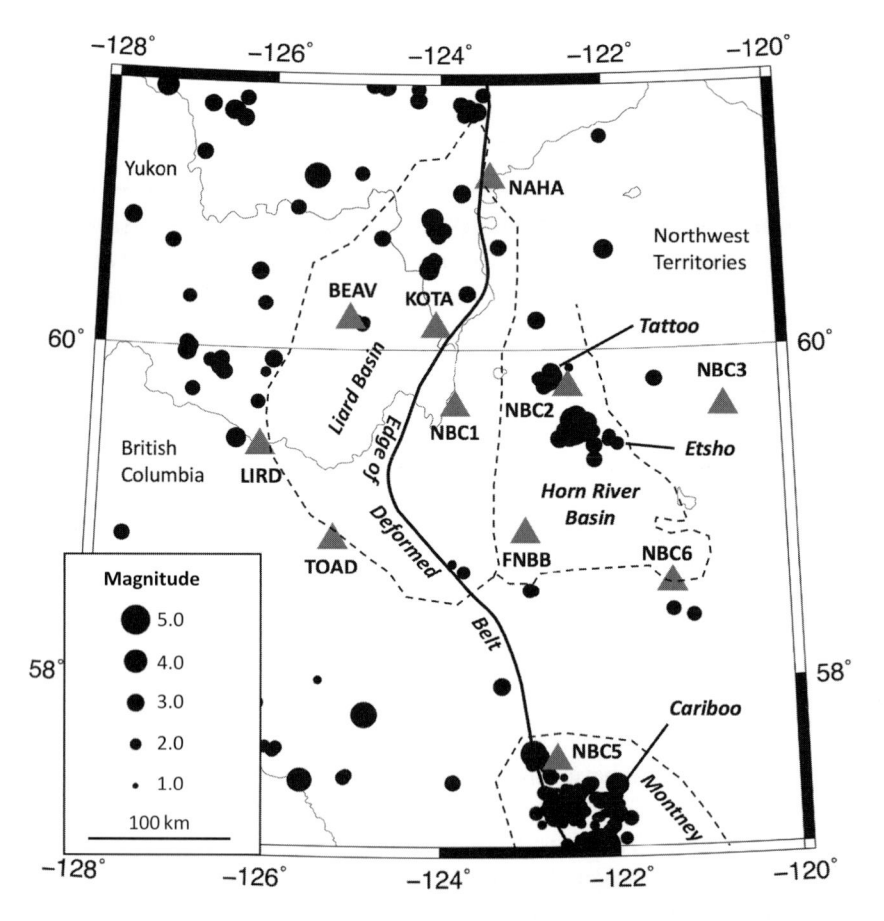

Fig. 5.10 Regional seismograph network example from northeastern British Columbia (NEBC), Canada and adjoining territories, highlighting specific areas of industry activity (Etsho, Tattoo, Cariboo) as well as important sub-basins (Liard, Horn River, Montney). Triangles indicate broadband seismograph stations, many of which were installed for the purpose of monitoring induced seismicity and/or to provide baseline seismicity data in areas that are prospective for future unconventional resource development.

at Fort Nelson (FNBB) and there was little indication of background seismicity in the region (Farahbod et al., 2015b). Following initial reports of injection-induced seismicity in the Tatoo, Etsho and Cariboo development areas (among others) (BCOGC, 2012, 2014), eight additional seismograph stations were installed as part of the NEBC regional seismic network. In 2016, seismograph network coverage was extended by the Yukon Geological Survey to cover the northern part of the Liard Basin, in anticipation of potential future activities for unconventional resource development in an area where considerable resource potential has been identified in shales of Mississippian to Devonian age (Ross and Bustin, 2008; National Energy Board, 2016). The current network configuration provides relatively uniform coverage over current and prospective development areas in the Horn River Basin and Liard Basin.

Unlike most microseismic acquisition systems, data from these seismograph networks is generally telemetered to data centres in near real time and most of the raw waveform data are freely available. A description of typical instrumentation and station design for a portable broadband telemetered seismograph station is given by Eaton et al. (2005).

5.3 Background Model Construction and Calibration

In order to determine reliable locations of seismic sources using wavefield observations, an accurate background model is required, which includes (but is not limited to) a *velocity model*. The background model needs to contain sufficient information to explain essential features of qP and qS wave propagation from the source region to the sensor array, including any attenuation and anisotropic effects that are expressed in the observed waveforms. Although development of the background model is primarily a data-processing task, a brief overview is included here due to the importance of recording calibration sources in order to enable model refinement and validation. Various calibration sources, such as perforation shots, string shots and sleeve-opening signals, are elaborated below.

The initial background model is typically constructed using well-log information. At a minimum, it must encompass a subsurface region that contains the entire seismically active volume, extending upward to the surface (i.e. not limited to the depth range of the target zone). Well-log data from the treatment well should be used, if available; for downhole surveys, data from the observation well should also be used to capture lateral heterogeneity within the intervening region. If necessary, logs from a nearby well can be used instead of the treatment well or observation well, where the concept of "nearby" depends on the degree of lateral heterogeneity for the field area. Although some areas have flat-lying stratigraphic sequences with sufficient continuity that well logs from 10 km away are useful, in other areas with greater structural complexity, more proximal well control is required. Sonic logs are relatively standard in typical log suites and provide acoustic transit-time measurements that can be easily inverted for the *P*-wave velocity. On the other hand, *S*-wave logs are less common, meaning that some other approach may be required such as a statistical analysis based on V_P/V_S measurements from other wells. If crossed-dipole logs are available, it is possible to determine fast and slow qS-wave velocities, as discussed in

§4.1.2. Ultimately, sufficient prior data is rarely available and a parsimonious, horizontally layered velocity model is often the best starting point.

5.3.1 Calibration Sources

Refinement of the initial velocity model using *calibration sources* is essential for obtaining reliable and accurate hypocentre locations. Calibration sources are seismic sources with known locations; the velocity model is calibrated using an iterative method, wherein hypocentre locations of these signals are computed using methods that are described in Chapters 6 and 7. Given the known locations of calibration sources, the background model (including anisotropic parameters, if necessary) is adjusted until the computed hypocentres match the known locations.

For hydraulic fracturing completions that use the perf-and-plug method (§4.2.1), perforation shots with known locations are available to use as calibration sources. Figure 5.11 shows a comparison of waveforms and time-frequency plots[7] for a typical microseismic event and a perforation shot recorded using a deep downhole array (Eaton et al., 2014d). The microseismic event exhibits distinct *P*- and *S*-wave arrivals with an apparent signal bandwidth from $\approx$ 20 Hz to 500 Hz. By comparison, the perforation shot lacks a clear *S*-wave arrival, but it has a distinct *P*-wave arrival with an apparent bandwidth that extends up to nearly 1000 Hz. For the microseismic event in Figure 5.11, the *S*-wave amplitude is nearly a factor of ten times greater than the *P*-wave amplitude, consistent with a radiation pattern for a double-couple source mechanism; conversely, for the perforation shot, the

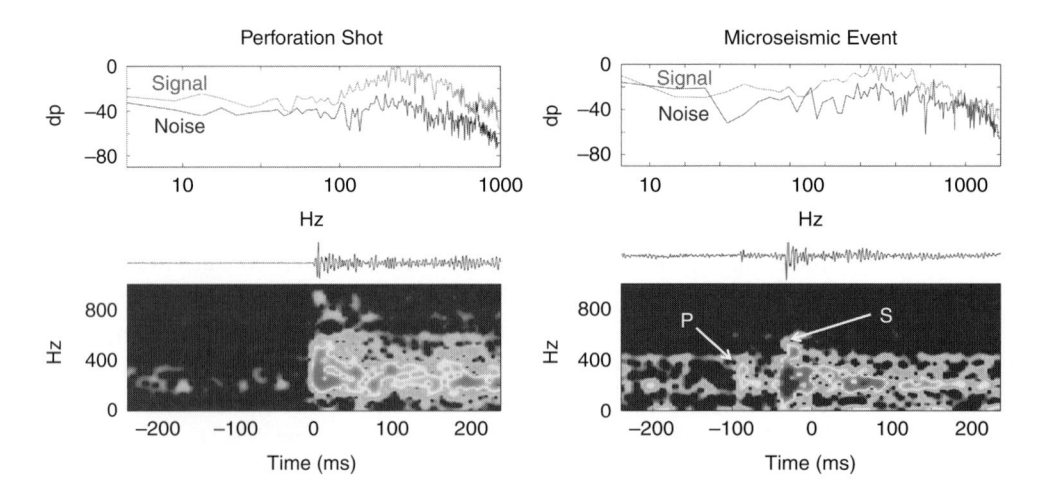

Fig. 5.11 Comparison of time-frequency spectra for a representative perforation shot and a microseismic event from a hydraulic fracturing program in the Montney trend of northeastern British Columbia (Eaton et al., 2014d). Upper panels show signal and noise spectra. Lower panels show recorded vertical-component waveforms and their corresponding spectragrams obtained using the short-time Fourier transform. Hot (red) colours indicate high amplitude. Used with permission from the Society of Exploration Geophysicists. (A colour version of this figure can be found in the Plates section.)

[7] Time-frequency analysis is discussed in Appendix B.

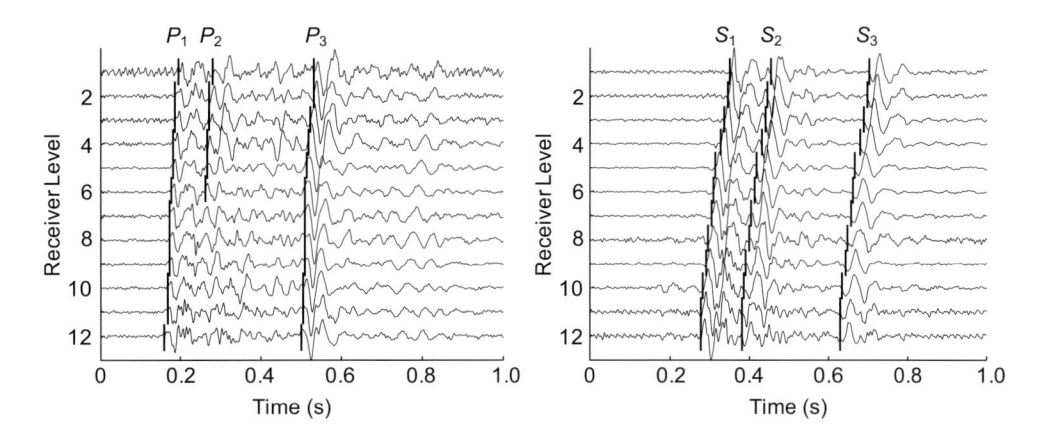

Fig. 5.12 Sleeve-opening event sequence from a hydraulic-fracture monitoring program in central Alberta, showing characteristic triplet of P- and S-wave arrivals. Traces are independently normalized and rotated into orthogonal horizontal components.

strong P wave and weak (or nonexistent) S wave is consistent with an isotropic (explosive) source mechanism. As in most monitoring experiments, the origin time of the perforation shot is too imprecise for direct velocity measurement. Special perforation-timing procedures have been developed (Warpinski et al., 2005b), but are not in common use.

For open-hole completions that use a sliding-sleeve system, perforation shots are not available, but waves generated by sleeve-opening events can be used for velocity-model calibration (Maxwell and Parker, 2012). Figure 5.12 shows a representative example of a sleeve-opening sequence from a microseismic monitoring program in central Alberta, Canada (Eaton et al., 2014c). The three-component recordings in Figure 5.12 have been rotated into ray-centered coordinates, as outlined in the next chapter, which effectively separates the P- and S-wave signals onto different records. As described by (Maxwell and Parker, 2012), due to the mechanical action of the assembly (Figure 4.10) the sleeve-opening event is characterized by multiple discrete signals within a timeframe of less than a second. In this example, all three of the arrivals have similar amplitude, but in other cases the first signal has the highest amplitude (Maxwell and Parker, 2012). Since the P and S first arrivals are clear and the position of the sliding sleeve assembly in the wellbore is known, these events make excellent calibration sources.

In some cases, neither perforation shots nor sleeve opening events are available. This scenario may occur if signals from the perforation shots are too weak to be detected. In addition, some hydraulic fracturing methods, such as a hydra-jet stimulation approach (Surjaatmadja et al., 2008), do not produce any detectable calibration signals. In such cases, one option is to deploy one or more *string shots* in the well (Figure 5.3). String shots are seismic detonators that are placed into the wellbore in order to generate an impulsive source. The use of string shots has the advantage that the precise time can be determined easily, in addition to the location.

As mentioned previously, for a downhole survey the orientations of the horizontal sensor elements in the toolstring are generally unknown. Thus, in addition to their use for calibration of the velocity model, calibration sources are also employed to determine the orientation of horizontal geophones, using a process that is described in Chapter 6. If no subsurface calibration source is available, surface sources with known locations are a viable alternative to determine the geophone orientations. In a worst-case scenario where no calibration sources whatsoever are detectable, it is sometimes necessary to assume, for each stage, that the first observed microseismic event coincides with the injection point. Since it is not possible to validate this approach, it should be regarded as a "last resort."

5.4 Survey Design Considerations

5.4.1 Sensor Types

Geophones are motion-sensitive transducers that use a damped mass-spring system to convert ground motion into an electrical signal that is proportional to the ground velocity (Knapp and Steeples, 1986). The basic design of a geophone uses a magnet positioned inside a wire coil. The magnet is fixed to the geophone case, while the coil and mass are inertially isolated by a spring such that relative motion between the magnet and coil generates an electromotive force as described by Faraday's Law of induction. Three-component geophones have three elements with mutually orthogonal orientations. The natural frequency of a geophone is given by

$$f_0 = \frac{1}{2\pi}\sqrt{k/M}, \tag{5.6}$$

where k is the spring constant and M is the mass. Although lower frequency geophones are available, most conventional geophones have natural frequencies of 10 Hz or more. The velocity *transfer function*[8] of a geophone can be written as (Wielandt, 2002)

$$H_g(f) = S_g \frac{(f/f_0)^2}{-(f/f_0)^2 + 2i\lambda_g f/f_0 + 1}, \tag{5.7}$$

where S_g is the geophone sensitivity in V/m/s and λ_g is the geophone damping factor. The geophone sensitivity is related to the spring constant and coil resistance R_c by (Knapp and Steeples, 1986)

$$S_g = k\sqrt{R_c}. \tag{5.8}$$

From Equation 5.7, it is clear that for an undamped geophone ($\lambda_g = 0$) there is a singularity in the transfer function associated with harmonic resonance of the system for $f = f_0$. This is not a desirable characteristic, so in practice geophones are always damped. For a *critically damped* geophone, $\lambda_g = 1/\sqrt{2}$ (Wielandt, 2002). Geophones are normally intended for use at frequencies well above the natural frequency, where they have a flat response.

A *seismometer* is a capacitive force-balance device. This type of device uses a mass-spring system, but rather than measuring the elongation of the spring a compensating

---

[8] A mathematical function that relates the output of a system to the input; see Appendix B.

force is applied to maintain the mass in a central position (Wielandt, 2002). The output of the device is calculated by measuring the strength of the compensating force. The resistance and capacitance parameters of the feedback loop can be adjusted with the mass/spring system to design a seismometer with a frequency response that covers the desired part of the signal bandwidth for a given application (Figure 5.1). For a seismometer with N zeros (denoted as a_j) and M poles (denoted as p_j), the transfer function can be written as

$$H_s(f) = S_s \frac{\prod_{j=1}^{N}(s - a_j)}{\prod_{j=1}^{M}(s - p_j)}, \qquad (5.9)$$

where $s \equiv 2\pi i f$ and S_s is the overall sensitivity factor, which is often expressed as a product of various intermediate parameters such as a gain correction. The response of a broadband seismometer (Trillium 240) and a geophone (SM-24 10 Hz) are compared in Figure 5.13.

Several other types of sensors are used for passive-seismic monitoring. *Accelerometers* respond to instantaneous acceleration of the ground, using piezoelectric ceramic transducers with a mass that provides a pressure in order to generate a voltage (Knapp and Steeples, 1986). *Force-balance accelerometers*, like seismometers, use the force-balance feedback principle, but have a resistive electronic feedback loop such that the output voltage is proportional to acceleration, rather than velocity (Wielandt, 2002).

Distributed acoustic sensing (DAS) is an emerging technology in which an optical fibre provides both the medium for signal transmission as well as the sensor (Daley et al., 2013). DAS systems use coherent optical time-domain reflectometry, in which short laser pulses of coherent light are transmitted along a fibre (Molenaar et al., 2012). Subsurface wave motion produces a strain along the fibre that can be detected by backscattered light waves, which

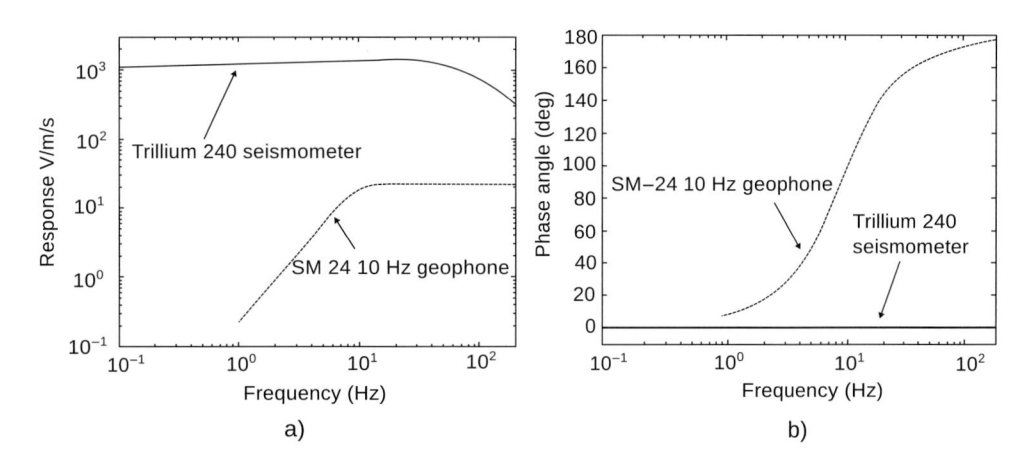

Fig. 5.13 Sensor frequency response curves for a Trillium T240 broadband seismometer and a SM-24 10 Hz geophone, showing a) amplitude and b) phase response.

are analyzed using an interrogation unit that segments the fibre with a spatial resolution of 1–10m. Although published results using downhole-deployed DAS systems exhibit higher noise levels than geophone arrays, the technology is rapidly evolving and is showing great potential for use as a passive monitoring tool during well completions (Molenaar et al., 2012).

5.4.2 Noise

Noise can be defined by exclusion as all undesirable components of ground-motion recordings that are not signal; the concept of signal depends upon the specific application, but is generally understood as the components of recorded waveforms that fit a conceptual model, such as coherent reflected wave energy (Kumar and Ahmed, 2011). Passive-seismic monitoring relies on detection of arrivals above the background noise, which varies by more than 60 dB in some frequency bands (Webb, 2002).

A few common types of noise are considered here. *Cultural* noise is generated by human activities such as vibrations caused by manufacturing and other industrial operaitons traffic, typically within a frequency band of 1–35 Hz (Boese et al., 2015). As previously mentioned, at a well pad where hydraulic fracturing operations are ongoing, strong vibrations are produced by pumps, blenders and vehicles, producing high-amplitude surface waves that radiate away from the pad. For example, Figure 5.14 shows noise levels computed for the shallow-well array depicted in Figure 5.9. The map was constructed using

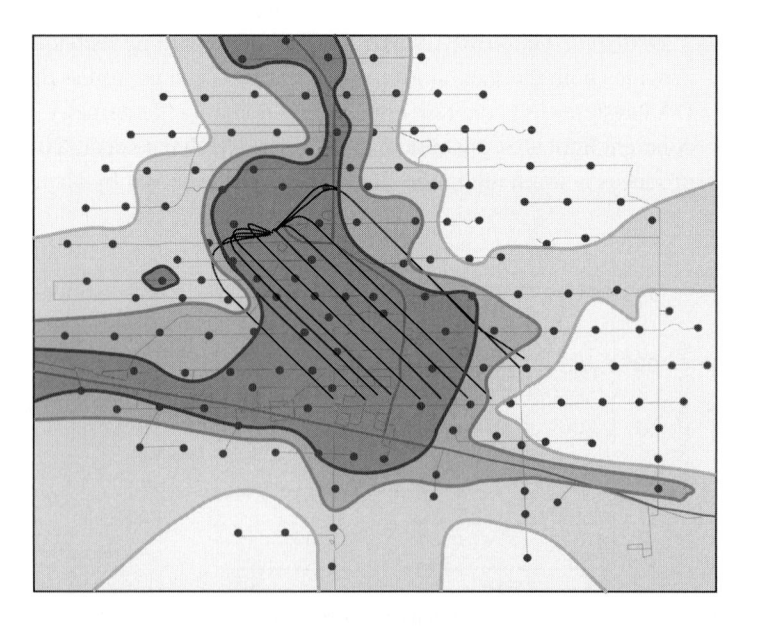

Fig. 5.14 Root-mean-squared (RMS) noise levels for the shallow wellbore array depicted in Figure 5.8. Dark gray represents high noise levels and light gray low noise levels. In general, the highest noise levels are observed near facilities and roads. A small increase in noise levels correlates with the position of a Holocene near-surface channel. Modified from Snelling and Taylor (2013).

the root-mean-squared (RMS) amplitude value $\bar{u}_{RMS}$, which is a type of average value calculated using the formula

$$\bar{u}_{RMS} = \sqrt{\frac{\sum_{i=1}^{N} u_i^2}{N}}, \qquad (5.10)$$

where $u_i, i = 1, 2, 3, \ldots N$ is the input series. The noise levels shown in Figure 5.14 were calculated for a 24-hour time window with no known seismicity. The highest noise levels are observed along roads used by vehicles accessing the well pad in the centre of the map, as well as a second well pad to the north. Slightly high noise values are found along a buried Holocene channel, where thick sediments are known to be present based on seismic data (Snelling and Taylor, 2013). This relationship could be anticipated because elevated noise levels are normally expected in any area of soft or unusually thick soil (Building Seismic Safety Council, 2003).

Wind is also a significant source of background noise within a frequency band that is similar to cultural noise. Wind noise is generally coupled into the ground through the root systems of plants and trees. The amplitude of wind noise decreases rapidly with depth (Carter et al., 1991), which is consistent with the use of a shallow-well array design for improvement of S/N. Like cultural noise, wind noise is effectively random on timescales of the seismic sources of interest (seconds). Random noise can be divided into various classes on the basis of frequency response. For example, *white* random noise has equal intensity for all frequencies, whereas *pink* random noise has intensity that scales as $1/f$. Pink noise therefore has greater intensity at lower frequency.

By analyzing noise records from 75 digital stations world-wide, Peterson (1993) developed a set of standard noise models in these units, known as the New High Noise Model (NHNM) and New Low Noise Model (NLNM). The measured noise level at NBC2 falls between these to limits within the period band from 0.3 s to 70 s, which means that this station is average with respect to noise levels in comparison with other stations from around the world. Figure 5.15 shows background noise calculated for station NBC2 (Figure 5.9). The noise is presented as a *power-spectral density* (PSD) plot, which is the Fourier spectrum of the noise autocorrelation function with respect to amplitude (Peterson, 1993).

The noise spectrum in Figure 5.15 contains a prominent noise peak at about 6 s period. This is known as the *microseismic noise peak*, which is caused by wave interference in the world's oceans (Webb, 2002). This noise peak is a global phenomenon and, as previously noted, has traditionally split seismology into short-period and long-period observational bands (Webb, 2002).

5.5 Survey Design Optimization

First-order survey-design goals for passive-seismic monitoring are, in general, to optimize the detection of microseismic events while minimizing errors in source locations (Maurer et al., 2010; Grechka, 2010; Zimmer, 2011). Various approaches and metrics have been

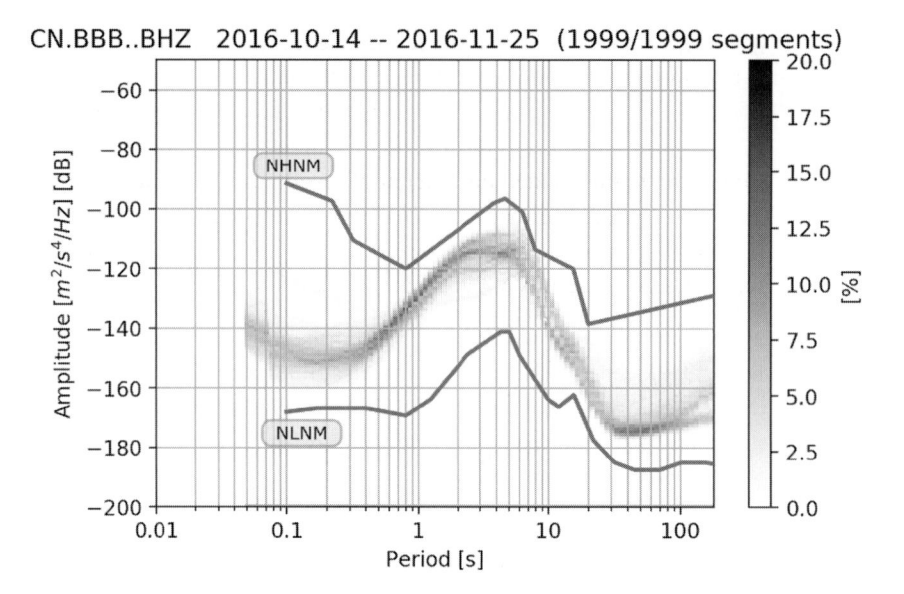

Fig. 5.15 Seismic noise models NLNM and NHNM (Peterson, 1993), compared with power-spectral density (PSD) calculated from a 40-day continuous noise sample from CNSN station BBB.

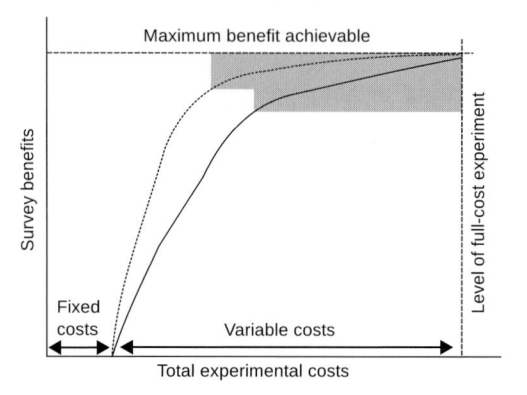

Fig. 5.16 Schematic graph depicting notional cost–benefit relationships for survey-design optimization. Shaded areas show diminishing returns. Dotted line shows optimized design, while solid line shows standard design. From Maurer et al. (2010), with permission from the Society of Exploration Geophysicists.

proposed to evaluate the performance of seismograph arrays for design optimization. A few of these approaches are discussed here.

Maurer et al. (2010) proposed a notional cost-benefit relationship for the impacts of design optimization, shown schematically in Figure 5.16. In the asymptotic limits, the concepts conveyed by this diagram are:

1 There will be fixed costs, such as mobilization of equipment, that are incurred before the collection of any data and thus before any survey benefits are realized.

2 Ultimately, for any survey design it is expected that there will be a point of diminishing returns. As this point is approached, the acquisition of more data mainly increases the redundancy of observations, rather than contributing new independent information.

According to this model, investment of resources into survey-design optimization can generally be expected to realize higher benefits for a given survey cost, or lower cost for the same benefit. This schematic depiction also implies that an optimized survey design will achieve higher benefits at the point of diminishing returns.

Martakis et al. (2006) recommended the use of *checkerboard tests* for design optimization of tomographic models from passive-seismic observations. A checkerboard test is a standard type of sensitivity analysis that is carried out by determining the expected response for an input model containing known anomalies (perturbations), usually on a regular grid. Synthetic data are determined by forward-modelling with actual sensor locations and the checkerboard model as input, and then inverting the results. The inversion results are then compared with the input checkerboard model to assess the resolving capabilities of the sensor array. A checkboard approach for assessing hypocentre uncertainty is presented in Figure 5.17, using four different sensor configurations. The configurations consist of: a) a single vertical observation well above the target zone; b) three vertical observation wells that span the target zone; c) a horizontal well within the target zone; and d) a surface cross-arm array. The model contains 12 microseismic source points. For each of these points, indicated by gray circles, 100 synthetic locations are computed by adding uniform random picking errors of ± 2ms for the P wave and ± 4ms for the S wave and uniform random azimuth errors of $\pm 15°$. This approach provides measures of location uncertainty that depend on sensor geometry as well as source location. The single monitor well performs best for the closest event, with uncertainty that increases with distance. The three-well array has more uniform errors, assuming that every event is recorded on all three arrays; in practice, this is unlikely to be the case. In the case of the horizontal well, the scatter increases toward the edge of the model domain. Finally, the surface array has relatively uniform response across the grid but a greater amount of scatter. This mainly reflects the much greater path length for this geometry than for the others. It should also be noted that the calculations shown here use a traveltime-based approach, whereas many algorithms for processing surface data use an imaging approach, as discussed in Chapter 7.

Peters and Crosson (1972) describe a statistical method known as prediction analysis, which they applied to estimate errors in hypocentre determination for a dense array of seismographs at the surface. The conventional approach for error analysis uses a least-squares approach assuming random errors in the input parameters. Prediction analysis enables the determination of errors without actually carrying out the inversion.

Grechka (2010) posed the optimization problem for survey design using the *Fréchet-derivative matrix*, which is the matrix of partial derivatives of the traveltime function with respect to the model parameters (x, y and z positional coordinates). Grechka (2010) showed how application of *singular-value decomposition* (SVD), a well-known technique for analyzing the behaviour of the solution to a linear inverse problem (Press et al., 2007), to the Fréchet-derivative matrix could be used to compare various survey designs.

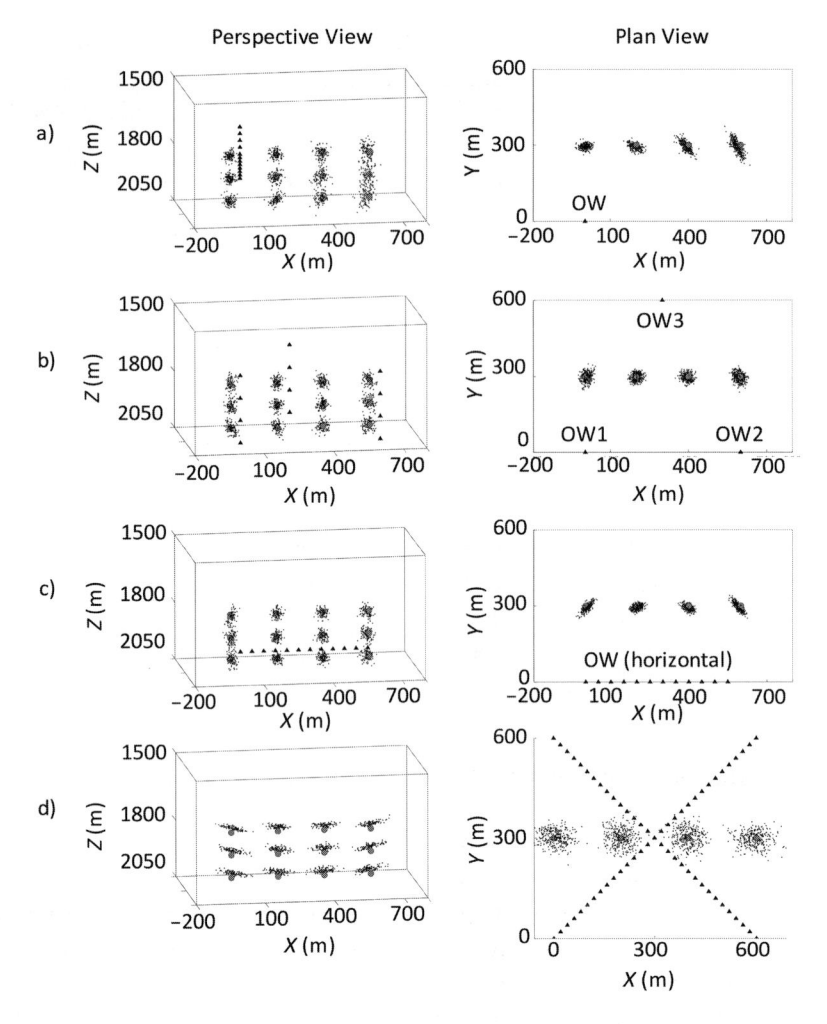

Fig. 5.17 Checkerboard approach to assess the impact of receiver geometry on microseismic event locations. Four receiver configurations are considered: a) a single vertical observation well (OW) located above the target zone; b) three vertical observations wells that span the target zone; c) a horizontal observation well within the target zone; and d) a surface cross-arm receiver array.

Rabinowitz and Steinberg (1990) considered the problem of optimizing sites for a seismograph network using a similar approach called the *D*-optimality criterion. This criterion maximizes the value of det(A^TA), where A is the Fréchet-derivative matrix; a station configuration that maximizes this is called *D*-optimal. They showed that, for a particular hypocentre, the *D*-optimal configuration must place all stations on concentric circles about the epicentre, with stations on each circle equidistant from each other.

Eaton and Forouhideh (2011) considered the problem of network design for moment-tensor inversion. The condition number of the generalized inverse matrix provides a simple measure for the stability of the inversion. Eaton and Forouhideh (2011) showed that the condition number varies approximately inversely with the solid angle subtended by the

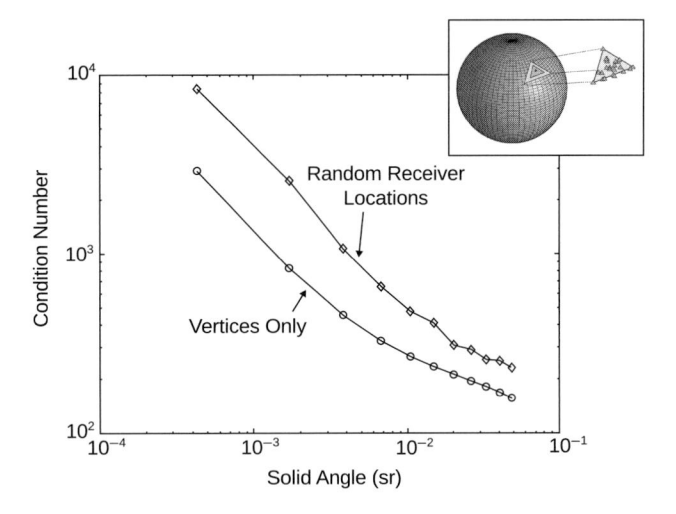

Fig. 5.18 Graph of condition number for moment-tensor inversion versus solid angle subtended by a receiver array within a triangular patch (inset). Top curve shows results for 20 random receivers plus three receivers at the vertices. Lower curve shows the results when receivers are positioned only at the vertices. In general, condition number decreases with increasing solid angle. Modified from Eaton and Forouhideh (2011), with permission of the Society of Exploration Geophysicists.

receiver array (Figure 5.18). For reference, solid angle is an angular measure that is used in 3-D geometry. It has units of steradians (sr), which are the 3-D equivalent of radians. For an entire sphere the solid angle is 4π sr.

The magnitude of completeness (M_c), which was defined early in this chapter as the magnitude above which there is a 90 percent chance for event detection, provides another measure of seismic-network performance (Schultz and Stern, 2015). Although methods for determining M_c are generally based on *a posteriori* catalog data (Woessner and Wiemer, 2005; Schorlemmer and Woessner, 2008), this parameter can also be used for optimizing seismograph network design using a simplified form of the seismic network evaluation through simulation (SNES) approach (Biryukov, 2016). This study used the ground-motion prediction equation (GMPE) of Atkinson et al. (2014) to compute the pseudoacceleration amplitude (PSA) of ground motion with period T at distance R:

$$\log_{10} \text{PSA}_T = C_T + 1.45M - \log_{10} Z(R) - \gamma_T R, \tag{5.11}$$

where C_T is a calibration constant, γ_T is the coefficient of anelastic attenuation and $Z(R)$ is the geometrical attenuation, given by

$$\log_{10} Z(R) = \begin{cases} 1.3 \log_{10} R & \text{if } R \leq 50; \\ 1.3 \log_{10} 50 + 0.5 \log_{10} R/50 & \text{if } R > 50. \end{cases} \tag{5.12}$$

In these expressions, R is expressed in km and PSA is in cm/s^2. A period of $T = 0.3$ s was used, which is typical for small-to-medium events (Atkinson et al., 2014). The criterion for a single-station detection was that the calculated PSA from Equations 5.12 and 5.13 had to

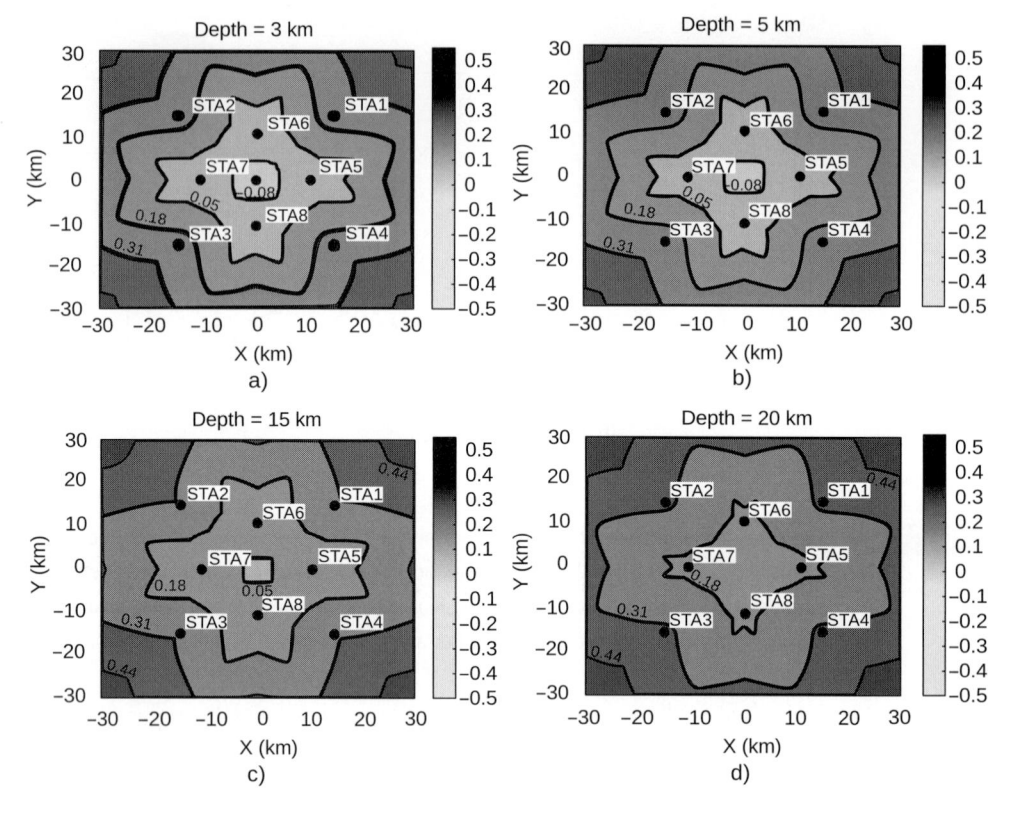

Fig. 5.19 The distribution of M_c as a function of the earthquake depth. Note the apparent shrinking of low M_c toward the network's centre of symmetry. From Biryukov (2016), with permission of the author.

be at least ten times greater than the PSA from Figure 5.15 at $T = 0.3$s. For an event to be deemed to be detected, this criterion had to be met at four or more stations.

Figure 5.19 shows the results of SNES calculations with varying focal depth from 3 km to 20 km. The distribution of stations is configured as a simple D-optimal network with nine stations. For a shallow focal depth (3 km), the magnitude of completeness shows a strong variability from the centre of the network ($M_c = -0.08$ to the edge of the grid where $M_c > 0.31$. This variability decreases with increasing focal depth, for the simple reason that the hypocentral distance is less variable for a deeper event. Figure 5.20 shows the results of SNES calculations under various values of the number of stations deemed to be required for hypocentre location.

5.6 Summary

The scientific and engineering foundations for hydraulic-fracture monitoring (HFM) using microseismic methods were established during the 1970s to 1990s by the Hot Dry Rock (HDR) experiment at Fenton Hill, New Mexico (Albright and Pearson, 1982), the M-Site project in the Piceance Basin, Colorado (Warpinski et al., 1998) and the

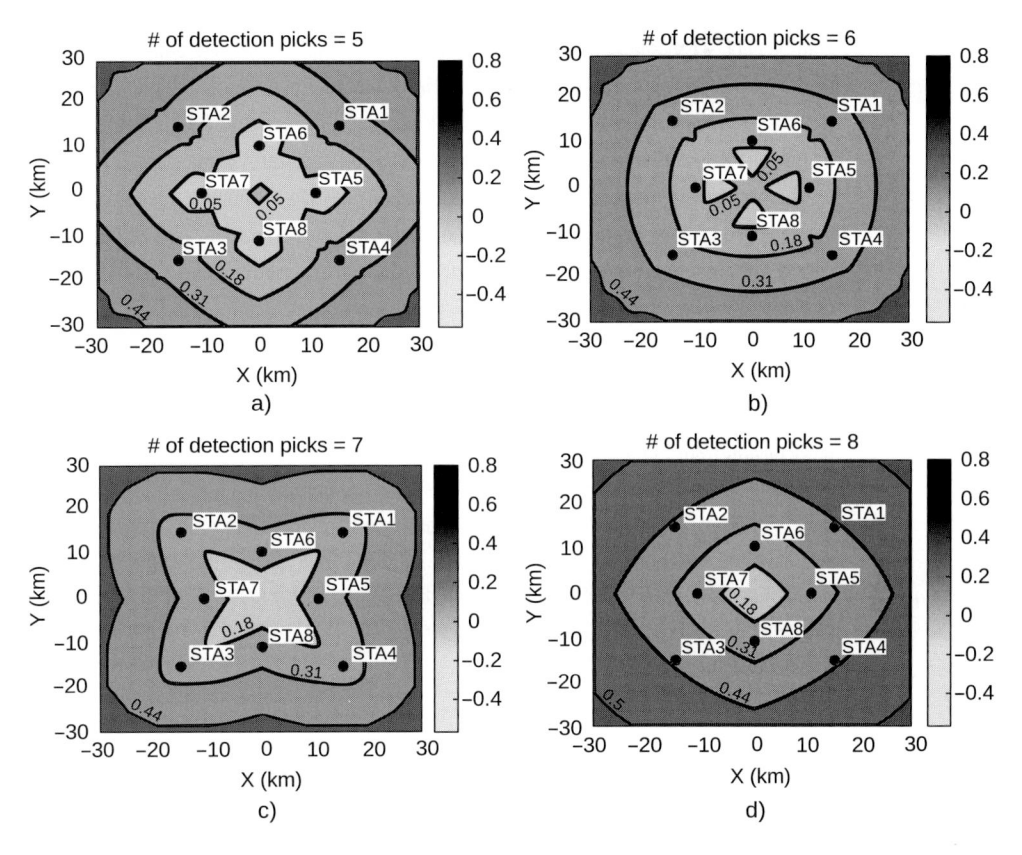

Fig. 5.20 Induced seismicity network design showing the distribution of magnitude of completeness as a function of the number of stations required for hypocentre location. From Biryukov (2016), with permission of the author.

Cotton Valley experiment in Texas (Walker, 1997; Rutledge and Phillips, 2003). Lessons learned from these early studies contributed to the initial large-scale commercial success of microseismic monitoring, in the Barnett play in Texas (Maxwell et al., 2002).

A number of distinct sensor configurations are in current use for HFM. The main objective of deep-downhole monitoring is to install sensors in close proximity to the injection zone, in order to minimize the effects of attenuation and surface noise. Deep-downhole arrays typically use 8–36 multicomponent geophones deployed in one or more observation wells. Surface arrays use a large number of sensors, often along radial arms centred on the well pad, in order to optimize noise reduction by stacking using linear arrays. The overall array diameter is usually designed to be about twice the target depth (Duncan and Eisner, 2010). In some cases, a patch design is employed at the surface, which enables 2-D array filtering of surface noise (Pandolfi et al., 2013; Roux et al., 2014). In the case of shallow-well arrays, sensors are installed below the overburden, which greatly reduces the effects of surface noise as well as high attenuation that typifies the near-surface environment. Regional seismograph networks use seismometers, rather than geophones, in order to obtain an instrumental response that is suited to the analysis of earthquakes with magnitude greater than zero.

Sensors used for passive-seismic monitoring have application-specific frequency response. Geophones are motion-sensitive transducers that use a mass-spring system. A seismometer is a capacitive force-balance device.

Traditionally, seismograph instrumentation has been split into short-period and long-period bands based on a global microseismic noise peak at about 0.15 Hz. Other types of instrumentation for passive seismic monitoring include accelerometers and distributed acoustic sensing (DAS) systems. The instrument response of sensors can be fully characterized by a sensitivity scalar as well as poles and zeros of the transfer function.

The background model that is used for processing passive-seismic data requires calibration using sources with known locations. Examples of calibration sources include perforation shots, string shots and sleeve-opening events. The background model may include density as well as anisotropic and attenuation parameters, in addition to P- and S-wave velocity models.

Random or coherent noise from different sources has a negative impact on microseismic recordings. Cultural and wind noise can be mitigated by appropriate placement of sensors. Reference noise spectra, known as the New High Noise Model (NHNM) and New Low Noise Model (NLNM) were developed by Peterson (1993) based on noise records from 75 digital stations worldwide.

Various approaches can be used for survey-design optimization. Checkerboard tests make use of a grid of source locations, which are inverted with various levels additive noise to investigate uncertainty based on the scatter in the inversion results (Martakis et al., 2006). Other optimization approaches include the application of singular-value decomposition (SVD) to the Fréchet-derivative matrix (Grechka, 2010), condition number of the generalized inverse matrix (Eaton and Forouhideh, 2011) and the D-optimality criterion for array design (Rabinowitz and Steinberg, 1990).

5.7 Suggestions for Further Reading

- Microseismic survey design and instrumentation: Maxwell (2014).
- Seismograph network design and instrumentation: Lee et al. (2002).

5.8 Problems

1. Use Equations 5.1–5.3 to estimate the minimum magnitude for a locatable event, at a distance of $r = 500$ m and a frequency of $f_0 = 15$ Hz. Start by calculating the seismic moment, assuming that the amplitude with 10% probability of detection (A_{10}) is 0.1μm; then calculate the corresponding moment magnitude using Equation 5.3. Use the following parameters: $v_P = 4000$ m/s, $v_S = 2000$ m/s, $\rho = 2500$ kg/m^3, $Q_P = 200$ and $Q_S = 100$.

2. Design a linear geophone array using Equation 5.4, for which the first notch rejects horizontally propagating waves such as ground roll with a frequency $f = 10$ Hz and an apparent horizontal velocity of 500 m/s.

3. Suppose that the mass element of a 10 Hz geophone is 100 g.

a) What is the spring constant, k?
b) Assuming that the geophone is critically damped and has a sensitivity of $S_g = 1$, what is the amplitude and phase of the transfer function at 2 Hz and 20 Hz?

4. A seismometer has a sensitivity of 1000 V/m/s, two poles at $+/-$ 4.4 Hz and two zeros at 0 Hz. Calculate the amplitude and phase of the instrument response at 2 Hz and 20 Hz.

5. Consider a four-station surface network, at locations that are 3.0 km to the north, south, east and west of an event with hypocentre at 4000 m depth. Assume that the medium is homogeneous and isotropic with velocity V.

a) Calculate the Fréchet derivative matrix of traveltimes, $A_{ij} = \frac{\partial t_i}{\partial x_j}$. Here, i denotes the station number and $j = 1, 2, 3$ denotes x, y and z location indices for the source.

b) Calculate the determinant $\det(\mathbf{A}^T\mathbf{A})$, which is the basis for the D-optimality criterion for survey design.

6. **Online exercises**. A set of Matlab tools is provided to aid in visualization of uncertainty for parameters associated with the design of a passive-seismic experiment. These tools can be used to create graphs of seismic-detection probability versus distance, as a function of injected fluid volume, stochastic diagrams of location uncertainty and condition number for moment-tensor inversion.

6 Downhole Microseismic Processing

> We are what we repeatedly do. Excellence, then, is not an act, but a habit.
> Aristotle, paraphrased by Durant (The Story of Philosophy, 1926)

Downhole microseismic data are distinct from surface and near-surface recordings inasmuch as the sensors are generally installed in relatively close proximity to the treatment zone. This characteristic is both a help and a hindrance. The close proximity to the source region, coupled with the avoidance of highly attenuating near-surface layers, typically results in high-fidelity waveforms from events that might otherwise be undetectable; but, the constraints imposed by limited availability of deep, nearby wellbores means that the acquisition geometry is rarely close to ideal. In many cases, an expedient but sub-optimal configuration is employed with a single vertical observation well; it then becomes a matter of doing more with less.

Determining hypocentre locations using downhole microseismic data involves phase picking and quality control (QC) of P and S arrivals,[1] estimation of hypocentral distance using the picked arrival times, and estimation of backazimuth based on waveform polarization. If more than one observation well is available, there are additional requirements for correlating observations between different wells (Warpinski et al., 2005a). While the need to pick P and S arrivals is shared by hypocentre location methods used for local or regional seismograph networks, there are important differences. For the most part, seismograph networks are deployed at the surface and sample a substantially greater aperture of the focal sphere. The techniques used for downhole microseismic are thus more akin to the seldom-used method of hypocentre location with a single seismograph station (Roberts et al., 1989; Farahbod et al., 2015b).

The goal of microseismic data processing is to transform continuous wavefield observations into precise and accurate estimates of event locations, magnitudes and other source characteristics. Figure 6.1 shows a simplified flow chart for the processing of downhole microseismic data after the monitoring program is complete. The basic processing sequence consists of two parallel workflows:

- a primary workflow, in which the waveform data are processed to produce an event catalogue;
- a secondary workflow that is used for estimation of sensor orientations, as well as construction and validation of a calibrated background model.

---

[1] The above quotation may resonate for anyone who has made a substantial time investment picking seismic waveforms.

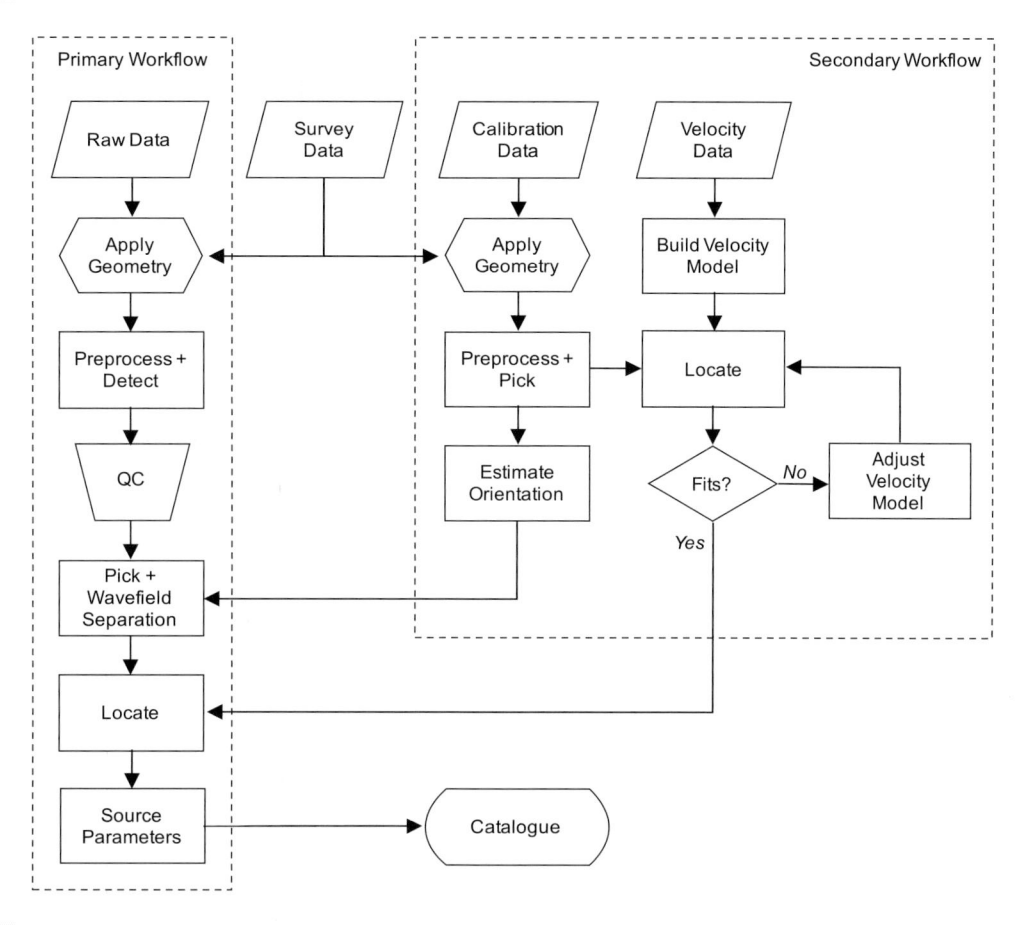

Fig. 6.1 Basic workflow for processing downhole microseismic data. QC denotes quality control.

There are four input data components for this workflow: raw digital waveform data, covering one or more time windows during the period of acquisition; survey data, containing positional information for the sensors, as well as observation and treatment wells; calibration data, which is often a subset of the raw data; and velocity data, consisting of well logs and/or other sources of information for the background medium. Raw microseismic time series require signal conditioning prior to the determination of hypocentres and source characteristics (Maxwell, 2014). The initial processing steps involve transformation of field coordinates into a fixed geographic reference frame, noise attenuation, signal detection, phase picking and rotation into ray-centred coordinates to produce a set of event files. In parallel with the primary processing workflow, it is necessary to construct a calibrated background model. Methods such as matched filtering (Eaton and Caffagni, 2015) or subspace detection (Harris, 2006) can be applied for detection of weak events with low S/N that may otherwise be missed, in which case automated methods can be used to determine relative hypocentre locations that are anchored to independently determined absolute locations of template events (Caffagni et al., 2016).

In the case of real-time processing carried out during hydraulic-fracturing operations, the basic workflow is essentially the same as shown in Figure 6.1 – although for real-time processing there is a greater emphasis on computational efficiency, in order to achieve sufficiently fast turnaround in each step. It is important, in this case, to have the velocity model prepared in advance. In practice, real-time processing can be carried out on-site, or via telemetry to a central processing facility. Real-time processing can be valuable as a decision-making tool, including identification of problems such as anomalous induced seismicity (Chapter 9) that may occur during well completion.

This chapter covers the principal steps in the overall workflow, from input data to generation of the final event catalogue, along with an alternative workflow that describes how template-based approaches for event detection can be incorporated into the data processing stream. Pertinent theoretical and practical aspects are summarized throughout. When processing is complete, event information is organized into a catalogue and used in conjunction with injection data, geomechanical simulation and other fracture diagnostics to interpret stimulation effectiveness. Evaluation of uncertainties in all parameters is a critical component of these processes.

6.1 Input Data

Raw digital waveform data from downhole microseismic monitoring of hydraulic fracturing are customarily stored in a record-based data format used by the petroleum industry for seismic reflection surveys, rather than a format designed for continuous waveform observations. Although there are well-established standards for exchange of data from earthquake networks, there is no standardized format for digital microseismic data and several common digital data formats are described in Appendix C. In general, commonly used data formats cannot accommodate long time-windows within a single file; consequently, raw microseismic data are typically stored in short-duration files (e.g. 1 minute) that must be organized and concatenated to produce a continuous time series.

The survey data include all applicable locations and deviation surveys for observation and treatment wells, sensor arrays and perforations, as well as information about the treatment schedule and target formation. Positions are commonly reported using a Universal Transverse Mercator (UTM) projection; care is required with UTM data, as the use of different ellipsoids (NAD27, NAD83, etc.) can lead to positional errors on the order of 10s m. The minimum data requirements for survey data are summarized in a set of guidelines for standard deliverables for microseismic monitoring of hydraulic fracturing prepared by the Canadian Society of Exploration Geophysicists.

The calibration data consist of waveforms, positional information and (approximate) timing for calibration sources, such as perforations, sleeve-opening events, string shots, surface sources, etc. As elaborated below, these data are used for two purposes: to determine the orientation of the downhole toolstring, and for calibration of the background model.

The velocity data are usually derived from well-log measurements. As mentioned in the previous chapter, ideally data from the observation well and treatment well should be incorporated into the model. In general, the background model is not strictly limited to P- and S-wave velocity, but may include attenuation and anisotropy parameters. In addition to well logs, seismic-reflection and refraction data may also provide useful sources of information.

6.2 Coordinate Systems and Transformations

Geometrical information about the monitoring program, such as injection locations and receiver locations and orientations, need to be established for use throughout the data-processing sequence. Three-component geophones measure ground velocity in mutually orthogonal directions. These directions form a local coordinate system, with one axis, denoted $\hat{\zeta}(\mathbf{x})$, oriented axially along the wellbore and two others, denoted $\hat{\psi}_1(\mathbf{x})$ and $\hat{\psi}_2(\mathbf{x})$, within the plane that is locally normal to the wellbore (Figure 6.2). In the case of a vertical observation wellbore, $\hat{\zeta}$ is always vertical, while $\hat{\psi}_1(\mathbf{x})$ and $\hat{\psi}_2(\mathbf{x})$ are horizontally oriented; however, in many cases the observation well is deviated. Furthermore, the azimuths of the unit vectors $\hat{\psi}_1$ and $\hat{\psi}_2$ are generally not known a priori, as the toolstring rotates during installation as a result of unavoidable twisting due to torque on the wireline. Consequently, except in rare cases where a special device (e.g. gyroscope) is installed in the sensor package, it is necessary to estimate the tool orientation using calibration sources.

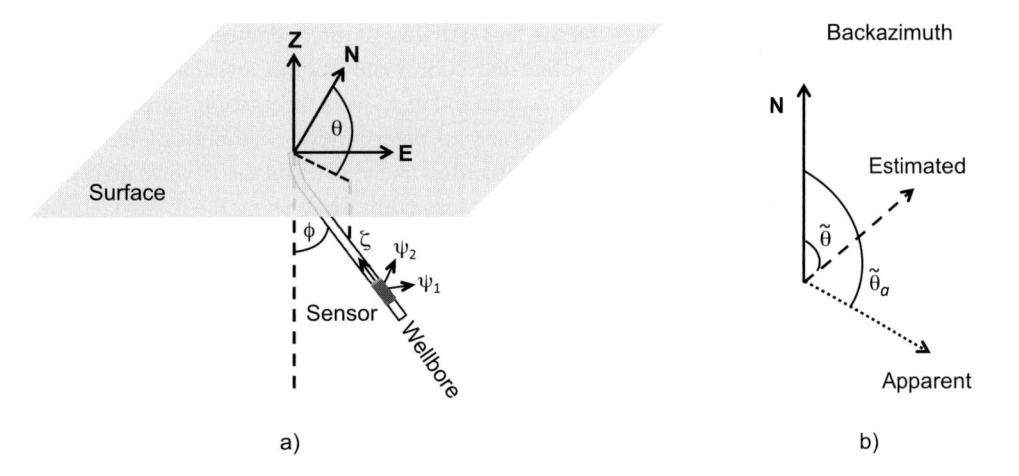

Fig. 6.2 Coordinate system and sensor orientation for downhole monitoring in a deviated wellbore, defined by angular parameters θ and ϕ. Geographic coordinates (E–N–Z) are used as a reference system. a) Three-component sensors have one component ($\hat{\zeta}$) oriented along the axis of the wellbore, with the other two ($\hat{\psi}_1$ and $\hat{\psi}_2$) in a plane that is locally perpendicular to the wellbore. Due to typically unknown rotation of the wireline during installation, the absolute orientations of the $\hat{\psi}_1$ and $\hat{\psi}_2$ components must be determined by calibration sources, unless a special device (e.g. gyroscope) is available to measure tool orientation. b) Map view, showing estimated and apparent backazimuth directions, as well as angles used for co-ordinate rotation.

Prior to further processing, a coordinate transformation is required in order to bring the measured ground motions into a geographic coordinate system, such as east–north–vertical (Figure 6.2). The wireline depth (distance along the borehole) of every sensor is determined during the installation procedure, including small corrections for wireline stretching. In general, every wellbore is surveyed after it is drilled and this deviation survey can be used to map wireline depth into geographic coordinates, $\{x_1, x_2, x_3\}$ or easting, northing and elevation. The orientation of the $\hat{\zeta}(\mathbf{x})$ component can thus be obtained unambiguously using the well deviation survey, since it falls along the axis of the wellbore.

Before any rotation operation or further processing is performed, it is good practice to preprocess (precondition) the data (Maxwell, 2014; Akram and Eaton, 2016b). This step typically includes the application of a *DC-shift* by independently subtracting the mean value from each component, which may have a large (and spurious) amplitude relative to the waveform amplitudes. It is then useful to perform a trend removal and to "de-glitch" the data by removing spurious large-amplitude spikes and interpolating any missing data samples within short time segments (Tapley and Tull, 1992).

Once preconditioning has been applied, a general transformation between Cartesian co-ordinate systems is performed using rotation operators (see Box 1.1) in a cascading sequence. If a wellbore is neither vertical nor horizontal, an initial transformation is required, in which a rotation is applied around a horizontal axis (e.g. north) in order to account for wellbore inclination. The resulting co-ordinate axes are denoted here as $\hat{h}_1(\mathbf{x})$, $\hat{h}_2(\mathbf{x})$ and $\hat{z}$, where the two provisional horizontal axes $\hat{h}_1$ and $\hat{h}_2$ have unknown orientation and are different for each receiver location. Here, this is referred to as an intermediate reference frame. In the case of a vertical observation well the raw data are acquired within this reference frame and the first transformation operation is not needed.

The next step is to rotate the coordinate system around the vertical axis in order to achieve a consistent north–south and east–west orientation for every receiver location. To perform this step, it is necessary to determine the azimuth of the $\hat{h}_1$ and $\hat{h}_2$ axes. This can be achieved based on polarization analysis using the *P*-wave arrival from a calibration source with a known location, or less commonly using the *S*-wave (Eisner et al., 2009a). For simplicity, only the *P*-wave method is considered here.

A method to perform polarization analysis is illustrated in Figure 6.3. A semi-automatic procedure has been developed by De Meersman et al. (2006), which uses a noise-weighted analytical signal method (see Appendix B). A simplified approach is described here, for purposes of illustration. First, the picked arrival times are used to select a time window following the phase arrival. The length of the time window, shown by the black bars in Figure 6.3, should be selected such that 1–2 pulse cycles are bracketed. The next step is to compute *hodograms* for each sensor, where a hodogram is simply a cross-plot of two orthogonal components of ground motion. In Figure 6.3, hodograms are shown in the horizontal and vertical planes. As described in Chapter 3, in an elastic medium the particle motion associated the passage of a plane body wave is linear and can be represented by eigenvectors of the Christoffel matrix. In the case of a *P* wave in an isotropic medium, the particle motion is perpendicular to the wavefront and parallel to the direction of the ray; in the case of an *S* wave in an isotropic medium, the particle motion is in the plane of the wavefront. For anisotropic media, the polarization of body waves is still expected to be

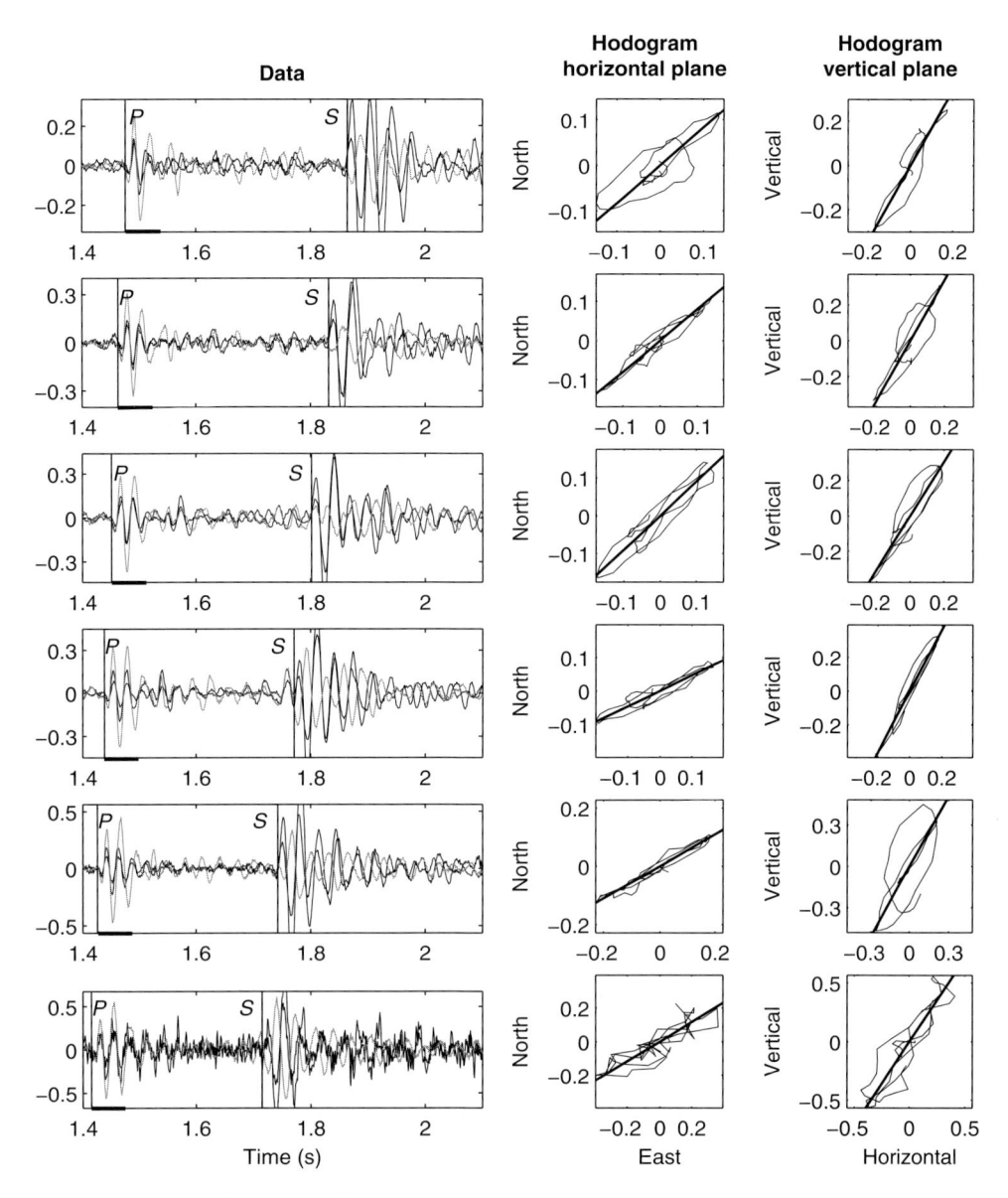

Fig. 6.3 Polarization analysis of three-component observations using a semi-automated approach. The panels on the left show three-component waveforms from a downhole array, with receivers numbered 1 to 6 from top to bottom. The hodogram panels on the right show crossplots of 60 ms of data, for the time windows marked by the bold horizontal lines below the P-wave picks on the left. The hodograms are overlain by automatically determined P-wave polarization vectors. Reproduced with permission from K. de Meersman, M. van der Baan and J.-M. Kendall, Signal Extraction and Automated Polarization Analysis of Multicomponent Array Data, *Bulletin of the Seismological Society of America*, Vol. 96, 2415–2430, 2006. Copyright Seismological Society of America.

linear but may not have exactly the same geometrical relationship to the ray as for isotropic elastic media. Furthermore, in the case of split shear waves, the interference between fast and slow shear waves can result in elliptical motion if the time separation between the two qS arrivals is less than the pulse length (Silver and Chan, 1991).

The best-fitting linear polarization vector within a selected time window can be obtained by constructing the *waveform covariance matrix*, given by

$$\mathbf{C} = \begin{bmatrix} C_{11} & C_{12} & C_{13} \\ C_{21} & C_{22} & C_{23} \\ C_{31} & C_{32} & C_{33} \end{bmatrix}. \quad (6.1)$$

The elements of $\mathbf{C}$ are the auto- and cross-variances of the three components of ground motion, after pre-conditioning and transformation into the intermediate reference frame. For a window with N time samples, the covariance matrix elements are defined as (Jurkevics, 1988)

$$C_{jk} = \frac{1}{N} \sum_{i=1}^{N} u_{ij} u_{ik}, \quad (6.2)$$

where u_{ij} denotes the ith time sample of the jth component in the time window. The eigenvectors of $\mathbf{C}$ define the three principal axes of the *polarization ellipsoid*, while the corresponding eigenvalues define the axis lengths (Jurkevics, 1988). In principle, for pure rectilinear motion there is only one nonzero eigenvalue, and for 2-D elliptical motion there are two nonzero eigenvalues. In practice, all three of the eigenvalues, $\lambda_k, k = 1, 2, 3$, are generally nonzero and the particle motion is ellipsoidal (Jurkevics, 1988). In the presence of random noise, the principal eigenvector of the covariance matrix provides an optimal estimate of the apparent linear polarization direction.

Let $\hat{\gamma}_{ij}$ represent the ith component of the unit eigenvector of the waveform covariance matrix, corresponding to the eigenvalue λ_j where the eigenvalues are ordered $\lambda_1 \geq \lambda_2 \geq \lambda_3$. The apparent polarization azimuth $\tilde{\theta}_a$ is given by

$$\tilde{\theta}_a = \arctan\left[\hat{\gamma}_{11}\mathrm{sign}(\hat{\gamma}_{31}), \hat{\gamma}_{21}\mathrm{sign}(\hat{\gamma}_{31})\right], \quad (6.3)$$

where the sign of the vertical component is introduced to eliminate the 180° ambiguity. The dip of the signal polarization $\tilde{\phi}_a$ is given by

$$\tilde{\phi}_a = \cos^{-1} \hat{\gamma}_{31}. \quad (6.4)$$

Once the apparent polarization azimuth of the P-wave polarization is known, or alternatively the S-wave backazimuth is determined (Eisner et al., 2009a), a transformation can be applied to go from the intermediate reference frame into a geographic reference frame. For each receiver, this transformation is applied as rotation through the angle $\tilde{\theta} - \tilde{\theta}_a$ (Figure 6.2b), where $\tilde{\theta}$ is the estimated backazimuth from the receiver to a calibration source location. If multiple calibration sources are available, the average value can be used.

The final rotation transformation thus requires an estimate of the true backazimuth. In general, if the background velocity model is known a priori, which unfortunately is a rare situation, then the expected backazimuth and/or polarization orientation can be obtained by ray tracing. A much simpler method is commonly used in practice, based on an assumption that the velocity model can be approximated as a horizontally stratified medium composed of isotropic layers. In this case, the estimated backazimuth is given by

$$\tilde{\theta} = \arctan\left(x_c - x_r, y_c - y_r\right), \quad (6.5)$$

where $\{x_c, y_c\}$ is the east–north geographic location of the calibration source and $\{x_r, y_r\}$ is the east–north geographic location of the receiver. Another consideration is the polarity of the sensor; for example, does positive amplitude on the $\hat{\zeta}$-component correspond to motion that is upwards or downwards? This polarity information can be obtained empirically during installation using a so-called "tap" test.

From the foregoing discussion, it should be clear that:

1 considerable care is required to transform the raw data into a proper geographic frame of reference;
2 propagation of errors in each step will introduce azimuthal uncertainty, which will impact the uncertainty in event locations (Feroz and Van der Baan, 2013).

6.3 Event Detection and Arrival-Time Picking

Event detection is a critical aspect of the processing workflow. In theory, event detectors can be viewed as the implementation of a binary hypothesis test for the presence or absence of a signal within a set of observations (Van Trees, 1968; Harris, 2006). In this paradigm, the test chooses between a hypothesis that signal is present and the null hypothesis that only noise is present.

Most hydraulic-fracture monitoring surveys produce catalogues containing hundreds or thousands of event detections. In view of this multiplicity of events, along with the typical low S/N of waveforms and the occurrence of other types of coherent arrivals that could be misidentified as microseismic events, the use of automated procedures rather than interactive picking has considerable practical value. In addition to overcoming the highly cumbersome nature of interactive picking, automated methods have the desirable qualities of consistency and repeatability. A few representative algorithms for automated event detection and arrival-time picking are described here; for a more extensive set of methods and a comprehensive literature review, see Akram and Eaton (2016b).

Although similar algorithms have been developed for the distinct tasks of event detection and arrival-time picking, it is often more efficient to perform these processes separately. Within the primary workflow for downhole microseismic processing presented in Figure 6.1, event detection, interactive quality control (QC) and arrival-time picking are carried out sequentially. In general, validation and selection of events for further processing can be a very time-consuming step. One reason to break these into separate steps is to avoid unnecessary effort picking arrival times for spurious detections, such as very noisy events or tube waves (see Chapter 5). Tube waves, in particular, are problematic for automated processing methods because they are repetitive and have high amplitude and high S/N (St-Onge and Eaton, 2011). These signals are easily recognized, however, based on their linear moveout and apparent velocity equal to the acoustic wavespeed in water, which is $\approx$ 1500 m/s.

In practice, prior to passing continuous waveform data on to modules for event detection or arrival-time picking, it can be helpful to apply an edge-preserving smoothing filter (Luo et al., 2002) and/or a bandpass filter as a preprocessing step. The underlying philosophy is to maximize the S/N by removing frequency components that fall outside the main bandwidth of the source. If a bandpass filter is applied to the data, good practice dictates that an unfiltered version of the data is archived and the filter has a minimum-phase response (see Appendix B) to ensure that arrival-time picks are not contaminated by precursory side-lobes due to the *Gibbs phenomenon* (Maxwell, 2014).

6.3.1 Single Receiver Methods

We begin by considering methods that use data from a single receiver, which may have one or more components. The *short-time average/long-time average* (STA/LTA) method (Earle and Shearer, 1994) is a straightforward and robust technique that is extensively used for event detection. For the ith data sample of a time series **u**, generalized expressions for the STA and LTA parameters are given by (Akram and Eaton, 2016b)

$$\text{STA}_i(\mathbf{u}) = \frac{1}{N_S} \sum_{j=i}^{i+N_S-1} \text{CF}_j(\mathbf{u}), \tag{6.6}$$

and

$$\text{LTA}_i(\mathbf{u}) = \frac{1}{N_L} \sum_{j=i-N_L+1}^{i} \text{CF}_j(\mathbf{u}), \tag{6.7}$$

where N_S and N_L denote the number of samples in the short and long windows, respectively. As elaborated below, $\mathbf{CF}(\mathbf{u})$ is a characteristic function that provides a measure of signal amplitude or energy within a time window. The value of the STA/LTA parameter at the ith sample is

$$\text{STA/LTA}_i(N_S, N_L) = \frac{\text{STA}_i}{\text{LTA}_i}. \tag{6.8}$$

The STA/LTA parameter is a robust signal detector, as it provides a time-dependent measure of S/N (Akram and Eaton, 2016b). As illustrated in Figure 6.4, exceedance of a threshold value provides a simple criterion for signal detection. Based on the principle of causality, the STA window should lead the LTA window; it is also desirable to avoid overlap between STA and LTA time windows in order to ensure statistical independence between these two measures (Taylor et al., 2010). Rather than using the maximum of the STA/LTA function at $t = t_{max}$, a good estimate of the arrival time is given by the maximum of the derivative of STA/LTA that occurs immediately prior to t_{max} (Akram and Eaton, 2016b).

Figure 6.4 shows a single-component waveform example for which the characteristic function is represented by the RMS amplitude (Equation 5.10). For multicomponent data, other options can be used; for example, Saari (1991) used the absolute value of the product

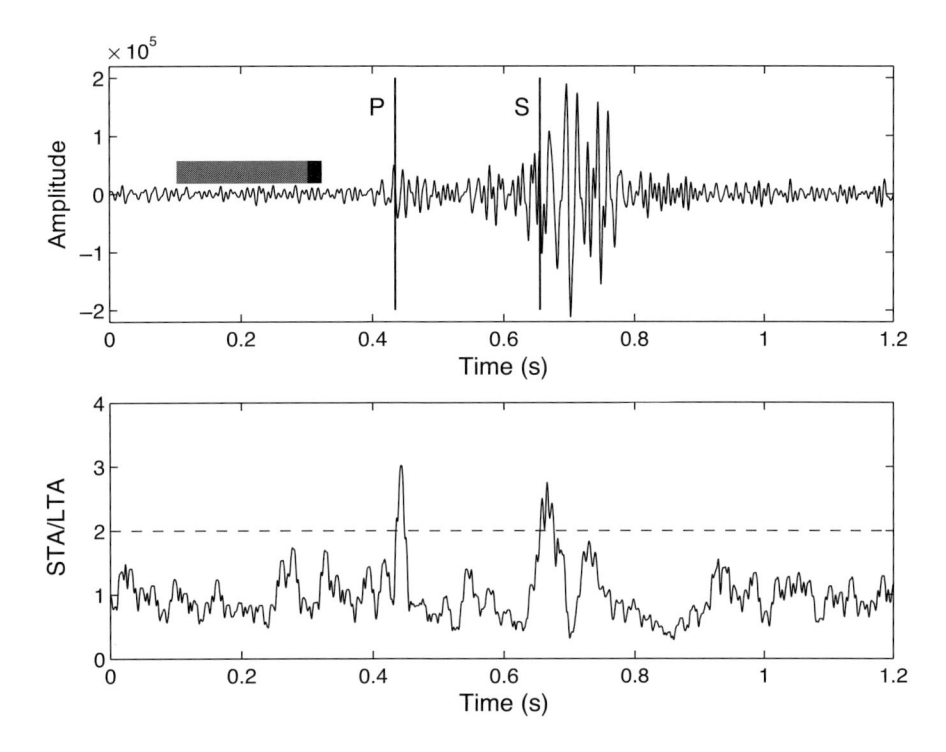

Fig. 6.4 Short-time average/long-time average method for automatic event detection and time picking. Upper panel shows a sample vertical-component waveform from a representative microseismic event. The grey and black regions mark the long-time average (LTA) and short-time average (STA) time windows, respectively, that are used to calculate the STA/LTA ratio. The lower panel shows the calculated STA/LTA time series for time windows of length 0.1000s and 0.0125s. The P- and S-wave picks are indicated based on a unitless detection threshold of 2.0.

of the 3-C amplitudes as a characteristic function, in order to increase the S/N by damping random noise, while Oye and Roth (2003) used the stack of absolute values within the STA and LTA time window.

Apart from the choice of characteristic function, performance of the STA/LTA algorithm depends upon the choice of three parameters, namely N_S, N_L and the detection threshold. Akram and Eaton (2016b) recommend a STA window that is 2–3 times the dominant period of the source pulse, and a LTA window that is 5–10 times longer than the STA window. The choice of detection threshold is dataset dependent and varies with S/N, so trials using a small subset of the data are recommended. The STA/LTA algorithm is well suited for real-time microseismic monitoring because it is robust, fast, and capable of simultaneously performing event detection and arrival-time picking (Akram and Eaton, 2016b).

A different single-component characteristic function proposed by Li et al. (2014) is based on the fourth statistical moment, called the *kurtosis*. For a N-sample time window of a waveform $\mathbf{u}$, the ith sample of the kurtosis function is given by

$$K_i(\mathbf{u}) = \frac{\sum_{j_1(i)}^{j_1(i)+N-1} (u_j - \bar{u}_i)^4}{N\sigma_i^4}, \tag{6.9}$$

where $\bar{u}_i$ is the mean of the distribution,

$$\bar{u}_i = \frac{\sum_{j_1(i)}^{j_1(i)+N-1} u_j}{N}, \qquad (6.10)$$

and σ_i^2 is the variance

$$\sigma_i^2 = \frac{\sum_{j_1(i)}^{j_1(i)+N-1}(u_j - \bar{u}_i)}{N}. \qquad (6.11)$$

Along the same lines as the STA/LTA method, the parameter N is assigned different values for a short window (N_S) or a long window (N_L); similarly, the summation index $j_1(i)$ is equal to i for a leading (short) window and $i - N$ for a trailing (long) window. Signal detection using this approach, called STK/LTK, is based on the expression

$$\mathrm{STK/LTK}_i(N_S, N_L, \epsilon) = \frac{\mathrm{STK}_i}{\mathrm{LTK}_i + \epsilon}, \qquad (6.12)$$

where ϵ is a small value used for numerical stability (Li et al., 2014). Kurtosis is essentially a measure of the "tailedness" of the probability distribution, that is the propensity to produce outliers (Westfall, 2014). If the background noise is random with a Gaussian amplitude probability distribution, then the noise would have a normative kurtosis level of 3; the use of STK/LTK as a signal detector thus amounts to a search for signals with distinctly different (non-Gaussian) kurtosis signatures (Li et al., 2014). Parameter choices for short- and long-time window lengths and detection threshold parallel the approach described above for STA/LTA.

STA/LTA and STK/LTK are two of many single-channel methods that can be used as triggers for event detection (Akram and Eaton, 2016b). A detection that is triggered on only one or a few receiver locations, however, is unlikely to be of interest for further processing. Automated procedures are therefore used to filter potential events, based on dastset-specific criteria such as a minimum number of triggers at different sensor locations within a prescribed time interval. QC inspection and further analysis, such as arrival-time picking, is then limited to potential events that meet the detection criteria.

The final single-receiver method considered here is the *Akaike Information Criterion* (AIC), a statistical model that was developed during the 1970s (Akaike, 1998) based on *autoregressive processes*. Like kurtosis, the use of AIC as a detection and arrival-time picking algorithm (Oye and Roth, 2003; St-Onge, 2011) assumes that observed waveforms can be treated as a series of discrete segments with distinct statistical properties. Consider the simple case of a time series with two distinct segments, here representative of noise and signal. For the kth data sample of a time series $\mathbf{u}$, the AIC value can be expressed as

$$\mathrm{AIC}_k(\mathbf{u}) = (k - M)\log(\sigma_1^2) + (N - M - k)\log(\sigma_2^2) + C, \qquad (6.13)$$

where N is the number of data samples, M is the order of the autoregressive model and C is a constant (Zhang et al., 2003). In addition, σ_1^2 and σ_2^2 are the variances calculated within the time series intervals (before and after the kth sample) not explained by the autoregressive model (St-Onge, 2011). M can be estimated by trial-and-error, using a time window containing noise (Akram and Eaton, 2016b). If $M \ll N$, however, Equation 6.13 can be approximated by

$$\text{AIC}_k(\mathbf{u}) \simeq k \log\left(\text{var}\left\{u(1,k)\right\}\right) + (N - k - 1) \log\left(\text{var}\left\{u(k+1, N)\right\}\right), \quad (6.14)$$

which is the form of the AIC criterion that is primarily used for seismic arrival-time picking. The optimal separation of the boundary between the two discrete stationary segments, coincides with the time index where the AIC function is minimum (Akram and Eaton, 2016b).

Figure 6.5 shows an example of AIC analysis applied to the same test waveform as in Figure 6.4. The time picks coincide with local minima of the AIC function and are earlier than the picks obtained using STA/LTA (Figure 6.4). The AIC method is regarded to give more accurate picks of signal onset, suitable for determining event locations (Oye and Roth, 2003). In most cases, however, there are multiple local minima in the AIC function and those that represent signal onsets of interest are not always clear. For that reason, AIC works best as a semi-automated approach, coupled with interactive quality control (QC).

6.3.2 Multi-Receiver Methods

We now turn our attention to multi-receiver methods, which exploit the waveform similarity of signals of interest across an array of receivers, in order to perform automatic arrival-time picking and/or refinement of picks generated using other methods. Methods

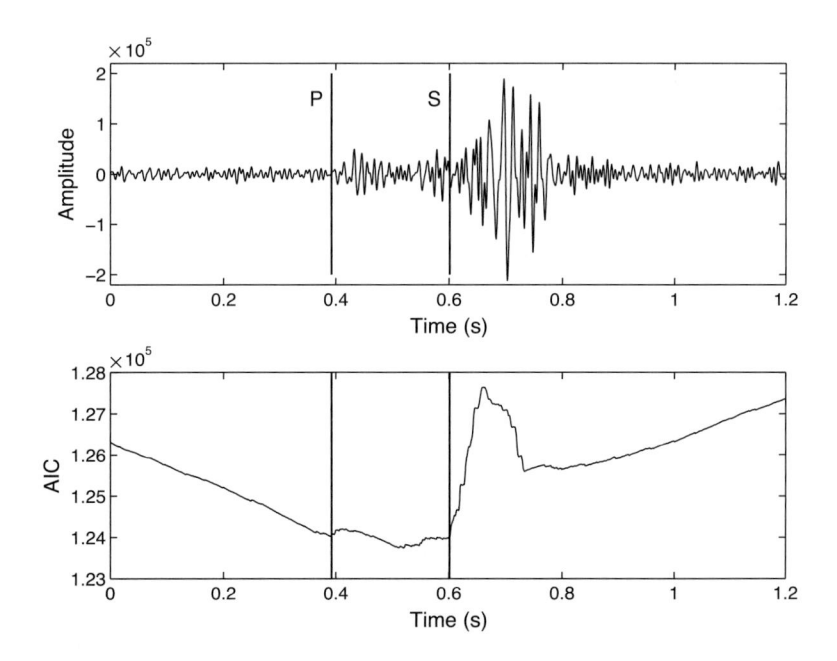

Fig. 6.5 Akaike information criterion method used for semi-automatic time picking. Upper panel shows the same waveform as in the previous figure. Lower panel shows the calculated AIC time series. The P- and S-wave picks are based on local minima.

that are considered here (Irving et al., 2007; De Meersman et al., 2009; Akram and Eaton, 2016a) are based on the signal-processing technique of *cross-correlation*, a tool to obtain a quantitative measure of similarity between two signals as well as an estimate of the lag time between the signals. The *j*th element of the cross-correlation of two discrete time series *a* and *b* is given by

$$\phi_j^{ab} = \sum_{i=-\infty}^{\infty} a_i b_{i+j}, \quad -\infty < j < \infty. \tag{6.15}$$

Similarly, the *j*th element of the *autocorrelation* of *a* is

$$\phi_j^{aa} = \sum_{i=-\infty}^{\infty} a_i a_{i+j}. \tag{6.16}$$

The normalized form of the cross-correlation of *a* and *b* is given by

$$\tilde{\phi}_j^{ab} = \frac{\phi_j^{ab}}{\sqrt{\phi_0^{aa} \phi_0^{bb}}}, \tag{6.17}$$

where ϕ_0^{aa} and ϕ_0^{bb} are the zero-lag autocorrelation coefficients of *a* and *b*, respectively. Normalization of the cross-correlation function facilitates quantitative evaluation of the similarity between two signals. For example, a value of +1 indicates that the signals are identical (perfectly correlated); conversely, a value −1 indicates that the two signals have opposite polarity and are anti-correlated.

The use of cross-correlation to refine time picks is based on a simple model for recorded seismic waveforms as the superposition of a signal component and a noise component. Based on this model, for a set of *M* recordings of a microseismic event, the *k*th seismogram $u_k(t)$ can be expressed as

$$u_k(t) = \alpha_k s(t) + n_k(t), \tag{6.18}$$

where α_k is a complex amplitude factor that incorporates variations between receivers in terms of source radiation pattern, attenuation and wellbore coupling, while $n_k(t)$ is the noise recorded at the *k*th receiver. According to this model, the noise is assumed to be uncorrelated between different receivers.

Based on this simple model, a number of different schemes can be used to determine the time lag between traces based on cross-correlation with a *pilot trace* (Bagaini, 2005). The basic principle for each scheme is that the lag time corresponding to the maximum value of the cross-correlation between two time series represents the optimal time shift to align the signals. Irving et al. (2007) developed an iterative approach for picking first arrivals of single-component waveforms from an active-source crosswell dataset. To pre-condition the signals, they normalized the input waveforms by the maximum signal amplitude. Using the highest S/N waveform as the first pilot trace, Irving et al. (2007) obtained the initial time picks by cross-correlation. In subsequent iterations, the pilot trace was formed by aligning every trace based on the current time pick to generate a stacked trace. The process was repeated until convergence.

De Meersman et al. (2009) used a similar approach to refine time picks for three-component passive-seismic observations. Their method is illustrated in Figure 6.6, which shows a *P*-wave time window for a microseismic event. The column on the left shows the three-component waveforms with no time shift and original time picks indicated. Unlike the method of Irving et al. (2007), the first pilot trace is formed by aligning and stacking the three-component waveforms using the initial time picks. Prior to stacking, the traces are normalized by the amplitude of the pre-event noise window. For each subsequent iteration, optimal time shifts were estimated using three-component cross-correlation to generate a new pilot trace. In this example, there is a clear improvement in the *P*-wave stack after cross-correlation alignment. Iterations continued until the time shifts reduced to zero. This method was used to refine time picks for both *P* and *S* waves, and was found to converge within three or fewer iterations.

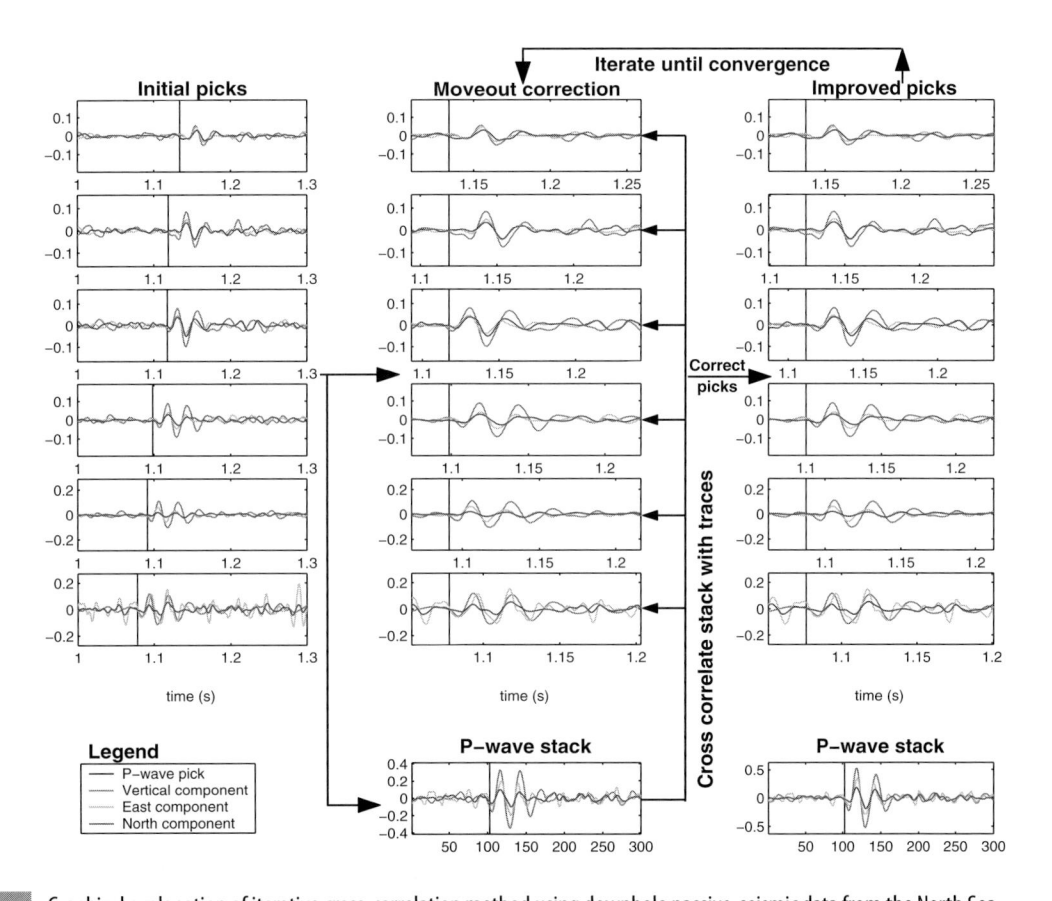

Fig. 6.6 Graphical explanation of iterative cross-correlation method using downhole passive-seismic data from the North Sea. Initial picks are used to align and stack the three-component traces. Picks are refined by cross-correlation with the stacked trace. The process is repeated until the arrival-time corrections converge to a small value. Reprinted from De Meersman et al. (2009), with permission of the Society of Exploration Geophysicists. (A colour version of this figure can be found in the Plates section.)

6.3.3 Wavefield Separation

Resolution of both P and S waves is important for downhole microseismic processing. Wavefield separation of the observed waveforms into P and S components has long been of interest (Dankbaar, 1985). Wavefield separation is a complicated process for surface sensors due to the effects of the free surface (Eaton, 1989), but in principle it can be readily accomplished for observations with downhole sensors that are not close to a high-contrast interface through transformation into ray-centred coordinates.

The concept of ray-centred coordinates is illustrated in Figure 6.7 for a receiver embedded within an anisotropic elastic continuum. As discussed in §3.2.1, for plane-wave propagation in a homogenous anisotropic medium there are three mutually orthogonal polarization vectors that emerge as eigenvectors of the Kelvin–Christoffel matrix. These are shown in Figure 6.7 in relation to the qP wavefront normal, indicated by the large arrow. For weakly anisotropic media, the unit vector $\hat{\gamma}_P$ is expected to be close to parallel with the P raypath (Thomsen, 1986). As discussed in §1.3, in general the unit vectors $\hat{\gamma}_{S1}$ and $\hat{\gamma}_{S2}$ are determined by the effective anisotropic symmetry of the medium. If the medium is isotropic, then the ray-centred coordinate axes are aligned parallel to the wavefront for the P wave, and coplanar with the wavefront such that the S_H axis is horizontal.

In practice, an empirical approach to determine the ray-centred coordinates can be performed by polarization analysis using the procedure described in §6.2 for determining receiver orientation. One approach used in previous studies (Caffagni et al., 2016; Akram and Eaton, 2016a; Van der Baan et al., 2016) to perform wavefield separation assumes that the sensors are located within a homogeneous medium and can be summarized as follows:

1. For each trace, begin by extracting three-component data within an approximately 100 ms time-window around the fast S wave (qS_1), using the previously determined arrival-time picks as the window start time. The S wave is analyzed first to construct the ray-centred coordinates since it typically has a higher S/N than the P wave.

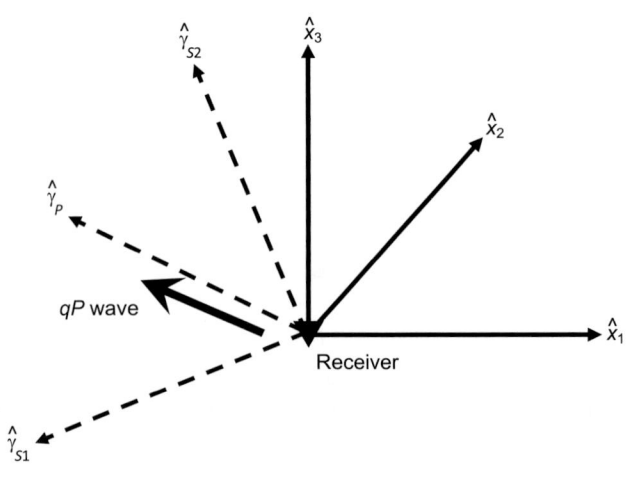

Fig. 6.7 Schematic illustration of ray-centred coordinate system $\{\hat{\gamma}_{S1}, \hat{\gamma}_{S2}, \hat{\gamma}_P\}$ in relation to east–north–elevation geographic coordinates $\{\hat{x}_1, \hat{x}_2, \hat{x}_3\}$.

2. Obtain the best-fitting linear polarization direction for the qS_1 wave using the covariance method described in §6.2. This defines the unit vector $\hat{\gamma}_{S1}$.
3. Use the same approach to obtain the polarization direction for the qP wave, using the P-wave pick times as the start of the time window. The long axis of the polarization ellipsoid in this window is taken as an approximate unit vector for $\hat{\gamma}_P$.
4. If a clear qS_2 wave is discernible in the waveforms, then the same approach can be used to determine an approximate unit vector for $\hat{\gamma}_{S2}$. Alternatively, $\hat{\gamma}_{S2}$ can be approximated using $\hat{\gamma}_P \times \hat{\gamma}_{S1}$.
5. The separated wavefields are obtained by projecting the three-component observations onto the ray-centred axes. For example, for each trace the qS_1 component can be approximated by $\mathbf{u}(t) \cdot \hat{\gamma}_{S1}$.

There are a number of qualitative metrics that can be used to evaluate to the results of this procedure. In practice, the empirical ray-centred coordinates obtained in this manner are not, in general, mutually orthogonal. This lack of orthogonality arises due to many complicating factors affecting field observations including noise, non-ideal sensor coupling, scattering from local heterogeneities, inaccurate picks, etc. The degree of orthogonality between $\hat{\gamma}_{S1}$ and $\hat{\gamma}_P$, however, can be used to assess the effectiveness of the wavefield separation. In addition, for two receivers that are closely spaced (relative to wavelength), large variations in the orientation of ray-centred coordinate axes are inconsistent with the assumption of a homogeneous medium. Thus, strong trace-to-trace variability in the inferred orientation of the axes provides a qualitative indicator that the results are unreliable. Finally, the effectiveness of the wavefield separation can be judged based on the apparent leakage between phases. Thus, if there is little evidence for S wave motion within the P wave window, and vice versa, this is an indicator that wavefield separation has been successfully accomplished.

Figure 6.8 shows a microseismic event on which this procedure has been applied. Here, the red traces have been projected onto the $\hat{\gamma}_P$ axis and approximate the qP wavefield, the green traces have been projected onto the $\hat{\gamma}_{S1}$ and approximate the qS_1 wavefield and similarly the blue traces have been projected onto the $\hat{\gamma}_{S2}$ axis and approximate the qS_2 wavefield. Close inspection shows that there is no strong apparent leakage between different wave components, which is an indication that the wavefield separation has been successful.

The arrow in Figure 6.8 highlights an inferred qP-wave polarity reversal that may be associated with a nodal plane (*cf.* Figure 3.11). In general, for most azimuths from a double-couple source, the radiated P-waves are weaker than radiated S waves. Consequently microseismic events with low S/N may not have a discernible P-wave arrival; such events are referred to as *single-phase*.

6.3.4 Matched Filtering and Subspace Detection

The use of cross-correlation for picking and refinement of arrival-times, as discussed above, takes advantage of the capability of this method to measure time lags between similar signals. Cross-correlation also provides a powerful tool to measure the similarity

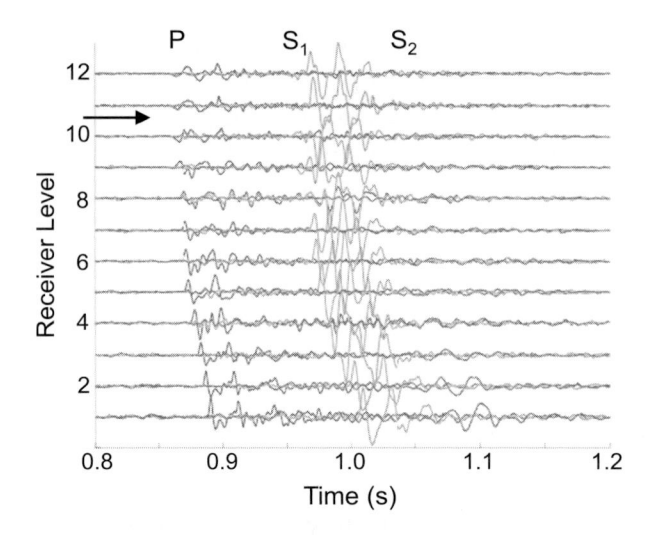

Fig. 6.8 Microseismic event recorded by a downhole geophone array, after rotation into ray-centered coordinates that separate qP (red), qS_1 (green) and qS_2 (blue) wavefields into distinct components. Arrow highlights interpreted qP-wave polarity reversal. (A colour version of this figure can be found in the Plates section.)

between two signals, irrespective of time lag. This characteristic of cross-correlation forms the basis for *matched filtering* (MF), a technique for enhanced event detection using template (or "parent") events (Van Trees, 1968). This approach is highly effective for the detection of weak, low S/N events with waveform characteristics that are similar to the template. The waveform similarity implies that the detected (or "child") events have a similar focal mechanism and a similar hypocentre location to the template event.

Matched filtering has been employed for event detection in a number of recent studies (Van der Elst et al., 2013; Goertz-Allmann et al., 2014; Skoumal et al., 2015b). In these studies, the template events are obtained from existing earthquake catalogues and have high S/N. In essence, the template waveforms are extracted and cross-correlated with continuous raw recordings to find matching signals. A specified value of the normalized cross-correlation function is used to establish a threshold for event detection.

Most previous applications of matched filtering have used data from sensor networks located at the surface. Caffagni et al. (2016) developed a MF method that is tailored for use with downhole microseismic data. Their event-detection method uses a stacked cross-correlation function (SCCF) that exploits the capabilities of the downhole sensor array by incorporating *beamforming*, a shift-and-sum procedure for waveforms (Ingate et al., 1985), into the detection process. Another novel aspect of this method is the use of an automated scheme to obtain relative hypocentre locations for child events, which are anchored to a predetermined absolute hypocentre location for the corresponding parent. A case study in which this method was applied to a hydraulic-fracturing monitoring program yielded a fourfold increase in the number of located events, which helped to resolve temporal clustering patterns and improved the delineation of the stimulated reservoir region (Eaton and Caffagni, 2015).

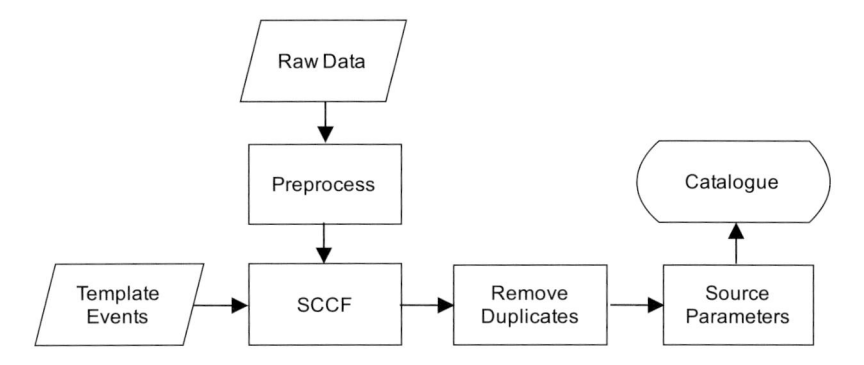

6.3 Event Detection and Arrival-Time Picking

Fig. 6.9 Flow chart illustrating the workflow used for event detection using matched filtering applied to downhole microseismic data. SCCF denotes stacked cross-correlation function.

The basic workflow for the MF method of Caffagni et al. (2016) is shown in Figure 6.9. Prior to cross-correlation with template events, the continuous raw data are pre-processed using a polarization-preserving three-component normalization procedure that is similar to *automatic gain control* (AGC), a widely used method to process seismic-reflection data (Yilmaz, 2001). For a given receiver, this process normalizes all three components using a modulating function that is given by

$$A_M(t) = A_E(t) * \Delta(t, t_\Delta), \tag{6.19}$$

where $*$ denotes convolution, $A_E(t)$ is the three-component amplitude envelope (Appendix B) and $\Delta(t, t_\Delta)$ is a triangular operator of duration t_Δ. If t_Δ is chosen to be ≈ 5 times the pulse duration, this procedure has the effect of enhancing the detectability of weak events compared with normalized cross-correlation (Caffagni et al., 2016), without altering the polarization. Event detection is based on a summed cross-correlation function, S_C, given by

$$S_C(t) = \sum_{i=1}^{3} \sum_{j=1}^{N} \phi_j^i(t), \tag{6.20}$$

where N is the number of receivers in the array and $\phi_j^i(t)$ is the cross-correlogram of the ith component of the template signal with the ith component of the jth receiver. This is a multi-component cross correlation, and the superscript notation used to denote the two time series is omitted for notational brevity. Due to the use of cross-correlograms rather than waveforms in the summation process, beam-forming is implicit in this scheme. To understand this, imagine two identical events separated by a fixed time lag Δt and recorded by a receiver array. Although the arrival times of the events will differ from one receiver to the next, as represented by the moveout across the array, the cross-correlation peak recorded by $\phi_{ij}(t)$ is aligned at Δt. Consequently, the cross-correlation peaks sum constructively by summing the cross-correlograms, without any requirement to apply a time shift or to measure the moveout. As with other detection methods, event detection is implemented based on exceedance of a user-specified threshold value.

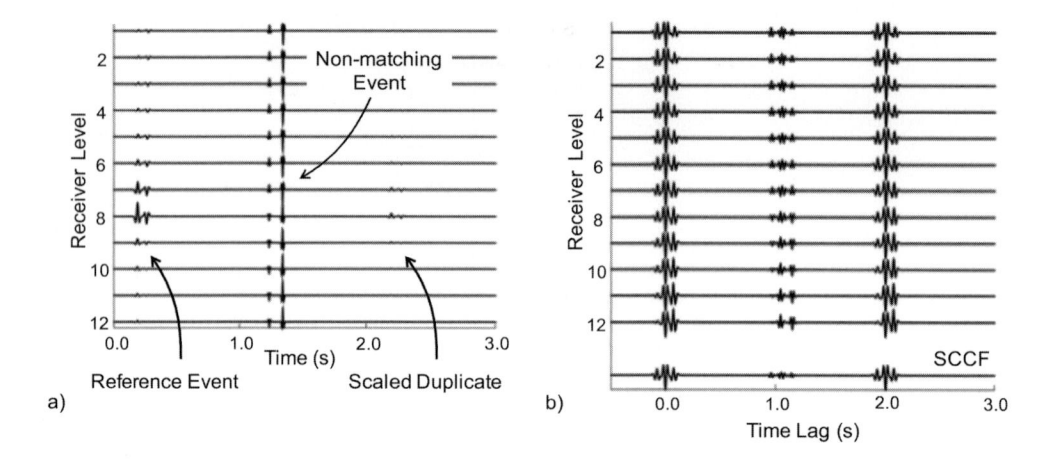

Fig. 6.10 Illustration of matched-filtering analysis based on the use of stacked cross-correlation function (SCCF). a) Synthetic waveforms, constructed using a reference event, a scaled duplicate event of lower amplitude and a non-matching high-amplitude event. b) Individual correlograms and SCCF. Modified from Caffagni et al. (2016), with permission from Oxford University Press.

The detection process is illustrated using synthetic data in Figure 6.10. The raw input data consists of 12 traces, each containing 3 events. The first event is a reference signal that provides the template. A scaled duplicate event is present with a time delay of 2.0 seconds. In addition, a high-amplitude non-matching event occurs with a time delay of one second. Although the non-matching event has a moveout pattern and P and S amplitude distribution that are distinct from the template, there is sufficient similarity in the waveforms that it generates a false detection with the standard MF approach. As shown in Figure 6.10, the false detection is avoided by preprocessing the input signals by normalizing with the amplitude modulation function in Equation 6.19 and then applying the stacked cross-correlation function in Equation 6.20.

Figure 6.11 shows a representative template event and three associated child events from the study by Eaton and Caffagni (2015), which used downhole microseismic data from a hydraulic-fracture monitoring program in western Canada. The template event (Figure 6.11a) has a magnitude of −1.95 and was detected using a conventional single-receiver method. The three weaker child detections have magnitudes ranging from −2.55 to −2.37. Careful inspection reveals a number of waveform similarities, such as the strong S-wave response seen in Figures 6.11b and 6.11c on the ψ_1 channel (Figure 6.2) for receiver level 12. There are also notable waveform differences between the child events that may reflect differences in source processes. For example, the child event in Figure 6.11b appears to show two arrivals, suggesting that two events occurred in fast succession, while the child event in Figure 6.11d is unusual as it appears to lack a visible S wave. These features highlight the robust detection capabilities of this method; child events are similar in character to the parent, but need not be identical.

In practice, the use of multiple parent events produces duplicate matching events, that is individual child events that match with multiple templates. Therefore, removal of duplicate child events is an important step in the MF workflow (Figure 6.9). Duplicate events can

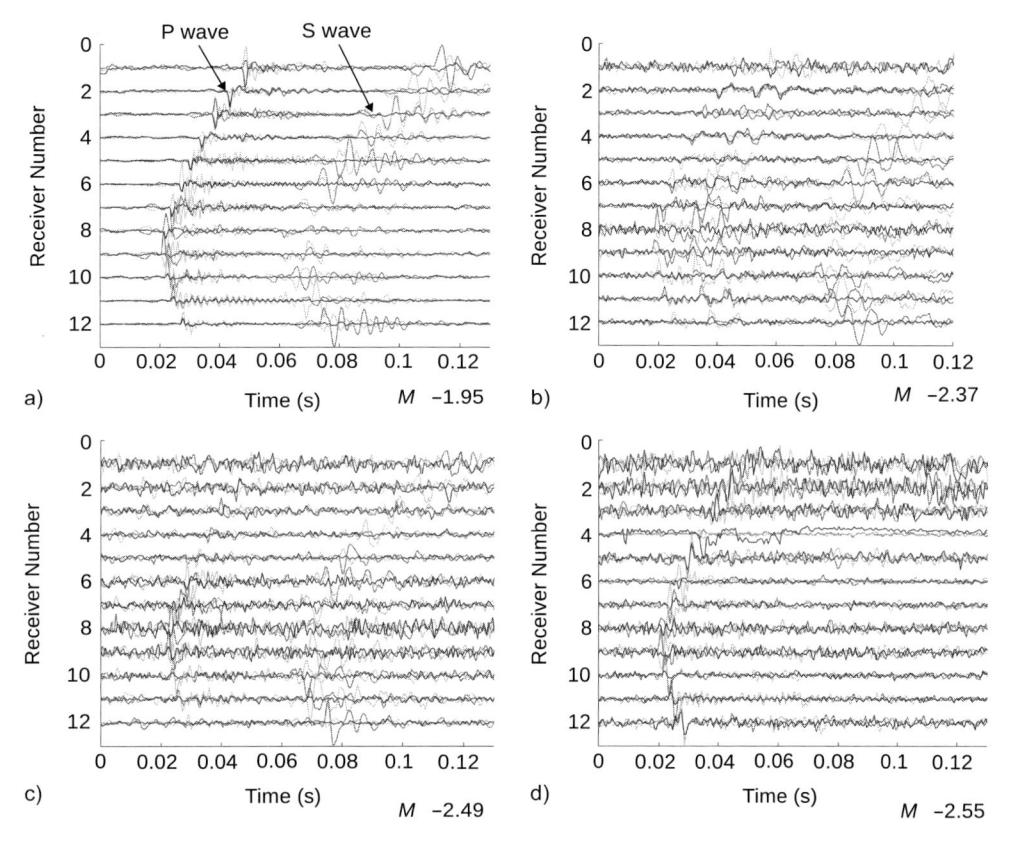

Fig. 6.11 Example of template and child events detected using matched filtering. Panels show unfiltered record sections of a) a template (parent) event and b) to d) a set of detected (child) events. Receiver levels are independently normalized and calculated magnitudes are indicated in the lower-right corner of each panel. Trace colours denote unrotated geophone orientations: red = ψ_1, green = ψ_2, blue = ζ. Modified from Eaton and Caffagni (2015) with permission from Oxford University Press. (A colour version of this figure can be found in the Plates section.)

be identified automatically based on common origin time. Since every parent–child pair is associated with a peak amplitude of the summed cross-correlation function, S_C; this parameter can be used for removal of duplicate child events by retaining only those with the highest value of S_C.

Caffagni et al. (2016) developed a procedure for determining relative locations of child events without any requirement to pick arrivals. In brief, this procedure works as follows:

1. First, the child event waveforms are projected onto the ray-centred coordinates of the corresponding parent event. This serves (approximately) to separate qP and qS waveforms.
2. An iterative beam-forming process is used to maximize a weighted stack based on a lookup table-driven search within a region centred on the parent event. The lookup table contains calculated qP and qS arrival times for every receiver.

3 The qP wavefield of the child event is aligned using the picked qP times for the parent. Then, for each node within the search region, the qP waveforms are shifted by the time difference between the calculated times for that node and the calculated time for the centre of the search region (i.e. the parent hypocentre). The same alignment method is applied to the qS waves.

4 The individual wavefields are stacked and combined into a single beamformed stack using user-defined weights for the individual wave types. The optimal hypocentre is defined by the position of the node with the highest stacked amplitude. If the location falls on the edge of the search grid, the event is discarded.

5 In the case of monitoring from a single vertical observation well, an additional step is required. In this case, a 2-D cylindrical coordinate system is used with its origin at the wellhead. The grid node positions are then parameterized based on focal depth and radial distance from the well. The optimum grid node is determined as before, but then an angular search is performed by applying a rotation over a small angular aperture around the parent hypocentre. The optimum rotation angle, combined with the optimum grid node in 2-D cylindrical coordinates, is used to compute the hypocentre.

Subspace detection provides a more general approach than the MF algorithm for detection of events within noisy observations, by representing the signals to be detected as a linear combination of orthonormal basis waveforms (Harris, 2006). The subspace algorithm operates by projecting the observed data into a subspace spanned by a design set of signals from a particular source. *Singular-value decomposition* (SVD), a generalization of the eigenvalue decomposition method for factorization of a matrix, provides an efficient approach to obtain an orthonormal representation that captures the important characteristics of the design set (Barrett and Beroza, 2014). For subspace detection, the detection statistic ranges from zero to one and is represented by the ratio of the squared norm of the projected vector to the squared norm of the original data vector (Harris, 2006). Like MF, a detection is obtained when the statistic exceeds a predefined threshold value.

Subspace detection is becoming popular as a way to improve computational efficiency relative to MF, while reducing the number of waveform templates required to fully characterize an earthquake sequence (Benz et al., 2015). For example, a subspace detection approach was applied to dual-array microseismic monitoring data by Song et al. (2014), leading to enhanced detection and improved viewing distance. Benz et al. (2015) implemented this method to analyze an induced earthquake sequence near Guthrie, Oklahoma, and area that was characterized by hundreds of robust detections per day over a seven-month period. Barrett and Beroza (2014) have developed an empirical approach that uses the stack and time-derivative of the stack as basis functions, as these strongly resemble the first two basis functions derived from SVD. They applied the empirical subspace method to observations of the 2003 M_W 5.0 Big Bear earthquake sequence and found that it provides substantial advantages over MF detection, including improved detection capabilities for overlapping events. All of these studies highlight the capabilities of subspace detection as a powerful approach for enhanced event detection.

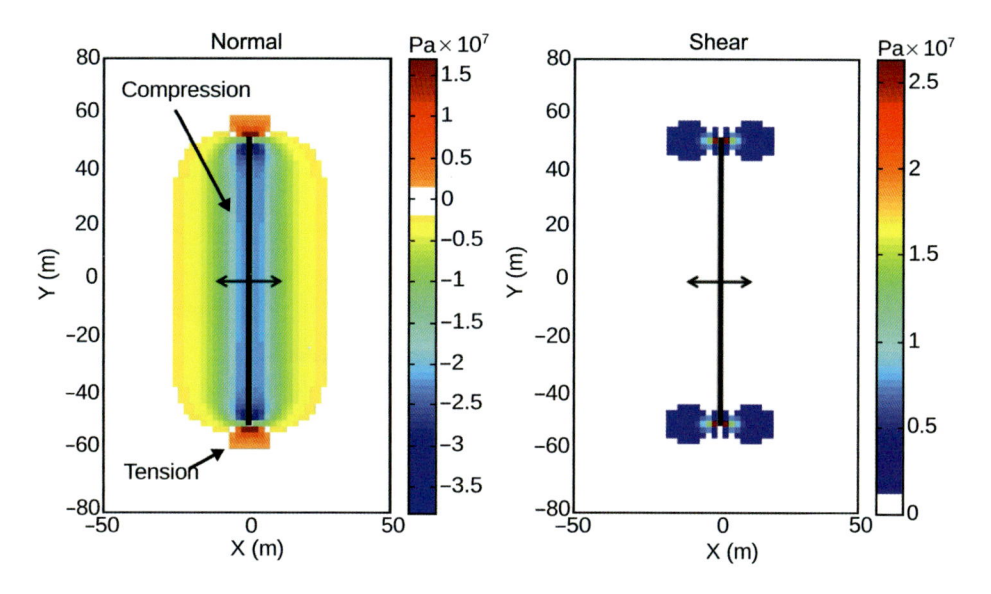

Fig. 2.4 Map view of normal stress and shear stresses around a vertical mode I crack. The crack is shown by a black line and is 100 m in length, 10 m in height, and located 3000 m below the surface of a homogeneous elastic half-space with $E = 2.25 \times 10^{10}$ Pa and $\nu = 0.28$. Fracture opening is 1.0 cm. Tensile stress is positive and localized near the crack tip. Stress calculations used the method of Okada (1992) and are resolved for the crack orientation.

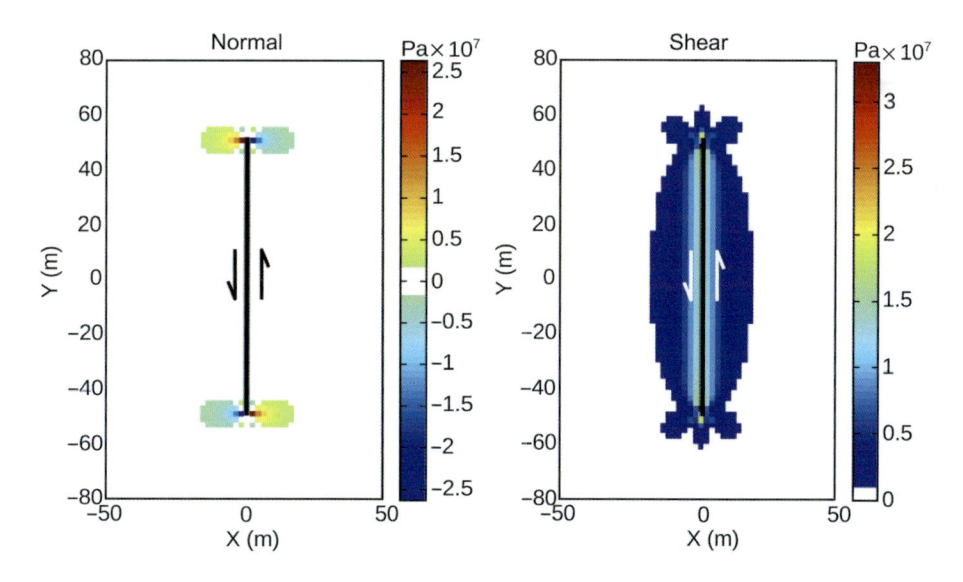

Fig. 2.5 Map view of normal stress and shear stress around a vertical mode II crack. The crack is shown by a black line and is 100 m in length, 10 m in height, and located 3000 m below the surface of a homogeneous elastic half-space with $E = 2.25 \times 10^{10}$ Pa and $\nu = 0.28$. Slip is 1.0 cm. Stress calculations used the method of Okada (1992) and are resolved for the crack orientation.

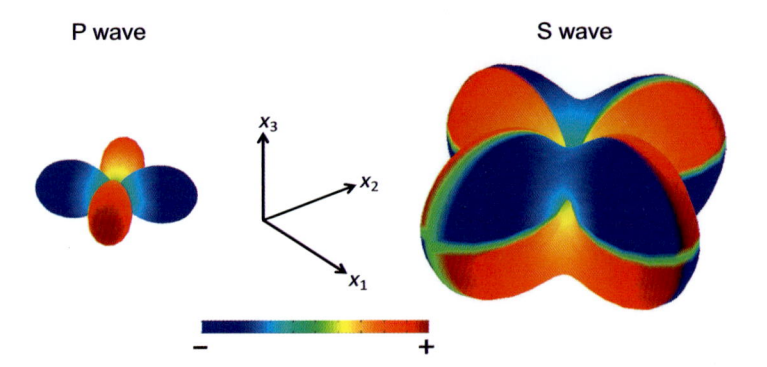

Fig. 3.11 Perspective view of the P-wave and S-wave radiation pattern for a double-couple source with $M_{12} = M_{21} = 1$. Colours denote amplitude, with positive outwards for the P wave and positive upwards for the x_3-component of the S wave. The P wave has two mutually orthogonal nodal planes, while the S wave has mutually perpendicular nodal axes.

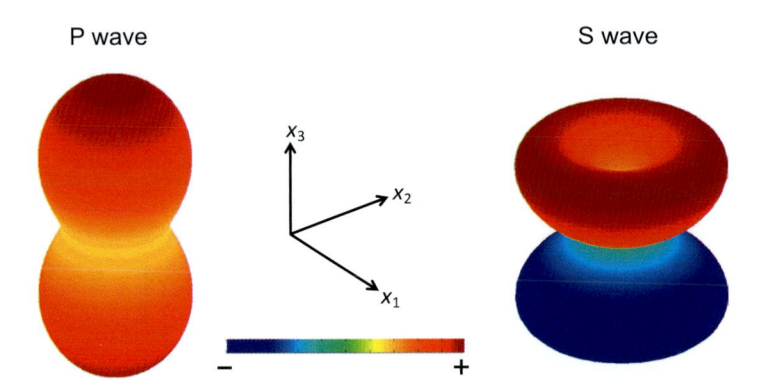

Fig. 3.14 Perspective view of the P-wave and S-wave radiation pattern for a tensile crack opening source with $M_{11} = M_{22} = \lambda$ and $M_{33} = \lambda + 2\mu$. Colours denote amplitude, with positive outwards for the P wave and positive upwards for the x_3-component of the S wave. The P wave has no nodal planes, while the S wave has a nodal plane and a perpendicular nodal axis.

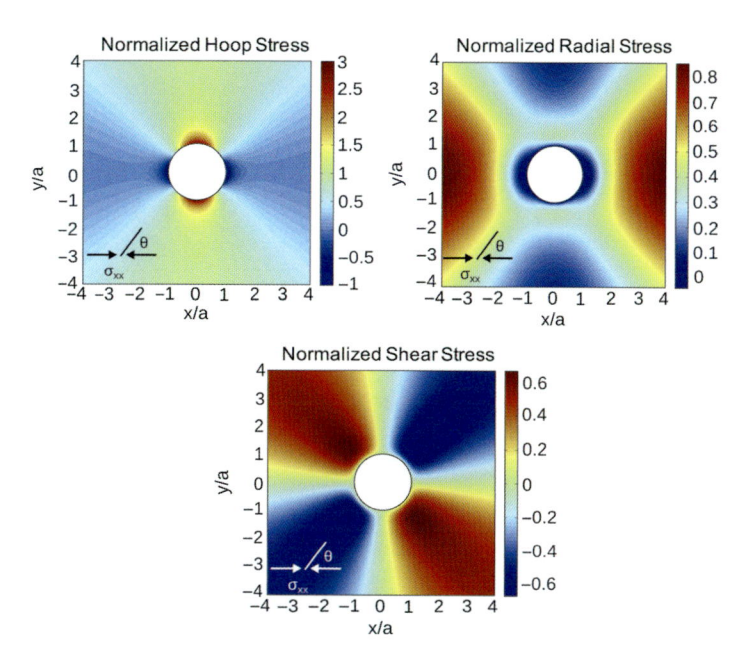

Fig. 4.1 Stress distribution in an elastic plate around an empty circular cavity of radius a, showing hoop stress, $\sigma_{\theta\theta}$, radial stress, σ_{rr} and shear stress, $\tau_{\theta r}$ computed using the Kirsch equations (Equation 4.1) and normalized based on the magnitude of the ambient uniaxial stress, σ_{xx}. The origin of the cylindrical coordinate system (r, θ) is at the centre of the cavity, with θ measured counterclockwise from the direction of the uniaxial stress. Stress amplitudes are plotted using the sign convention that tensile stress is negative.

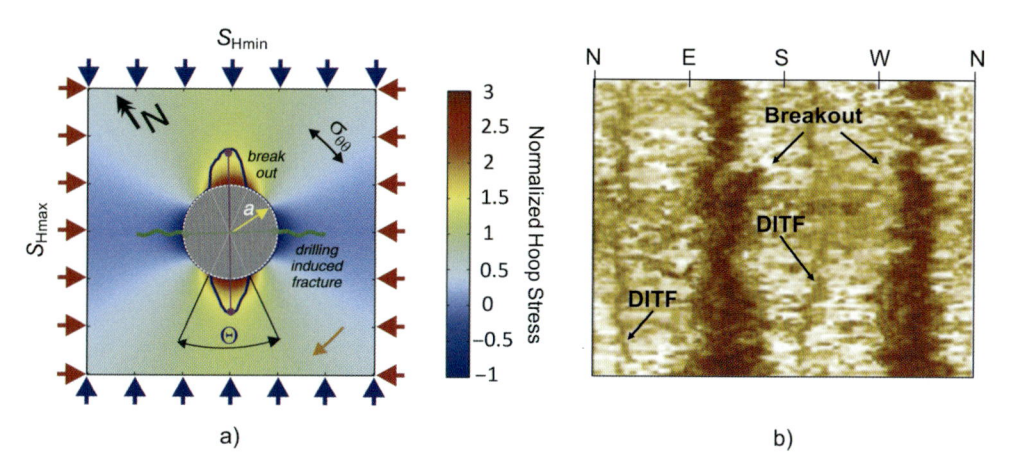

Fig. 4.2 a) Schematic illustration of regions of borehole breakout and drilling-induced tensile fracture (DITF) in relation to the generated hoop stress, $\sigma_{\theta\theta}$ around a wellbore of radius a. Angular breakout width is shown by Θ. Reprinted from *Tectonophysics*, Vol 580, Douglas R. Schmitt, Claire A. Currie and Lei Zhang, Crustal stress determination from boreholes and rock cores: Fundamental principles, Pages 1–26, copyright 2012, with permission from Elsevier. b) Borehole image data showing spatially oriented observations of borehole breakout and DITF. Geographic directions of north (N), east (E), south (S) and west (W) are indicated. Reprinted from *International Journal of Rock Mechanics & Mining Sciences*, Vol 40, M. D. Zoback et al., Determination of stress orientation and magnitude in deep wells, Pages 1049–1076, copyright 2003, with permission from Elsevier.

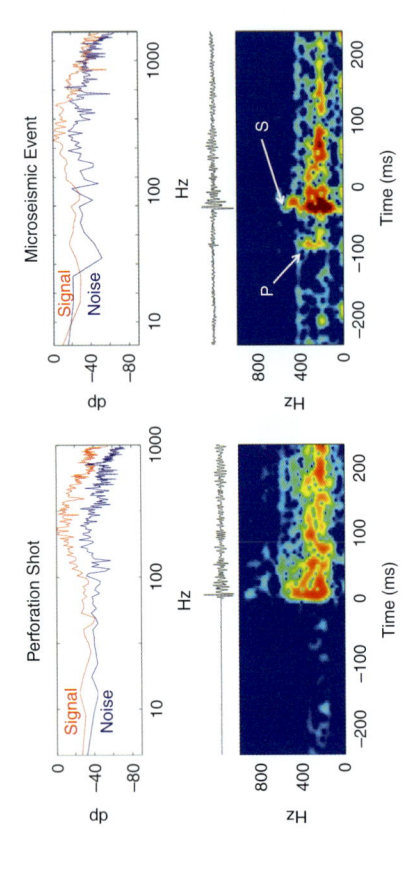

Fig. 5.11 Comparison of time-frequency spectra for a representative perforation shot and a microseismic event from a hydraulic fracturing program in the Montney trend of northeastern British Columbia (Eaton et al., 2014d). Upper panels show signal and noise spectra. Lower panels show recorded vertical-component waveforms and their corresponding spectragrams obtained using the short-time Fourier transform. Hot (red) colours indicate high amplitude. Used with permission from the Society of Exploration Geophysicists.

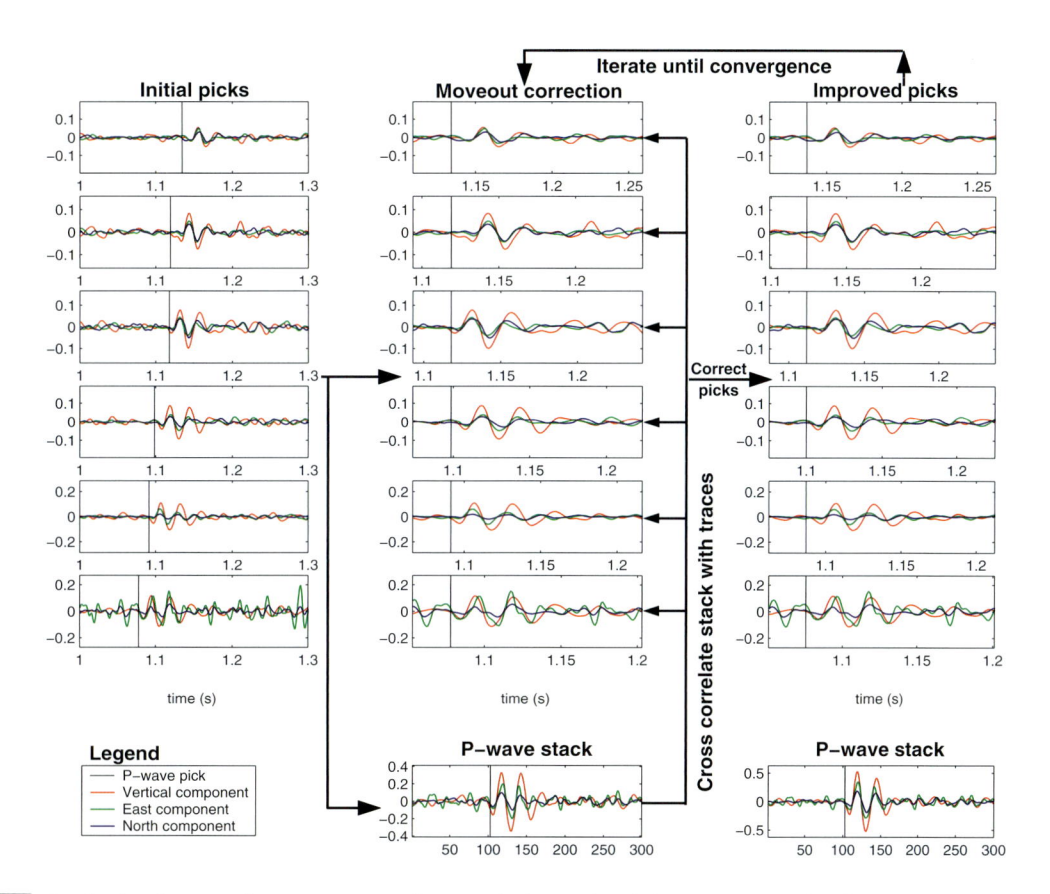

Fig. 6.6 Graphical explanation of iterative cross-correlation method using downhole passive-seismic data from the North Sea. Initial picks are used to align and stack the three-component traces. Picks are refined by cross-correlation with the stacked trace. The process is repeated until the arrival-time corrections converge to a small value. Reprinted from De Meersman et al. (2009), with permission of the Society of Exploration Geophysicists.

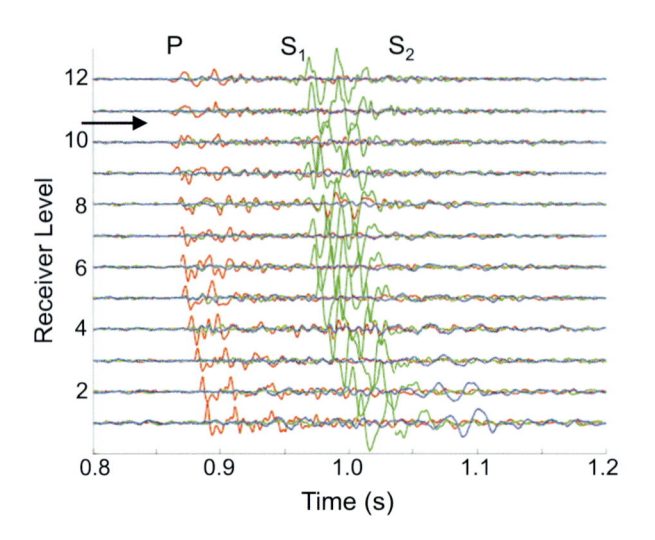

Fig. 6.8 Microseismic event recorded by a downhole geophone array, after rotation into ray-centered coordinates that separate qP (red), qS_1 (green) and qS_2 (blue) wavefields into distinct components. Arrow highlights interpreted qP-wave polarity reversal.

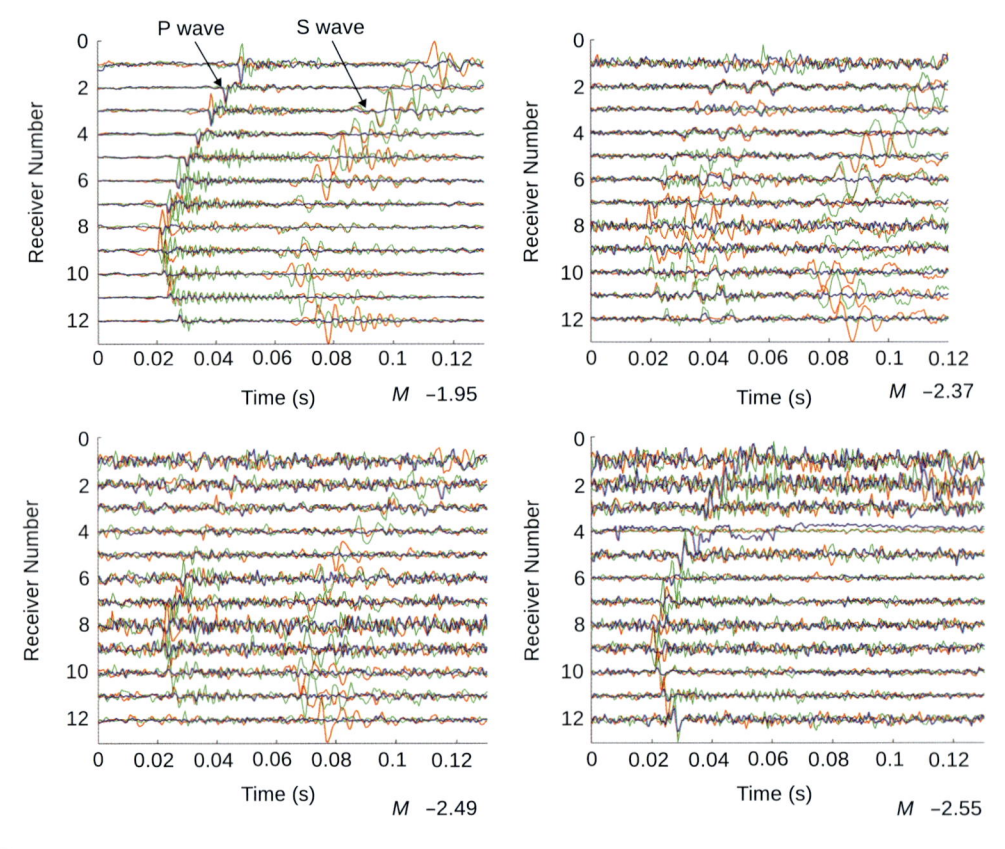

Fig. 6.11 Example of template and child events detected using matched filtering. Panels show unfiltered record sections of a) a template (parent) event and b) to d) a set of detected (child) events. Receiver levels are independently normalized and calculated magnitudes are indicated in the lower-right corner of each panel. Trace colours denote unrotated geophone orientations: red = ψ_1, green = ψ_2, blue = ζ. Modified from Eaton and Caffagni (2015) with permission from Oxford University Press.

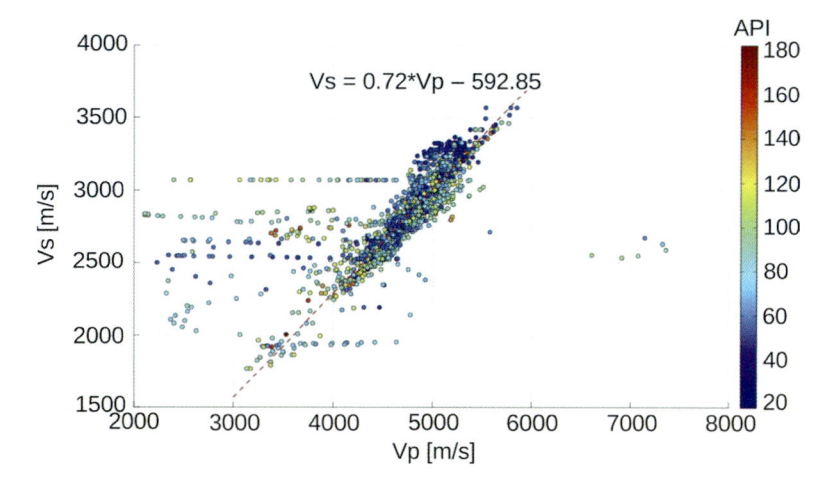

Fig. 6.15 Cross plot of v_P versus v_S within the depth interval of the microseismic events showing the line of best fit. Data points are coloured by gamma-ray value in API (American Petroleum Institute) units. Many anomalous values come from coal intervals where the S-wave picks were clipped in the dipole sonic log. From Pike (2014), with permission of the author.

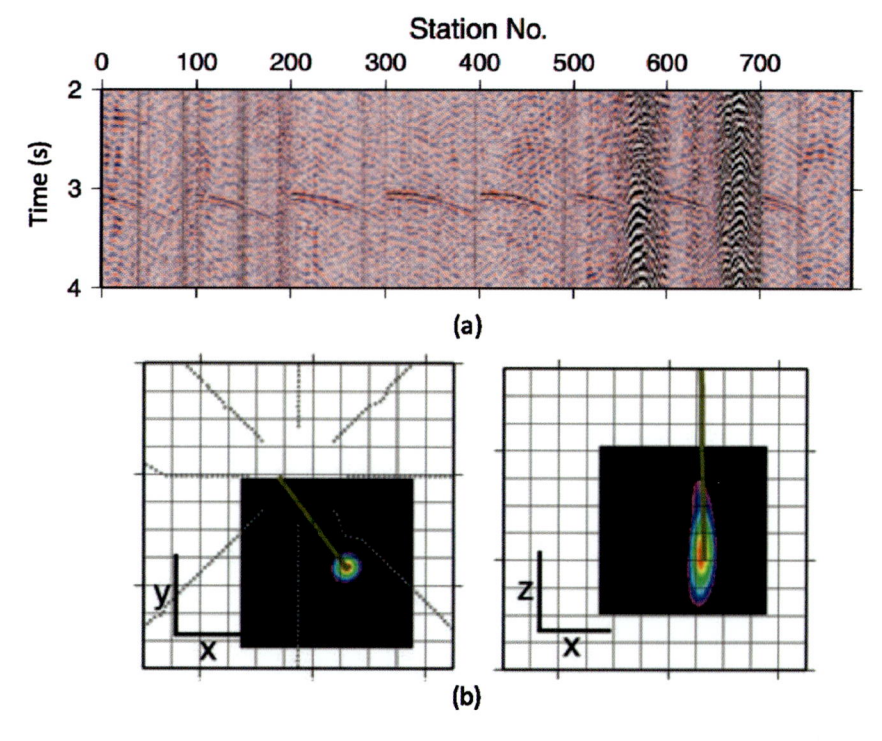

Fig. 7.5 Field data example of beamforming process. a) Raw waveform recordings of a perforation shot. b) Colour density plots of beamformed stack of the perforation shot. Left: map view, showing configuration of surface geophones. Right: cross section view. Reproduced from Chambers et al. (2010), with permission from Wiley.

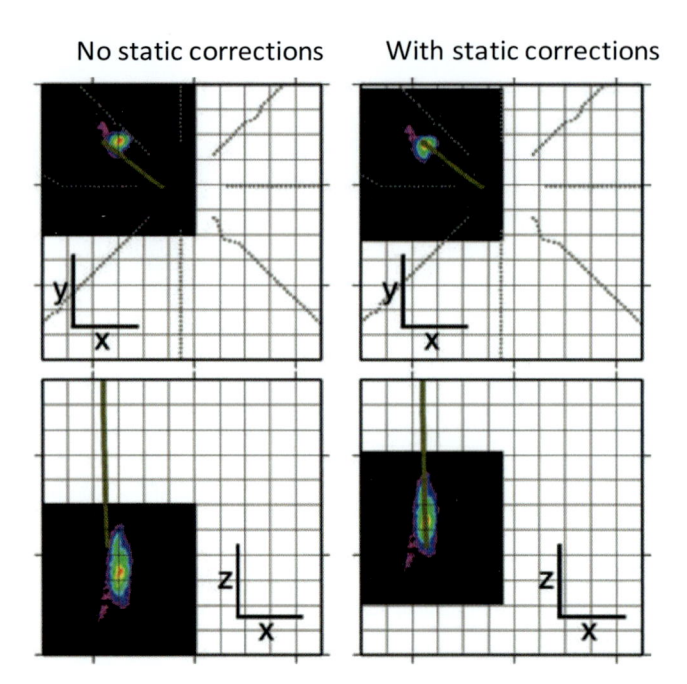

Fig. 7.6 Effect of statics on calculated location of a perforation shot, based on colour-density plots. Without the application of statics, the perforation shot is located off the wellbore in both map view (upper panel) and cross section (lower panel). After application of statics, the perforation shot is centred at the correct location. Grid lines are spaced at 152.4 m intervals. Reproduced from Chambers et al. (2010), with permission from Wiley.

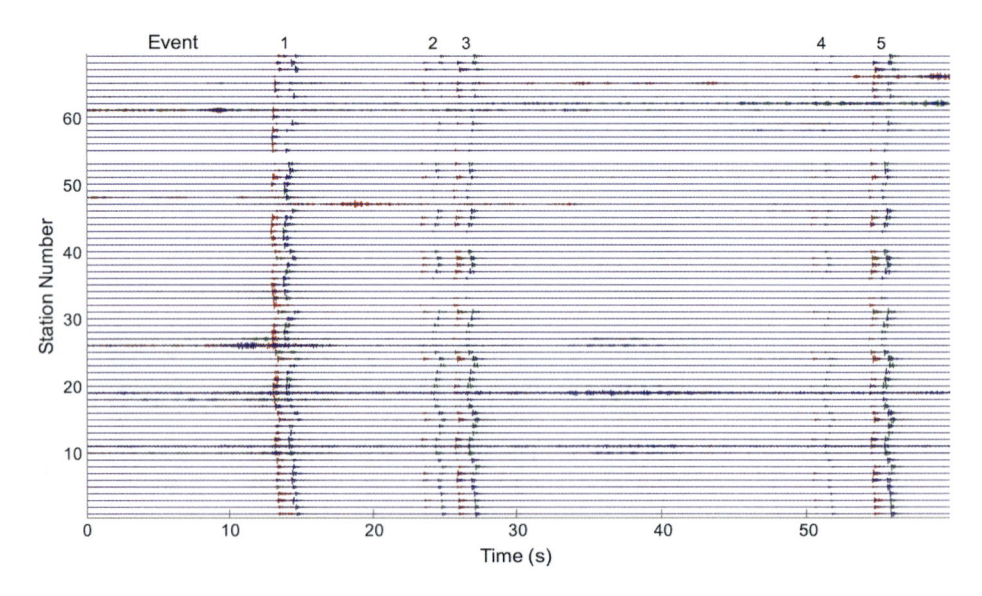

Fig. 7.8 One-minute raw multicomponent data record from the ToC2ME passive monitoring project. The array configuration is shown in Figure 7.7. This data record contains five easily discernible induced events, which is unusual as most events that are detectable using surface arrays are not visible in raw data (Chambers et al., 2010). The P-wave arrival is evident on the vertical component (red), while the S-wave arrival is most easily discernible on a horizontal component (blue).

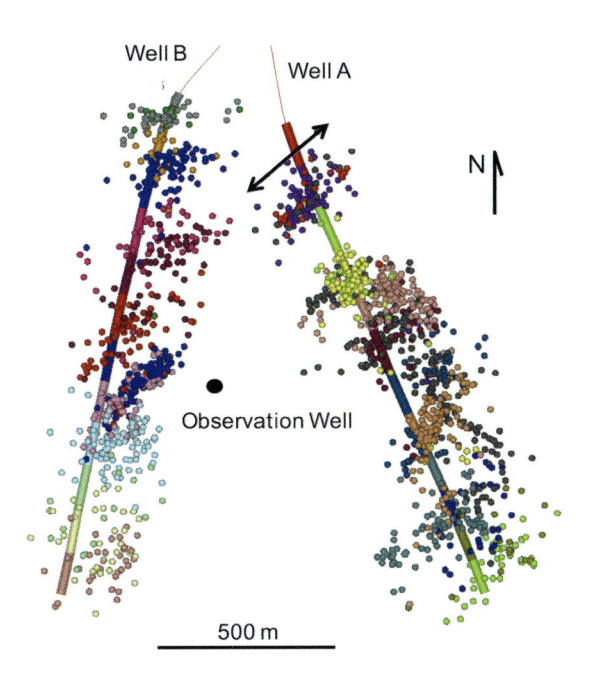

Fig. 8.1 Hydraulic-fracture monitoring (HFM) example. These microseismic observations are from the Hoadley experiment, a two-well open-hole completion in western Canada that was monitored with a 12-level downhole array deployed in a vertical observation well (Eaton et al., 2014c). Treatment stages are shown as coloured bars along the horizontal well paths. Microseismic events are coloured by treatment stage. Typical attributes inferred from microseismicity included fracture dimensions and orientation, as marked by the double arrow near one of the event clusters along well A.

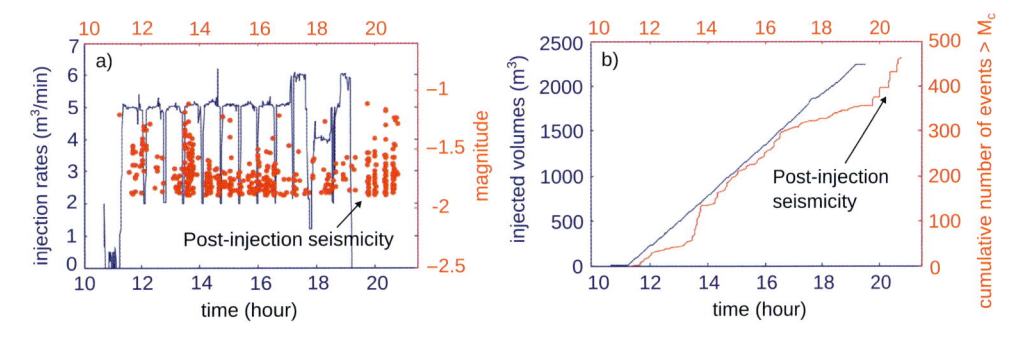

Fig. 8.3 Comparison of induced microseismicity with fluid injection schedule. a) Seismicity above the magnitude of completeness and fluid injection rate corresponding to 12 treatment stages for the horizontal well A in Figure 8.1. Stages are separated by brief periods of lower injection rates. b) Cumulative injected fluid volume and number of microseismic events above the magnitude of completeness. Reproduced from Maghsoudi et al. (2016), with permission of Wiley. Copyright 2016 by the American Geophysical Union.

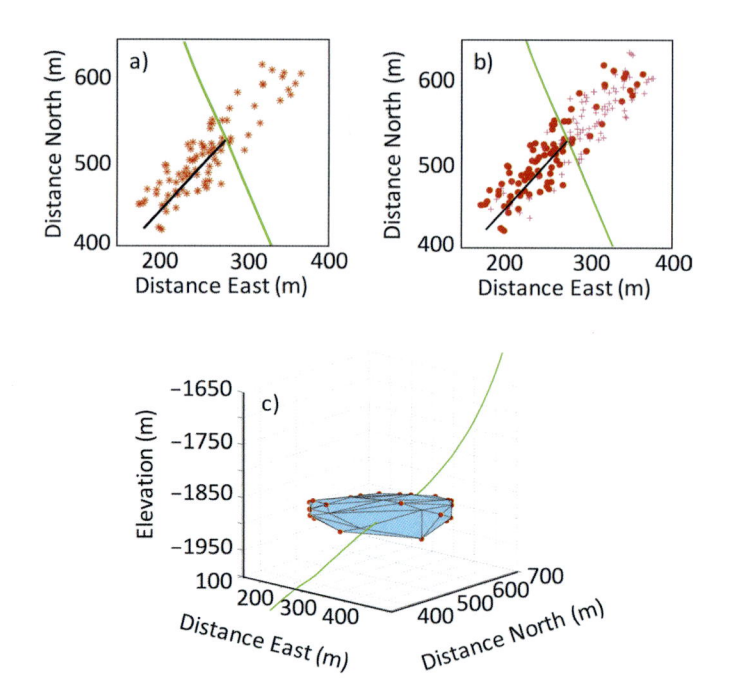

Fig. 8.7 Convex-hull method for estimated stimulated volume (ESV) determination, based on a microseismic cluster from Figure 8.1 that is indicated by the double arrows. a) Map view of the observed microseismicity. Distance is relative to the observation well. Green line shows treatment wellbore and black line shows line of best fit through the hypocentres. b) Augmented microseismic point cloud, with mirror events added (shown by plus symbols). c) Convex hull, fit to the point set in (b).

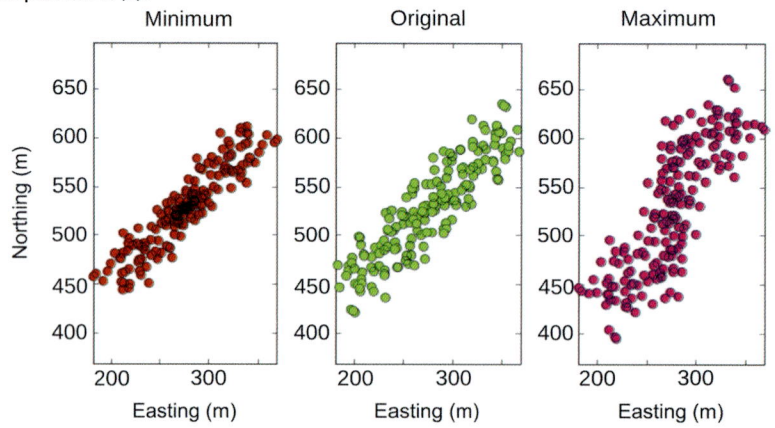

Fig. 8.8 Method for calculating uncertainty of estimated stimulated volume (ESV) volume using the convex hull approach. The original cluster shown in the central panel is the augmented point cloud from Figure 8.7b. The convex-hull that encloses this point cloud has a volume of 7.70×10^5 m^3. The minimum ESV (6.13×10^5 m^3) is determined by displacing every hypocentre in the augmented set toward the injection point, whereas the maximum ESV (1.41×10^6 m^3) is determined by displacing away from the injection point. In both cases, the displacement distance is defined by the location uncertainty.

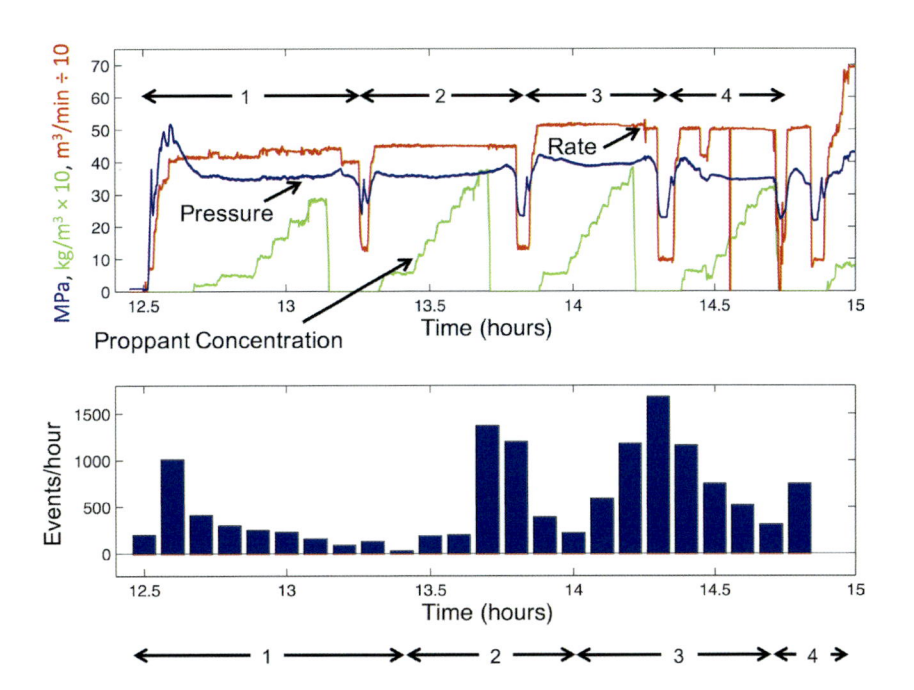

Fig. 8.4 Comparison between pump curves (upper panel) with seismicity rate (lower panel) determined using matched filtering analysis, showing temporal clustering during open hole completion. Stages 1–4 are marked by arrows at the top of the figure. Due to the short time interval between stages, microseismic activity exhibits temporal overlap from one treatment stage to the next. Temporal clusters are nevertheless discernible, as indicated by numbers at the bottom of the figure. Modified from Caffagni et al. (2016), with permission from Oxford University Press.

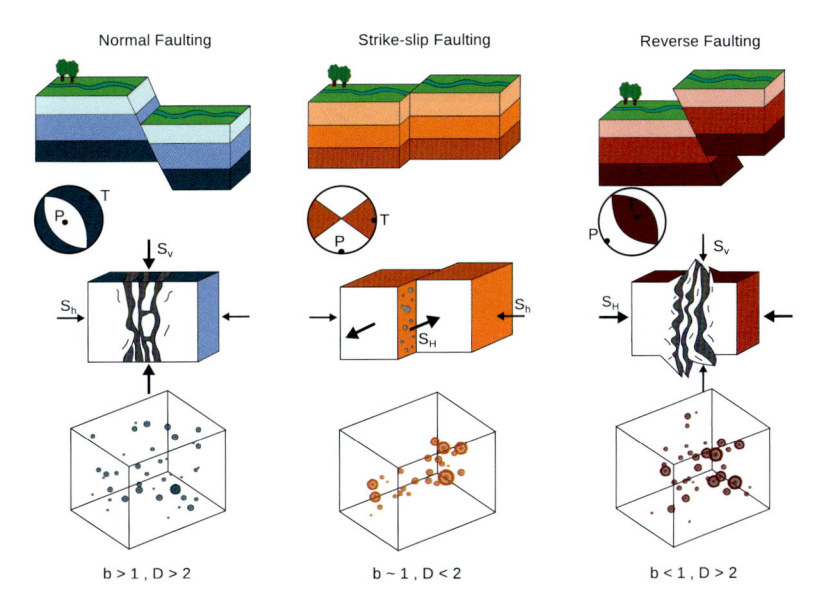

Fig. 8.9 Inferred relationships between stress regime, focal mechanism, b and D parameters. P and T denote the pressure and tension axes, respectively (Equation 3.58). In terms of Anderson's fault regimes, the extensional regime is associated with normal faulting and a greater proportion of small-magnitude events; the strike-slip regime is associated with planar features with a b value of close to unity; and the compressive stress regime is associated with reverse faulting and higher proportion of large-magnitude events. From Grob and Van der Baan (2011). Reprinted with permission of the Society of Exploration Geophysicists.

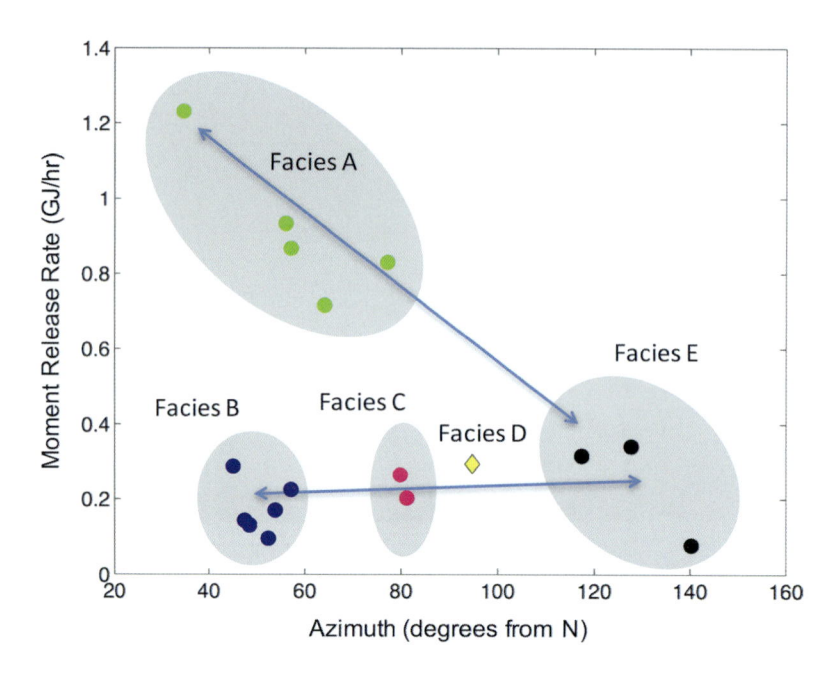

Fig. 8.10 Cross-plot of two microseismic attributes, moment release and cluster azimuth, measured from the Hoadley downhole microseismic dataset shown in Figure 8.1. Each dot represents the average value for an event cluster. Grouping of event clusters coupled with trend analysis, as illustrated here, is used as a basis to infer microseismic facies. From Rafiq et al. (2016), with permission of the Society of Exploration Geophysicists.

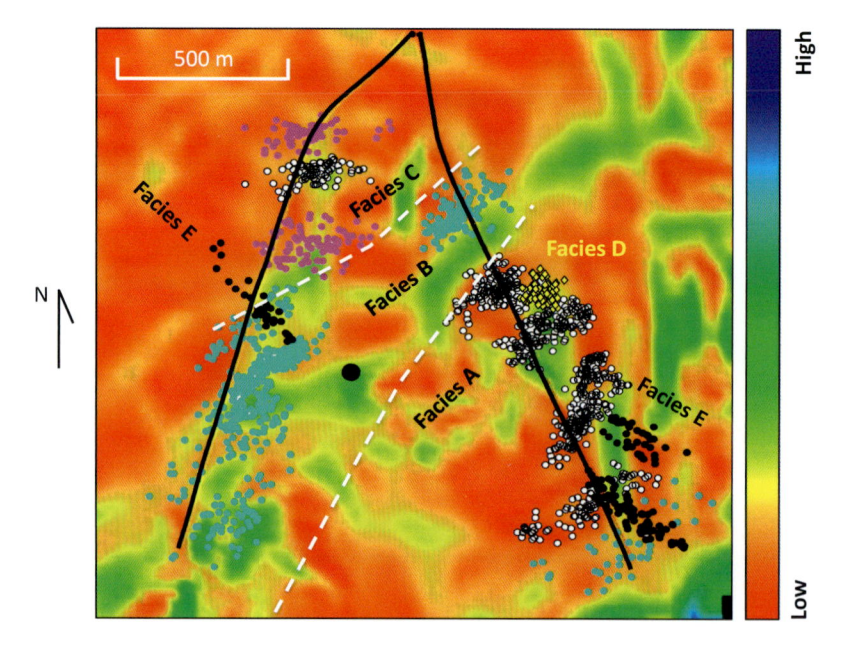

Fig. 8.11 Most-positive curvature attribute derived from 3-D seismic data, overlain with microseismicity from the Hoadley experiment. Northeast–southwest-trending anomalies (green) are parallel to the strike of a barrier-bar complex. Symbols for microseismic events represent facies, as shown in Figure 8.10. Modified from Rafiq et al. (2016), with permission of the Society of Exploration Geophysicists.

Fig. 8.12 Source mechanisms showing crack opening and closing during hydraulic fracturing, based on moment-tensor inversion. a) Scatter plot (left) and density contour plot (right) showing Hudson source-type diagram. These plots contain events that occurred during the time window indicated by the yellow region in the underlying graph, which shows pump curves for pressure and proppant volume. The majority of events cluster close to a tensile crack opening (TCO) mechanism. Subplots b) – d) contain similar graphs for subsequent time windows during the treatment stages. As time progresses during the treatment, the source mechanism shifts toward tensile crack closure (TCC) in the lower right part of the source mechanism diagram. Modified from Baig and Urbancic (2010), with permission of the Society of Exploration Geophysicists.

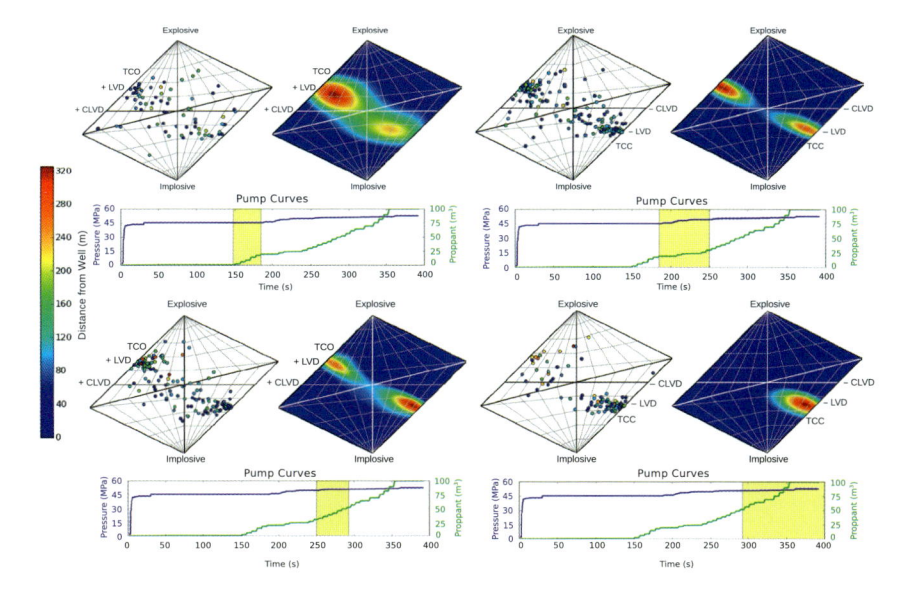

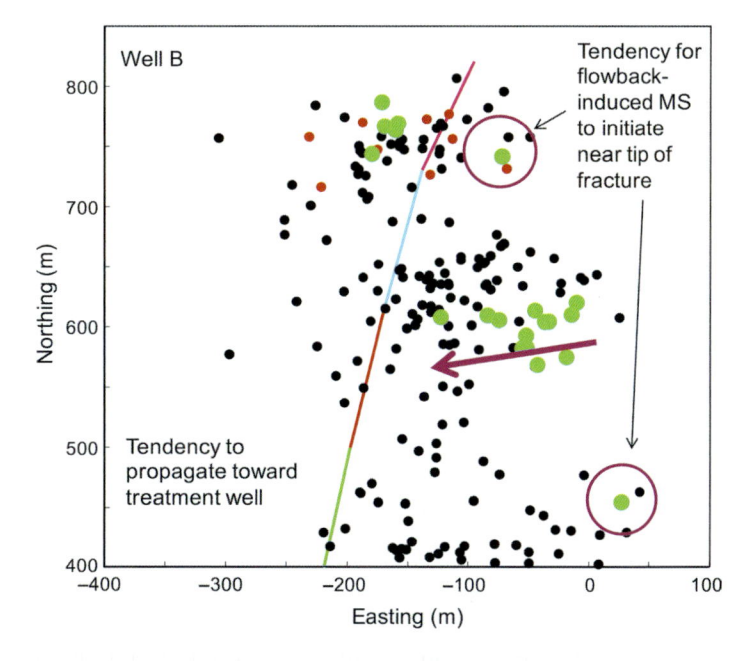

Fig. 8.14 Flowback-induced microseisms recorded near the heel of well B during the Hoadley experiment (Eaton et al., 2014c). Events shown in black occurred during hydraulic fracturing. Events in red occurred after treatment but prior to flowback. Those shown in green occurred during flowback. Large arrow shows progression of one set of flowback-induced events.

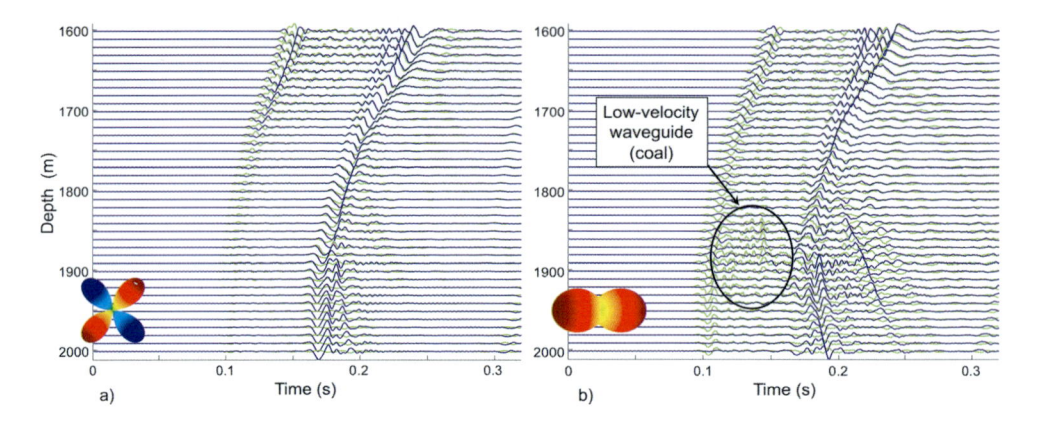

Fig. 8.15 2-D finite-difference wavefield simulation of shear and tensile mechanisms. Blue traces show vertical displacement, green traces show horizontal. a) Bedding-plane shear-slip mechanism. b) Tensile crack-opening (TCO) mechanism. The geological model and geophone array are based on the Hoadley experiment. The source radiation pattern is shown in the lower left of each panel. Low-velocity coal layers produce a waveguide, which is more obvious for the TCO mechanism due to stronger excitation by P waves.

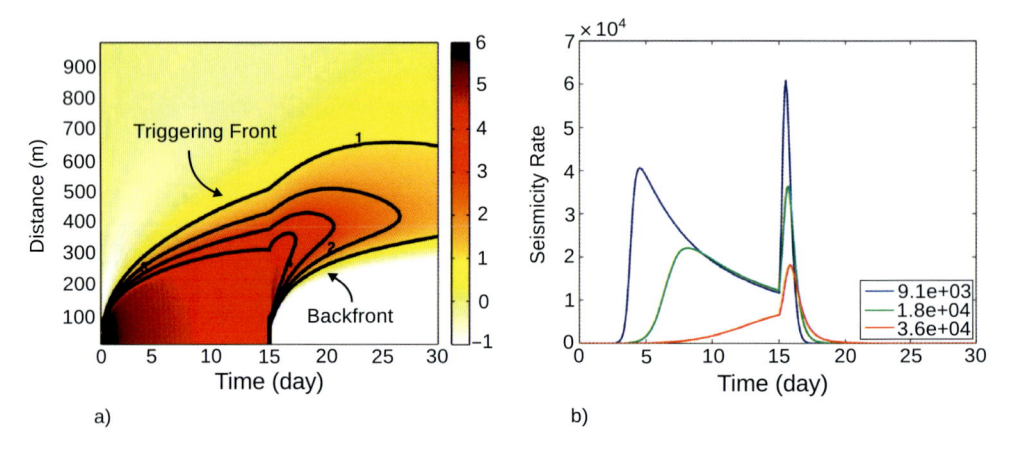

Fig. 9.4 a) Distance-time ($r - t$) plot showing triggering front, backfront and calculated seismicity rate for a constant-flux point injection source that operates for 15 days. The model contains faults with a rate-state constitutive relationship, and orientation as described in Figure 9.3. Colour scale and contours represent $\log_{10}$ of seismicity rate. b) Seismicity rate at a position 250 m from the injection point, for various values of the rate-state characteristic time parameter, t_a. Note sharp increase in seismicity rate after shut-in. Modified from Segall and Lu (2015), with permission from Wiley.

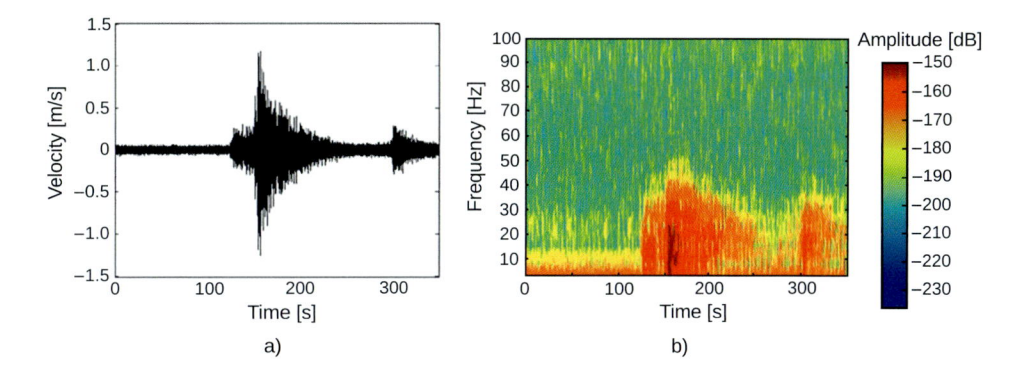

Fig. 9.6 Induced earthquakes at regional distance recorded using a 15 Hz geophone installed at 2 km depth. a) Filtered (10–80 Hz) horizontal-component trace containing regional waveforms from small earthquakes of M_L 3.1 and M_L 2.8 on 29 June 2013. b) Spectrogram for the unfiltered trace. Time window is 350 s and the spectral amplitudes are referenced to 1 m/s. These earthquake-generated signals show a strong similarity to reported waveforms linked to previously proposed models for long-period long-duration (LPLD) events, as described by Das and Zoback (2013). Modified from Caffagni et al. (2014).

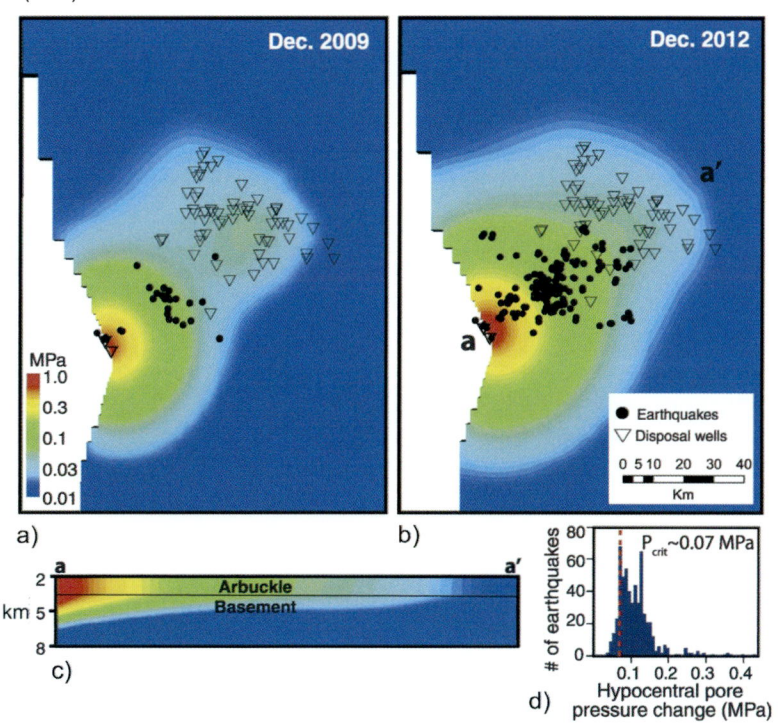

Fig. 9.15 Hydrogeologic model of diffusive expansion of pore-pressure perturbation away from high-rate injection wells in central Oklahoma. a) Modelled pressure perturbation in December 2009, showing earthquakes from 2008 to 2009. b) Modelled pressure perturbation in December 2012, showing earthquakes from 2008 to 2012. Hydraulic diffusivity is 2 m^2/s. The largest injected fluid volumes correspond with four high-rate wells, located within the area of highest pressure perturbation. c) Vertical cross section showing modelled pore pressure perturbation in 2012 along profile $a - a'$. There is a strong pore-pressure signal in the Arbuckle Group as well as the uppermost basement. d) Histogram of pore-pressure increase at all earthquake hypocentres in the catalog. A pore-pressure increase of ∼0.07 MPa is inferred to be the triggering threshold. Reproduced from Keranen et al. (2014), with permission of the American Association for the Advancement of Science.

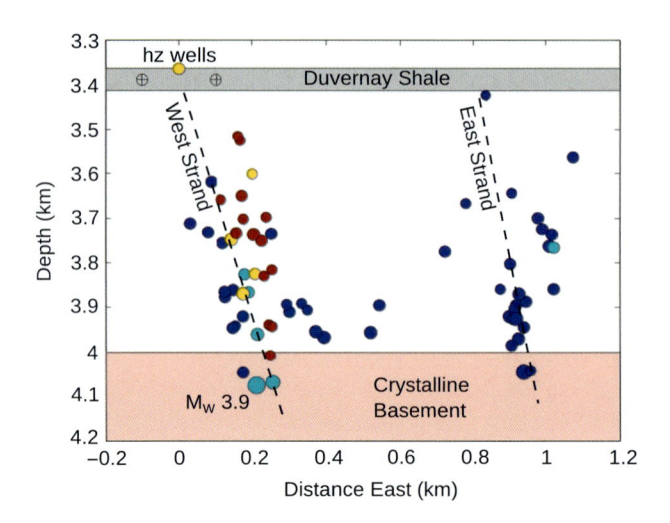

Fig. 9.17 Example of fault activation from hydraulic fracturing, presented as an east–west cross section showing hypocenters of induced events from January to March 2015. Dark blue symbols show events that occurred during hydraulic fracturing in two horizontal (hz) wells. Light blue, yellow and red symbols show post-injection events over a two-month time period. Two fault strands are evident, with contrasting temporal activation patterns. Reproduced from Bao and Eaton (2016), with permission of the American Association for the Advancement of Science.

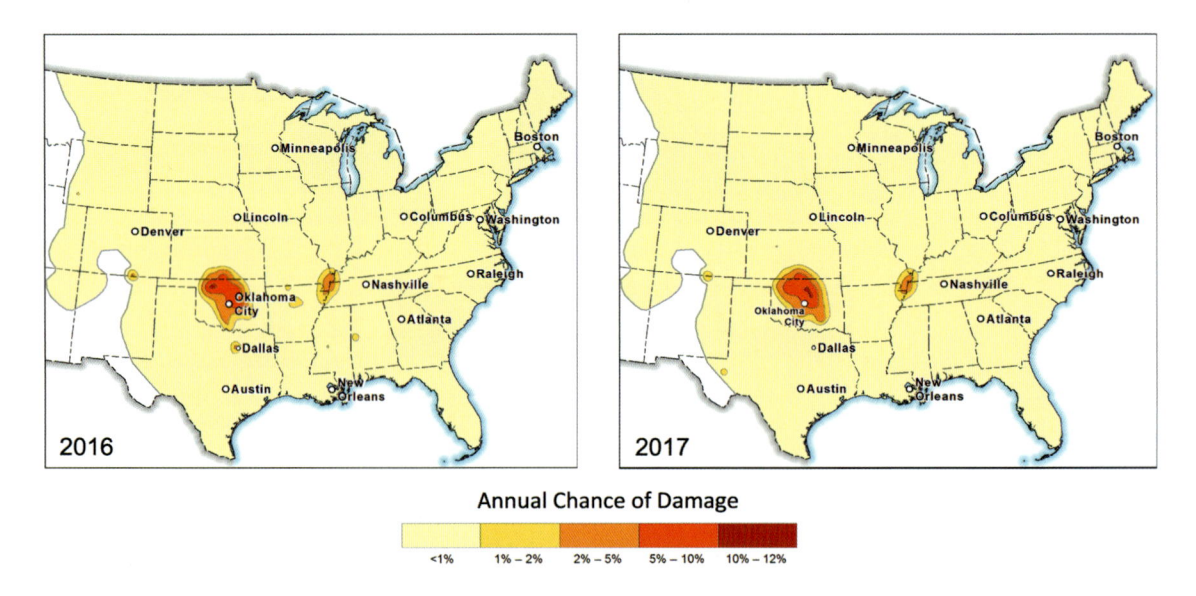

Fig. 9.20 Comparison of USGS one-year hazard forecast maps for 2016 and 2017. From Petersen et al. (2016) and Petersen et al. (2017), with permission of the USGS.

6.4 Hypocentre Estimation

In general, there are two classes of hypocentre-location methods used for downhole microseismic monitoring (Maxwell, 2014). The first approach is more common and involves inversion of traveltime observations combined with event backazimuth information derived from hodogram analysis. The second approach, called *coherency scanning*, uses beam-forming or semblance-weighted stacking (Eaton et al., 2011) to determine event locations. Like the procedure described above to obtain relative hypocentre locations for child events, the coherency-scanning approach has the practical advantage that arrival-time picking is not required. As such, this approach is particularly well suited to real-time microseismic processing. Related methods, such as reverse-time migration, are also gaining popularity in recent years. This section focuses on the first class of methods, as the coherency-scanning and reverse-time migration methods are more commonly used for processing of surface and shallow-array microseismic data. These topics are therefore covered in Chapter 7.

All methods for hypocentre estimation rely upon an accurate background model, which is constructed as part of the secondary workflow (Figure 6.1). Validation of the background model requires an algorithm to determine the hypocentres of calibration events, which should be identical to the method used to locate the detected events. These topics are organized with location methods covered in this section and background model construction in the next. Simultaneous inversion of the velocity model and hypocentre locations is also possible (Zhang et al., 2009; Tian et al., 2017) but, for simplicity, we focus here on the workflow described in Figure 6.1.

We begin with the scenario of a receiver array deployed in a single vertical observation well, which imposes a natural cylindrical symmetry on the problem as depicted in Figure 6.12. Thus far in the processing workflow, we have discussed arrival-time picking methods and techniques to determine backazimuth based on hodogram analysis of the qP arrival. Hence, given a set of N observed arrival-time picks, $\tilde{t}_i$ and M inferred event backazimuth directions, $\tilde{\theta}_j$, the best-fitting event hypocentre can be determined by minimizing an objective function, E_0, such as

$$E_0 = \sum_{i=1}^{N}(t_i(\mathbf{m}) - \tilde{t}_i)^2 + w \sum_{j=1}^{M}(\theta_j(\mathbf{m}) - \tilde{\theta}_j)^2 , \qquad (6.21)$$

where t and θ denote modelled arrival times and backazimuth orientations, respectively, while the parameter w is an adjustable weight factor that controls the relative importance of fitting the observed times versus fitting the backazimuth directions. One approach to determine this factor uses a ratio of the uncertainty in time picks to the uncertainty in directional data (Maxwell, 2014). The model parameter vector $\mathbf{m} = \{x, y, z, \tilde{\tau}\}$ includes the source position and the origin time, $\tilde{\tau}$.

Different approaches and variants on the objective function have been developed for hypocentre estimation, as summarized by Maxwell (2014). Taking into consideration the inherent cylindrical symmetry, the problem in its most basic form can be reduced to two

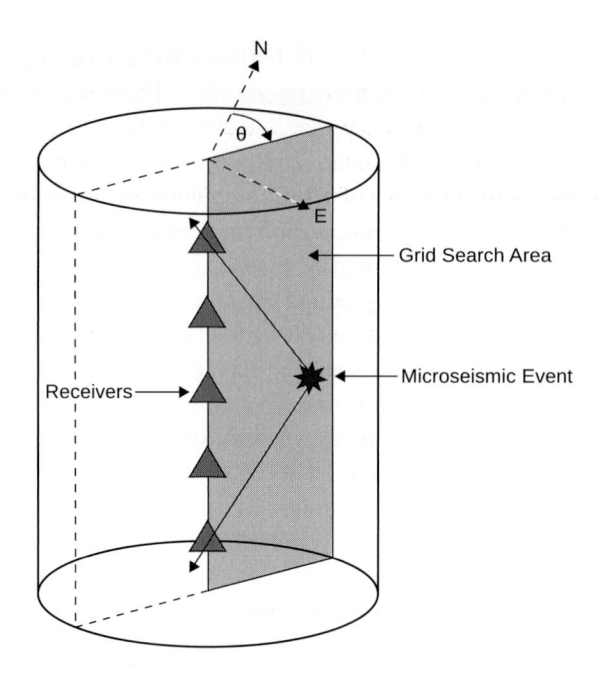

Fig. 6.12 Cylindrical symmetry and grid search area for a single vertical observation well. Modified from Jones et al. (2014), reprinted with permission from Wiley. Copyright 2013 European Association of Geoscientists & Engineers.

steps. The first step is to estimate the 2-D hypocentre location in cylindrical coordinates $\{\rho, z\}$ by minimizing a reduced form of the objective function,

$$E_\rho = \sum_{i=1}^{N} \left(\Delta t_i - \Delta \tilde{t}_i \right)^2, \qquad (6.22)$$

where $\Delta t_i = (t_S - t_P)_i$ denotes the modelled time difference between the qS and qP arrivals at the ith receiver and $\Delta \tilde{t}_i$ denotes the observed time difference. The use of the $qS - qP$ time difference, rather than absolute arrival times, decreases the number of model parameters by eliminating the origin time, $\tilde{\tau}$. A solution to Equation 6.22 was obtained by Pike (2014) using an exhaustive search on a 2-D grid, where the modelled qP and qS arrival times were obtained by ray tracing. Jones et al. (2010) obtained a solution using Geiger's method, a classical least-squares method that is covered in Chapter 9. Geiger's method uses an initial estimate of the hypocentre parameters to obtain the solution based on the generalized inverse matrix (Equation 9.6). If the errors in the arrival time picks are normally distributed, the variance in the model parameters (Equation 9.7) can be calculated in order to determine the error ellipsoid at a specified level of confidence. This confidence region encapsulates uncertainty in arrival time picks, but does not provide information about uncertainty in the velocity model.

The next step is to determine the backazimuth, which defines the azimuthal direction from the receiver to the source. As previously discussed, this can be determined from the measured particle motion of the qP wave using a covariance method. In the simplest case of a vertical observation well and a horizontally layered isotropic medium, the backazimuth is expected to be the same for every receiver. Therefore, in principle a simple backazimuthal average over all receivers provides a direct estimate of the azimuth to the source, while the variance in the backazimuth can be used to estimate uncertainty. One problem with this approach is the presence of a 180° ambiguity in the measured backazimuth, due to the fact that the covariance method provides the direction of the principal axis of the polarization ellipsoid but not a polarity. Moreover, the polarity of the qP wave is unknown, so first motion could be pointing either toward the source or away from it.

One way to overcome this directional ambiguity is to invoke a priori information about the source location. Jones et al. (2010) developed a data-driven approach that uses two or more receivers along with the measured inclination of the qP polarization. In essence, their method is based on the principle that rays from the source will tend to converge in the direction of the source, but diverge in the opposite direction. Jones et al. (2010) also developed a statistical t-test that can be used to quantify the confidence in the inferred orientation.

Combining the above two steps, the estimated hypocentre location and its uncertainty can thus be determined in cylindrical coordinates $\{\rho, z, \varphi\}$. It remains only to transform the hypocentre locations into a more convenient Cartesian reference frame. One important aspect of this transformation is that uncertainties in backazimuth lead to increasingly large positional uncertainties with distance from the well, as shown in the survey-design tests depicted in Figure 5.16.

Using a Monte Carlo simulation approach, Jones et al. (2014) noted that, when the hypocentre problem is cast in terms of $S - P$ times as in Equation 6.22, small errors in time picks or in the background model can lead to large errors in depth. Artifacts are often manifested as the clustering of events near boundaries with large velocity contrasts. To improve the results, they proposed two different approaches. One approach uses the objective function in Equation 6.21 to fit the absolute times, rather than the time difference. This approach doubles the number of observation parameters, at the expense of an increase in the number of model parameters from three to four. Jones et al. (2014) also developed a geometrical approach that they call the *equal distance time* (EDT) method. This technique defines the hypocentre as the point where the maximum number of arrival-time surfaces intersect.

The dependence of the hypocentre estimate on backazimuth, and the resulting increasing uncertainty with distance from the receiver array, is relaxed if multiple observation wells are available (Maxwell, 2014). For events that are recorded on two or more wells, there is sufficient coverage of the focal sphere that a well-resolved location can be determined using arrival-time picks alone. Typically, however, only a subset of events are well recorded at multiple observation wells; weak events are recorded only at the nearest well.

Castellanos and Van der Baan (2013) describe a case study of microseismic event locations observed using downhole receivers in multiple wells at a deep underground mining operation. The favourable recording geometry enabled the use of a powerful double-difference relocation method (Castellanos and Van der Baan, 2013) to refine the solutions in order to improve precision. The double-difference method is described in §9.2.2.

Ultimately, three distinct but often confused aspects of estimated hypocentre locations that need to be considered are (Maxwell, 2014):

1. *Accuracy*, which quantifies the closeness of an estimated location to its true location, in absolute terms. Inaccurate locations may come from systematic errors in a velocity model or neglect of anisotropy in the background model.
2. *Precision*, which quantifies the degree of scatter for a set of estimates of a location. Lack of precision may arise from data uncertainty, which is linked to the S/N level.
3. *Ambiguity/non-uniqueness*, which refers to the potential for multi-modal topology of the error surface that can lead to multiple locations that fit the observations. Linearized approaches such as Geiger's method are prone to becoming trapped in a local minimum near the initial hypocentre; hence, direct global-search methods (Lomax et al., 2014) are more robust.

6.5 Background Model Determination

In order to determine hypocentre locations it is necessary to construct a background model, which comprises the most significant part of the secondary workflow in Figure 6.1. The basic steps are:

1. Construct an initial a priori velocity model that defines $v_P(x, y, z)$ and $v_S(x, y, z)$. Typically, the initial model uses well log data, ideally from the treatment well(s) and/or observation well(s). In cases where well log data is missing or unreliable, it is sometimes necessary to construct a lithologic model first, which is then converted into a velocity model based on other information.
2. If applicable, the velocity model may be generalized to incorporate the effects of anisotropy and attenuation. In the case of a medium characterized by vertical transverse isotropy (VTI), horizontal transverse isotropy (HTI) or tilted transverse isotropy (TTI), rather than specifying elastic stiffness values it is often more convenient to parameterize the background model by augmenting v_P and v_S using the orientation of the symmetry axis as well as Thomsen parameters (Thomsen, 1986), $\delta_T(x, y, z), \varepsilon_T(x, y, z)$ and $\gamma_T(x, y, z)$. Similarly, the effects of attenuation can be incorporated by specifying quality factors $Q_P(x, y, z)$ and $Q_S(x, y, z)$.
3. Using data from calibration sources, the initial background (velocity) model is adjusted until the correct hypocentre locations are retrieved. Various strategies can be used to adjust the model, ranging from a nonlinear inversion to manual adjustment of parameters. In general, the background model is underdetermined (i.e. there are typically not enough observations to uniquely determine the model).

To illustrate the process of velocity-model construction, we consider a case study that uses a simplified approach that neglects the effects of anisotropy and attenuation and assumes horizontal layering. The study area is located in the Deep Basin of Alberta, Canada (Pike, 2014). Within this region, the term "Deep Basin" (Masters, 1979) refers to a

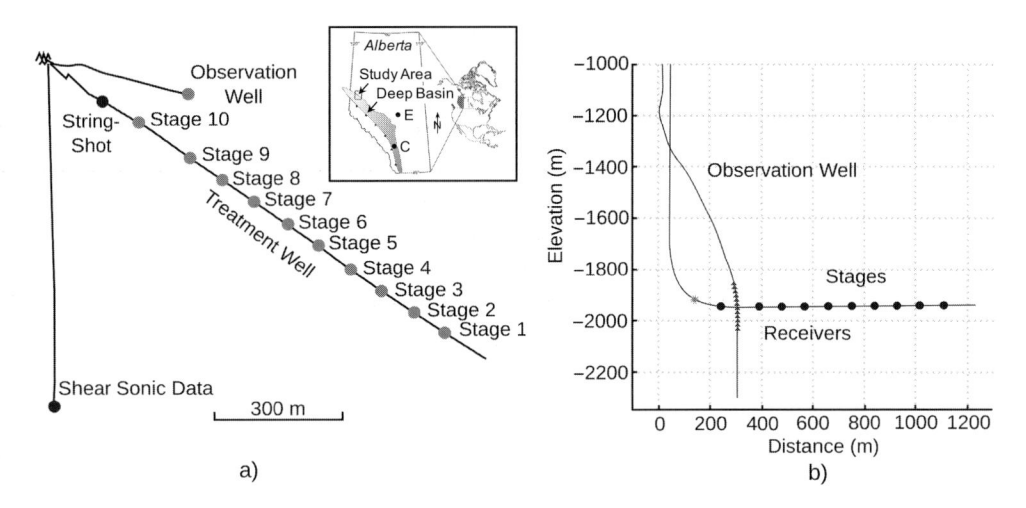

Fig. 6.13 Case study in the Deep Basin area of Alberta, Canada. a) Map view, showing that hydraulic fracturing was performed in ten stages within a horizontal treatment well. The treatment well was drilled into the Falher Member of the Cretaceous Spirit River Group. A velocity model was constructed using well log data from a third well. In the inset map, E denotes Edmonton and C denotes Calgary. b) Cross section showing location of receivers (small triangles) relative to the treatment stages. Modified from Pike (2014), with permission of the author.

clastic wedge of low-permeability, gas-bearing Cretaceous strata that reaches a maximum thickness of over 4.5 km within the foreland basin in western Canada. Figure 6.13 shows the layout of the hydraulically fractured horizontal well and the microseismic monitoring program, where the receivers were deployed in an offset deviated well. The horizontal well was drilled into gas-bearing shales of the Falher member of the Spirit River Group, which were deposited in a coastal-plain environment. Overlying strata in the Spirit River Group are characterized by shales, siltstones and coals (Pike, 2014). The horizontal well was completed with a process that utilized hydra-jets to perforate the well casing for fracture initiation; consequently, no perforation shots were available for velocity calibration. Consequently, a string shot was used as a calibration source.

Both P- and S-wave velocities are required to build the velocity model; in practice, however, S-wave logs are not available for all wells. In this example, a sonic log was available in the observation well, which was converted to velocity and upscaled from 0.1 m to 1.0 m sampling using a running median filter. It should be noted that various methods can be used to upscale from logging-tool resolution to seismic wavelengths (Rio et al., 1996; Lindsay and Van Koughnet, 2001). The observation well lacks S-wave velocity information, which requires a dipole sonic log. In order to build the initial velocity model, S-wave velocities from a nearby well were used (Pike, 2014). Figure 6.14 shows the velocity data used for construction of the initial model. Prominent low-velocity layers visible in these logs represent coal beds, which have strong elastic contrasts with the surrounding clastic units, resulting in complex seismic propagation paths and interference patterns.

The initial velocity model was built under the assumption that the medium is horizontally stratified. This is a reasonable approximation in this region, where sedimentary units are

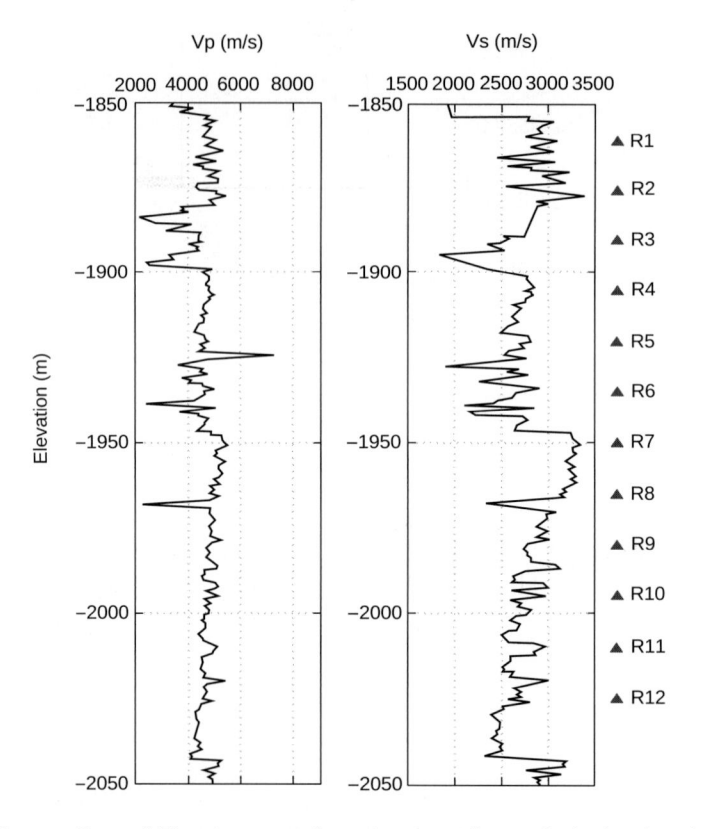

Fig. 6.14 v_P and v_S logs from an offset well. These logs are used in conjunction with v_P and other logs from the observation well to construct the initial velocity model. Receiver depth levels are shown as R1,...R12. Modified from Pike (2014), with permission of the author.

largely undeformed. Using this approximation, correlative layers were identified in both the observation well and the offset well using γ-ray log data common to both wells. A cross-plot of v_P vs. v_S is shown in Figure 6.15, coloured according to the γ-ray value in API units. Points with low γ-ray values correspond to clean sands, whereas higher values correspond with shales. Based on all of the measurements, the average v_P/v_S ratio is 1.65, although many anomalous values were found in coal intervals (Pike, 2014).

A calibration event, in the form of a string shot in the treatment well (Figure 6.13), was available to refine the velocity model. The initial velocity model was used to calculate the P- and S-wave traveltimes from the string shot location to every sensor. These calculated times were then compared with the observed times, as shown in Figure 6.16. Since the times for the initial model are not in agreement with the observed values, the velocity model was iteratively adjusted to achieve a satisfactory fit.

The final velocity model used a Vp/Vs ratio of 1.75 within the coal layers, and a Vp/Vs of 1.6 within the siliciclastic (sand and shale) sediments.

The static velocity model obtained using this iterative adjustment process was used to determine microseismic event locations. The interpretation of these results, including the profound influence of the coal layers on wave propagation in these units, are discussed

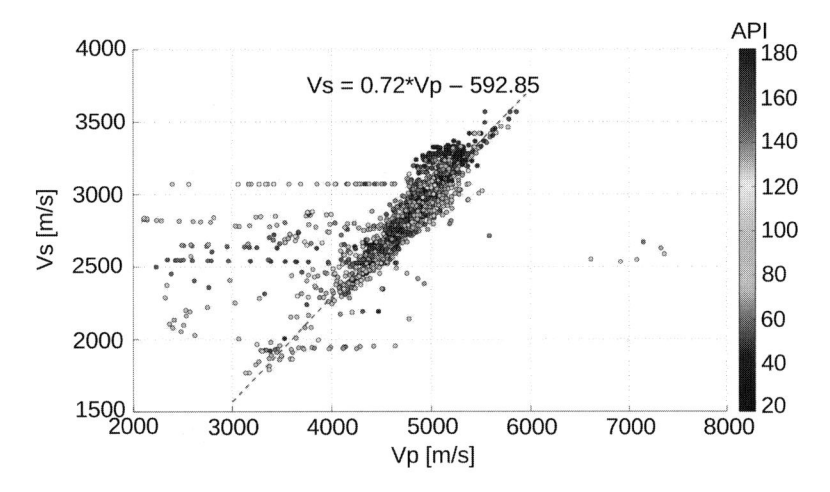

Fig. 6.15 Cross plot of v_P versus v_S within the depth interval of the microseismic events showing the line of best fit. Data points are coloured by gamma-ray value in API (American Petroleum Institute) units. Many anomalous values come from coal intervals where the S-wave picks were clipped in the dipole sonic log. From Pike (2014), with permission of the author. (A colour version of this figure can be found in the Plates section.)

in Chapter 8. In this case, only one calibration event was available, due to the nature of the completion process used here. In many cases, multiple calibration sources are available, especially if a perf-and-plug completion method is used (§4.2.1). In this scenario, it may be advantageous to update the velocity model during the completion in order to monitor velocity changes that are caused by the induced fracture network (Akram, 2014).

6.6 Source Characterization

Magnitude is a fundamental source parameter that provides a first-order estimate of event size based on observations of radiated seismic waves. In particular, for monitoring of induced seismicity, the accuracy of magnitude holds special significance as a metric used for traffic light systems (§9.4). A number of different magnitude scales have been developed and calibrated for moderate to large earthquakes. Similar methods are used to estimate event magnitudes using downhole microseismic data, under the general assumption that magnitude calibration for earthquakes can be extrapolated over orders-of-magnitude and applied to microseismicity. As a further cautionary note, large discrepancies in magnitude have been reported for different processing of the same dataset (Shemeta and Anderson, 2010); hence, prior to analysis of information from a microseismic catalogue it is important to be aware of the method used to determine event magnitudes.

Stork et al. (2014) calculated seismic moment and magnitude using downhole microseismic waveforms recorded during the Cotton Valley experiment (Walker, 1997; Rutledge

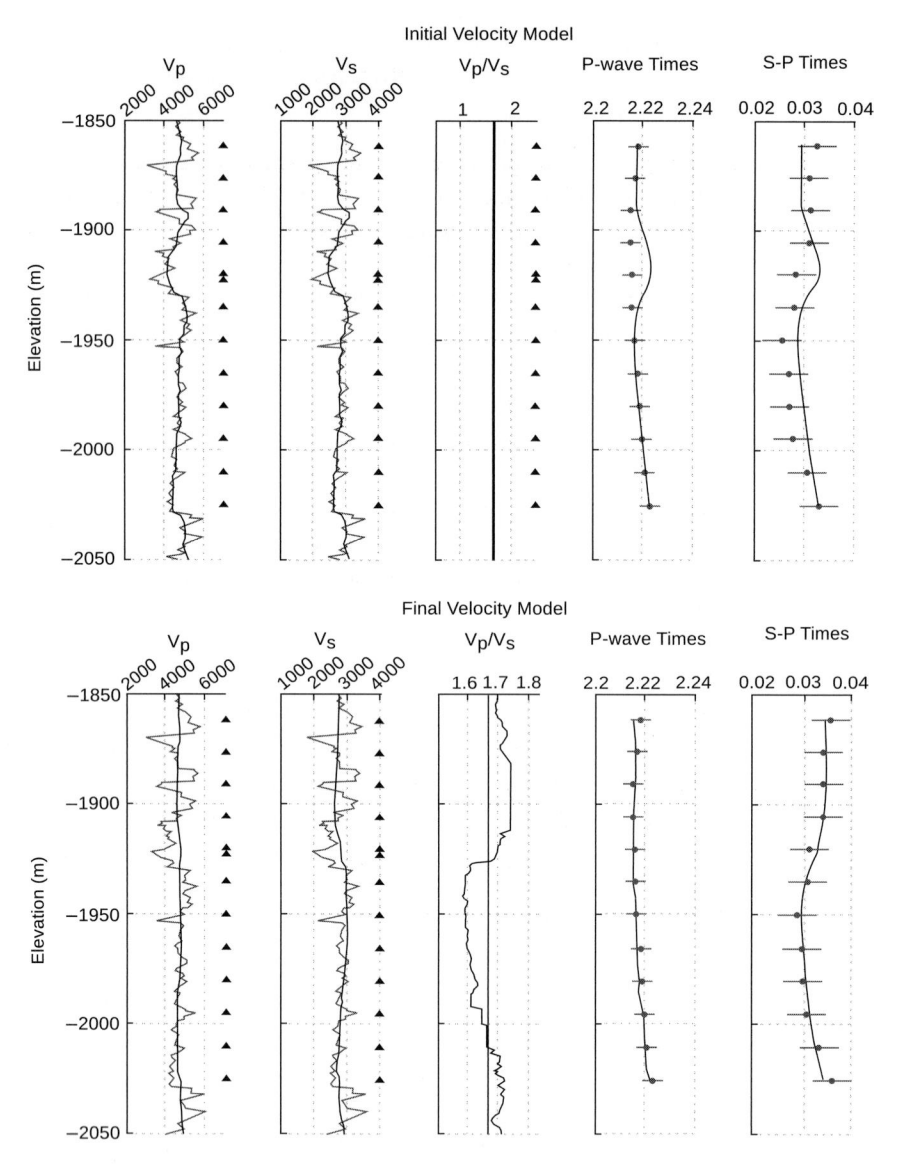

Fig. 6.16 Example of velocity-model refinement using data from a calibration source. In both the initial and final velocity models, the original v_P and v_S logs are shown as thin curves, while corresponding re-sampled and upscaled logs are shown as thick curves. On the traveltime graphs, the solid lines represent the computed times, while the picked times are indicated by circles with error bars. After model refinement, the resultant v_P/v_S is approximately 1.75 within the coal intervals (marked by C) and 1.6 within the clastic interval. Modified from Pike (2014), with permission of the author.

and Phillips, 2003) discussed in §5.1. Figure 6.17 shows an example of their analysis for one event. As illustrated in Figure 6.17a, seismic moment can be determined using a time-domain calculation expressed using an integral derived from the far-field elastodynamic Green's function,

$$M_0 = \frac{4\pi\rho v^3 r}{R} \int_{t_1}^{t_2} u(t)dt, \qquad (6.23)$$

where $u(t)$ is the displacement seismogram after correction for instrument response (Appendix B), while velocity v and average radiation amplitude R have different values for P and S waves. Once the seismic moment is calculated, the moment magnitude is easily determined from Equation 3.53.

The magnitude can also be measured based on the low-frequency asymptote of the displacement spectrum, as discussed in §3.7. Figure 6.17b shows spectral fits using several different spectral models. The two models that are considered (Brune and Boatwright models) converge at low frequency, and therefore predict the same magnitude value. The models diverge at higher frequency, however, which affects the determination of corner frequency.

Stork et al. (2014) compared magnitudes calculated using different approaches and with different assumptions and found discrepancies as large as 0.6 magnitude units. In order to reduce uncertainty in magnitude determination, they provided a series of recommendations, including:

- Focal-mechanism solutions should be used to determine the radiation pattern corrections, averaged over P- and S-phases and the available sensors.
- Noise should be estimated and removed from the displacement spectra.
- An attenuation correction should be applied.
- A minimum of four receivers should be averaged.
- Measurements should be made using recordings with SNR > 3.

Various other source parameters can be obtained from downhole microseismic observations, including spectral parameters (Eaton et al., 2014d) and seismic moment tensors (Baig and Urbancic, 2010).

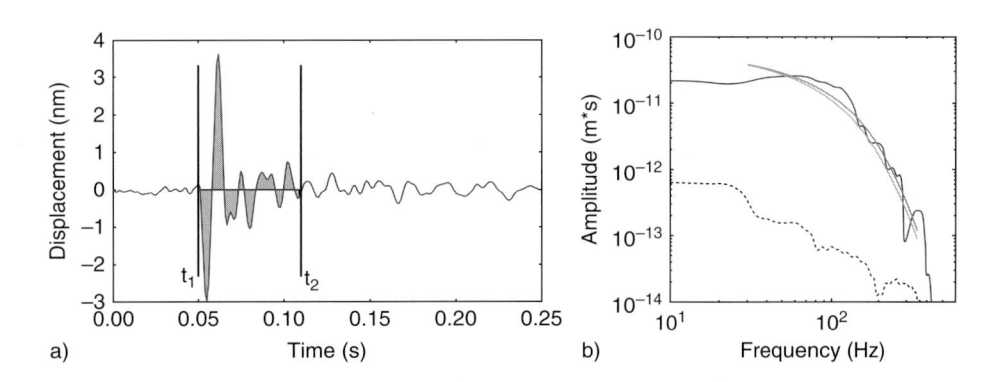

Fig. 6.17 a) Example S-wave seismogram (displacement) used for magnitude calculation. b) Smoothed S-wave amplitude spectrum of time window in a) along with the best-fitting Brune model and Boatwright model with the same Ω_0 and Q. The noise spectrum is given by the black dashed line. From Stork et al. (2014).

6.7 Summary

The main goal of microseismic processing is to produce a catalogue of event locations, times and magnitudes. The processing is organized into primary and secondary workflows. The primary workflow includes event detection, picking P- and S-wave arrivals, performing wavefield separation, computing hypocentre locations and estimating source parameters such as magnitude. The secondary workflow, which includes the determination of receiver orientation and construction of the background model, operates in parallel with the primary workflow. Determination of a background model is of core importance, as errors in the velocity model map into hypocentre location errors. The initial velocity model is usually constructed using available well log data. Calibration sources are required in order to refine the velocity model. A case study from western Canada illustrates some of the practical difficulties in dealing with low-velocity layers, in this case coal.

Event locations are generally reported using Cartesian geographic coordinates. For deviated observation wells, correct determination of the sensor orientation requires considerable care, particularly since the orientation of the sensor within the wellbore is not known a priori due to twisting of the wireline during installation of the array. Polarization analysis can be performed using a covariance matrix method, which yields the major axis of the polarization ellipsoid as an estimate of the linear polarization of an observed pulse.

Event detection can be performed using single-receiver or multi-receiver methods. The most widely used single-receiver approach is based on the ratio of the short-time average/long-time average amplitude (STA/LTA) values within specified time windows. Single-receiver methods for arrival-time picking include variants of STA/LTA using parameters such as kurtosis, or the Akaike Information Criterion (AIC) method. Multi-receiver methods for phase picking make use of cross-correlation to refine time picks. Wavefield separation, in which the three-components of ground motion at the receiver are projected onto ray-centred coordinates corresponding to qP, qS_1 and qS_2 arrivals, is highly effective for improving and validating arrival-time picks.

Enhanced event detection can be achieved using matched filtering and/or subspace detection. Matched filtering applies cross-correlation of the continuous waveform data using template ("parent") events in order to detect similar "child" events. For downhole microseismic data, matched filtering can be facilitated using stacked cross-correlation functions. Subspace detection is a more general approach in which signals are represented as a combination of orthonormal basis waveforms.

Hypocentre locations can be obtained from downhole microseismic data by inverting the qP and qS arrival times, combined with polarization analysis to determine the backazimuth of the event with respect to the receiver array. Microseismic sources can then be characterized on the basis of magnitude, moment tensor and spectral parameters.

6.8 Suggestions for Further Reading

- Event detection and arrival-time picking: Akram and Eaton (2016b).
- Subspace detection: Harris (2006).
- Magnitude and scaling relations: Kanamori and Anderson (1975).

6.9 Problems

1. Suppose that recorded ground motion as a function of time t ($0 < t < T = N\pi$) for an event is defined by
$u_x = a \sin t$
$u_y = b \cos t$
$u_z = 0$.
Construct a covariance matrix using Equation 6.1 and show that the two nonzero eigenvalues of this matrix are proportional to a and b.

2. Consider a time series given by random white noise with a root-mean-squared amplitude of A_N and a unit impulse at $t = 0$. Draw a sketch of the characteristic functions using STA/LTA, kurtosis and AIC approaches for signal detection.

3. Consider a shale layer that contains thin coal intervals, as defined in problem 4b in chapter 1. In addition, the shale density is 2200 Kg/m³ and the coal density is 1800 Kg/m³. Determine the layer parameters (isotropic v_P and v_S, or VTI parameters as appropriate).

a) Take the thickness-weighted average of the v_P and v_S of the constituent beds. This is a commonly used approach for blocking velocity models.
b) Take the *time-weighted* average of the v_P and v_S of the constituent beds.
c) Take the time-weighted average of the slownesses (reciprocal of velocity) of the constituent beds.
d) Calculate the equivalent VTI medium using Backus averaging (as in Problem 4b). Express your results using Thomsen (1986) parameters of $\alpha, \beta, \gamma, \delta$ and ϵ.

4. Use the Brune spectral method to estimate the magnitude of the event depicted in Figure 6.17, assuming v_S = 2500 m/s, ρ = 2600 Kg/m³ and r = 500 m.

5. **Online exercise**: A Matlab tool is provided that enables parameter-sensitivity testing using short-time average/long-time average (STA/LTA), modified energy ratio (MER) and Akaike Information Criterion (AIC) methods. Waveform examples of microearthquakes recorded during hydraulic fracture monitoring are provided as test signals.

7 Surface and Shallow-Array Microseismic Processing

> Indeed, we often mark our progress in science by improvements in imaging.
>
> Martin Chalfie (Nobel Lecture, 2008)

Although the goals of surface-based microseismic monitoring are the same as those for downhole monitoring, the data-processing methods are quite distinct. For acquisition programs that are deployed at or near the surface, the relatively large number of sensors (up to thousands), lower S/N environment and larger aperture lead to a processing approach that is typically more akin to imaging methods used for controlled-source seismic exploration than to methods used for earthquake seismology (Duncan and Eisner, 2010). In order to detect weak events, surface microseismic monitoring exploits the power of stacking and other signal-processing methods for noise reduction and enhancement of signal coherence. Shallow-well arrays constitute a hybrid approach; with this acquisition design, the seismic wavefield is recorded with sparser spatial sampling in a higher S/N environment, but the aperture and overall recording geometry – that is viewing from above – remain almost the same as for surface arrays.

The theoretical groundwork for source imaging using surface arrays was laid by McMechan (1982), who observed that seismic sources can be imaged by propagation of the observed wavefield backwards in time. This approach, in which the observed seismograms are treated as time-dependent boundary values, provided a foundation for reverse-time imaging (RTI) (Artman et al., 2010). Kao and Shan (2004) proposed an alternative method, which they called a source-scanning algorithm (SSA). This method operates by summing absolute amplitudes observed at all stations at predicted arrival times for a set of trial hypocentre locations and origin times. SSA belongs to a class of source-imaging methods that are based on a form of *Kirchhoff migration*.

The emergence, starting in 2004, of surface-based microseismic methods for monitoring hydraulic fracturing (Duncan, 2005; Lakings et al., 2006) was accompanied by vigorous technical debate (Eisner et al., 2011b, a). Given then-established low-magnitude characteristics of microseismicity induced by hydraulic fracturing (Warpinski et al., 1998; Maxwell et al., 2002; Rutledge and Phillips, 2003), the debate centred on the ultimate detection capabilities of sensors deployed in noisy surface environments. Skeptics argued that S/N improvements from stacking saturate at an upper limit, as demonstrated by Cieslik and Artman (2016), and suggested that amplitude loss from geometrical spreading and inelastic attenuation are insurmountable, irrespective of array design.[1] This argument was

---

[1] These arguments are reminiscent of the differences in sensitivity of astronomical observations from orbiting platforms versus surface-based astronomical observations, which suffer from the obscuring effects of Earth's atmosphere.

falsified, for example, by well-documented detection of calibration events using surface arrays (Chambers et al., 2010), as well as benchmarking comparisons between near-surface and downhole microseismic monitoring that demonstrate overall consistency in event locations (Maxwell et al., 2012). The issue was ultimately laid-to-rest by an experiment that employed simultaneous, time-synchronized monitoring with a variety of sensor configurations, including horizontal and vertical near-reservoir downhole arrays, shallow-well arrays, surface lines and 2-D patches combining single component and multicomponent sensors (Peyret et al., 2012). Figure 7.1 highlights one of the reported results from this experiment, showing the amplitude variation with depth for two relatively large events compared with a modelled amplitude trend using elastic finite-difference wavefield simulations (Maxwell et al., 2012). While a decrease in S/N from ≈ 1000 for the deepest receivers to ≈ 1 at the surface was reported, some microseismic events are discernible at the surface without stacking and many more could be directly correlated after processing (Maxwell et al., 2012).

This chapter describes the basic theory and practical aspects of full-wavefield imaging methods that are used to process continuous recordings from surface- and shallow-well monitoring arrays. A number of relevant topics are not covered because they are dealt with elsewhere. For example, although matched filtering has been developed for near-surface

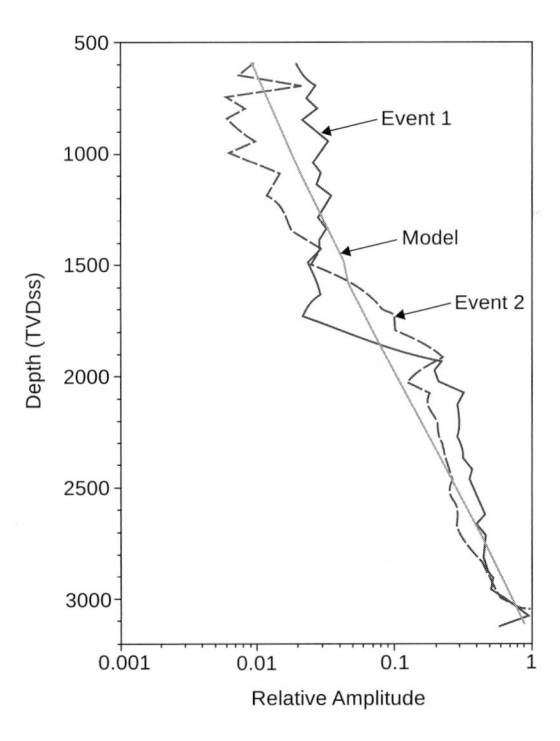

Fig. 7.1 Tracking of amplitude from the reservoir to the surface for two relatively large microseismic events, compared with modelled amplitude trend. Republished with permission of Society of Exploration Geophysicists, from *Microseismic Imaging of Hydraulic Fracture: Improved Engineering of Unconventional Shale Reservoirs*, S. Maxwell, copyright 2014; permission conveyed through Copyright Clearance.

monitoring (Eisner et al., 2008), this approach is not discussed here as it is essentially the same as template-based methods discussed in §6.3.4 for downhole microseismic processing and §9.2.1 for induced seismicity. In addition, construction of a velocity model is important, as in the downhole microseismic processing, but the methods described in the previous chapter are applicable for surface monitoring also.

The chapter begins with a description of wave-amplification effects in the near-surface environment, which explain in part why monitoring from the surface is feasible. Next, beamforming and vespagram methods used for global-seismic arrays are considered, as these methods are easily adaptable for analysis of data from surface microseismic arrays. Following this, a basic workflow for surface-microseismic processing is described, including a theoretical framework for elastic imaging based on variants of the Kirchhoff migration method and reverse-time imaging method. The elements of resolution and uncertainty are discussed, from the perspectives of accuracy and precision. Finally, case examples are presented for both surface and shallow-well configurations.

7.1 The Free-Surface Effect and Wave Amplification

Wavefield amplification that occurs in the vicinity of the *free surface* is a significant factor that leads to enhancement of signal amplitudes for surface and near-surface recording (Eisner et al., 2011b). A free-surface boundary condition is one in which stress components vanish along the surface (Eaton, 1989). A consequence of this boundary condition is that the surface acts as a near-ideal reflection interface, in which upcoming *P* and *S* wavefields generate a phase-shifted reflection and a mode-conversion. In the case of a vertically incident *P*-wave, the downgoing reflection has negative polarity and the superposition of the upgoing and downgoing wavefields at the surface results in doubling of the signal amplitude. The interaction of upgoing elastic waves with the surface is analogous to acoustic sea-surface reflections, which generate well-known ghost events for marine seismic data (Parkes and Hegna, 2011). For sensors installed in a shallow well, a time lag exists between the upgoing and downgoing wavefield that produces a characteristic series of notches in the observed frequency spectrum, similar to the spectral notches in Figure 5.7. For the general case of oblique incidence, the interaction of superposed seismic waves at the free surface is more complex, in accordance with plane-wave reflection coefficients from the Zoeppritz equations (Dankbaar, 1985).

Observed interactions of the seismic wavefield with the free surface are illustrated in Figure 7.2, which shows unfiltered waveform data from a M_W 1.6 microearthquake recorded by a vertical array of geophones cemented into a 27.4 m-deep well. Three-component (3C) 15 Hz geophones are situated at the surface and at the bottom of the shallow well, in addition to three single-component (vertical) 15 Hz geophones at 12 m, 17 m and 22 m depth and coincident 3C 15 Hz geophone at the surface and a direct-burial broadband (BB) seismometer. The pre-signal noise is not discernible in this record, due to the large amplitude of the signal for this event compared with the amplitude of the background noise. A slight moveout of the upgoing *P* and *S* waves can be used to estimate the near-surface velocities at this shallow-well geophone array.

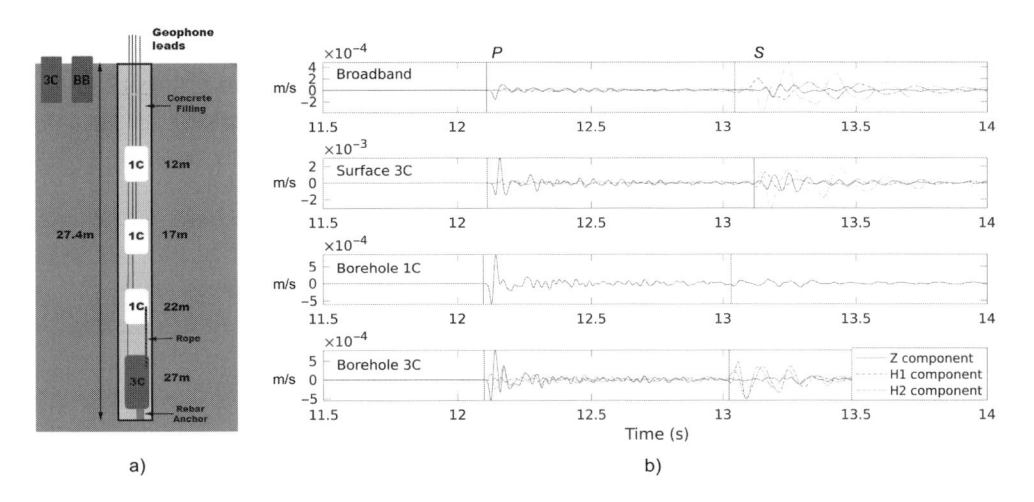

Fig. 7.2 Waveforms from a M_W 1.6 microearthquake recorded using a shallow-well geophone array, along with a surface three-component (3C) geophone and direct-burial broadband (BB) seismometer. a) Sensor configuration. b) P and S waveforms. The borehole vertical component (1C) waveform is at 22 m depth. Note amplification of the geophone recordings at the surface due to the free-surface effect as well as phase and amplitude differences between the surface geophone and broadband seismometer.

On the surface geophone the P-wave pulse appears to be nearly symmetric; this is caused by superposition of the upgoing incident wave and reverse-polarity downgoing reflected P wave, both of which have asymmetric pulses with approximately minimum-phase characteristics (Appendix B). The incident S wave has lower dominant frequency; it is apparent on the (unrotated) H_1 and H_2 horizontal components of the deep geophone, but is not evident in the intermediate vertical-component geophone. The incident S wave is recorded on the vertical component of the surface geophone, likely due to complex wavefield interactions at the free surface. The surface geophone waveforms exhibit a fourfold increase for the P-wave amplitude relative to the deep geophone, and sixfold increase for the S-wave. As elaborated below, these amplification values exceed the approximate amplitude doubling predicted by the free-surface effect. The broadband signals show less amplification and a different amplitude and phase response than the geophones, due to the differences in their instrument response.

Near-surface wavefield interactions are impacted in various ways by the presence of overburden. An overburden layer is nearly ubiquitous in most sedimentary basins and is composed of weathered material that is typically characterized by low P-wave velocity (800–1800 m/s) and extremely low S-wave velocity (100–400 m/s) (Park et al., 1999), as well as high attenuation ($Q < 50$). For example, in Figure 7.2, all of the geophone traces reveal a P-wave *coda*, which consists of scattered waves that manifest with diminishing amplitude after the primary phase. This coda is partly a result of wave energy that is trapped in a near-surface low-velocity wave guide, together with scattering from near-surface heterogeneities such as irregular bedrock topography. Furthermore, in accordance with Snell's Law the presence of a low-velocity layer at the surface leads to bending of rays into nearly vertical incidence, meaning that vertical-component geophones, which are predominantly used in practice for surface-monitoring arrays, are ideally suited to record the incident P wave. The presence of a low-velocity layer also leads to further amplification of ground motion, both signal and noise, relative to sites where bedrock is exposed (Building Seismic Safety Council, 2003). The increased signal amplification arises from the conservation of energy (§3.2.3), due to reduction of velocity and density of the near-surface material.

7.2 Beamforming and Vespagrams

Beamforming is a shift-and-sum procedure to process seismic waveforms from an array (Ingate et al., 1985). The arrival times (or moveout) at different receivers depend on the source location and origin time, as well as the subsurface velocity structure. If time shifts are applied such that phase arrivals at every station are aligned, coherent signal will be amplified while suppressing incoherent noise.

In global seismology, the term beamforming[2] assumes an incident plane wave and refers to application of a shift-and-sum process for fixed values of slowness and backazimuth (Rost and Thomas, 2002). If the slowness and backazimuth are unknown, they can be estimated using the *vespa process* (velocity spectral analysis) (Davies et al., 1971), by picking the horizontal slowness value, p, that corresponds with the maximum stacked amplitude on a *vespagram*

$$V(t, p) = \frac{1}{M} \sum_{i=1}^{M} u_i(t - \tau_i), \quad (7.1)$$

where $u_i(t)$ is the seismogram for station i, $\tau_i(p)$ is the relative traveltime and M represents the number of stations in the array. This simple waveform-stacking approach is effective for teleseismic arrivals (i.e. arrivals observed at source-receiver distances of greater than ~ 3000 km) since the pulse shape is relatively invariant across the aperture of the array. A similar expression can be written for estimating backazimuth (Rost and Thomas, 2002).

Nonlinear variants of the vespa process are used to improve resolution of teleseismic array processing; these methods are discussed briefly here, due to their utility for surface and near-surface microseismic processing. For example, the Nth-root process (McFadden et al., 1986) ($N = 2, 3, 4, \ldots$) can be used to enhance the resolving power. This is calculated by summing the Nth root of the absolute values of the data samples, while preserving the sign for each sample:

$$V'_N(t, p) = \frac{1}{M} \sum_{i=1}^{M} |u_i(t - \tau_i)|^{1/N} \operatorname{sign}\{u_i(t)\}. \quad (7.2)$$

The Nth root vespagram, $V_N(t, p)$, is obtained by taking the Nth power, while again preserving the sign of each data sample:

$$V_N(t, p) = |V'_N(t, p)|^N \operatorname{sign}\{V'_N(t, p)\}. \quad (7.3)$$

---

[2] In the passive-seismic monitoring literature, the term beamforming is sometimes used rather loosely in reference to any array-based coherency-scanning approach.

The Nth root process is useful for analysis of weak signals, but it introduces waveform distortion (Rost and Thomas, 2002).

Another nonlinear variant on the vespa process is the *phase-weighted stack* (PWS) method (Schimmel and Paulssen, 1997). This method uses the instantaneous phase $\varphi(t)$ from complex-trace analysis (Appendix B). For a set of M traces, a stacked function can be defined using the instantaneous phase,

$$c(t) = \frac{1}{M} \left| \sum_{j=1}^{M} e^{i\varphi_j(t)} \right| , \qquad (7.4)$$

which defines a parameter $c(t)$ that represents a measure of stack coherency (see discussion below). The maximum value of this parameter is unity for an identical set of traces, and zero for an incoherent stack. In order to reduce noise and detect weak, coherent signals, the PWS approach can be combined with vespagram analysis as follows:

$$V_{PWS}(t,p) = \frac{1}{M} \sum_{j=1}^{M} u_j(t - \tau_j) \left| c(t - \tau_j) \right|^{\nu} , \qquad (7.5)$$

where sharpness of the coherency filter is controlled by the exponent, ν. A similar measure of stack coherency is the *semblance*, which can be defined as (Neidell and Taner, 1971)

$$S_c(t) = \frac{\left[\sum_{j=1}^{M} u_j(t) \right]^2}{\sum_{j=1}^{M} u_j(t)^2} , \qquad (7.6)$$

which measures the power of the sum of M data samples, normalized by the total power in the set of samples. Like the previous coherency parameter, semblance falls within the range $0 \leq S_c \leq 1$ and a semblance value of unity implies ideal trace-to-trace coherency.

7.3 Basic Processing Workflow

A basic workflow for processing surface microseismic data (P. M. Duncan, *pers. comm.*, 2017), listed below, provides a step-by-step framework for processing continuous recordings to detect and locate microseismic events using surface and shallow-well arrays.

1 Data formatting and removal of instrument response
2 Editing of noisy traces
3 Data preconditioning
4 Velocity analysis (calibration)
5 Application of receiver statics corrections
6 Focal-mechanism estimation
7 Imaging and event detection

As described in Chapter 5 and Appendix B, the instrument-response correction in the first step of this workflow accounts for the transfer function of the sensors, thus converting the recorded signals into units of ground motion. An approximate instrument scaling can be achieved by dividing the raw recorded signals by the scalar correction term ξ_c,

$$\xi_c = S_i \, 10^{g_a/20} \,, \tag{7.7}$$

where S_i is the sensitivity of the sensor in units of V/m/s and g_a is the amplifier-*gain* setting (in dB) used by the digitizer. This scaling correction provides a reasonable approximation for the complete instrument response at frequencies where the transfer function is "flat" (see Figure 5.13), such as at frequencies that are well above the natural frequency of a geophone. A trace editing procedure in the second step is used to remove traces containing high-amplitude noise, such as transient noise from a vehicle driving past a station.

Various data preconditioning processes can be used in the third step. Commonly, a bandpass filter is applied to isolate the useful (highest S/N) bandwidth of the recorded data. Amplitude normalization and spectral balancing can also be applied to improve the stacking (beam-forming) process. The construction of a background model using a priori information such as well logs follows the same approach as described in Chapter 6. Velocity analysis is a procedure to refine and validate the velocity model to optimize the traveltime fit to the moveout of the observed arrivals. In this context, the effects of anisotropy can be significant at large offsets, where the moveout may deviate from the predicted traveltime curve for an isotropic medium (Tsvankin and Thomsen, 1994).

Variable time shifts occur at receiver locations that can negatively impact the beam-formed image. These are caused by surface topography coupled with spatially variable overburden thickness and velocity. A simplified approach to correct for the kinematic effect of these time shifts is through the use of *statics corrections*. This type of correction is widely used to process exploration seismic data, as exemplified by Cary and Eaton (1993). In exploration seismology, use of the term "statics" emphasizes that, unlike moveout, statics corrections do not vary with time. Statics corrections can be applied as simple time shifts to the recorded waveforms. This avoids the need for explicit incorporation of near-surface complexities into the velocity model. As described by Chambers et al. (2010), short-wavelength statics can produce wavefield incoherence that diminishes the effectiveness of beamform stacking, whereas long-wavelength statics can lead to bias in event locations.

Well designed surface and shallow-well arrays usually provide sufficient coverage of the focal sphere that the focal mechanism of large events can be determined with high confidence (Eaton and Forouhideh, 2011). Indeed, estimation of the source mechanism can be incorporated directly into the event-localization procedure (Zhebel and Eisner, 2014).

The imaging and event-detection step is arguably the most important element of the processing sequence and is described in detail in the next section. As an alternative approach, this processing step can also be carried out using processing tools that are described in Chapter 6, including matched filtering and/or subspace detection coupled with absolute and relative hypocentre methods. Various other aspects of the processing workflow, particularly receiver statics and focal mechanism determination, are highlighted in case studies at the end of this chapter. These case studies cover both surface and shallow-array approaches for data acquisition.

7.4 Elastic Imaging

We begin in a very general manner by briefly considering the *Kirchhoff–Helmholtz* formula for anisotropic elastic media (Pao and Varatharajulu, 1976), which expresses a fundamental representation theorem in elastodynamics (Aki and Richards, 2002). In a general anisotropic medium,[3] the mth displacement component of the seismic wavefield at a point P at a location $\mathbf{x}'$ that is interior to a closed surface, Σ (Figure 7.3) is given by

$$u_m(\mathbf{x}') = \int_{\Sigma} \hat{n}_j c_{ijkl} \left[u_i(\mathbf{x}) G_{lm,k}(\mathbf{x}'|\mathbf{x}) - G_{lm}(\mathbf{x}'|\mathbf{x}) u_{l,k}(\mathbf{x}) \right] d\Sigma, \tag{7.8}$$

where $\hat{\mathbf{n}}(\mathbf{x})$ is the normal to the surface Σ and $G_{lm}(\omega, \mathbf{x}'|\mathbf{x})$ denotes the elastodynamic Green's function describing the displacement at $\mathbf{x}'$ for a point source at $\mathbf{x}$. In addition, $G_{jm}(\omega, \mathbf{x}'|\mathbf{x})$ is a solution to

$$c_{ijkl}(\mathbf{x}') G_{lm,ik} - \rho(\mathbf{x}') \omega^2 G_{jm} = -\delta_{jm} \delta(\mathbf{x}' - \mathbf{x}), \tag{7.9}$$

where δ_{jm} is the Kronecker delta and $\delta(\mathbf{x}' - \mathbf{x})$ is the Dirac delta. The surface integral in Equation 7.8 can be considered as a statement of *Huygens' principle*, as it implies that every point on Σ radiates a secondary wavefield that propagates, via the elastodynamic Green's function $\mathbf{G}(\omega, \mathbf{x}'|\mathbf{x})$, to an interior point at $\mathbf{x}'$ (Pao and Varatharajulu, 1976). As a theoretical framework for imaging, this formula is powerful: it assures us that the elastic wavefield at any point within the interior of a subsurface region can, in principle, be completely reconstructed through a weighted integration of the wavefield and its normal derivative measured around an arbitrary bounding surface.

Since we are free to choose any bounding surface, we begin by tailoring the geometry of Σ to suit the problem of interest by representing it as a lower hemisphere that is split into

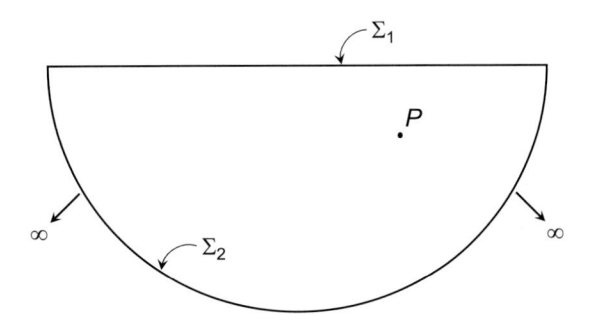

Fig. 7.3 The Sommerfeld radiation condition for a hemispherical surface Σ composed of two parts, Σ_1 and Σ_2. This condition implies that the contribution of the wavefield on Σ_2 tends to zero as the radius of Σ_2 goes to ∞. This can be used with the elastic Kirchhoff–Helmholtz formula (Equation 7.8) to calculate the wavefield at a point P in the subsurface.

---

[3] Strictly speaking, the Kirchhoff–Helmholtz formula in Equation 7.8 assumes a source-free region interior to Σ; for the sake of simplicity we are therefore careful to identify sources as localized regions of high-amplitude waves.

two parts, as shown in Figure 7.3. Here, Σ_1 represents a circular region of the horizontal plane where wavefield measurements are made, while Σ_2 forms the enclosing lower part of the surface. As the radius of Σ_2 goes to ∞, the *Sommerfeld radiation condition* holds that the contribution to the summation from this part of the surface can be neglected. By invoking the Sommerfeld radiation condition, the surface of integration is thus reduced to upper surface of a half-space, rather than a complete bounding surface.

It is useful at this point to consider *Kirchhoff migration*, an imaging method introduced by Schneider (1978) that is now routinely used to process seismic-reflection data. The theoretical foundation is based upon the Kirchhoff–Helmoltz formula for acoustic media, with an appropriate choice of Green's function to eliminate the normal derivative of u at the surface. This method is used to obtain a migrated image, $\psi_m(\mathbf{x}', \omega)$, from the zero-offset wavefield, $\psi(\mathbf{x}, \omega)$, by evaluating the integral (Carter and Frazer, 1984)

$$\psi_m(\mathbf{x}', \omega) = 2 \int_\Sigma \psi(\mathbf{x}, \omega) \partial_n G \, d\Sigma, \tag{7.10}$$

where $\mathbf{x}$ gives the position on Σ ($z = 0$). By using a ray-theoretical form of the Green's function, Kirchhoff migration can be expressed as a weighted integration along a traveltime curve. An equivalent representation of Kirchhoff migration views it as a weighted backprojection of each data point onto an *isochron surface* (surface of constant traveltime) that passes through the scattering object (Esmersoy and Miller, 1989).

The latter viewpoint is incorporated into a class of Kirchhoff-style methods for wavefield source imaging (Kao and Shan, 2004; Chambers et al., 2010; Pesicek et al., 2014; Chambers et al., 2014; Hansen and Schmandt, 2015; Vlček et al., 2016; Trojanowski and Eisner, 2017) that have their roots in the Kirchhoff–Helmholtz formula. These methods define a grid of potential source locations within the expected source region. The backprojection process can be expressed as a generalized type of beamforming with the use of an image function, $F(\mathbf{x}, t)$,

$$F(\mathbf{x}, t) = \sum_{k=1}^{N} R_P^k \left(t + \tau_P^k(\mathbf{x}) \right) + R_S^k \left(t + \tau_S^k(\mathbf{x}) \right), \tag{7.11}$$

where $R_P^k(t)$ and $R_S^k(t)$ are characteristic functions for P and S waves, which are derived from the observed wavefield at the kth receiver, while $\tau_P^k(\mathbf{x})$ and $\tau_S^k(\mathbf{x})$ are the traveltimes (or moveout correction) to the kth receiver location from $\mathbf{x}$ for P and S waves. Various types of characteristic functions are described below. The source-detection process is usually based on identification of local maxima of the image function that meet user-defined detection criteria. In this sense, wavefield imaging algorithms for event detection and hypocentre location can be characterized as "locate, then pick," in contrast to downhole microseismic processing where the procedure can be characterized as "pick, then locate." The picking process can be formalized using detection theory (Johnson and Dudgeon, 1992) in which a maximum likelihood approach is used to choose between the null hypothesis, H_0 = signal absent, versus the alternative process, H_1 = signal present (Thornton and Mueller, 2013).

Returning to the backprojection formula, numerous variants exist. For example, many published algorithms consider only one type of wave, for the sake of simplicity. Alternatively, in order to compensate for errors in the predicted traveltime, the method used by

Pesicek et al. (2014) includes an additional summation over a time window of size $2M$ centred on the predicted traveltime

$$F(\mathbf{x}, t) = \sum_{k=1}^{N} \left\{ \sum_{m=-M}^{M} R_P^k \left(t + \tau_P^k(\mathbf{x}) + m\delta t \right) + R_S^k \left(t + \tau_S^k(\mathbf{x}) + m\delta t \right) \right\}. \quad (7.12)$$

The simplest form of the characteristic function is the wavefield itself,

$$R_P^k(t) = R_S^k(t) = u_k(t). \quad (7.13)$$

This characteristic function leads to the classic *diffraction-stack* algorithm. Unlike teleseismic beamforming methods discussed above, this simple approach is impacted by waveform incoherency between receivers that can degrade the stack. Causes of stack degradation include varying path effects as well as polarity changes across the array due to the radiation pattern of the source (§3.5.1). As noted by Schisselé and Meunier (2009) and Chambers et al. (2014), since the source-radiation pattern is neglected, this method is expected to produce a complex image that is likely to yield a null at the hypocentre, which defeats the purpose of source imaging. A simple approach to mitigate this problem is to use the absolute value of the wavefield; this method defines the source-scanning algorithm of Kao and Shan (2004). This approach eliminates the destructive interference caused by summing positive and negative amplitude values at the source point, but does not always perform well (Trojanowski and Eisner, 2017). A similar philosophy underlies the use of *envelope stacking* (Liang et al., 2009), where the characteristic function is the amplitude envelope (§6.3.4) of the observed seismograms.

Another type of characteristic function is the STA/LTA function discussed in §6.3.1. This approach combines the event detection capabilities with the backprojection algorithm, and was used by Hansen and Schmandt (2015) to locate microearthquake sources in a passive-seismic monitoring experiment at the Mount St. Helens volcano. The semblance along the diffraction path (Equation 7.6) was used as a characteristic function by Chambers et al. (2010). Trojanowski and Eisner (2017) compared a number of different characteristic functions, combined with a scheme for polarity weighting assuming that the source-radiation pattern is known a priori. They showed that methods that account for the source radiation pattern have superior performance for detecting weak events. Chambers et al. (2014) developed a set of weights for the characteristic function that can be used for simultaneous determination of moment tensor and source location.

The combination of generalized beamforming to construct the image function in Equation 7.11 or 7.12, together with picking events based on localized maxima (or a more sophisticated maximum-likelihood approach) is referred to as *coherency scanning*. This process is illustrated with a toy example in Figure 7.4. The first three panels show a set of synthetic traces, corresponding to five receivers located symmetrically above a point source at the origin. Here, the *j*th synthetic trace, $s_j(t)$, was generated by convolving a delayed impulse with a simple source wavelet,

$$s_j(t) = w(t) * \delta(t - \tau_j), \quad (7.14)$$

where $\delta(t-\tau)$ is the Dirac delta, a *Ricker wavelet* was used for $w(t)$ and $\tau_j(r)$ is the moveout time at distance r for a source at the origin in a uniform isotropic medium with velocity V,

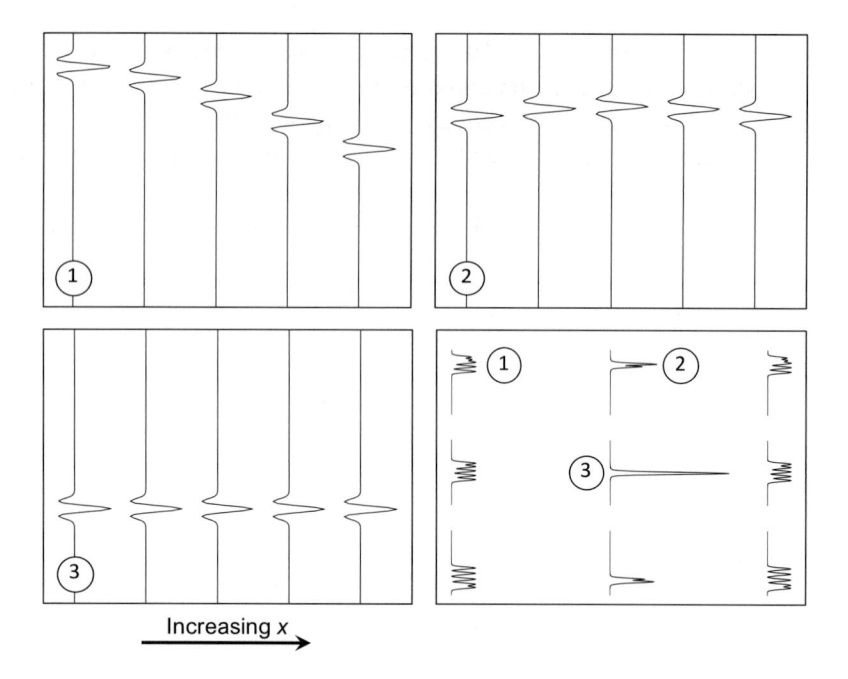

Fig. 7.4 Schematic illustration of the coherency scanning process used for event detection with surface microseismic data. Panels 1, 2 and 3 show synthetic waveforms after application of a moveout correction for a trial hypocentre location. Panels 1 and 2 show the case for a test source that is mislocated, while for panel 3 the trial hypocentre coincides with the actual hypocentre. The stacked envelope traces are shown in the lower right panel.

$$\tau_j(r) = \frac{r_j}{V}. \tag{7.15}$$

A Ricker wavelet is the second derivative of a Gaussian and is commonly used to compute simple synthetic seismograms; it has the following time function:

$$w(t) = (1 - 2\pi^2 f_M^2 t^2) e^{-\pi^2 f_M^2 t^2}, \tag{7.16}$$

where f_M is the wavelet peak frequency. In Figure 7.4, the first two panels show moveout-corrected traces with an incorrect trial hypocentre location. For panel 1, the trial hypocentre position is above the origin and laterally offset in the direction of negative x. The moveout is undercorrected, leaving residual time delays. For panel 2, the trial hypocentre location is above the origin and the moveout is again undercorrected. For panel 3, the trial hypocentre location is at the origin and the moveout-corrected traces are aligned. The lower right panel shows the image function obtained using the amplitude envelope as the characteristic function. The peak amplitude corresponds with the correct hypocentre location.

An alternative approach to Kirchhoff-style migration is to perform a numerical back-propagation of the recorded seismograms through a known velocity model. This approach was initially proposed by McMechan (1982) and was further developed by Gajewski and Tessmer (2005) using the reverse-time method of Fink (1999). Artman et al. (2010) extended this approach further by introducing physically meaningful elastic *imaging conditions*.

The reverse-time modelling approach is, in principle, conceptually straightforward; it entails reversing the time axis of the observed seismogram and injecting this as a source wavefield at receiver locations. Following Artman et al. (2010), the extrapolation direction is defined as z and the depth axis is built by recursive propagation using

$$\psi'(k_x, z + \Delta z, \omega) = \psi'(k_x, z, \omega) e^{-i\Delta z k_z}, \qquad (7.17)$$

where Δz is the depth sampling interval, ψ' denotes the time-reversed wavefield and the exponential term represents a downward propagation operator. In addition,

$$k_z = \sqrt{\omega^2 s^2 - k_x^2} \qquad (7.18)$$

is the vertical wavenumber, where $s(z)$ is the slowness of the medium and k_x denotes horizontal wavenumber. The recursive wavefield extrapolation process defined by Equation 7.17 is initialized with $\psi'(k_x, 0, \omega)$, which is the 2-D Fourier transform of the time-reversed data, $[u(x, -t)]_{z=0}$.

In essence, the propagating wavefields from each receiver are aggregated, producing a stacked reverse-time image. As elaborated below, localized sources are identified by focusing of the wavefield at the source locations. This procedure is more computationally efficient than a Kirchhoff approach but, as noted by Chambers et al. (2014), it does not enable the flexible assignment of characteristic functions as in the Kirchhoff methods described above. In addition, the implementation of the reverse-time imaging approach outlined above requires the data to be sampled on a regular grid, which in general will require a method for interpolation from the acquisition grid and may not be feasible for all configurations.

7.4.1 Imaging Condition

A key component of a migration algorithm is the imaging condition, which is a condition that is applied to extract sources from the migrated data volume. In the case of Kirchhoff methods, the imaging condition is either a local maximum of the image function that satisfies a user-defined threshold criterion, or a maximum-likelihood test that is designed to reduce false positive detections. In the case of the reverse-time imaging method, Artman et al. (2010) proposed the use of the zero-lag of the autocorrelation of the wavefield as it accumulates over time. This is given by

$$\psi_m(x, t) = FFT^{-1} \left[\sum_\omega \psi'(k_x, z, \omega) \psi'(k_x, z, \omega)^* \right], \qquad (7.19)$$

where ψ_m is the migrated image and * denotes complex conjugate. This imaging condition is chosen because large values of the zero-lag autocorrelation indicate the presence of an event and because waveform complications are smoothed. This approach only locates the source in space; a coarse estimate of the origin time is available, however, by selection of time windows for processing.

7.4.2 Resolution and Uncertainty

Although the two concepts are closely related, resolution and uncertainty have distinct meanings (Thornton, 2012). In the context of migration imaging, resolution refers to the ability to distinguish two closely spaced events and thus relates to the size and shape of the focused event image. The resolution is determined by aperture of the recording array as well as the bandwidth of the data. The data bandwidth is important because the dominant wavelength corresponds to the measured width of the source pulse and forms a bound on the best achievable resolution. The role of array aperture derives from antenna theory; the use of a larger aperture results in better theoretical resolution. In practice, the diameter of the array is usually chosen to be twice the target depth.

As discussed in §6.4, factors that contribute to location uncertainty include accuracy, precision and non-uniqueness. Precision quantifies the experimental uncertainty that is inherent to a given dataset or assumed Earth model. For example, uncertainty that derives from random errors is called *aleotoric* uncertainty, from the Greek *aleo*, which means rolling of a dice (Der Kiureghian and Ditlevsen, 2009). This is the type of uncertainty that is typically most emphasized in reports and publications. On the other hand, accuracy quantifies the closeness of estimated event locations to their true locations, and is related to the validity of model assumptions. For example, if a background model is incorrect, it may lead to a very close-spaced distribution of events characterized by high precision but insufficient accuracy. This type of uncertainty is called *epistemic* uncertainty, from the Greek *episteme*, which means knowledge.

Eisner et al. (2010) developed a simple probabilistic model based on aleotoric uncertainty, by assuming that the errors at different receivers are uncorrelated and follow a Gaussian distribution. Since the probability distribution for each parameter and each receiver are assumed to be independent, this approach leads to a probability density function for the observed data of the form

$$p(t_P, t_S) = N e^{-\sum_R (\tilde{t}_P - t_P)^2 / 2\sigma_P^2} e^{-\sum_R (\tilde{t}_S - t_S)^2 / 2\sigma_S^2}, \qquad (7.20)$$

where $\tilde{t}_P$ and $\tilde{t}_S$ are the measured arrival times, t_P and t_S are the modelled arrival times and σ_P and σ_S are the standard deviations of the Gaussian distributions for P and S time picks. The normalization constant N is used to ensure that the integral over all possible locations is equal to unity (Eisner et al., 2010).

A similar probabilistic model was used by Eisner et al. (2009b) to compare location uncertainties for three scenarios: downhole monitoring in a single vertical well, downhole monitoring with dual observation wells, and surface microseismic monitoring with an areal distribution of receivers. The results of their analysis are summarized in Table 7.1. In general, within areas where adequate coverage is available (considering the distance bias discussed in Chapter 6) and if the depth extent of the receiver array straddles the zone of interest, downhole monitoring provides superior depth resolution. The use of dual monitoring arrays can increase the spatial extent of the area of coverage, but can it can also lead to artifacts if a plane of symmetry exists such as a plane containing two vertical or two horizontal observation wells.[4] On the other hand, the use of a surface or shallow array

---

[4] A plane of symmetry is unlikely if one or both of the observation wells is deviated.

Table 7.1 Summary of Uncertainties for Different Acquisition Scenarios[1]

Geometry	Vertical Position	Horizontal Position	Velocity-Model Sensitivity
Single vertical borehole	~10s m for most common scenarios	Significantly better in radial direction, azimuthal uncertainty several 10s m	All coordinates affected (both vertical and horizontal)
Dual downhole arrays	Similar to single monitoring array, if good velocity model	Significantly dependent on position relative to plane of symmetry	Very sensitive and creates artifacts near plane of symmetry
Surface or shallow array	~10s to 100s m for most common scenarios	No specific bias in any direction, ~10s m for most common scenarios	Vertical position very sensitive, horizontal position is very robust.

[1] Modified from Eisner et al. (2009b)

typically results in more uniform lateral resolution that is also very robust in the presence of velocity-model uncertainty.

Other approaches can be used to assess location uncertainty and non-uniqueness. Thornton and Mueller (2013) used a synthetic modelling approach to assess the impact of random noise, but adding Gaussian random noise to the synthetic data prior to location using a beamforming method. Their results showed that S/N is a key indicator for both resolution and location uncertainty. Pesicek et al. (2014) applied two standard statistical tests (bootstrap and jacknife) to compare location uncertainty using double-difference relocation with template event detection and wavefield source imaging with a Kirchhoff method. They found that location uncertainties for both approaches are comparable, but the waveform imaging method was more robust with respect to station geometry.

7.5 Case Examples

Several salient aspects of the foregoing discussion as well as elements of the basic processing flow are highlighted in the following case studies. Figure 7.5 shows a field data example from Chambers et al. (2010) that illustrates the concepts of uncertainty. This example uses a perforation shot; since this event has a known source location, it enables the evaluation of location uncertainty. As expected, vertical resolution is inferior to horizontal resolution, as a geometrical consequence of the use of a surface array for imaging. Based on Kirchhoff theory, optimal resolution would require receivers that entirely surround the target zone.

The application of receiver statics, as outlined above under the basic processing workflow, is important to fine-tune the velocity model and to ensure that the events are well focused. An example using signals from a perforation shot is shown in Figure 7.6.

An example of a shallow-well array configuration is shown in Figure 7.7. Passive seismic data for this project, known as the Tony Creek Dual Microseismic Experiment (ToC2ME),

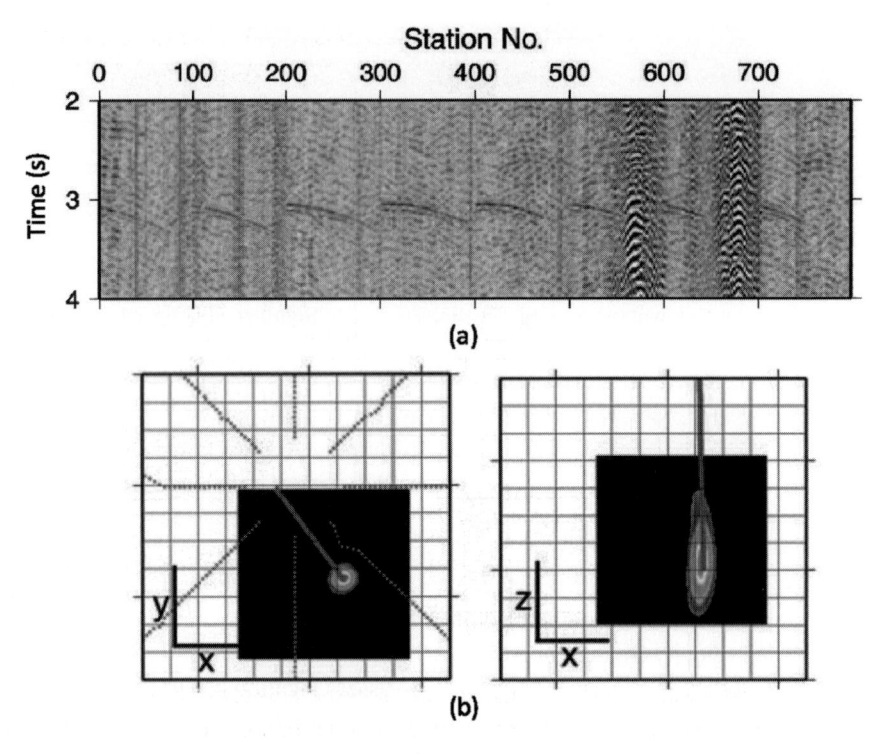

Fig. 7.5 Field data example of beamforming process. a) Raw waveform recordings of a perforation shot. b) Colour density plots of beamformed stack of the perforation shot. Left: map view, showing configuration of surface geophones. Right: cross section view. Reproduced from Chambers et al. (2010), with permission from Wiley. (A colour version of this figure can be found in the Plates section.)

were acquired by the University of Calgary in October–November 2016. Acquisition used a six-channel OYO GSX-3 digitizer at each station, with a sampling rate of 2 ms. The continuous data were formatted into SEG-2 files with a record length of 60 s.

An example of a 60-second recording is shown in Figure 7.8. This record contains five events that are discernible in the raw data, which is unusual for hydraulic-fracture monitoring since most events are typically too weak to be visible without beamforming (e.g. Chambers et al., 2010; Maxwell et al., 2012). Each event is characterized by a distinct *P*-wave arrival, primarily visible on the vertical component, followed by *S*-wave signals that are mainly confined to the two horizontal components. The moveout is evident from the varying arrival times, but does not show a simple coherent trend due to the positioning and associated numbering of the shallow array stations. It is possible to infer, even from the raw recordings, that event 1 has a different focal mechanism than the other events based on the apparent position of the *P*-wave nodal plane.

As previously discussed, an advantage of the use of large-aperture surface arrays, compared with downhole monitoring, is that the focal mechanism can often be determined with confidence by direct measurement of the polarity of the *P*-wave arrival. Knowledge of the focal mechanism can be used to design polarity weights for the characteristic function that is used for coherency scanning, which can significantly improve event resolution

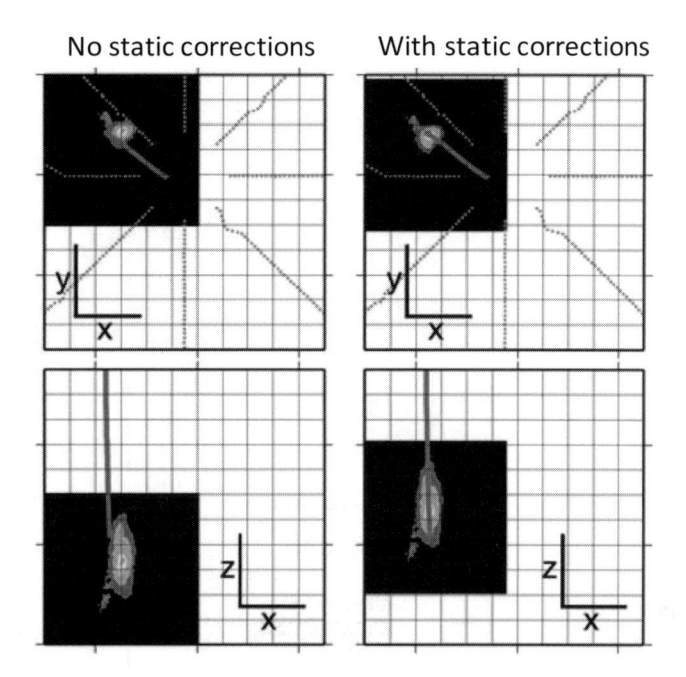

Fig. 7.6 Effect of statics on calculated location of a perforation shot, based on colour-density plots. Without the application of statics, the perforation shot is located off the wellbore in both map view (upper panel) and cross section (lower panel). After application of statics, the perforation shot is centred at the correct location. Grid lines are spaced at 152.4 m intervals. Reproduced from Chambers et al. (2010), with permission from Wiley. (A colour version of this figure can be found in the Plates section.)

(Trojanowski and Eisner, 2017). As an example, Figure 7.9 shows the P-wave polarity for a M_W 1.6 event on November 10, 2016, which has a clear strike-slip mechanism.

It is important to recognize that, despite the close visual similarity, a first-motion map derived from surface or near-surface recordings, as in Figure 7.9, is not directly equivalent to a focal-mechanism diagram as discussed in §3.5.2. In particular, focal-mechanism diagrams are based on lower-hemisphere projections of the polarity of the P-wave first-motion data, whereas such maps show the polarity on a horizontal surface located above the hypocentre. Thus, a projection of the observations onto the lower hemisphere is needed before nodal planes and other parameters can be accurately determined.

7.6 Summary

Ground motions recorded by sensors at or near the surface are strongly affected by the free-surface boundary condition and low-velocity overburden. These factors lead to amplification of upgoing wavefields and also introduce complexity arising from near-surface scattering and mode conversions. This recording environment tends to have a lower S/N compared with deep downhole recording, but this can be compensated through the use of stacking with a large number of sensors.

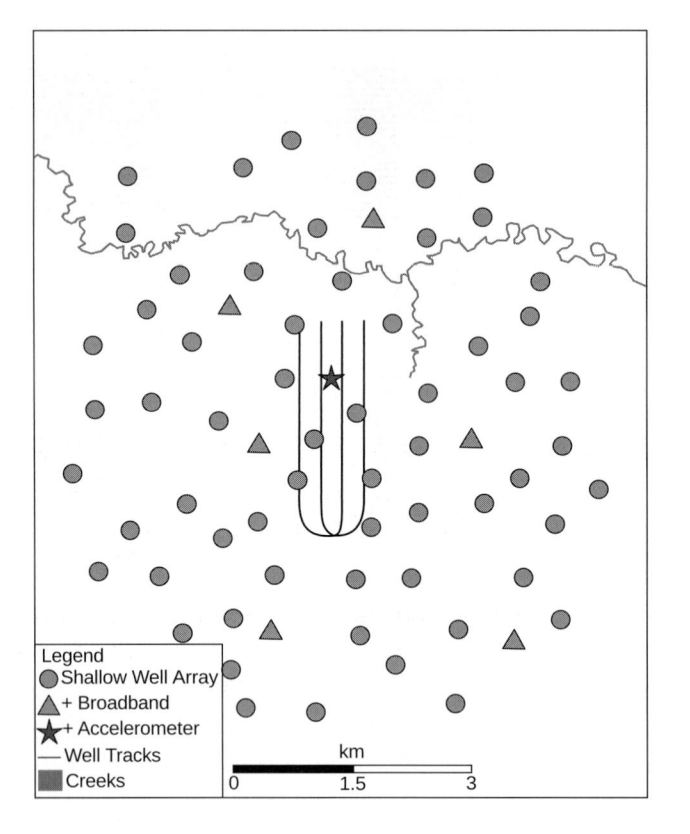

Fig. 7.7 Acquisition layout for the ToC2ME experiment. The triangles indicate locations where there was both a broadband station and a shallow borehole array station. The configuration of the geophone array installed in the shallow wellbores is shown in Figure 7.1a. Modified from Igonin and Eaton (2017).

Beamforming is a shift-and-sum procedure that can be used to localize and characterize events by aligning them based on moveout characteristics. Nonlinear variants of this procedure, such as Nth root stacking, phase-weighted stacking and semblance-weighted stacking, can be used to enhance the S/N.

The basic processing workflow for surface and shallow-well passive seismic data includes instrument-response correction, data preconditioning, velocity analysis, statics corrections, focal-mechanism determination and elastic imaging. In practice, microseismic processing with surface arrays or shallow wellbore arrays commonly uses a locate-and-detect approach, in contrast to hypocentre location methods used with deep downhole arrays, which employ methods that are generally similar to earthquake seismology (i.e. detect-then-locate).

Downward wavefield extrapolation used for elastic imaging can be formulated in several ways, including reverse-time imaging and the Kirchhoff migration. A flexible framework for Kirchhoff migration is described in this chapter, based on the use of characteristic functions. The imaging problem has also been formulated based on a reverse-time modelling procedure. A key component of this method is the application of an imaging condition, which is used to extract sources from the migrated data volume.

The resolving power of surface and shallow-well monitoring is determined by the array geometry and bandwidth of the data. Important factors that contribute to hypocentre location uncertainty include precision, accuracy and non-uniqueness, which can be expressed using the concepts of aleatoric and epistemic

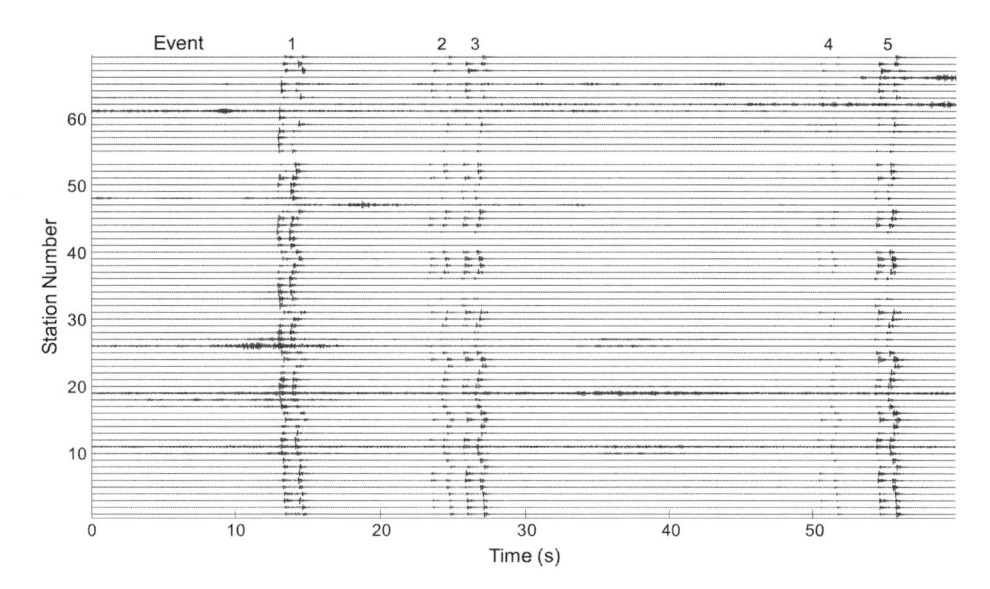

Fig. 7.8 One-minute raw multicomponent data record from the ToC2ME passive monitoring project. The array configuration is shown in Figure 7.7. This data record contains five easily discernible induced events, which is unusual as most events that are detectable using surface arrays are not visible in raw data (Chambers et al., 2010). The P-wave arrival is evident on the vertical component (red), while the S-wave arrival is most easily discernible on a horizontal component (blue). (A colour version of this figure can be found in the Plates section.)

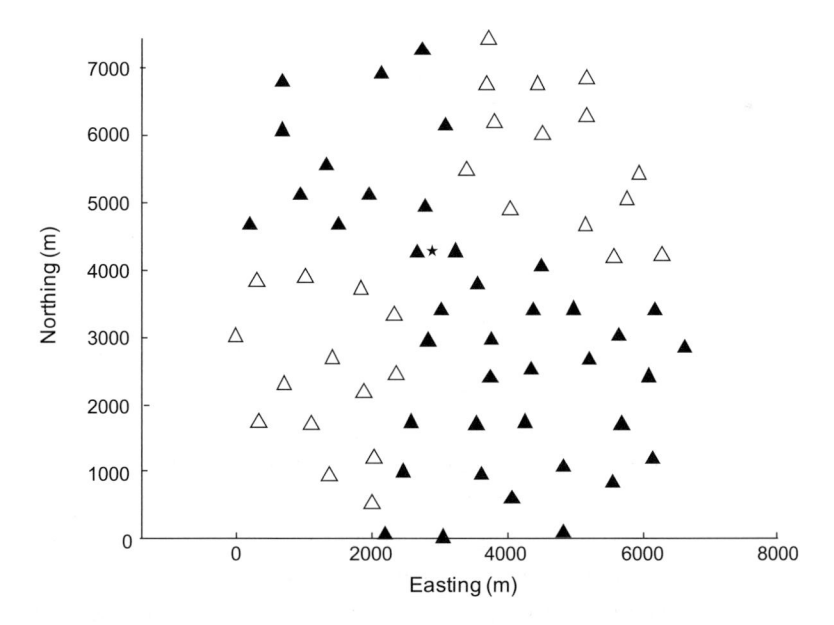

Fig. 7.9 P-wave first motion polarities from a M_W 1.6 microearthquake on November 10, 2016 recorded by the ToC2ME shallow-well array. Filled triangles denote positive-polarity first motion, while open triangles denote negative polarity. Star shows the inferred epicentre. The polarity distribution is indicative of a strike-slip focal mechanism with approximately N–S and E–W nodal planes.

uncertainties. Case examples illustrate some of the practical considerations covered in this chapter, including statics corrections and determination of focal mechanism based on *P*-wave first-motion polarity.

7.7 Suggestions for Further Reading

This chapter provides a framework and introduction to the theory and practice of passive-seismic data processing using recordings from surface and shallow-well arrays. Additional information on some subjects is available in the following sources.

- Array processing and beamforming techniques: Rost and Thomas (2002)
- Imaging methods and conditions: Claerbout (1985)
- Aleatory and epistemic uncertainty: Der Kiureghian and Ditlevsen (2009)

7.8 Problems

1. Raw microseismic data requires conversion into units of velocity. If the background noise falls within the range of ± 0.2 raw amplitude units, what is the corresponding range of displacement in m/s, given a sensitivity of $S_i = 85.8$ V/m/s and an amplifier gain of $g_a = 72$ dB?

2. Consider the shallow borehole station depicted in Figure 7.2a. Given the known depths, determine the near-surface *P*-wave velocity, given the arrival times of 2.451, 2.455, 2.459 s for the three 1C components (from the top), and 2.465 s for the 3C receiver at the bottom. What is the near-surface velocity? Does this value fall into the range of expected near-surface *P*-wave velocity?

3. Consider the shallow buried array in Figure 7.7. If hydraulic fracturing is performed at a depth of 3.0 km in the eastern horizontal well and 3C geophones are used to record the resulting microseismic events, on which components do you expect to record the *P*-wave, and on which do you expect to record the *S*-wave? Next, consider an event originating at depth outside the limits of the array to the northwest. On which components would you then expect to see the dominant *P*- and *S*-wave motion? Assume that the v_p and v_s of the near-surface unconsolidated layer is much less than the velocities in the underlying layers.

4. Sketch the wavefronts for both cases in Problem 3. How would you characterize the moveout that would be expected for both events, if the receivers were reordered by distance from the central well?

8 Microseismic Interpretation

> Take care of the sense and the sounds will take care of themselves.
> Lewis Carroll (Alice's Adventures in Wonderland, 1865)

Interpretation is the science of finding meaning in data. The primary purpose of microseismic monitoring is to develop insights into the nature and extent of brittle deformation processes in the subsurface. Passive-seismic monitoring has many applications, including monitoring of fluid stimulation for enhanced geothermal systems and mine-safety protocols. The main focus of this chapter is passive seismic monitoring of $M_W < 0$ induced microseismicity associated with hydraulic fracturing, although the interpretation workflow is more broadly applicable and could be used, for example, in passive seismic monitoring of geothermal stimulation programs. Interpretation of passive-seismic data for induced seismicity, considered here as events with $M_W \geq 0$, is the subject of the next chapter. In the case of hydraulic-fracture monitoring, microseismic data are generally acquired for the following purposes:

- surveillance of the growth of fracture networks;
- estimation of the spatial extent of the stimulated region within the reservoir;
- characterization of the complexity of induced fracture networks;
- diagnosing potential operational issues during well completion.

There is considerable research potential for development of more advanced microseismic interpretation methods, such as quantifying the state of stress, imaging discrete fracture networks, time-lapse studies, understanding possible slow-slip processes and reservoir characterization, including facies analysis.

In this chapter, a systematic approach to microseismic interpretation is introduced. This approach is guided by an underlying philosophy that the interpretation of any experimental data should be repeatable, observation-driven and, to the greatest extent possible, quantitative. As part of this approach, applicable uncertainties in measured quantities should be reported, along with metadata describing how reported uncertainties were obtained. The proposed interpretation procedure is built upon four key concepts:

1 *Attributes*, following Taner (2001), are physical characteristics derived from observations, either by direct measurements or by logical or experience-based reasoning. There are two fundamentally different types of attributes: measured and inferred (indirect). Examples of measured attributes are the length of a microseismic cluster and the dimensions of the estimated stimulated volume (ESV); examples of inferred attributes are fracture length and SRV.

2 *Methods* are the tools used for interpreting microseismic data. Examples of methods that are discussed in this chapter include determination of the convex hull or smallest-enclosing ellipsoid for a point set, and microseismic facies analysis (MFA).
3 *Applications* of microseismic data, which can be classified as either visualization-based or model-based (Cipolla et al., 2012). Examples of visualization-based applications include estimation of SRV (Mayerhofer et al., 2010) and microseismic-depletion delineation (MDD) (Dohmen et al., 2017). Examples of model-based applications include construction of a discrete fracture network (Virues et al., 2016) and mapping of proppant placement into hydraulic fractures based on microseismic observations (McKenna, 2014).
4 *Validation* of measured and inferred attributes, which is based on the use of different measurement technology to verify characteristics derived from microseismic observations (Warpinski and Wolhart, 2016).

This chapter begins by outlining a systematic interpretation workflow built upon the concepts described above. We then consider the elements of the workflow in sequence, namely data preconditioning, event attributes, clustering analysis, cluster and global attributes, and applications. The chapter finishes with a brief look at a number of common interpretation pitfalls.

8.1 Interpretation Workflow

The following basic workflow is proposed for systematic interpretation of microseismic data:

1 data preconditioning, to mitigate the effects of observational inconsistencies and biases;
2 determination of event attributes and their associated uncertainties;
3 attribute-driven event clustering;
4 determination of cluster and global attributes;
5 applications of microseismic interpretation to achieve specific engineering objectives.

The input to the interpretation workflow is often an event catalogue, which contains a list of events and their respective attributes derived from microseismic data processing, such as hypocentre parameters (location and time), magnitude values and other source parameters. Prior to undertaking quantitative interpretation, it is important to precondition these data to ensure that results are not skewed by any observational bias. A number of useful additional attributes can then be obtained, either from the raw waveform data or from the event catalogue, that are not normally determined during data processing. Once a complete set of attributes (along with associated uncertainties) is available, some form of attribute-driven clustering can be applied in order to group events that have similar characteristics. For multi-stage hydraulic fracturing, temporal grouping based on stage number is a straightforward, but not necessarily optimal, example of event clustering. Statistical measures of attributes, such as mean value, standard deviation or fractal dimension, can then be extracted from event clusters. A final step in the workflow involves application and validation of microseismic interpretation to achieve specific engineering objectives. In most cases, the most powerful results are achieved through integration of microseismic data with independent data from other sources. Each of these steps in the interpretation workflow is considered in the following sections.

8.2 Data Preconditioning

As with interpretation of any type of geophysical remote-sensing data, it is important to be wary of inherent uncertainties, limitations and pitfalls associated with the dataset under consideration. As illustrated in Figure 8.1, microseismicity typically occurs in spatially distributed event clouds. This is a representative example of microseismic data recorded during hydraulic fracturing (Eaton et al., 2014c). In this example, two horizontal wells are completed, with 12 stages in each well. Microseismic monitoring was performed using a 12-level array of three-component geophones deployed in a vertical observation well. The treatment intervals for each stage are shown by coloured bars and the microseismic events during the injection are plotted with the same colour as the corresponding stage.

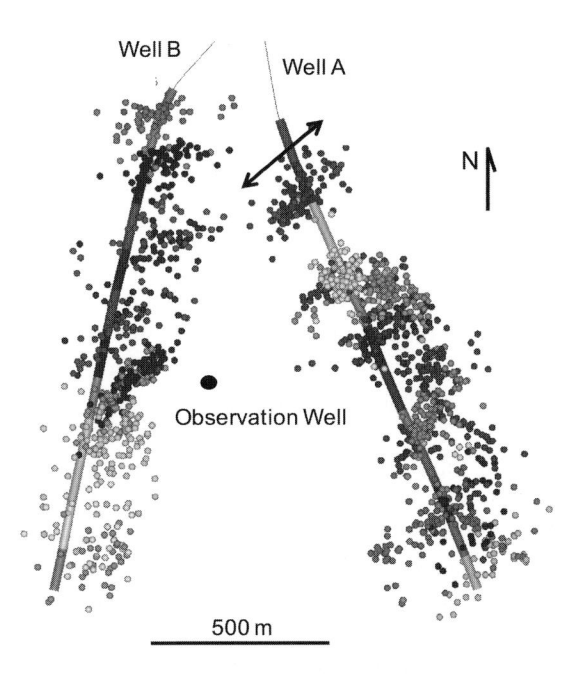

Fig. 8.1 Hydraulic-fracture monitoring (HFM) example. These microseismic observations are from the Hoadley experiment, a two-well open-hole completion in western Canada that was monitored with a 12-level downhole array deployed in a vertical observation well (Eaton et al., 2014c). Treatment stages are shown as coloured bars along the horizontal well paths. Microseismic events are coloured by treatment stage. Typical attributes inferred from microseismicity included fracture dimensions and orientation, as marked by the double arrow near one of the event clusters along well A. (A colour version of this figure can be found in the Plates section.)

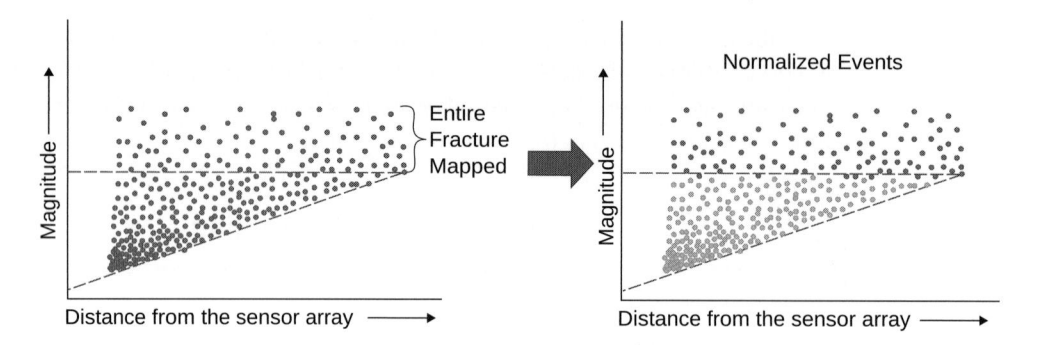

Fig. 8.2 Data preconditioning for downhole microseismic observations, by normalizing the catalogue through removal of events below a user-determined threshold magnitude indicated by the dashed line. This step ensure that consistent attributes are derived for event clusters, regardless of distance from the downhole receiver array(s). Modified from Cipolla et al. (2011), with permission of the Society of Petroleum Engineers.

The treatment took place over two days and, for each well, proceeded from the toe of the well (at the south end, where the lateral segments have the greatest separation) to the heel of the well. This dataset contains evidence for activation of natural fracture networks, as described by Eaton et al. (2014c).

Observed microseismic event distributions invariably contain scatter, some of which arises from uncertainty in hypocentre locations rather than the geometry of the underlying fracture system. Moreover, different acquisition methods impose characteristic biases on the event distribution. As discussed by Cipolla et al. (2011), downhole microseismic monitoring from a vertical observation is more sensitive to events that occur near the sensor array. On the other hand, as discussed in Chapter 7 surface-microseismic monitoring typically has a greater uncertainty in focal depth (Eisner et al., 2009b) that depends largely on aleatoric uncertainty from S/N of the data (Thornton and Mueller, 2013) and epistemic uncertainty from inadequate knowledge of the velocity model.

Figure 8.2 illustrates a procedure for normalizing the event distribution from downhole microseismic monitoring (Cipolla et al., 2011). Due to the inherent viewing bias of downhole microseismic observations (*cf.* Figure 5.6), this data preconditioning step removes (or filters) observed events with magnitude below a user-determined threshold. This step ensures the consistency of attributes derived for every cluster, regardless of distance from the observation well. The cut-off magnitude is an example of metadata that should be included as part of the interpretation. Similar procedures can be applied to surface-microseismic observations, such as a S/N cut-off.

8.3 Event Attributes

Once preconditioning has been applied, the next step in the interpretation workflow is determination of event attributes, which are quantitative characteristics of events that can be used as a basis for clustering. A list of basic event attributes is given in Table 8.1; this is

Table 8.1 Event Attributes

Attribute	Category[a]	Type	Metadata
Pre-injection flag	M	Logical flag	Description
During injection flag	M	Logical flag	Description
Post-injection flag	M	Logical flag	Description
Stage	M	Integer	Description
Date	M	Text	Year/Month/Day
Julian date	M	Integer	Year
Event start time	M	Scalar	Units and time zone[b]
Origin time	I	Scalar	Units, time zone and method
Event location	M	Vector	Units and map projection
Magnitude	M	Scalar	Magnitude type and method
Seismic moment	M	Scalar	Units and spectral model
Corner frequency	M	Scalar	Units and spectral model
Stress drop	I	Scalar	Units and spectral model
Seismic moment tensor	M	Tensor	Units and method
Double-couple (DC) %	M	Scalar	Method
CLVD %	M	Scalar	Method

[a] M = measured, I = inferred
[b] Universal Coordinated Time (UTC) is strongly recommended

not intended to be exhaustive, but rather a guideline to illustrate the types of event attributes that can be used to support the interpretation. Most of these attributes are obtained during data processing (Chapters 6 and 7) and are stored in the event catalogue.

Some of these attributes, such as source characteristics (stress drop, moment tensor) are considered as advanced processing attributes (Cipolla et al., 2011). Such a distinction is arbitrary, as what is deemed to be "advanced" clearly depends on the technological maturity of the field. Here, the primary discriminator is not application-driven, but relies instead on whether an attribute applies to a single event or a set of events. The fundamental principles and methods used to obtain these attributes are discussed in previous chapters of this book.

8.4 Clustering Analysis

Clustering relies on attributes of observations and is a fundamental tool for analysis of data in many different disciplines. It can be defined as unsupervised classification of observations into groups (clusters) (Jain et al., 1999). Clustering of microseismicity has been used, for example, to extract fine details within a cloud of events (e.g. Moriya et al., 2003).

There is a great diversity of existing clustering techniques (Jain et al., 1999). Most methods can be classified as either agglomerative and divisive. *Agglomerative* clustering is a bottom-up approach that initially bins each event into a separate cluster and then merges the clusters until a stopping criterion is met. Conversely, a *divisive* clustering algorithm

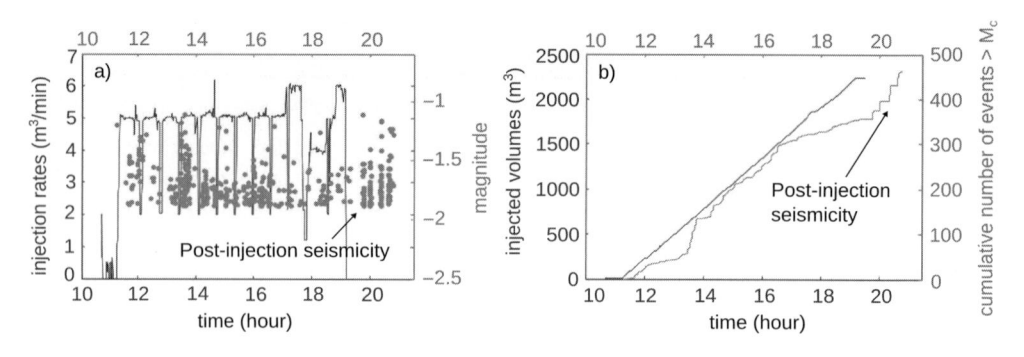

Fig. 8.3 Comparison of induced microseismicity with fluid injection schedule. a) Seismicity above the magnitude of completeness and fluid injection rate corresponding to 12 treatment stages for the horizontal well A in Figure 8.1. Stages are separated by brief periods of lower injection rates. b) Cumulative injected fluid volume and number of microseismic events above the magnitude of completeness. Reproduced from Maghsoudi et al. (2016), with permission of Wiley. Copyright 2016 by the American Geophysical Union. (A colour version of this figure can be found in the Plates section.)

starts with all observations grouped into a single cluster and performs splitting until a stopping criterion is met. Both methods have been applied to passive seismic observations.

The simplest and most common type of microseismic-event clustering is binning based on stage number – that is events are grouped together if they occurred within a specified time window after the initiation of a given treatment stage and prior to the end of the stage (or within a short time window thereafter). In the lexicon of clustering analysis, this approach is called a *monothetic* method. Binning-by-stage is often very effective, especially for well completions such as perf-and-plug, where there is a lengthy time interval between stages. A shortcoming of this simple approach, however, is highlighted in Figure 8.1. For example, a set of events near well A beside the double arrows fall within a discernible localized region of space, but this region contains events from two different stages. Other examples of overlapping stages are apparent in this figure, reflecting the nearly continuous style of open-hole completion that was used for this hydraulic fracturing program (Figure 8.3).

Another clustering-related example of temporal overlap between stages is discussed by Eaton and Caffagni (2015) and illustrated in Figure 8.4. In that study, a matched-filtering method (§6.3.4) was used to extract a more complete set of events. Figure 8.4 shows the first four stages of the treatment program. A sliding-sleeve procedure in an open hole was used here, such that the time interval between successive stages is relatively brief. The initial formation breakdown, marked by the elevated pressure near the start of stage 1, was accompanied by a microseismic response. However, there was no response observed at the start of stage 2. Rather, a build-up of seismicity rate is observed that overlaps with the beginning of stage 3. The same pattern is repeated from stage 3 to stage 4. Eaton and Caffagni (2015) and Caffagni et al. (2016) argued that clustering of the microseismicity by partitioning it into time windows marked by local minima in observed seismicity rate was more effective than simply binning the events based on stage number. The interpreted time windows for the microseismicity clusters are indicated at the bottom of Figure 8.4.

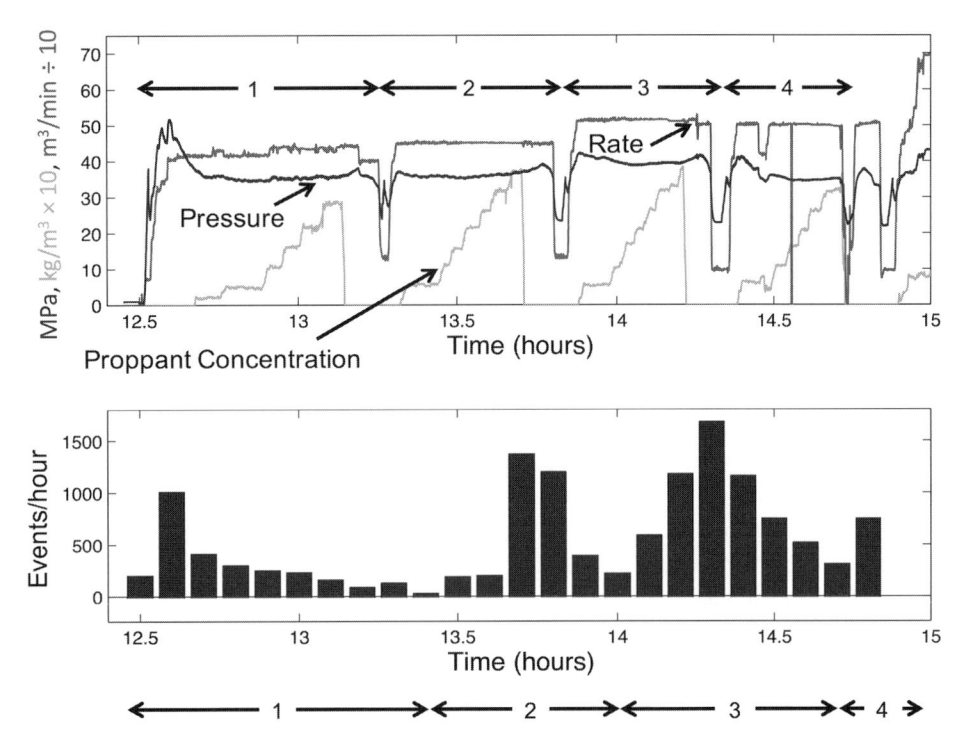

Fig. 8.4 Comparison between pump curves (upper panel) with seismicity rate (lower panel) determined using matched filtering analysis, showing temporal clustering during open hole completion. Stages 1–4 are marked by arrows at the top of the figure. Due to the short time interval between stages, microseismic activity exhibits temporal overlap from one treatment stage to the next. Temporal clusters are nevertheless discernible, as indicated by numbers at the bottom of the figure. Modified from Caffagni et al. (2016), with permission from Oxford University Press. (A colour version of this figure can be found in the Plates section.)

Several different clustering schemes have been used, all of which are based on some measure of similarity defined by event attributes (including the stage number). For example, Skoumal et al. (2015b) used a hierarchical agglomerative clustering algorithm based on temporal and spectral characteristics of observed waveforms in order to group events detected using a template method into distinct clusters, which they call swarms. Eaton et al. (2014b) discuss the use of the k-means algorithm, a divisive clustering approach using a two-phase iterative algorithm to partition the distribution by minimizing the Euclidean norm (E) of the event-to-centroid distance (Seber, 2009). Vasudevan et al. (2010) applied a definition of recurrences for record-breaking process within directed-graph theory to perform clustering analysis of intraplate seismicity patterns. The same approach was used by Maghsoudi et al. (2016) to perform spatial-temporal clustering analysis of the dataset in Figures 8.1 and 8.3, revealing nontrivial relationships within clusters. Ultimately, whether a set of events that is localized in space and time is classified as a single cluster, or as multiple clusters, will impact the interpretation of the results. Binning by stage is not necessarily optimal for engineering analysis, whereas the use of an ad hoc approach to make this determination is subjective, challenging to document, prone to error and unlikely to be repeatable. Further research is needed for advance the development of rigorous, data-driven, objective clustering algorithms for microseismic-monitoring datasets.

8.5 Interpretation Methods

Once event clusters have been determined, whether through binning based on stage number or through the use of a more sophisticated approach, collective attributes of event clusters can be determined. Some of these cluster attributes can be aggregated to determine global attributes that characterize the entire dataset. Table 8.2 provides a partial list of cluster or global attributes of microseismic observations. Many of these attributes are discussed in greater detail by Eaton et al. (2014b) and Rafiq et al. (2016).

We will now consider several specific methods used to determine some of these attributes. The most basic interpretative measurements using microseismic observations are the physical dimensions and azimuth of the microseismic cloud (Maxwell, 2014). Implicit in such an interpretation is an assumption that the microseismic cloud is a reliable proxy for the dimensions of the associated hydraulic fracture network. Warpinski and Wolhart (2016) give a comprehensive review of studies that have attempted to validate this assumption, and conclude that available data, while sparse, support the use of this spatial proxy for high-quality microseismic observations. In particular, fracture azimuth inferred from the measured azimuth of microseismic clusters agrees with independent observations such as drill-through of a fracture and "fracture hits" at offset wells. Moreover, the observed agreement between fracture length from microseismicity and independent observations implies that at least some of the microseisms are generated in the vicinity of the fracture tip (Warpinski and Wolhart, 2016). Similarly, inferred fracture height is useful for assessing the potential for out-of-zone fracture growth. Out-of-zone fracture growth is undesirable for a number of reasons: it falls outside of the target reservoir and therefore does not contribute to hydrocarbon production; and, moreover, it may open up communication pathways to potable water supplies or geohazards (Fisher and Warpinski, 2012).

While seemingly straightforward, there are a number of nuances that need to be considered to assure accurate measurement of these attributes as well as repeatability of the results. One important consideration is the method used to perform clustering analysis, particularly with respect to how outliers are handled. For extraction of azimuth and fracture length, a method for calculating the smallest enclosing ellipsoid based on linear programming (Welzl, 1991) can be used. The length and azimuth of the cluster are then uniquely determined by the major axis of the ellipsoid. Rich and Ammerman (2010) discussed the effects of stress anisotropy and showed that a complex ellipsoidal volume is expected when horizontal stress anisotropy is low.

8.5.1 Estimated Stimulated Volume

The concept of stimulated-reservoir (or rock) volume, SRV, holds that the effective volume of a reservoir within which the permeability is enhanced by a fracture network induced by

Table 8.2 Cluster and Global Attributes

Attribute	Category[a]	Type	Metadata
Cluster length	M	Scalar	Units and method
Cluster height	M	Scalar	Units and method
Cluster azimuth	M	Scalar	Units and method
Cluster duration	M	Scalar	Units and method
Fracture length	I	Scalar	Units and method
Fracture height	I	Scalar	Units and method
Fracture azimuth	I	Scalar	Units and method
Propped fracture length	I	Scalar	Units and method
Fracture aperture	I	Scalar	Units and method
Fracture complexity	I	Category[b]	Descriptive rationale
Estimated Stimulated Volume (ESV)	M	Scalar	Units and method
Stimulated Reservoir Volume (SRV)	I	Scalar	Units and method
Mean magnitude	M	Scalar	Magnitude type
Magnitude variance	M	Scalar	Magnitude type
Mean stress drop	M	Scalar	Units and spectral model
Stress-drop variance	M	Scalar	Units and spectral model
Average DC %	M	Scalar	Method
Average CLVD %	M	Scalar	Method
b value	M	Scalar	Method
D value	M	Scalar	Method
Net seismic moment	M	Scalar	Units
Seismic-moment density[c]	M	Scalar	Units
Maximum-moment rate	M	Scalar	Units
Transience[d]	M	Scalar	Units and method
Apparent permeability	I	Scalar	Units and method

[a] M = measured, I = inferred.
[b] See Figure 8.5.
[c] Net seismic moment normalized by cluster volume.
[d] Ratio of net seismic moment during a time window at the start of a cluster to the net seismic moment for the entire cluster.

a stimulation program. Early research on this topic was published by Fisher et al. (2002) and Maxwell et al. (2002), whose work documented the creation of large fracture networks in the Barnett shale, leading to the notion of fracture complexity as illustrated in Figure 8.5. These studies also highlighted an initial relationship between the hydraulic-fracture treatment size and the production response. This subsequently led to the establishment an empirical relation between production data, the hydraulic-fracture fluid volume injected, and the observed SRV from the microseismic observations. Mayerhofer et al. (2010) linked the estimated SRV to the dimensions of the microseismic point cloud; the general concept is illustrated in Figure 8.6. There is evidence, however, that use of the microseismic point cloud as a direct proxy for SRV is not universally applicable in terms of forecasting reservoir productivity (Cipolla and Wallace, 2014). SRV calibration is complicated by

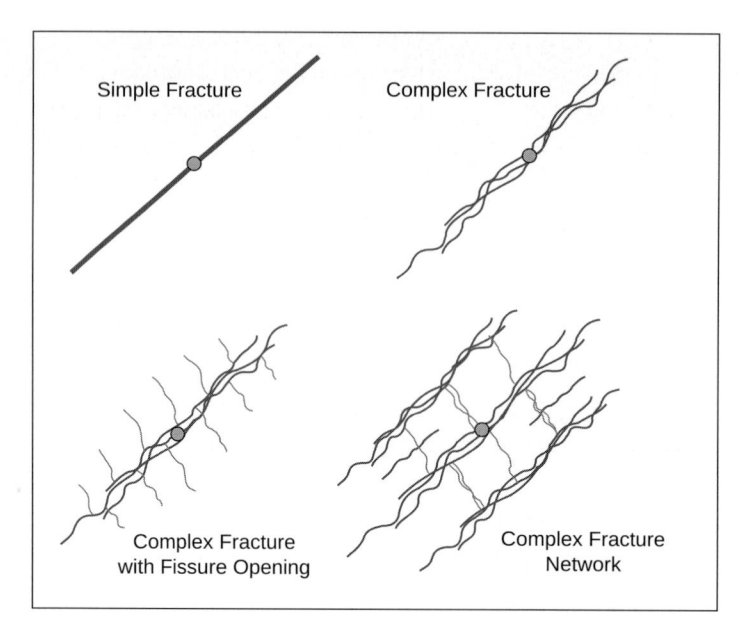

Fig. 8.5 General categories of fracture complexity inferred from microseismic observations. For each panel, injection location is indicated by the grey circle. Reprinted from Warpinski et al. (2009), with permission of the Society of Petroleum Engineers.

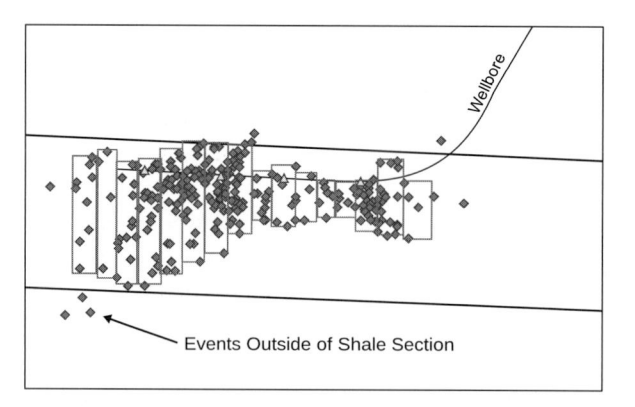

Fig. 8.6 Cross section showing height calculation to determine stimulated reservoir volume (SRV) using microseismic observations. Modified from Mayerhofer et al. (2010), with permission of the Society of Petroleum Engineers.

factors such as: 1) observational bias, 2) uncertainties in microseismic event locations and 3) incomplete geomechanical understanding of the spatial relationship between induced microseismicity and propped fracture networks.

Due to uncertainties for geomechanical application of SRV derived from microseismic data, a preferred alternative term for an interpretive subsurface volume bounded by contiguous microseismic point cloud is *estimated stimulated volume*. An objective and quantitative approach for volumetric calculation is outlined in Box 8.1, which describes

8.5 Interpretation Methods

> **Box 8.1** **Convex Hull Method**
>
> A *convex hull* is defined as the smallest convex set containing all of the points from a point cloud (Barber et al., 1996). The volume is formed from triplets of points that create a tessellated convex volume comprised of triangular surface elements and the quickhull algorithm (Barber et al., 1996) enables rapid calculation of convex hull volume. An example of a convex hull produced using a microseismic point set is given in Figure 8.7c. Informally, a convex hull can be viewed as a shrink-wrapped surface around the exterior of the point cloud. In contrast, Mayerhofer et al. (2010) used an ad hoc binning approach coupled with fracture height estimation to measure SRV from microseismic data (Figure 8.3). Other common methods include measurement of the volume enclosed within a user-defined threshold by a 3-D density field of microseismic events, or fitting a simple geometrical shape such as a rectangular prism to a defined subset of microseismic events. Since these approaches tend to require user selection of parameters that are qualitative, it is unclear what quantitative relationship may exist between SRV estimates obtained using these different approaches. In practice, use of the quickhull method has some desirable characteristics: 1) it is unique for a given point cloud, and 2) it provides an inherently conservative estimate of volume, since it is the smallest convex volume that contains the points.

the method of convex hulls. This technique is borrowed from the field of computational geometry and provides a unique minimum volume estimate for a point cloud. Figure 8.7 uses data from the cluster marked with a double arrow in well A of the Hoadley experiment (Figure 8.1) to illustrate an approach for ESV calculation that accounts for observation well bias. In this case, the observation well is situated west of the treatment well and microseismic events are less abundant with increasing distance. The observational bias arises because of the reduction in wave amplitude with distance from the source; as a consequence, events located farther from the observation well are more difficult to detect. This known source of bias can be accounted for using a simple process of augmenting the point cloud by pairing each observed microseismic event with a virtual mirror point that is reflected across the centroid of the injection zone in the wellbore (Figure 8.7b). The ESV can then be determined using the convex hull algorithm, using either the original point cloud or the augmented point cloud as shown in Figure 8.7c. Underlying this approach is an assumption that the hydraulic fracture has a symmetrical bi-wing geometry, consistent with semi-analytical geomechanical models outlined in Chapter 4.

Another important factor to consider for ESV calculation is the uncertainty in hypocentre locations of observed events. Volumetric uncertainty arising from this factor can be accounted for in a simple way, as illustrated in Figure 8.8. In this example, each microseismic event has a reported location uncertainty that incorporates probabilistic uncertainty in the background velocity model, the geometry of the sensor array and traveltime picking errors (Maxwell, 2009). An upper bound on the ESV can be obtained by generating a synthetic inflated microseismic point cloud, by displacing each point in the augmented point cloud away from the injection centroid by a distance equal to the location uncertainty for that event. The direction of displacement is given by the unit vector directed from the centroid to the event. Conversely, a lower bound on the ESV can be

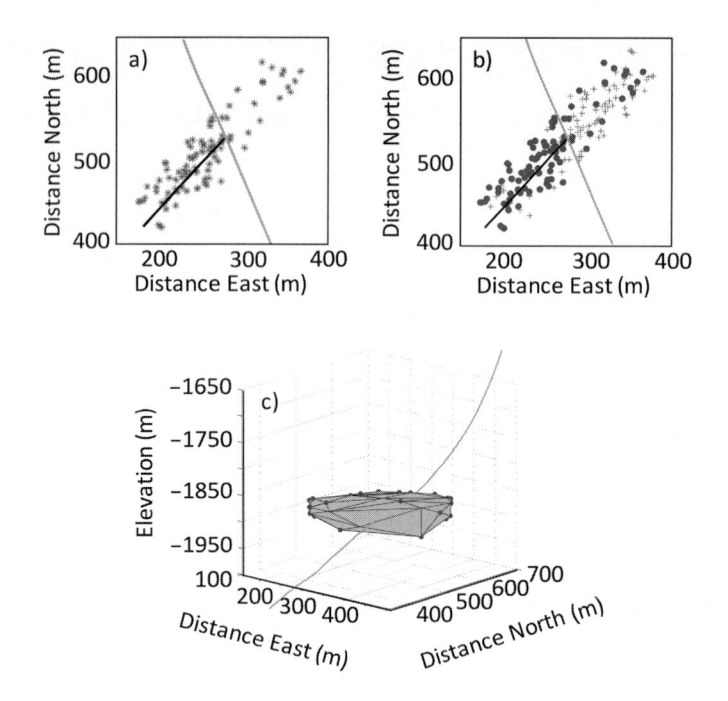

Fig. 8.7 Convex-hull method for estimated stimulated volume (ESV) determination, based on a microseismic cluster from Figure 8.1 that is indicated by the double arrows. a) Map view of the observed microseismicity. Distance is relative to the observation well. Green line shows treatment wellbore and black line shows line of best fit through the hypocentres. b) Augmented microseismic point cloud, with mirror events added (shown by plus symbols). c) Convex hull, fit to the point set in (b). (A colour version of this figure can be found in the Plates section.)

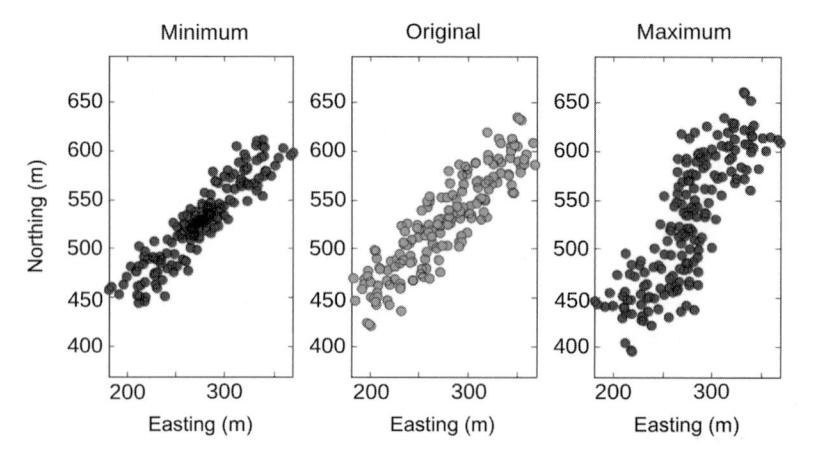

Fig. 8.8 Method for calculating uncertainty of estimated stimulated volume (ESV) volume using the convex hull approach. The original cluster shown in the central panel is the augmented point cloud from Figure 8.7b. The convex-hull that encloses this point cloud has a volume of 7.70×10^5 m^3. The minimum ESV (6.13×10^5 m^3) is determined by displacing every hypocentre in the augmented set toward the injection point, whereas the maximum ESV (1.41×10^6 m^3) is determined by displacing away from the injection point. In both cases, the displacement distance is defined by the location uncertainty. (A colour version of this figure can be found in the Plates section.)

obtained by generating a synthetic deflated microseismic point could, where each point is displaced toward the centroid by one uncertainty unit. This approach provides a quantitative and objective measure of uncertainty in seismically stimulated volume that may be of direct value for reservoir modelling.

8.5.2 Frequency–Magnitude Distributions and Fractal Dimension

It is usually assumed that the frequency-magnitude distribution derived from microseismic catalogues is well described by the Gutenberg–Richter relationship for earthquakes

$$\log_{10} N = a - bM , \qquad (8.1)$$

where N is the number of events with magnitude $\geq M$. The slope of this relationship, called the b value, is a measure of the relative abundance of large events versus small events in the catalogue. Methods for estimating b and its statistical uncertainty from an event catalogue are discussed in §3.8. In general, most earthquake fault systems produce seismicity that is characterized by a b value of close to unity (e.g. El-Isa and Eaton, 2014). The logarithmic form of the relationship implies, assuming $b = 1$, that a unit increase in magnitude is associated with a tenfold reduction in the number of earthquakes. Based on energy-scaling relationships for earthquakes, however, a unit increase in moment magnitude implies a 30-fold increase in energy (see Equation 3.74). When expressed in terms of seismic moment, the Gutenberg–Richter relationship takes the form of a power law,[1] which implies scale invariance of the underlying rupture processes.

The scale-invariant property of earthquakes has been linked to the concept of *self-organized criticality* (SOC) by Bak and Tang (1989), who showed that the Gutenberg–Richter relation can be interpreted to emerge from critical behaviour of fault systems. The term "self-organized" implies that a dynamic system naturally evolves to a critical state, with little or no dependence on the initial conditions (Bak et al., 1988). In addition to earthquakes, a classic example of this behaviour is a growing pile of granular material, whose slope evolves toward a material-dependent angle of repose. The addition of a single grain is unlikely to destabilize the pile, but it can trigger an avalanche of virtually any size. The concept that Earth's crust is in a critically stressed state (Townend and Zoback, 2000) is another proposed example of SOC behaviour.

For natural seismicity, the b value is thought to depend upon earthquake fault regime (Schorlemmer et al., 2005) and, more generally, upon the state of stress and its temporal variations (El-Isa and Eaton, 2014). Consequently, due to the potential for its use as a diagnostic parameter, a great deal of attention has been placed on characterizing and mapping the b value associated with injection-induced seismicity. The fractal dimension, D, is another potentially diagnostic parameter that has been used to characterize the spatial distribution of hypocentres (Grob and Van der Baan, 2011). For a point cloud, the fractal dimension has the following values:

- $D = 0$ for a single point;

---

[1] Other notable examples of power-law distributions include the diameter of lunar craters, individual net worth and the number of citations for scientific papers; see Newman (2005).

- $D \sim 1$ for a set of colinear points;
- $D \sim 2$ for a set of coplanar points;
- $D \sim 3$ for points that are uniformly distributed within a sphere.

As a cautionary note, in the application of these canonical distributions to the interpretation of the D value of a microseismic point cloud, it is important to bear in mind the scatter in locations arising from the uncertainty in event locations. For example, an event distribution that is intrinsically coplanar would appear to have a D value > 2 in the presence of scatter arising from location uncertainty.

For a cluster of points, the D parameter can estimated using the correlation-integral method of Grassberger and Procaccia (1983). The first step is to determine the correlation integral, $C_F(r)$, which is given by

$$C_F(r) = \frac{2}{N_P(N_P - 1)(N_R(r))}, \tag{8.2}$$

where N_P is the total number of events within the cluster and $N_R(r)$ is the number of pairs of events separated by a distance $< r$. For a fractal distribution, the correlation integral is characterized by a power-law relationship with respect to r,

$$C_F(r) \propto r^D. \tag{8.3}$$

Using this approach, the fractal dimension of a distribution can therefore be estimated from the slope of a logarithmic plot of C versus r.

The D value provides a simple scalar measure that can be used to monitor changes in the geometrical distribution of events. Moreover, some statistical models for earthquakes predict a linear relationship between the D and b (Hirata, 1989). If such a relationship can be validated for a specific area, then measuring one of these parameters is sufficient to characterize them both, and measuring both parameters provides statistical redundancy. These interconnected relationships are discussed by Grob and Van der Baan (2011) and summarized for different earthquake fault regimes in Figure 8.9.

In marked contrast to natural seismicity, the b value reported for microseismic catalogues recorded during hydraulic fracturing is typically ~ 2 (Eaton et al., 2014a). An abrupt drop from this value to $b \sim 1$ has been postulated to be indicative of fault activation during hydraulic fracturing (Maxwell et al., 2009; Kratz et al., 2012). To gain insights into the underlying reason for this change in b value, a number of scenarios were considered by Eaton and Maghsoudi (2015) to explain the unusually high values during treatment, as follows:

- instrument response is incorrectly scaled;
- too few events in some catalogues to obtain a statistically robust determination of b-value;
- superposition of two or more distinct populations with different b values;
- magnitude saturation due to the use of high-frequency sensors (i.e. conventional geophones);
- the frequency-magnitude distribution exhibits preferred scaling.

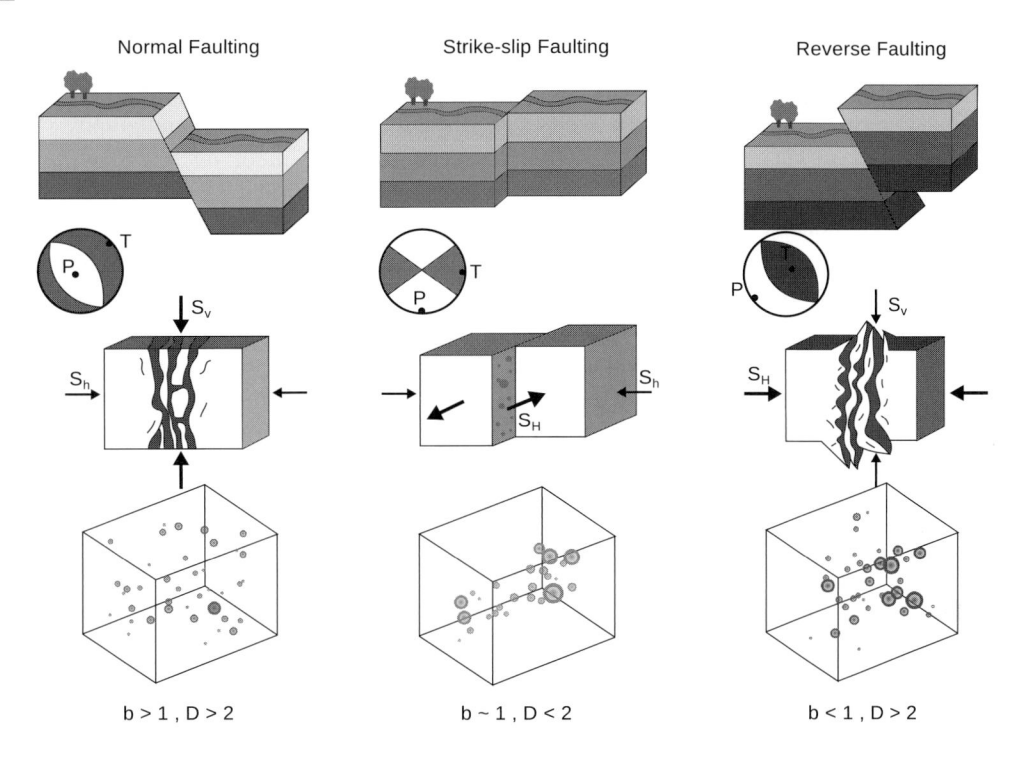

Fig. 8.9 Inferred relationships between stress regime, focal mechanism, b and D parameters. P and T denote the pressure and tension axes, respectively (Equation 3.58). In terms of Anderson's fault regimes, the extensional regime is associated with normal faulting and a greater proportion of small-magnitude events; the strike-slip regime is associated with planar features with a b value of close to unity; and the compressive stress regime is associated with reverse faulting and higher proportion of large-magnitude events. From Grob and Van der Baan (2011). Reprinted with permission of the Society of Exploration Geophysicists. (A colour version of this figure can be found in the Plates section.)

Using random synthetic catalogues that included realistic effects of viewing bias, instrument response and noise characteristics, Eaton and Maghsoudi (2015) falsified all but the last hypothesis. In practice, the finite dimensions of earthquake fault systems means that the scale-invariant behaviour of earthquakes ultimately breaks down at very large magnitude. This behaviour is represented by the tapered Gutenberg–Richter formula in Equation 3.84, which is characterized by a rollover in the frequency-magnitude distribution at magnitudes approaching the upper limit. Based on the tapered distribution, estimation of the b value using a small range of magnitude values that are close to the upper limit will yield uncharacteristically large b values. Eaton and Maghsoudi (2015) argued that microseismic catalogues characterized by $b \gg 1$ can be interpreted as manifestations of systems operating close to their upper magnitude limit. Viewed from this perspective, activation of a fault system produces events that exceed this limit, thus restoring the natural scale invariance of the system. A similar conclusion was reached by Eaton et al. (2014a), who interpreted preferred scaling of microseismicity as an expression of the activation of a *stratabound* fracture network, a type of fracture network for which vertical fractures are systematically terminated at horizontal bedding boundaries.

8.5.3 Microseismic Facies Analysis

By analogy with the well-established method of seismic stratigraphy (Mitchum et al., 1977), Rafiq et al. (2016) introduced the concept of *microseismic-facies analysis* to describe a systematic, facies-based approach to microseismic interpretation. A microseismic facies is a body of rock with distinctive empirical characteristics that can be extracted from microseismicity. The goal of microseismic facies analysis is to highlight subtle stratigraphic details, structural deformation, fracture orientation and stress compartmentalization within a reservoir.

This approach makes use of the suite of attributes of microseismic clusters listed in Table 8.2, in order to enable statistical exploration of empirical relationships and elucidation of fine structure within microseismicity clouds. Inferred inter-cluster relationships are used to partition an unconventional reservoir into distinct facies units on the basis of microseismic response. Integration of the microseismic facies interpretation with other types of data, such as 3-D seismic data and regional structural and stratigraphic architecture from well logs, can provide important insights for models of unconventional reservoirs.

Figure 8.10 illustrates how microseismic facies can be defined by data clustering in microseismic attribute space. The graph shows a cross-plot of two attributes, cluster azimuth and moment release rate. In the study area in western Canada, S_{Hmax} is oriented close to 45° (Heidbach et al., 2010), so clusters with this azimuth are aligned parallel to the expected orientation for a simple mode I fracture. The seismic moment release rate characterizes the intensity of microseismic activity. Microseismic clusters along well B appear to have a relatively uniform moment release rate of < 0.4 GJ/hr. In contrast, well A clusters that are well aligned for mode I fractures have a significantly higher moment release rate. The two wells exhibit distinct trends on this graph, which may be indicative of intersection of reservoir facies with different geomechanical response to fracture treatment. Alternatively, the differing response may be related to the horizontal well azimuth, which differs for these two wells (Figure 8.1).

Facies were identified by spatial grouping of points on this graph, resulting in the classification of five facies units labelled A through E. The locations of microseismic events from each of these clusters is plotted on top of a 3-D seismic image in Figure 8.11. The microseismic events are coloured, based on the facies grouping from Figure 8.10. The seismic background image depicts the most-positive curvature attribute (Rafiq et al., 2016). This attribute is obtained from the migrated 3-D seismic volume by picking the time surface of the reflection horizon marking the top of the reservoir unit, which is the Glauconitic sand member of the Cretaceous Mannville Formation. After converting from two-way time to depth, reflection curvature can be determined from this empirically derived surface by fitting a quadratic polynomial of the form (Roberts, 2001)

$$z(x, y) = ax^2 + by^2 + cxy + dx + ey + f , \qquad (8.4)$$

where the mean curvature is given by

$$k_{mean} = \frac{[a(1 + e^2) + b(1 + d^2) - cde]}{(1 + d^2 + e^2)^{3/2}} , \qquad (8.5)$$

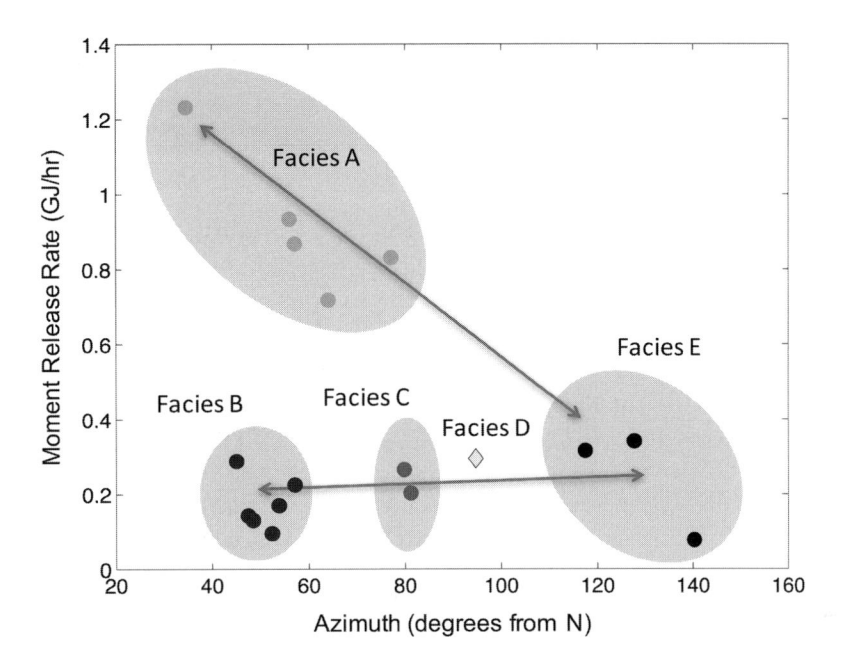

Fig. 8.10 Cross-plot of two microseismic attributes, moment release and cluster azimuth, measured from the Hoadley downhole microseismic dataset shown in Figure 8.1. Each dot represents the average value for an event cluster. Grouping of event clusters coupled with trend analysis, as illustrated here, is used as a basis to infer microseismic facies. From Rafiq et al. (2016), with permission of the Society of Exploration Geophysicists. (A colour version of this figure can be found in the Plates section.)

and the Gaussian curvature is given by

$$k_{Gauss} = \frac{4ab - c^2}{(1 + d^2 + e^2)^2}. \tag{8.6}$$

The *most-positive curvature* is given by a combination of these two parameters,

$$k_1 = k_{mean} + (k_{mean}^2 - k_{Gauss})^{1/2}. \tag{8.7}$$

Zones of high k_1 may be indicative of minor faults, other localized structures or stratigraphic deformation (Rafiq et al., 2016). A quasi-linear feature, marked by a high relative curvature anomaly that extends from from near the southwest of the image in Figure 8.11 to the northwest corner of the image, is parallel to the depositional axis of the barrier-bar reservoir complex (Hayes et al., 1994). This feature is also aligned with a flexural hinge/reverse fault identified using regional well log information (Rafiq et al., 2016). Microseismic facies B appears to correlate with this feature, and is juxtaposed with facies C to the north and facies A to the south. Facies E makes a high angle with the structure trends and the regional S_{Hmax} direction; this microseismic facies is interpreted by (Rafiq et al., 2016) to represent events that are controlled by a reactivated natural-fracture network. Taken together, the microseismic facies obtained by clustering in microseismic

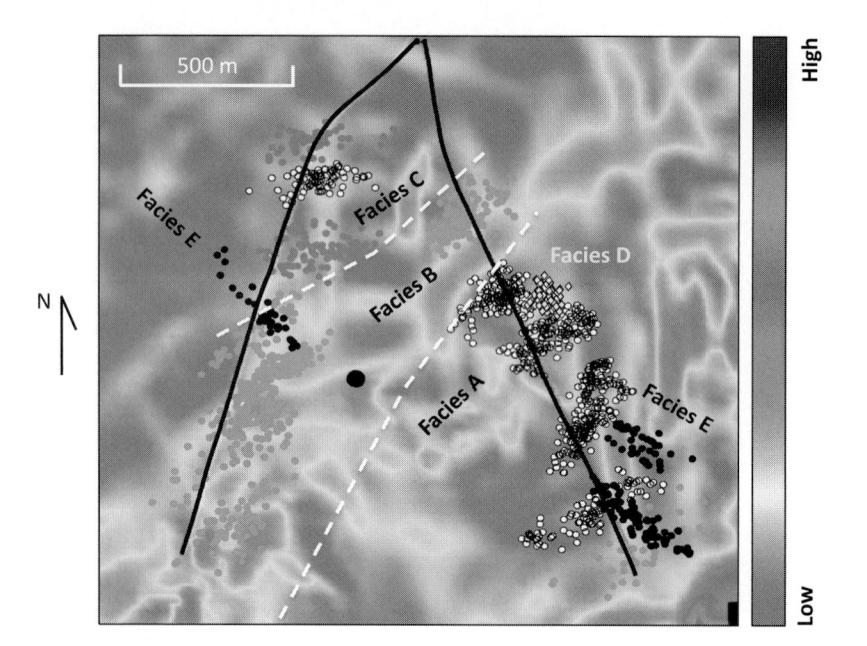

Fig. 8.11 Most-positive curvature attribute derived from 3-D seismic data, overlain with microseismicity from the Hoadley experiment. Northeast–southwest-trending anomalies (green) are parallel to the strike of a barrier-bar complex. Symbols for microseismic events represent facies, as shown in Figure 8.10. Modified from Rafiq et al. (2016), with permission of the Society of Exploration Geophysicists. (A colour version of this figure can be found in the Plates section.)

attribute space (Figure 8.10) exhibit a coherent pattern that appears to correlate with 3-D seismic attribute analysis. Insights from this analysis are useful for characterizing this complex unconventional reservoir, and highlight the importance of data integration using microseismic data in conjunction with other complementary types of data.

8.6 Source Mechanism Studies

As discussed in Chapter 3, a variety of different methods can be used to characterize microseismic sources, including:

- full moment-tensor solutions;
- decomposed moment-tensor solutions, represented as a linear combination of double-couple, compensated linear-vector dipole (CLVD) and isotropic (explosion/implosion) components;
- double-couple focal mechanisms, often represented as beachball diagrams;
- spectral parameters, such as corner frequency (f_c), Brune source radius and coseismic stress drop ($\Delta \tau$).

Seismic moment tensors provide an idealized mathematical representation of a seismic source as set of force couples that act simultaneous at a point within a continuous medium. The 3×3 moment tensor is symmetrical and can be fully characterized by six independent components. Most natural earthquakes can be well represented by a double-couple source mechanism, which only requires three independent components; these components are often specified using the strike and dip of the fault plane, plus the direction of the slip vector (rake). In essence, moment tensors capture the P- and S-wave amplitude radiation patterns, of which there are characteristic patterns for a double-couple event (Figure 3.12), a tensile crack opening (Figure 3.14), an explosion (uniform P-wave radiation for all azimuths with no S wave), etc. Various approaches can be used to interpret moment information. This includes decomposition of moment tensors into components that reflect shear-slip (i.e. double couple, DC) and non-DC mechanisms including mode I failure expressed as tensile crack opening.

Figure 8.12 shows source mechanisms from moment-tensor inversion during a hydraulic-fracturing treatment (Baig and Urbancic, 2010). The estimated source mechanisms during the treatment are presented using a Hudson source-type diagram. This type of diagram provides a convenient visual representation of the type of seismic source projected onto a skewed diamond-shaped grid (see Figure 3.16). Grid coordinates of explosive, implosive, tensile crack opening (TCO) and tensile crack closure (TCC), CLVD and linear vector dipole (LVD) types of sources are indicated on the graphs. The double couple mechanism plots at the origin in the centre of the diamond pattern. The diagram shows time-window snapshots at four times during the treatment, as indicated by the yellow-shaded rectangle in the pump-curve graphs. For example, the panel in the top left of Figure 8.12 shows a time window starting at $\sim$ 150 s, when the injection of the pad was complete and proppant was beginning to be mixed into the slurry. One of the principal advantages of the Hudson source-type plot is that the projection has equal-area characteristics, which means that the event density can be meaningfully contoured as shown by the colour-density Hudson plots, where red colours indicate a high density of events and blue colours indicate low density. In the earliest time window, the moment-tensor analysis reveals a preponderance of events that are positioned in the parameter space close to the tensile-crack opening mechanism. As the treatment stage proceeds, the bulk of activity shifts toward the tensile-crack closure (TCC) position. The sequence of activity can be interpreted as indicative of opening and closing of fractures in response to the fluid injection. For this dual-well downhole monitoring program, Baig and Urbancic (2010) report that the condition number for the matrix inversion is small ($<$ 30), which means that the inversion process is numerically stable and thus should not lead to artifacts caused by excessive amplification of noise (Vavryčuk, 2007; Eaton and Forouhideh, 2011).

Nevertheless, a few important caveats warrant discussion. First, there are significant trade-offs between moment tensor, velocity model and source location; thus, the best-fitting moment-tensor solution obtained using an incorrect velocity model or hypocentre will differ from the true solution. Warpinski and Wolhart (2016) emphasize the current lack of validation of moment-tensor solutions. In addition, Warpinski and Wolhart (2016) cite several studies demonstrating that the radiated seismic energy during treatment is a negligible fraction of the input energy for hydraulic fracturing (Maxwell and Cipolla, 2011;

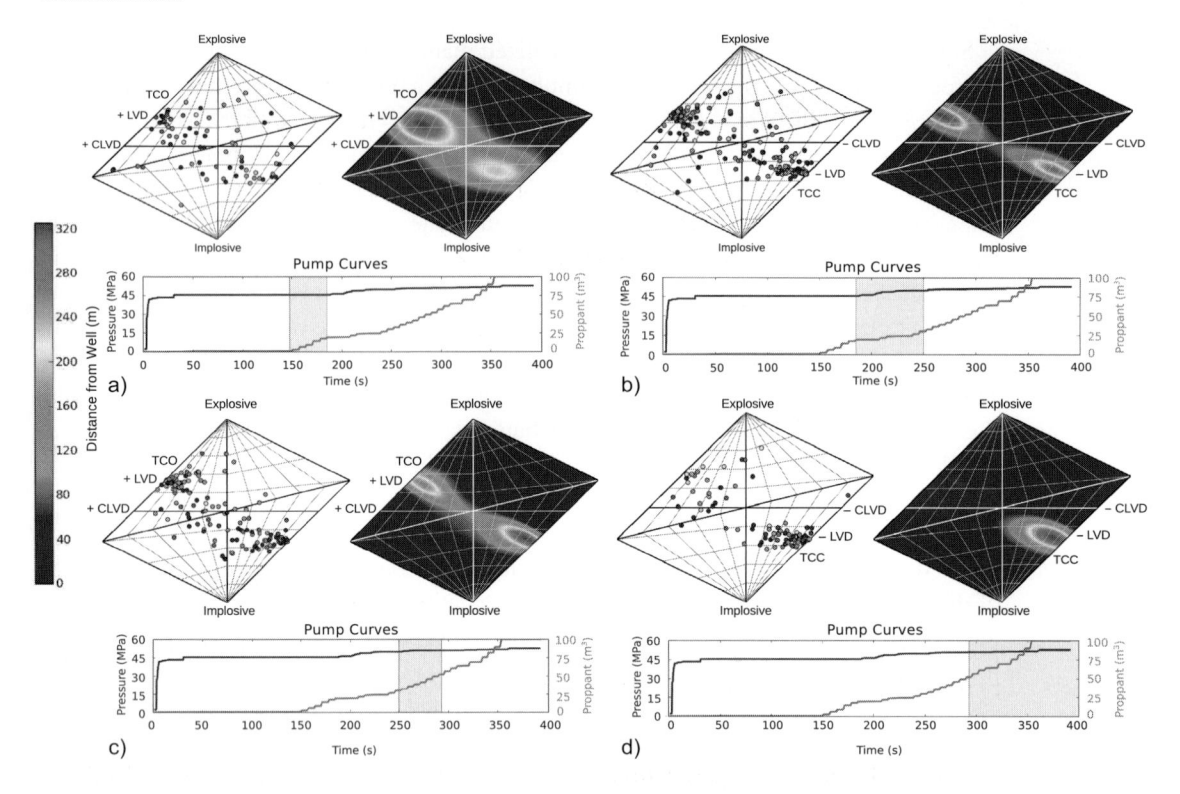

Fig. 8.12 Source mechanisms showing crack opening and closing during hydraulic fracturing, based on moment-tensor inversion. a) Scatter plot (left) and density contour plot (right) showing Hudson source-type diagram. These plots contain events that occurred during the time window indicated by the yellow region in the underlying graph, which shows pump curves for pressure and proppant volume. The majority of events cluster close to a tensile crack opening (TCO) mechanism. Subplots b) – d) contain similar graphs for subsequent time windows during the treatment stages. As time progresses during the treatment, the source mechanism shifts toward tensile crack closure (TCC) in the lower right part of the source mechanism diagram. Modified from Baig and Urbancic (2010), with permission of the Society of Exploration Geophysicists. (A colour version of this figure can be found in the Plates section.)

Boroumand and Eaton, 2012) as evidence that the physical connection between rupture processes of microseismic events and fracture geomechanics is tenuous. There are also certain ambiguities in the interpretation of a moment-tensor solution; for example, it is not possible to determine, for a double-couple solution (focal mechanism) alone, which of the two nodal planes is the correct fault plane. These and other issues are considered in the pitfalls section of this chapter.

As discussed in Chapter 7, surface and shallow-well microseismic monitoring typically provides a sufficient aperture that the focal mechanism from P-wave first motions, though not the full moment tensor, can be retrieved in a direct and robust manner. Several of these studies highlight the presence of two distinct yet consistent populations of focal mechanisms (Detring and Williams-Stroud, 2012; Snelling et al., 2013); one event population is characterized by an unusual dip-slip mechanism, with vertical and horizontal nodal

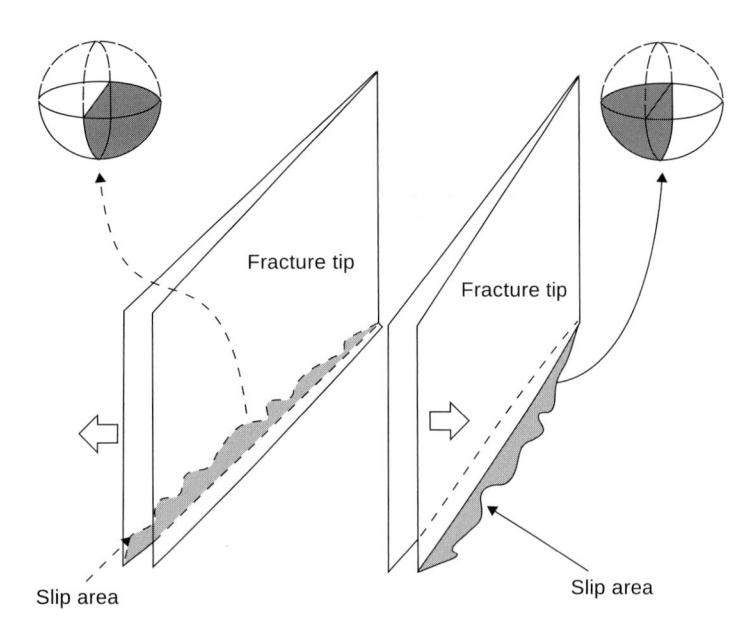

Fig. 8.13 Layer-parallel slip during hydraulic-fracture opening, producing double-lineament microseismic events. From Tan and Engelder (2016). Reprinted with permission of the Society of Exploration Geophysicists.

planes, while the second population of events is characterized by strike-slip mechanisms. In these studies, the orientation and mechanisms of the strike-slip events are consistent with activation of fault systems below the treatment level. A number of studies have proposed a bedding-slip model for the other population of events (Rutledge et al., 2013; Stanek and Eisner, 2013; Tan and Engelder, 2016). The model is illustrated in Figure 8.13. According to this model, although the opening of a vertical tensile fracture is generally an aseismic process, as the fracture opens the lateral displacement of the walls is accommodated by a slip along bedding planes.

Seismic slip on bedding planes is postulated to produce the observed focal mechanisms indicated in Figure 8.13. The slip surfaces may be situated at the top, bottom or stepover of the hydraulic fracture (Tan and Engelder, 2016). This mechanism is important, as it suggests that the bedding slip events are directly linked to opening of the hydraulic fracture. Furthermore, the double-lineament microseismic events could delineate the full extent of the SRV (Tan and Engelder, 2016). There is potential to investigate this model further by analyzing polished-slip surfaces, which have been documented in drillcore data (Soltanzadeh et al., 2015; Glover et al., 2015). These features are interpreted as slip surfaces that have developed along horizontal planes of weakness in response to natural deformation.

Microseismic events are sometimes observed during flowback operations. Figure 8.15 shows an example of microseismicity during flowback, documented by Eaton et al. (2014c). That study showed that an abrupt drop in pressure recorded during flowback coincided with a sequence of microseismic events. The flowback-induced events initiated at the distal parts of microseismic clusters and correlate in time with an abrupt drop in flowback pressure (Eaton et al., 2014c). The events observed during flowback progress in a retrograde fashion back toward the wellbore, suggesting that they record a process that initiated near the fracture tips. Recalling that the fracture closure can be considered as negative polarity of TCO mechanism, the characteristics of the observed microseismicity, including amplitude ratio $S/P < 5$ suggest that these events may record fracture closure that accompanied the flowback stress drop. If so, this observed region of fracture closure provides an indication of fractures that did not receive enough proppant to remain open.

8.6.1 Waveform Modelling

The previous methods have focused on visualization-based interpretation. We now look at some examples of model-based interpretations. Simulations of wave propagation can be very effective for understanding the expression of different moment tensors mechanisms based on specific array geometries. Figure 8.14 shows numerical simulations for two different source types, calculated using a *finite-difference* method, a numerical scheme in which the continuous differential equations are replaced with a difference equations (Virieux, 1986). The finite-difference method used modified from Boyd (2006) to provide the ability to use any moment-tensor source. Figure 8.15a shows 2-D finite-difference simulation for the case of a source with layer-parallel slip with hypocentre at 1950m, recorded using a vertical array of sensors from 1600 m to 2000 m at an epicentral distance of 400 m. The

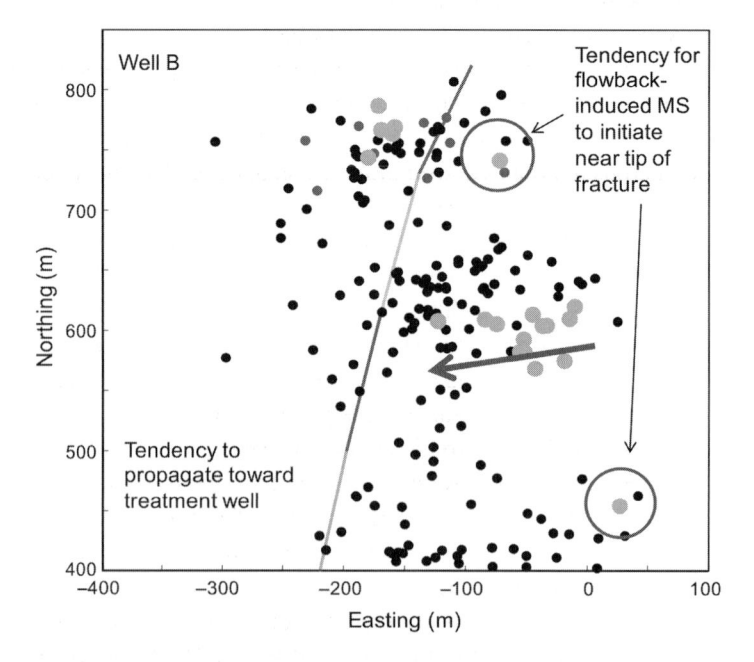

Fig. 8.14   Flowback-induced microseisms recorded near the heel of well B during the Hoadley experiment (Eaton et al., 2014c). Events shown in black occurred during hydraulic fracturing. Events in red occurred after treatment but prior to flowback. Those shown in green occurred during flowback. Large arrow shows progression of one set of flowback-induced events. (A colour version of this figure can be found in the Plates section.)

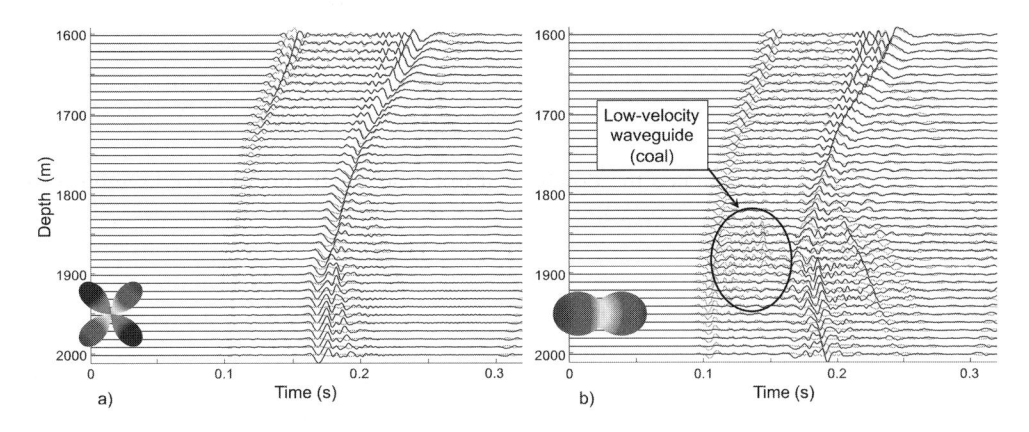

Fig. 8.15 2-D finite-difference wavefield simulation of shear and tensile mechanisms. Blue traces show vertical displacement, green traces show horizontal. a) Bedding-plane shear-slip mechanism. b) Tensile crack-opening (TCO) mechanism. The geological model and geophone array are based on the Hoadley experiment. The source radiation pattern is shown in the lower left of each panel. Low-velocity coal layers produce a waveguide, which is more obvious for the TCO mechanism due to stronger excitation by P waves. (A colour version of this figure can be found in the Plates section.)

velocity model is an isotropic, horizontally stratified model obtained by 1 m median blocking the well log from the observation well in Figure 8.1. The finite-difference simulation shows that P-wave arrival is discernible above ≈ 1800 m, but is weak or absent below this depth due to the effects of the horizontal nodal plane, which can be recognized from the radiation pattern, shown in the lower left corner of the record section. Note the upgoing $P - S$ conversion for depths above ≈ 1700 m. These are expected due to the presence of numerous interfaces at stratigraphic contacts. For the double-couple source in Figure 8.15a, the S wave arrival is strong for all depths.

Figure 8.15b shows a similar synthetic record section, for the case of a tensile-crack opening (TCO) source mechanism with a vertically oriented fracture. In this case the P wave has a higher amplitude than seen in the double-couple simulation. Eaton et al. (2014d) showed that in a binary scenario (i.e. either double couple or TCO) an average amplitude ratio of $S/P < 5$ is indicative of tensile failure at a 90 percent level of confidence. Low-velocity coal layers create a waveguide Pike (2014), which is more strongly excited by the TCO mechanism due to its radiation pattern.

McKenna (2014) describes a model-based interpretation of a hydraulic fracturing completion in the Eagleford shale in Texas. The focus of that study was to carefully exam microseismic data for proppant signatures, with the goal of tracking the placing of proppant in the reservoir. The study incorporated fluid mass-balance through the use of: 1) McGarr's formula relating seismic moment to net injected volume; 2) estimation of fracture dimensions from the microseismic data; 3) use of DFIT-derived infiltration coefficient, which determines the fraction of injected volume that enters into the reservoir; and 4) a calibrated empirical relationship between fracture length and aperture, which allows seismic-derived fracture surface area to be converted into a volume estimate. This combination was used to construct a calibrated discrete fracture network (DFN) model. The centroid of microseismic deformation was tracked by averaging event locations within a time window (McKenna, 2014). This analysis showed that the centroid oscillated in a systematic fashion back and forth away from the treatment well during injection. This change in the locus of microseisms suggests that the presence of proppant within the existing hydraulic fractures, creating local stresses that resulted in new failure mechanisms. During the injection of the pad at the start of each stage, the centroid migrated progressively away from the treatment well, in a pattern consistent with tip-induced microseismicity. At the start of the first proppant build-up the centroid jumped back to a position close to the treatment well. The treatment program included two phases of proppant injection for each stage, and when the first proppant phase was complete the microseismicity centroid jumped back out toward the tip of the fracture. This pattern repeated itself during the second proppant phase. The DFN model that was constructed during the initial mass-balance calculation was populated with areas where the interpreted proppant-related seismicity had occurred. This yielded a DFN model for the propped fracture network, calibrated using microseismic.

Boroumand and Eaton (2015) conducted a model-based interpretation in which a numerical hydraulic fracture model was tuned to fit observed microseismicity during a hydraulic fracturing stage. The results of this modelling exercise are presented in Figure 8.16. The numerical model used a lumped-parameter approach, which enabled efficient analysis of model sensitivity to a few key parameters. The assumption was made in this study that the spatio-temporal evolution of the microseismic cloud provides a proxy for the growth of the hydraulic fracture. Consequently, the parameters were adjusted to obtain a good fit. A particularly important geomechanical parameter was the critical energy release rate, G_{cr}, which determines the energy required to create new fracture surface. There are few laboratory constraints on this parameter. A value at the lower end of published values of G_{cr} was required in order to achieve a satisfactory fit.

Spectral characteristics of radiated wavefields can carry significant clues about source mechanism. For example, a source that produces a positive and negative radiation in rapid succession will generate periodic spectral notches (Walter and Brune, 1993). These are caused by superposition of two closely spaced events of opposite polarity, similar to the surface ghost that was discussed in Chapter 7. Eaton et al. (2014d) documented periodic spectral notches for microseismic events and argued that this spectral characteristic is best explained by opening and closing a tensile crack, noting the low likelihood of forward and reverse (retrograde) shear slip on a fracture, as this would require a near-instantaneous reversal in the polarity of the shear traction acting on the fracture. Van der Baan et al. (2016) developed this concept further by introducing a geomechanical model for the process zone near the fracture tip, which they called *stick-split* failure. They postulated that tensile fracture growth is episodic. During the opening phase, the tensile fracture tip grows too fast for the fluid to keep up. This results in a transient open fracture area near the tip that does not contain fluid, which closes very quickly after opening in the same manner as a hand clapping. Van der Baan et al. (2016) argued that this stick-split mechanism could explain ribs seen on exposed surfaces of magmatic dikes, which are a natural form of fluid-induced tensile failure. Figure 8.17 illustrates their model and also shows a source spectrum containing periodic notches.

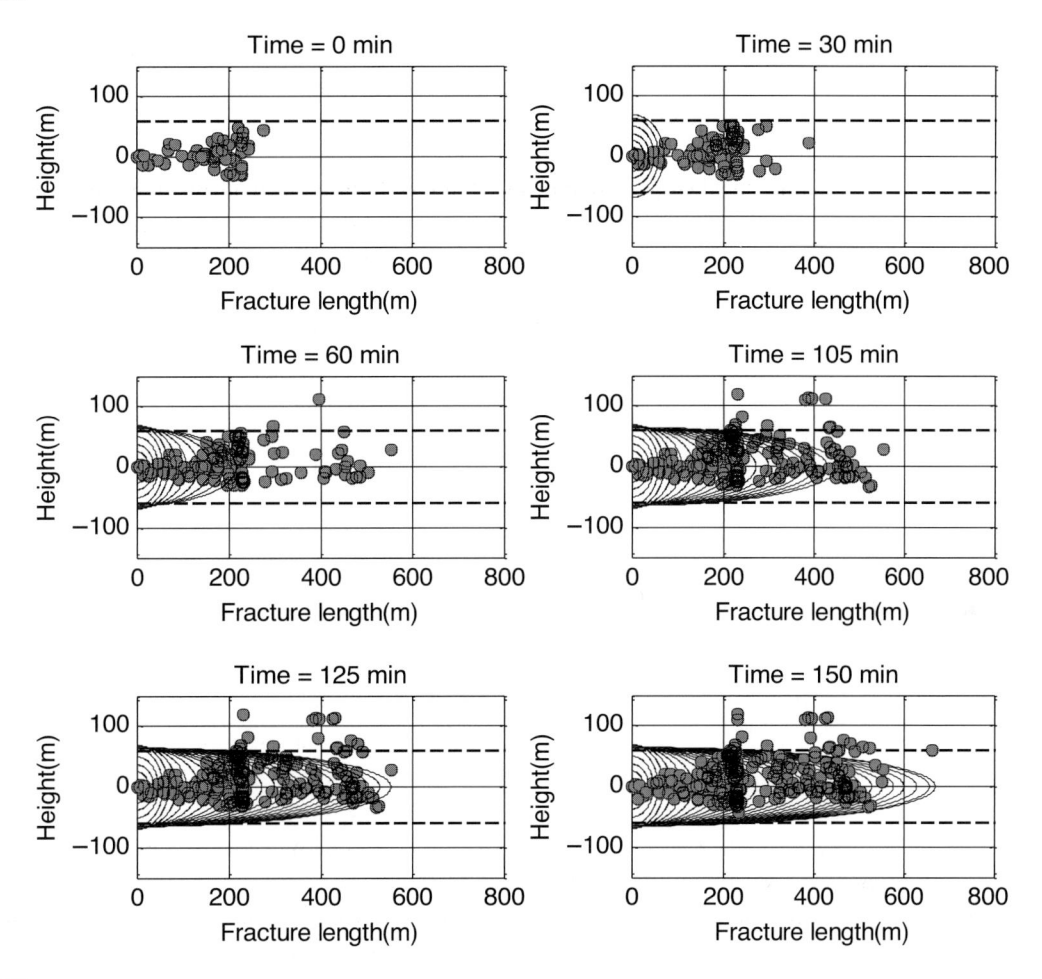

Fig. 8.16 Example of model-based interpretation, illustrating adjustment to a hydraulic-fracture model in order to fit the spatiotemporal evolution of microseismicity. Images show snapshots of the simulated fracture ellipses and their correlation to microseismic events, which are indicated by circles. In this example, fracture parameters such as the critical energy release rate, G_{cr}, were adjusted until fracture growth matched the expanding microseismic cloud. From Boroumand and Eaton (2015). Reprinted with permission of the Society of Exploration Geophysicists.

8.7 Applications

There are a number of value-added applications of microseismic interpretation that have been well documented to provide key insights for optimizing hydraulic-fracturing completions and well performance (Cipolla et al., 2012; Maxwell, 2014; Warpinski and Wolhart, 2016). It is also becoming increasingly recognized that the highest value from the use of microseismic observations comes from calibration of geomechanical models, by capturing the overall characteristics of microseismic clusters rather than individual events. In general,

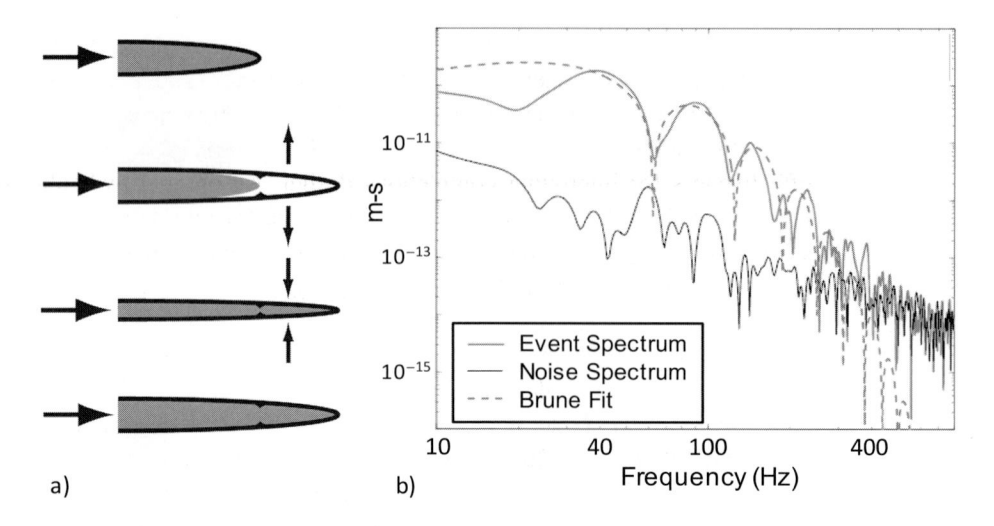

Fig. 8.17 Proposed split-stick fracture mechanism. a) Schematic representation of the process. From top to bottom, this mechanisms involves pressure buildup, tensile failure and fluid-pressure drop due to fracture-volume increase, leading to partial fracture closure. b) Displacement spectrum for the qS_1 wavefield (upper solid curve). Dashed curve shows the best-fitting tensile opening/closing model. Lower solid curve shows background noise, based on a pre-event noise window. Signal amplitude is above the noise level from 10 to 400 Hz. Republished with permission of the Geological Society of America from Stick-split mechanism for anthropogenic fluid-induced tensile rock failure, *Geology*, volume 44, number 7, copyright 2016; permission conveyed through Copyright Clearance Center, Inc.

applications and interpretation tools can be subdivided into three broad categories (Cipolla et al., 2012):

- providing process-control feedback in near-real-time;
- evaluating completions effectiveness;
- optimizing of field design for hydrocarbon recovery.

One example of real-time process-control feedback involves early detection of screen out, a proppant blockage that prevents continued fluid injection during treatment (§4.2). Incipient screen-out can be manifested by the development of a new cluster of events in close proximity to the injection point (Rodinov et al., 2012). Early identification of screen out enables preventative measures that may avoid delays from remediation, such as cleaning out the well (Maxwell, 2014). Near-real-time microseismic monitoring is also the best available approach to evaluate the effectiveness of *diversion* (Warpinski and Wolhart, 2016), a well-completions method that uses mechanical, rate-and-pressure modification, or chemical agents to divert fracturing fluid into regions that may not otherwise be stimulated due to stress conditions, rock properties, etc. (Van Domelen, 2017). The development of microseismicity within the intended regions of the reservoir provides process-control feedback for diversion (Daniels et al., 2007). Similarly, geohazards such as faults have near-real-time signatures that include a sudden increase in event magnitude or source mechanism, a change in *b* value and initiation of a planar event cloud that is oblique to the hydraulic fracture (Kratz et al., 2015). In addition to induced seismicity, which is covered in Chapter 9, the intersection of a hydraulic fracture with a fault can lead to problems such as fluid loss or casing failure (Warpinski et al., 2012).

Another key application of microseismic interpretation is the evaluation of completions effectiveness for improving completions design. An optimal hydraulic-fracturing treatment maximizes the area and conductivity of the fracture network within the reservoir, while minimizing wasted resources and energy associated with out-of-zone fracture growth (Maxwell, 2014). For example, fracture-height growth can be correlated with injection rate (Inamdar et al., 2010); thus, the application of microseismic monitoring to assess fracture height containment can provide valuable feedback that can be used to make adjustments to treatment design. The value of information increases significantly if near-real-time microseismic data are available. For example, in a case study in the Wolfcamp formation of west Texas, Ejofodomi et al. (2010) illustrate how staging and perforating can be modified using near-real-time microseismic. There is also considerable interest in the use of microseismic interpretation to map proppant placement, as documented by McKenna (2014) or, conversely, stimulated regions that did not receive proppant as shown in Figure 7.14. Finally, as discussed in §9.1.3, the *triggering front* is an idealized surface in distance-time $(r - t)$ space that represents an envelope to the cloud of microseismicity as it expands away from the injection point. Some studies suggest that permeability can be retrieved from the shape of the triggering front. For example, Shapiro and Dinske (2009) have developed a nonlinear model in which linear diffusion and hydraulic fracturing are end members.

The application of microseismic interpretation to field-development optimization focuses on calibrating reservoir-simulation models in order to understand drainage architecture (Cipolla et al., 2012). For example, Dohmen et al. (2017) describe a new application of microseismic monitoring for optimizing field development during *refracturing*.[2] Their study shows that stimulation of a well next to a producing well generates a strong microseismic response in the producing well, called *microseismic-depletion delineation* (MDD). Their proposed geomechanical model is shown in Figure 8.18. According to this model, production-related drop in pore pressure causes the Mohr circle to shift to higher effective normal stress. This is accompanied by an increase in deviatoric stress due to poroelastic effects, as marked by an increase in diameter of the Mohr circle. Fluid injection into an adjacent well leads to a pore-pressure increase, which moves the Mohr circle across the failure line resulting in strong microseismic response. This response is also accompanied by a low b value. Dohmen et al. (2017) postulate that the microseismic response delineates the *depleted zone*, and that the strength of the response (reflected by the b value) is indicative of the degree of depletion.

Some new and emerging challenges at the forefront of microseismic interpretation include:

- Robust discrimination between "wet" and "dry" events, where wet microseismicity is triggered by a change in pore pressure while dry microseismicity is caused by a change

---

[2] Refracturing is a process of hydraulically fracturing a well after the initial well completion and production cycle have been completed.

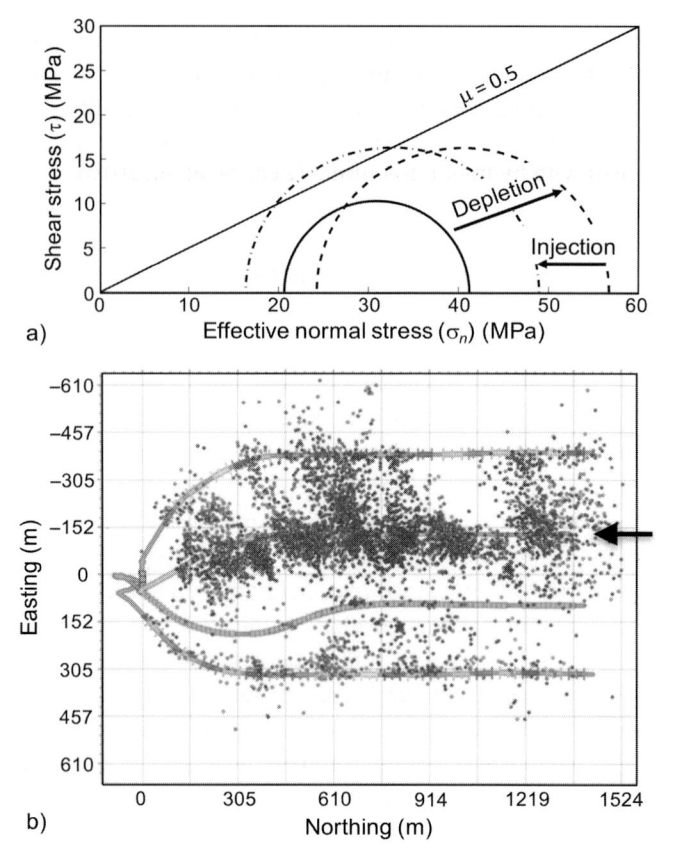

Fig. 8.18 Example of microseismic depletion delineation in the Bakken play of North Dakota. a) Poroelastic model for effective-stress evolution during production and restimulation, using a Biot coefficient of $\alpha = 0.75$. Production-driven depletion is represented by a pore-pressure decrease of 20.7 MPa, which results in increased deviatoric stress. Subsequent injection decreases the pore pressure by 10.3 MPa, pushing the state of stress beyond the Mohr–Coulomb failure line. b) Map view of microseismicity, showing only larger events in the range of $-2.2 \leq M \leq -0.6$. The two outer wells were hydraulically fractured. The arrow marks a producer well that was not stimulated. Induced microseismicity is interpreted to show reservoir regions that have been depleted. Modified from Dohmen et al. (2017), with permission of the Society of Petroleum Engineers.

in total stress (Nagel et al., 2012). This distinction is important to avoid overestimation of ESV by including dry events (Maxwell et al., 2015).
- Development of Bayesian methods for microseismic processing and interpretation, including the incorporation of reflected waves for hypocentre determination (Belayouni et al., 2015).
- Imaging reservoir boundaries, including further development of microseismic facies analysis (Rafiq et al., 2016).
- Better integration of microseismic interpretation with seismic petrophysics, geomechanical and reservoir models, and other fracture-diagnostic methods such as tiltmeter and distributed acoustic sensing (DAS) (Warpinski et al., 2014).

- Development of validated scaling relations that are calibrated to observed microseismicity, rather than relying upon earthquake scaling relations that are extrapolated over many orders of magnitude from their calibration range.
- Identification of a new observational paradigm for potential slow-slip processes associated with hydraulic fracturing (Zecevic et al., 2016).

8.8 Interpretation Pitfalls

A number of pitfalls for microseismic interpretation have been identified by Cipolla et al. (2012), who argue that misapplications primarily arise from flawed assumptions about the nature of physical links between microseismic events, hydraulic fracture geometry and well productivity. For example, although source mechanisms derived from moment-tensor inversion are representative of small-scale deformation processes associated with individual microseismic events, the link between these mechanisms, existing natural fractures and the growth of the hydraulic fracture remains unclear, particularly considering the negligible fraction of the total energy in the system that is radiated by seismic waves (Warpinski and Wolhart, 2016). Instead, Cipolla et al. (2012) advocate the use of microseismic data for calibration of geomechanical models, such as one developed by Weng et al. (2011) that accounts for complex interactions between the hydraulic fractures and natural fractures including interfacial friction, stress anisotropy and stress shadowing. On the other hand, double-lineament microseismic events with bedding-plane slip mechanisms (Tan and Engelder, 2016) seem to imply a direct physical link between hydraulic-fracture growth and microseismicity. Resolution of this issue is critical for improving our understanding of the interactions between aseismic and seismic deformation processes.

Another contentious issue is the concept of stimulated reservoir volume (SRV). Although measured values of SRV correlate with production in some areas (Mayerhofer et al., 2010), this correlation breaks down (and even reverses) elsewhere (Cipolla and Wallace, 2014). Part of the inconsistency stems from the aforementioned incomplete understanding of complex physical links between microseismicity and hydraulic-fracture growth in a fractured medium. In addition, hypocentre location uncertainty, misinterpretation of overlapping stages, failure to account for viewing bias (Cipolla et al., 2012), distinguishing between wet versus dry microseismicity (Maxwell et al., 2015) and uncertainties in mapping the extent of propped fractures (McKenna, 2014) further obscure the meaning of SRV. In view of these sources of ambiguity, use of the term *estimated stimulated volume* (ESV) is recommended.

Despite vast differences in spatial scale and rupture mechanism, scaling relations applicable to shear failure on large-scale fault systems have been applied to build discrete fracture networks for reservoir modelling, based on microseismic observations (e.g. McKenna, 2013). This issue is brought into focus by considering a study of microseismic monitoring at an underground rock laboratory in Canada, which demonstrates large discrepancies between scaling predicted by classic source models and observations of rock fractures near underground openings (Cai et al., 1998).

8.9 Summary

This chapter begins by introducing a new systematic approach for microseismic interpretation, guided by a philosophy that interpretation should be repeatable, data-driven and quantitative. The basic interpretation workflow consists of data preconditioning, determination of event attributes, event clustering, microseismic facies analysis and application of microseismic interpretation to achieve specific engineering objectives.

Data preconditioning involves careful filtering to mitigate observational bias and to remove events with inaccurate microseismic locations, in order to ensure consistency of inferred attributes and to avoid spurious complexity associated with low-confidence locations (Cipolla et al., 2011). In general, S/N can be used as an indicator of reliability, with higher S/N usually associated with more robust event parameters (Cipolla et al., 2012).

Clustering is defined as unsupervised classification of observations into groups (Jain et al., 1999), and can be classified as agglomerative or divisive. Agglomerative clustering is a bottom-up approach that initially bins each event into a separate cluster and then merges the clusters until a stopping criterion is met. Divisive clustering starts with all observations grouped into a single cluster and performs splitting until a stopping criterion is met. A convex-hull method is introduced to perform repeatable determination of estimated stimulated volume (ESV) for a point cloud (cluster). The method includes a deterministic approach to compute volumetric uncertainty.

There is a virtually unlimited number of cluster attributes. Examples include basic fracture geometry (e.g. apparent dimensions and azimuth), Gutenberg–Richter b value, fractal dimension (D), maximum moment release rate, transience and apparent permeability. Cluster attributes provide the basis for microseismic facies analysis (Rafiq et al., 2016), a systematic approach to microseismic interpretation. This method is based on the concept of microseismic facies, a body of rock with distinctive empirical characteristics that can be extracted from microseismicity. The overall goal of microseismic facies analysis is to highlight subtle stratigraphic details, structural deformation, fracture orientation and stress compartmentalization within a reservoir.

Various model-based approaches for microseismic interpretation are described in this chapter. An important tool is full-waveform modelling of source radiation and wave propagation for a realistic background model. This can be accomplished using a finite-difference numerical simulation approach.

Applications of microseismic interpretation to achieve engineering objectives are split into three basic categories: providing process-control feedback in near-real time, evaluating completions effectiveness and optimizing field design for hydrocarbon recovery. Notable pitfalls for microseismic interpretation include assuming that microseismic moment tensors provide a direct proxy for hydraulic fracture deformation processes, over-interpretation of the concept of SRV and scaling relations derived from large-scale fault processes.

8.10 Suggestions for Further Reading

- Practical guidelines for microseismic interpretation: Maxwell (2014)
- Review of methods used for clustering analysis: Jain et al. (1999)
- Quantitative seismic interpretation and petrophysics: Avseth et al. (2010)
- Empirical relations in seismology: Kanamori and Anderson (1975)

8.11 Problems

1. The table below summarizes the number events within magnitude bins of width one magnitude unit and centred at values shown by M. Suppose that these events were detected within the same study area, during three different time windows of varying length.

M	1 day	7 days	28 days
0	48	346	1329
1	4	28	110
2	0	2	9
3	0	0	1

a) Estimate the b value by linear regression. Assume that the table is complete to the lowest magnitude bin.
b) Based on this simple example, how many observations do you think are necessary for reliable determination of b value?

2. The table below shows the locations for 10 coplanar events. Locations denoted by (x', y', z') contain noise, whereas those denoted by (x, y, z) are noise-free. Using equation 8.2, calculate the D value for both of these sets of events. What does this tell you about the effects of location uncertainty on the calculated D value?

x	y	z	x'	y'	z'
162	451	759	158	465	919
794	84	525	772	78	613
311	229	502	331	216	309
529	913	1637	551	909	1928
166	152	314	165	132	588
602	826	1542	601	807	1741
263	538	941	255	560	852
654	996	1824	674	1019	1685
689	78	464	683	82	376
748	443	1041	729	421	1070

3. Supplementary Table 1 in the online materials contains locations, times and magnitudes for a series of events that could be classified variously into clusters. Discuss various approaches that could be used to cluster these events.

4. For the tabulated events in Problem 3:

a) make an $r - t$ graph to determine the apparent diffusivity;
b) evaluate attributes listed in Table 8.2.

5. Use Equation 8.7 to determine the most-positive curvature corresponding to the following analytically defined surfaces.

a) A sphere of radius a.
b) A paraboloid defined by $z = x^2/a^2 + y^2/b^2$.

Note that the calculations can be considerably simplified by consulting a textbook on elementary differential geometry to find parametric expressions for Gaussian curvature (k_{Gauss}) and mean curvature (k_{mean}).

6. **Online exercises**: A Matlab tool is provided to determine the estimated stimulated volume for a cloud of points, using the method of convex hulls (including volumetric uncertainty). A sample dataset from a hydraulic fracturing stage is included.

9 Induced Seismicity

> Logarithmic plots are a device of the devil.
> Charles F. Richter (Earthquake Information Bulletin, 1980)

The term *induced seismicity*, as used in this book, refers to earthquakes or other seismic events that have a clear anthropogenic association. Human activities with the potential to produce appreciable seismic activity include: underground mining (Gibowicz and Kijko, 1994) and construction of tunnels (Husen et al., 2013); impoundment of surface water reservoirs (Simpson et al., 1988); development of engineered geothermal systems (Breede et al., 2013); detonation of underground explosions (Massé, 1981); and, subsurface fluid injection or withdrawal (National Research Council, 2013). Among these different causal mechanisms for induced seismicity, this chapter focuses on injection-induced seismicity from energy technologies, including deep-subsurface saltwater disposal (SWD), hydraulic fracturing (HF) of low-permeability unconventional reservoirs and engineered (or enhanced) geothermal systems (EGS). These three causal mechanisms of induced seismicity occur in different settings and arise from distinct industrial applications, yet they share broadly similar underlying physical mechanisms.

The material in this chapter builds upon previous sections of the book, especially the fundamental theory discussed in the first four chapters. The chapter begins with a review of basic concepts, commencing with definitions for a few key terms. Pioneering work on injection-induced seismicity is then considered, which frames our current understanding of this phenomenon, including criteria for discriminating between induced and natural events. A fluid/stress systems approach is then used to explore necessary conditions for induced seismicity, such as nucleation of dynamic slip on a fault. Seismological tools of the trade that are employed for observations of induced seismicity are then covered, followed by a case histories for various causal mechanisms of deep saltwater disposal, hydraulic fracturing and geothermal applications. Finally, methods used for risk assessment and hazard mitigation are presented, along with a brief discussion of important societal factors.

9.1 Background

We set the stage for this chapter by considering several important semantic issues. First, some investigators reserve the adjective "induced" for seismicity caused by human activities that lead to shear stress preturbations on a fault that are comparable in magnitude to the shear stress from the ambient (unperturbed) stress field (McGarr et al., 2002). According to this viewpoint, the term "triggered" is then used if the human-induced stress change on a fault is only a relatively small fraction of the ambient stress level (Simpson, 1986). This category applies to faults that are prone to failure, such that the timing of rupture nucleation may be advanced by human-induced stress changes, but the magnitude is controlled by tectonic stresses (Dahm et al., 2013). A shortcoming of this distinction is the implied necessity to determine the magnitudes of ambient stresses and human-induced stresses that act on a fault – estimates that may be fraught with uncertainty (Passarelli et al., 2013). Furthermore, within the seismological community, the term "triggered" is now primarily used to described earthquakes that are caused by dynamic stress transfer from other earthquakes (Freed, 2005). Hence, following Rubinstein and Mahani (2015), we adopt the simpler definition of induced seismicity, given at the start of this chapter, to avoid any such confusion.

Moreover, from this definition, it follows that $M < 0$ microseismicity, which typically occurs during hydraulic-fracturing operations, should be considered as a form of induced seismicity – albeit of low magnitude. This may seem counter-intuitive, as induced seismicity is generally viewed as an uncommon – and undesirable – byproduct of human activities, whereas this type of microseismicity is expected and desirable, insofar as it can provide an effective means for remote surveillance of hydraulic fractures. Consequently, for the sake of clarity, seismicity that is an expected and intended outcome of a human activity is designated here as *operationally* induced, whereas events that are not normally associated with the activity are designated as *anomalous* induced seismicity.

9.1.1 Pioneering Studies

Induced seismicity that is caused by fluid injection has been extensively studied. Since the concept of effective stress was introduced by King Hubbert and Rubey (1959), it has been recognized that pore pressure plays a fundamental role in reducing the effective normal stress acting on a fault, thus reducing the frictional sliding resistance. One of the earliest and best-documented examples of a probable link between fluid injection, pore pressure and induced seismicity comes from a sequence of earthquakes that took place from 1962 to 1968 near Denver, Colorado (Healy et al., 1968). The injection source that is inferred to have produced the earthquake sequence was a disposal well at the Rocky Mountain Arsenal, where large volumes of chemical fluid waste were being injected into crystalline basement rocks at a depth of 3.7 km. Although considerable low-magnitude seismicity occurred during injection operations, the largest earthquake (M_W 4.85) occurred more than a year after injection had ceased (Herrmann et al., 1981). The "smoking-gun" evidence for induced fault activation in this case includes a high degree of temporal correlation between injection and seismicity rates, as well as an approximately coplanar distribution of earthquake hypocentres in the basement close to the injection site (Healy et al., 1968).

These observations from Rocky Mountain Arsenal prompted an experiment, undertaken from 1969 to 1974 by Raleigh et al. (1976), that was designed as a direct test of the effective-stress hypothesis. This test was conducted at the Rangely oil field in Colorado, which at that time had been on waterflood (the injection of water for secondary oil recovery) for over a decade and had produced a multitude of small earthquakes of up

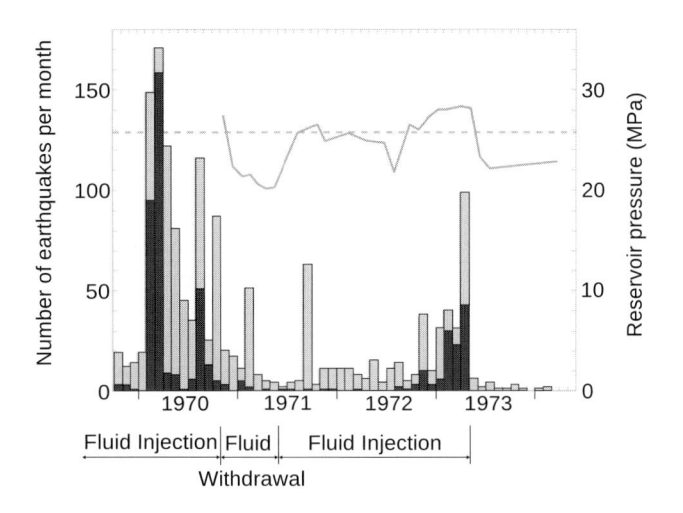

Fig. 9.1 Seismic response of a subsurface reservoir to change in pore pressure, showing good agreement with a simple model of effective stress and Mohr–Coulomb failure. Pressure history (solid black line) and earthquake frequency (bars) at the Rangely field in Colorado, during more than one full cycle of fluid injection and withdrawal. The dashed line shows the calculated critical pressure required for fault activation. Black bars show earthquakes within 1 km of the injection wells. Modified from Raleigh et al. (1976), with permission of the American Academy for the Advancement of Science.

to M_L 3.1. The experiment involved several cycles of raising and lowering of the reservoir pressure with the use of injection wells in which the pumping flow direction could be reversed. Observational data were provided by concurrent measurement of reservoir pressure in nearby wells, along with locations and timing of seismic activity obtained using a local seismograph network. The Mohr–Coulomb failure criterion for faults that transect the Rangely Field was determined based on the ambient stress state measured using a combination of methods, including mini-fracture tests and earthquake focal mechanisms, coupled with laboratory measurement of the coefficient of static friction ($\mu_s = 0.81$) from wellcores. The observed pattern of onset and cessation of seismicity (Figure 9.1) shows remarkable agreement with the model predictions, confirming the basic principles of effective stress and Mohr–Coulomb failure at the field scale.

9.1.2 Distinguishing Between Natural and Induced Seismicity

Davis and Frohlich (1993) proposed a set of criteria for distinguishing between induced and natural earthquakes, based on the following questions:

1. Are the events the first known earthquakes of this character in the region?
2. Is there a clear (temporal) correlation between injection and seismicity?
3. Are epicentres near wells (within 5 km)?
4. Do some earthquakes occur at or near injection depths?
5. If not, are there known geologic features that may channel flow to the sites of earthquakes?

6 Are changes in bottomhole pressures sufficient to encourage seismicity?
7 Are changes in fluid pressure (or stress conditions) at hypocentral locations sufficient to encourage seismicity?

Although a number of other approaches have been developed, this set of criteria continues to be used as a basis for discriminating induced events from background seismicity (e.g. Lamontagne et al., 2015).

Cross-correlation analysis (Appendix B) provides another simple tool, which can provide a quantitative measure of temporal correlation between disparate types of signals such as injection volume and seismicity rate (Schultz et al., 2015a). Oprsal and Eisner (2014) discuss the importance of subtracting the mean value from both time series to prevent bias, whereas Telesca (2010) developed a methodology to estimate statistical confidence for cross-correlation between different time series using random reshuffling tests.

Dahm et al. (2013) provided a set of recommendations for distinguishing between induced, triggered and natural seismic events, including strategies that fall into three general categories:

1 Physics-based probabilistic models, in which Mohr–Coulomb failure criteria (§2.2.1) and rate-and-state frictional rheology (§2.3.1) are applied. This approach requires extensive knowledge of fault architecture and geomechanical properties, but has an advantage over other strategies in that it can provide deeper insights into underlying physical processes.
2 Statistics-based seismicity models, in which an ETAS model (§3.9) is fit to a pre-injection time window in order to define the characteristics of the natural seismicity. Considering uncertainties in ETAS parameters, the probability that post-injection seismicity can be explained by ETAS-type natural activity can be used as the basis for distinguishing between natural and induced seismicity. Seismogenic index (see below) provides another statistical approach.
3 Source-parameter models, in which events are classified on the basis of moment tensors, focal depth and/or spectral-derived parameters such as stress drop. For example, non-double-couple (non-DC) moment-tensor components of tectonic earthquakes are generally relatively small, so an earthquake with a significant isotropic component could be interpreted as unlikely to be of tectonic origin. The converse is not true, however; the absence of non-DC components does not provide a basis to conclude that an event is natural (Dahm et al., 2013).

The latter approach was applied by Cesca et al. (2013) to compare full moment-tensor solutions for a set of natural and induced earthquakes in central Europe. Their results revealed distinct non-DC components for mining-induced events resulting from cavity collapses and pillar bursts, but did not show significant differences between other types of induced events and natural earthquakes based on robust moment-tensor inversion. Similarly, Zhang et al. (2016) determined source parameters including full moment tensors for a set of induced and natural earthquakes from western Canada. According to their analysis from western Canada, the focal depths of induced events are shallower than for natural earthquakes (Zhang et al., 2016). They found that stress-drop parameters for the analyzed set of induced events fall within the expected range for tectonic earthquakes, in contrast to previous findings by Hough (2014), suggesting that induced events may have systematically lower stress drop than tectonic earthquakes. Care is needed in interpreting these findings, as the stress-drop parameter is higher in the east than in the west, so induced events in the east may have stress-drop parameter comparable to that of deeper natural events in the west (Atkinson and Assatourians, 2017). Zhang et al. (2016) also observed significant non-DC components for injection-induced events, and suggested that these may arise from elastic moduli changes due to the brittle damage in the source, dilatant jogs created at the overlapping areas of multiple fractures, or non-planar pre-existing faults. Volcanic earthquakes can also have significant non-DC components, for example, due to rupture on a ring-shaped fault around the caldera (Nettles and Ekström, 1998). Skoumal et al. (2015a) argued that injection-induced and natural earthquake sequences can be distinguished on the basis of "swarminess," which represents the ratio of the number of events in a sequence above a threshold magnitude to the maximum magnitude in the sequence. They showed that induced sequences have a high degree of swarminess, since for a given maximum magnitude they contain many more events.

9.1.3 Activation Mechanisms

Figure 9.2 highlights three different activation mechanisms for injection-induced seismicity. The first mechanism is representative of seismicity induced by saltwater disposal (SWD), wherein fluids are injected into a permeable, laterally and vertically continuous storage unit such as the Cambrian and Lower Ordovician Arbuckle Group carbonate sequence (Fritz et al., 2012). If the storage unit is in hydrological contact with a fault, either directly or through an fracture network, there is potential for pore pressure from injection to influence the stress state of the fault. It is important to keep in mind that the process involves diffusive transport of fluid pressure, not advective transport of the injected fluids (Rubinstein and Mahani, 2015).

The second mechanism depicted in Figure 9.2 is representative of extraction of hydrocarbons (Baranova et al., 1999; Van Thienen-Visser and Breunese, 2015), which is sometimes combined with fluid injection during secondary hydrocarbon recovery. There is no hydrological connection between the reservoir and the fault in this case; based on the theory of poroelasticity, the activation mechanism is a large-scale stress changes from extraction or injection of fluids (Segall, 1989).

The third mechanism depicted in Figure 9.2 represents hydraulic fracturing (HF) that is in sufficiently close proximity to a critically stressed fault (hundreds of m) that pore pressure and/or poroelastic stress changes can trigger slip (Bao and Eaton, 2016). In terms of induced seismicity, this differs from the SWD scenario in several fundamental respects. First, in the absence of natural or induced fracture networks, diffusive transport of fluid or pore pressure is inhibited by the low permeability of the injection zone (Atkinson et al., 2016). Moreover, HF fluids are injected at relatively higher pressure (above breakdown) but at relative lower volumes – on the order of tens of thousands of m^3 per well (Bao and Eaton, 2016), which is relatively small in comparison with sustained injection (years) in high-rate class II well (Rubinstein and Mahani, 2015).

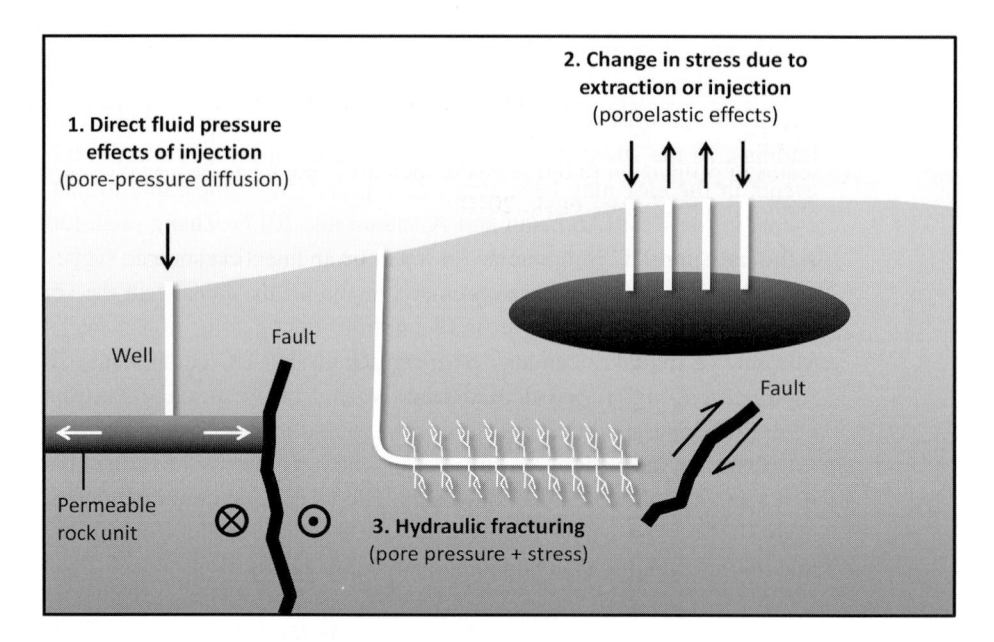

Fig. 9.2 Triggering mechanisms for three types of fluid-induced seismicity. A proximal, critically stressed fault is required in each case. Modified from Eaton (2016), with permission of the Canadian Energy Technology and Innovation Journal.

By analogy with a basic model for a petroleum system, consisting of a hydrocarbon source, a migration pathway, a reservoir and a seal, a fluid-systems approach provides a framework to describe coupled processes that give rise to induced seismicity. The main components of this system are:

- a source that causes a sufficient change in pore pressure and/or stress;
- a proximal fault that is in a state of incipient failure;
- a pressure diffusion or stress pathway from the source to the fault.

The occurrence of injection-induced seismicity requires all three of these components to co-exist. For example, statistical analysis of induced seismicity from SWD has demonstrated that high-rate injection wells ($> 50,000$ m^3 per month) are much more likely to be associated with earthquakes than lower-rate injection wells (Weingarten et al., 2015). Moreover, a relatively low fraction of disposal and HF wells are associated with induced seismicity in both western Canada – 1.4% and 0.3%, respectively (Atkinson et al., 2016) and Ohio – 1.5% and 0.4%, respectively (Skoumal et al., 2015a). Induced-seismicity susceptibility to fluid injection is generally observed to be strongly localized into discrete regions (Ghofrani and Atkinson, 2016), suggesting that one or more of the necessary components identified above is absent in many areas.

The *seismogenic index*, Σ, is a parameter that was introduced by Shapiro et al. (2010) to describe induced-seismicity susceptibility to fluid injection. It can be expressed in terms of net injected volume ($V(t)$) and b value of the Gutenberg–Richter distribution as (Dinske and Shapiro, 2013)

$$\Sigma = \log_{10} N_M(t) - \log_{10} V(t) + bM, \qquad (9.1)$$

where $N_M(t)$ is the number of injection-induced earthquakes of magnitude $\geq M$. The seismogenic index derives from a principle that the probability of failure of a defect (crack) scales in proportion to the pressure increase, which can be argued based on the following assumptions (Shapiro et al., 2010):

- the pressure field is generated by a point-like injection source;
- cracks are randomly oriented and homogeneously distributed within the stimulated region;
- failure of a crack is characterized by a critical pressure value that is uniformly distributed between a maximum and minimum value.

Based on these assumptions, the seismogenic index is a site-specific parameter that is time-independent; hence, Equation 9.1 can be recast in the form of the Gutenberg–Richter formula (Equation 3.80) with $a(t) = \Sigma + \log_{10} V(t)$.

Segall and Lu (2015) performed a similar analysis for a constant-flux point source embedded in a poroelastic medium (§1.5) containing a distribution of parallel fractures characterized by a rate-state frictional rheology (§2.3.1). Their model provides some basic insights about the pressure-diffusion and stress pathways from the source to the fault, as well as time-dependent characteristics of fault activation. The pore-pressure field at a distance r that is produced by a point source with constant flux q satisfies Equation 1.57 (the diffusion equation) and is given by (Rudnicki, 1986)

$$P(\mathbf{x}, t) = \frac{q}{4\pi \rho_F r} \frac{\eta}{\kappa} \, \mathrm{erfc}\left(\frac{r}{2\sqrt{ct}}\right), \qquad (9.2)$$

where position $\mathbf{x}$ is relative to the point source located at the origin, ρ_F and η are the fluid density and dynamic viscosity and κ is the permeability of the medium. In addition, the parameter c is given by (Segall and Lu, 2015)

$$c = \frac{\eta}{\kappa} \frac{(\lambda_u - \lambda)(\lambda + 2\mu)}{\alpha^2 (\lambda_u + 2\mu)}, \qquad (9.3)$$

where λ and λ_u denote the drained and undrained Lamé parameters, μ is the shear modulus of the elastic frame and α is the Biot coefficient. A closed-form expression for the stress field, similar to Equation 9.2, was obtained by Rudnicki (1986).

Segall and Lu (2015) generated a suite of numerical models to explore parameter sensitivity in the case of a simple poroelastic model. Figure 9.3 shows a graph of the effects of pore pressure, given by $\mu_s P$, and stress, given by $|\tau| + \mu_s \sigma_n$ for a particular model realization at $t = 5$ days following the start of injection. The graph shows a cross-over distance beyond which the effects of poroelastic stress on the Coulomb failure function are greater than the effects of pore pressure. At distances shorter than this cross-over point, the pore-pressure is dominant. While this example applies to a specific set of model parameters, as a general rule the effect of poroelastic stress is greater than pore pressure at large distance (Deng et al., 2016).

The onset of fluid-injection induced seismicity is characterized by an expanding cloud of events that is bounded in space and time by a *triggering front* (Shapiro, 2015). Conversely,

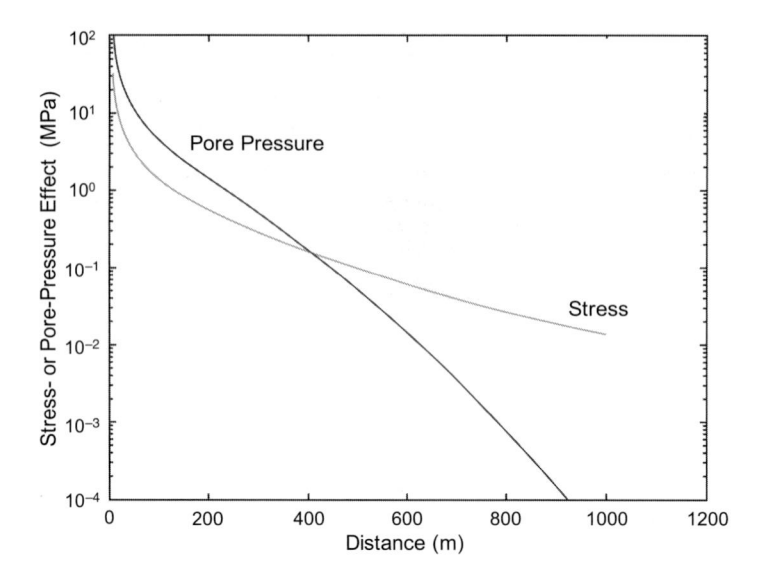

Fig. 9.3 Contributions of stress and pore-pressure to Coulomb failure in a homogenous poroelastic medium with a constant-flux point-source of fluid, five days after the start of injection. The background model contains normal faults that strike perpendicular to the direction of the profile and dip at 60°. As illustrated here, considering the effects of poroelastic coupling, in general there is a distance beyond which the influence of stress exceeds the influence of pore pressure. Modified from Segall and Lu (2015), with permission from Wiley.

termination of injection produces a *backfront*, which marks an abrupt cessation of induced seismicity in space and time (Parotidis et al., 2004). In general, for a finite-duration injection these two surfaces intersect to form a closed region in space and time, which contains all of the events induced by the injection source.

The triggering front and backfront are evident in Figure 9.4a, which shows a time-distance ($r - t$) plot of seismicity rate derived from a poroelastic numerical simulation reported by Segall and Lu (2015). The simulation incorporates a constant-flux source from $t = 0$ to $t = 15$ days. It also incorporates the effects of rate-state friction, based on a constitutive model developed by Dieterich (1994). The triggering front is manifested as the onset of sharply increased seismicity rate, from a nonzero background seismicity rate. Close to the source (< 200 m) there is a pronounced peak in seismicity rate that persists for a few days after passage of the triggering front. As a consequence of the rate-state constitutive model used for this simulation, another sharp increase in seismicity rate is evident after injection-source shut-in at $t = 15$ days. The apparent synchronous (at the time scale of the plot) onset of the post-shut-in seismicity rate increase suggests that this behaviour results from poroelastic stress and is not a diffusion phenomenon. This type of behaviour has been documented for some cases of injection-induced seismicity, such as for the Basel geothermal project (discussed below). Figure 9.4b shows graphs of seismicity rate at a distance of 250 m for different choices of the parameter $t_a \equiv a\bar{\sigma}/\dot{\tau}$, where a is a rate-state friction parameter (see §2.3.1). The parameter t_a represents a characteristic rate-state decay time

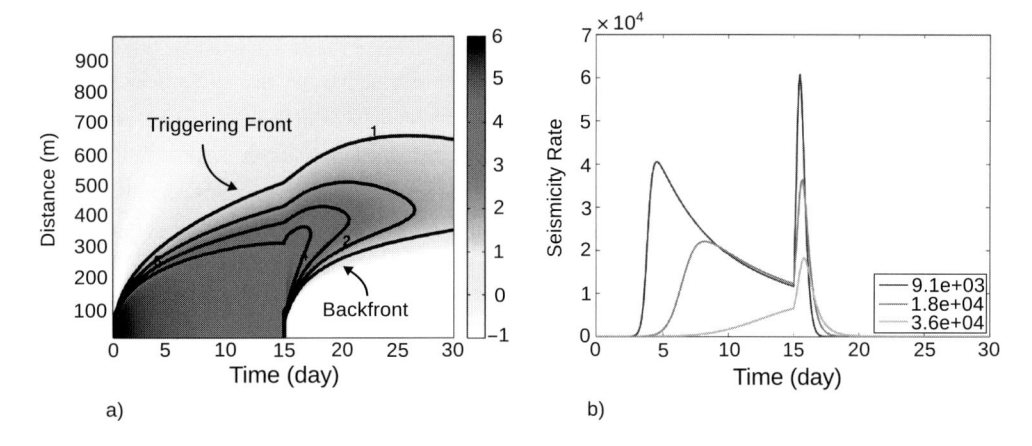

Fig. 9.4 a) Distance-time ($r - t$) plot showing triggering front, backfront and calculated seismicity rate for a constant-flux point injection source that operates for 15 days. The model contains faults with a rate-state constitutive relationship, and orientation as described in Figure 9.3. Colour scale and contours represent $\log_{10}$ of seismicity rate. b) Seismicity rate at a position 250 m from the injection point, for various values of the rate-state characteristic time parameter, t_a. Note sharp increase in seismicity rate after shut-in. Modified from Segall and Lu (2015), with permission from Wiley. (A colour version of this figure can be found in the Plates section.)

(Segall and Lu, 2015) and is directly determined by the stress perturbation through $\bar{\sigma}$ and background stressing rate $\dot{\tau}$.

These numerical simulations predict seismicity rate, but they do not explicitly consider magnitudes of induced events. The question of the maximum expected magnitude for a sequence of induced earthquakes is of considerable importance for various reasons, including pre-injection risk assessment, mitigation of hazards and public outreach and communication. There are a number of conjectural approaches that have been proposed to forecast the maximum magnitudes. These different approaches for estimation of maximum (plausible) magnitude are discussed in Box 9.1.

These numerical simulations provide key insights based on a relatively simple numerical model. In practice, $r - t$ plots from field observations often reveal more complex interactions, as illustrated in Figure 9.5. These graphs depict the space-time evolution of microseismicity from a two-well open-hole HF stimulation of a tight sand reservoir in western Canada (Eaton et al., 2014b). Both wells were completed in 12 stages. Composite diagrams were produced for each well by clustering microseismic events based upon injection stage and aligning the clusters using the stage initiation time. Parameters used for triggering fronts were computed using the cubic-parabolic approximation of Shapiro and Dinske (2009) with a spherical fracture domain and increase in effective porosity in the range from 2×10^{-7} (west well), 3×10^{-7} (east well) and 8×10^{-7} (back front for both wells). The backfront is less clear than the triggering front due to the variable duration for each stage. Activation of fractures is marked by non-diffusive, near-synchronous seismicity over a range of distances, forming vertical streaks in the diagrams (e.g. arrow in Figure 9.5a). Post-injection fracture activation persisted for more than eight hours after termination of the HF stimulation (Figure 9.5c).

> **Box 9.1** **Maximum Magnitude for an Injection-Induced Earthquake Sequence**
>
> Three different approaches for estimation of maximum magnitude are considered here, along with their implications for managing risk. First, a deterministic limit on seismic moment was proposed by McGarr (1976) for a change in volume, for example due to underground mining excavation. This approach has since been reformulated for application to fluid injection (McGarr, 2014). In essence, this method assumes that the upper limit for seismic moment release is constrained by the pressure-induced stress change. For a medium with shear modulus μ, the deterministic limit is
>
> $$M_0^{max} = \mu \Delta V,$$
>
> where M_0^{max} is the maximum seismic moment in an earthquake sequence induced by injection of a net volume ΔV. This formula is based on the assumption that the medium is fully saturated; it further assumes that the crust is in a state of incipient failure, such that on average a pressure increase of $\Delta P = \frac{\Delta \tau}{2\mu}$ is sufficient to induce slip, where $\Delta \tau$ is a characteristic earthquake stress drop. An alternative geometrical approach was proposed by Shapiro et al. (2011), who postulated that the rupture area for an induced earthquake falls entirely within the stimulated volume. This reduces the task to one of estimating the largest potential slip surface area within a given stimulated volume. Finally, Van der Elst et al. (2016) proposed that, statistically speaking, the maximum observed magnitude ($\widehat{M}_{max}$) is the expected maximum value for a finite sample drawn from an unbounded Gutenberg–Richter distribution:
>
> $$\widehat{M}_{max} = M_c + \frac{1}{b} \log_{10} N,$$
>
> where N is the number of events in the sequence for which $M \geq M_c$. If induced earthquakes occur on pre-existing faults, as commonly assumed, then M_{max} for induced seismicity is the same as that for a natural earthquake in the same area (Van der Elst et al., 2016).
>
> These three models imply different approaches for risk management. The deterministic method proposed by McGarr (2014) implies that a ceiling on the maximum magnitude can be imposed by limiting the net injected volume, whereas the approach developed by Shapiro et al. (2011) implies that the time-dependent maximum magnitude is governed by the spatial size of the microseismic event cloud. Finally, the sample-size hypothesis of Van der Elst et al. (2016) implies that the best available estimate of the maximum magnitude is based upon observed seismicity rate – see also discussion by Segall and Lu (2015). The latter two approaches suggest that real-time monitoring is essential for effective management of risk.
>
> A reliable estimate of maximum plausible magnitude would clearly be beneficial for quantitative risk assessment of injection-induced seismicity. At present this goal remains elusive and may be best considered in a probabilistic context.

9.2 Tools of the Trade

The methods used for acquisition and processing of induced-seismicity monitoring datasets are broadly similar to those used for previously discussed monitoring applications. The

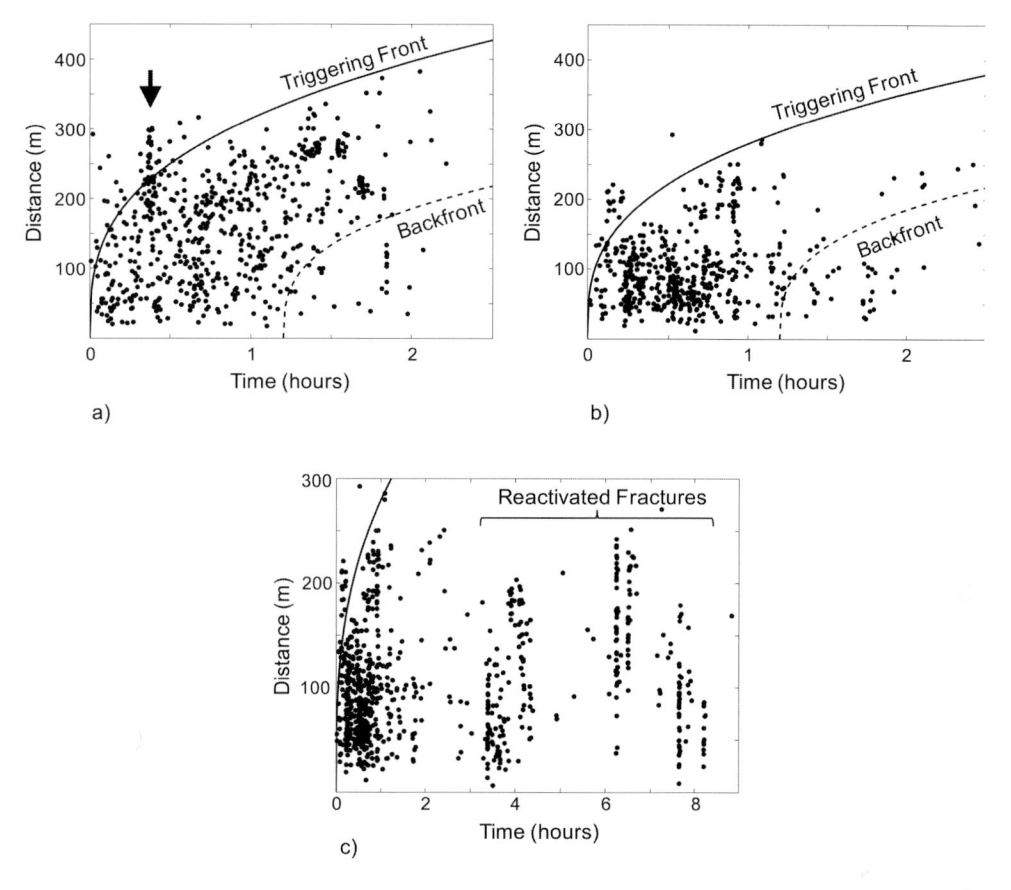

Fig. 9.5 Composite $r - t$ plots for open-hole well completions in a tight sand reservoir (Figure 8.1), showing estimated triggering front and backfront for horizontal wells on a) the west and b) the east side of the observation well. Near-synchronous activation of a fracture system during hydraulic fracturing is indicated in a) by the arrow. c) Expanded time interval for the east well; in this case, reactivation of fractures persists for more than 7.5 hours after completion of hydraulic fracturing. Modified from Eaton et al. (2014b).

primary objectives, however, are different: while hydraulic-fracture monitoring is usually undertaken in an effort to image hydraulic fractures, determine ESV and thus optimize well-completion design and operations (Maxwell, 2014), induced-seismicity monitoring is typically performed for risk management and/or regulatory compliance. The instrumentation is also somewhat different. Although geophones can be used for detection of larger induced events (Figure 9.6), a key difference for induced-seismicity monitoring is the need to acquire lower-frequency signals to prevent magnitude saturation (Eaton and Maghsoudi, 2015). As discussed in Chapter 5, seismological instrumentation used for induced seismicity monitoring includes:

- *broadband* seismometers capable of measuring ground motions over most of the frequency spectrum that is radiated by earthquakes;

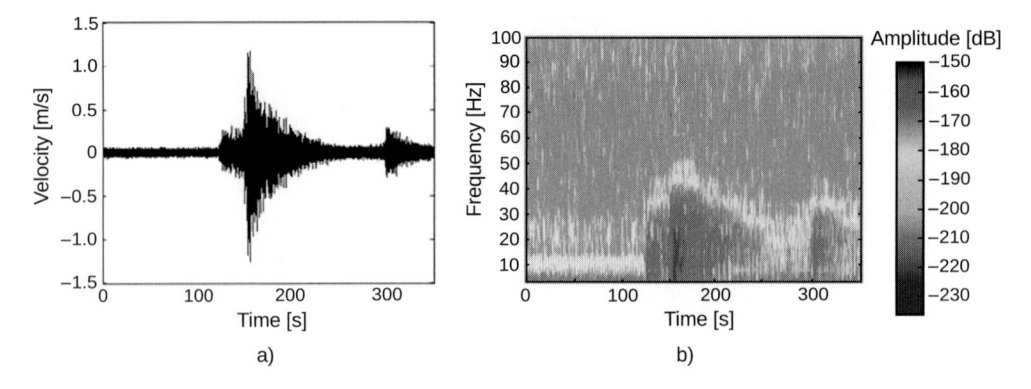

Fig. 9.6 Induced earthquakes at regional distance recorded using a 15 Hz geophone installed at 2 km depth. a) Filtered (10–80 Hz) horizontal-component trace containing regional waveforms from small earthquakes of M_L 3.1 and M_L 2.8 on 29 June 2013. b) Spectrogram for the unfiltered trace. Time window is 350 s and the spectral amplitudes are referenced to 1 m/s. These earthquake-generated signals show a strong similarity to reported waveforms linked to previously proposed models for long-period long-duration (LPLD) events, as described by Das and Zoback (2013). Modified from Caffagni et al. (2014). (A colour version of this figure can be found in the Plates section.)

- low-frequency geophones, which do not cover the full spectrum of ground motion, but nevertheless extend the range of traditional geophones to a frequency band that is important for characterizing local seismicity (i.e. to frequencies of < 10 Hz);
- *accelerometers*, which measure ground acceleration rather than particle velocity and have different sensitivity and noise characteristics than geophones and seismometers. Due to the different sensitivity, accelerometers do not saturate in amplitude for high ground motion, unlike broadband seismometers; however, broadband seismometers are more sensitive to weak ground motion.

In this section, a number of methods are revisited, with emphasis on induced-seismicity monitoring applications. These approaches draw heavily from material covered in previous chapters, particularly seismic sources (§3.5), magnitude scales (§3.6), ETAS models (§3.7) and microseismic data-processing methods described in Chapters 6 and 7.

9.2.1 Event Detection

The first step for induced seismicity monitoring involves event detection. In virtually all cases of practical interest, the sheer data volume means that automatic methods for event detection are essential. Moreover, the raw dataset is comprised of continuously recorded waveforms, which often contains problems that need to be addressed prior to event detection. These problems include gaps due to missing data, high-amplitude noise bursts ("glitches"), spurious long-period trends, etc.

In general, first-pass methods, used to detect potential events from continuous waveform data, resemble those discussed in previous chapters. Popular methods include the short-time average/long-time average (STA/LTA) method and other approaches described in §6.2. Implementation of these methods for induced-seismicity monitoring differs

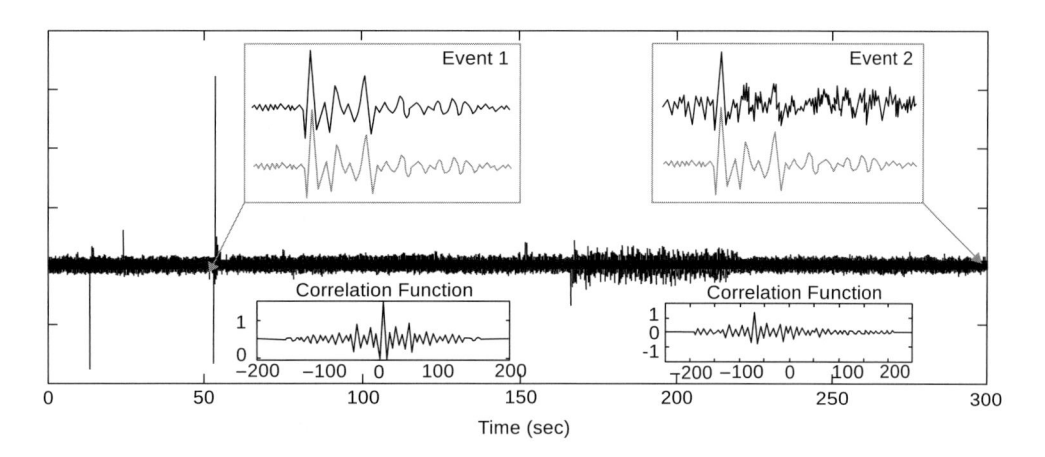

Fig. 9.7 Cross-correlation method using a template event with an example of two detections during 5 min of continuous z-component data. Insets compare waveforms of the detected child events (black) with the template (parent) event (gray). In this case, event 1 is the template and event 2 is a weaker event during this time period. The respective cross-correlation functions are shown below. From Goertz-Allmann et al. (2014), with permission.

somewhat from microseismic monitoring, in particular due to the necessarily sparser station distribution compared with downhole, surface and near-surface systems that are typically used for microseismic monitoring. A typical example of an induced-seismicity monitoring network is a system deployed around the Liard Basin in northwestern Canada (Figure 5.2). In general, a wider spatial separation between stations is more challenging since it reduces inter-station waveform coherency.

The use of template methods for enhanced detection is becoming routine for induced-seismicity monitoring. The theoretical basis for these methods, which include matched filtering and subspace detection, are described in §6.2.1. Figure 9.7 shows how a basic single-channel cross-correlation approach (see Appendix B) can be employed for template matching. The dataset is from the In Salah CO_2 sequestration project in Algeria (Goertz-Allmann et al., 2014). This example illustrates how a low signal/noise ratio (S/N) event that is characterized by a waveform that is highly similar to a template event can be detected even in the presence of noise and other types of signals. A match-and-locate method has been developed that uses stacked cross-correlation functions (SCCFs) and a hypocentre search around the location of the template event (Zhang and Wen, 2015); this is similar to the SCCF approach developed for downhole microseismic data by Caffagni et al. (2016).

For this type of analysis, it is important to consider how template events are selected. A simple approach is to select, amongst the potential events obtained from a first-pass detection run, those that exceed a defined threshold value of S/N. This approach has clear drawbacks, since it is difficult to assure, by any measure, that the detected events comprise a complete set for the dataset being analyzed. In order to avoid ad hoc selection of template events, Skoumal et al. (2016) developed an efficient repeating signal detector, which exploits the aforementioned swarmy nature of induced seismicity. Their method identifies signals of interest above a threshold S/N and then groups them based on spectral and time-domain characteristics, using an agglomerative clustering algorithm. Application of this approach using waveforms from a single seismograph station has demonstrated significant improvements on catalogs generated using more time-intensive approaches (Skoumal et al., 2016).

Another promising method is called the Fingerprint And Similarity Thresholding (FAST) algorithm (Yoon et al., 2015). This method uses feature extraction to construct waveform fingerprints, which contain key discriminative features and serve as condensed waveform proxies. A widely used technique for high-dimensional approximate nearest-neighbour search, called locality-sensitive hashing (LSH), is then applied in order to reduce computational burden of comparisons between dissimilar pairs. An existing fingerprinting algorithm is then used, which combines computer-vision techniques and large-scale data processing methods to detect similar waveforms. A test of the FAST method using a dataset from California produced results similar to autocorrelation for detection of repeating events, but with significantly greater computational efficiency.

9.2.2 Hypocentre Estimation

Earthquake-location methods used for induced seismicity monitoring and analysis are identical to those used for natural earthquake studies. The goal is to determine the location of earthquake hypocentres, that is the estimated locations of earthquake foci, together with location uncertainty. Taken together with the *origin time* and magnitude, this information provides a concise description of the basic characteristics of an earthquake. In some cases, an absolute earthquake location is determined, in which case the location is referenced to a geographic coordinate system and a fixed time base (e.g. Universal Time Coordinated, UTC). In other cases, relative earthquake locations are determined, in which case the locations are specified with respect to another event (Lomax et al., 2014).

Like the hypocentre-location problem for downhole microseismic arrays (§6.3), an earthquake location is usually determined by minimizing the misfit between predicted and observed *P*- and *S*-wave arrival times. In most cases, arrival times are required at four or more stations, but in unusual situations where waveforms are available at only one seismograph station the *single-station* method (SSM) (Roberts et al., 1989) can be used. For example, Farahbod et al. (2015b) used this approach to investigate background seismicity of the Horn River Basin prior to regional shale-gas development, when only one seismograph station, located near Fort Nelson, was deployed in the region. Much like downhole microseismic processing, this method uses the backazimuth and incidence angle estimated using a three-component waveform window containing the direct *P*-wave arrival. The hypocentre is determined by shooting a reciprocal ray in a reverse direction to the incoming *P* wave. An imaging condition is applied, such that the hypocentre coincides with the point along the raypath where the calculated $S - P$ time matches the observed $S - P$ time.

For the typical case where data are available from four or more stations, Geiger (1912) introduced an iterative least-squares method that still forms the basis for routine absolute earthquake locations using regional seismograph networks (Lay and Wallace, 1995). For an assumed velocity model, the problem is first linearized by expressing the data, represented by time residuals (Δd) of all picked phases, in terms of model perturbations:

$$\Delta \mathbf{d} = \mathbf{G}\Delta \mathbf{m}, \tag{9.4}$$

where $\mathbf{m} = [x, y, z, \tilde{\tau}]$ contains the hypocentre parameters and $\tilde{\tau}$ is the origin time. The elements of the matrix $\mathbf{G}$ are partial derivatives,

$$G_{ij} = \frac{\partial d_i}{\partial m_j}. \tag{9.5}$$

Since the problem is almost always *overdetermined* (i.e. there are more observations than model parameters), a generalized least-squares inverse solution is used:

$$\Delta \mathbf{m} \simeq \left[\mathbf{G}^{\mathbf{T}}\mathbf{G}\right]^{-1} \mathbf{G}^{\mathbf{T}}\Delta \mathbf{d}. \tag{9.6}$$

If the errors in the data are normally distributed with variance σ^2, the corresponding model errors are then given by:

$$\sigma_m^2 \simeq \mathbf{G}^{-1}\sigma^2 \left[\mathbf{G}^{-1}\right]^T. \tag{9.7}$$

This and other linearized approaches to determine an optimal solution (a point in space and time) with associated uncertainties are subject to limitations arising from intrinsic nonlinearity of the hypocentre-inverse problem (Lomax et al., 2014). Indeed, an earthquake's location can be represented in a more complete way using a Bayesian approach to determine the a posteriori probability-density function (PDF) over all possible solutions in a four-parameter space, $[x, y, z, t]$. Lomax et al. (2014) summarize direct, global-search methods that can be used to compute PDF solutions to the hypocentre-inverse problem, which are useful to quantify the goodness of fit between predicted and observed arrival times in relation to all uncertainties.

Figure 9.8 is a schematic representation of a hypocentre PDF, illustrating a moderately complex model that can be problematic for linearized methods. In this case, there is a sharp boundary across which there is a large velocity contrast. Depending upon the model, a linearized solution may become "trapped" in a local minimum of the misfit function that falls on the incorrect side of the sharp boundary. In such a case, the formal uncertainty estimates obtained using Equation 9.7 are not meaningful, whereas a global direct-search method is always assured of finding the maximum-likelihood hypocentre. This pathology is particularly relevant for analysis of hydraulic fracturing induced seismicity for the common scenario in which the target zone is located close to a large velocity contrast.

We now consider the double-difference method (hypoDD), which was developed by Waldhauser and Ellsworth (2000) and has been used extensively to calculate precise relative hypocentre locations. The hypoDD algorithm takes advantage of similar source-receiver raypaths in order to cancel errors arising from uncertainties in the velocity model. Successful application of this algorithm requires that the separation between events is small compared to the path length of rays from the sources to the stations, as well as the scale-length of velocity heterogeneity in the medium (Waldhauser, 2001). In addition, good azimuthal coverage of the source region by the sensor network is important (Ma and Eaton, 2011). The algorithm makes use of two time differences, one being the difference in observed time at two stations and the other being the difference in calculated times based

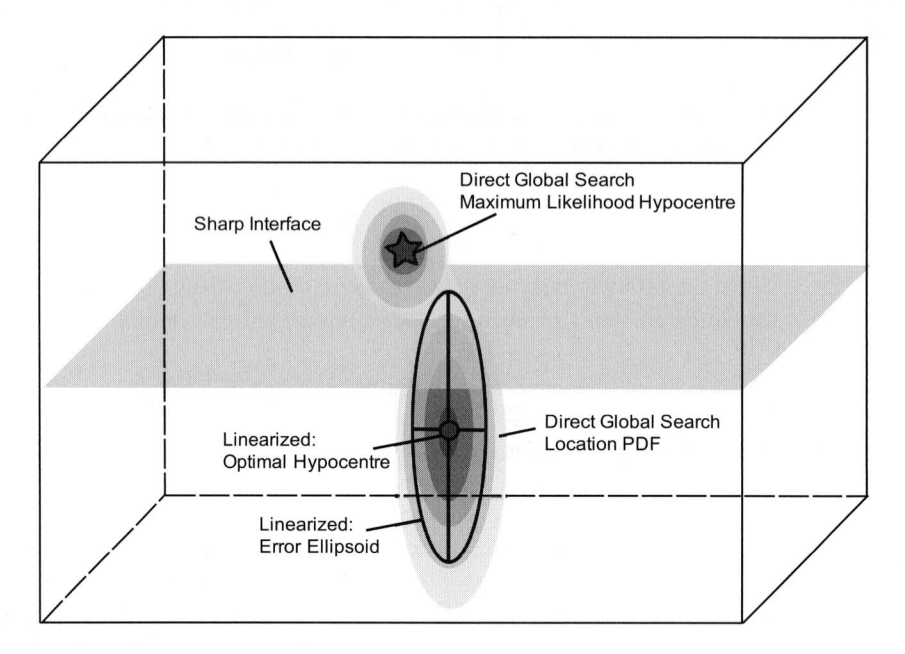

Fig. 9.8 Schematic illustration of hypocentre Probability-Density Function (PDF) for a scenario with two local maxima due to the presence of a sharp interface between low-velocity and high-velocity media. In this case, the direct global search result correctly locates the global maximum, whereas a linearized method may fail. Modified from Lomax et al. (2014), with permission.

on the velocity model. For the ith earthquake observed at the jth station, the traveltime residual (difference between observed and modelled time) may be written as

$$\Delta t_{ij} = t_{ij}^{ob} - t_{ij}^{0} \simeq \sum_{l=1}^{4} \frac{\partial t_{ij}}{\partial m_l^i} \|_{\mathbf{m}_0^i} \Delta m_l^i , \qquad (9.8)$$

where $\mathbf{m}_i = [x, y, z, \tilde{\tau}]$ defines the model parameters for the ith event and the initial model is represented by $\mathbf{m}_0^i = [x_0, y_0, z_0, \tilde{\tau}_0]$. Similarly, for the kth station we have:

$$\Delta t_{ik} = t_{ik}^{ob} - t_{ik}^{0} \simeq \sum_{l=1}^{4} \frac{\partial t_{ik}}{\partial m_l^i} \|_{\mathbf{m}_0^i} \Delta m_l^i . \qquad (9.9)$$

Subtracting Equation 9.9 from Equation 9.8 yields:

$$\Delta t_i^{jk} = \left(t_{ij}^{obs} - t_{ik}^{obs} \right) - \left(t_{ij}^{0} - t_{ik}^{0} \right) \simeq \sum_{l=1}^{4} \left(\frac{\partial t_{ij}}{\partial m_l^i} \|_{\mathbf{m}_0^i} - \frac{\partial t_{ik}}{\partial m_l^i} \|_{\mathbf{m}_0^i} \right) \Delta m_l^i . \qquad (9.10)$$

In the above equations, for one pair of events, there are eight unknown variables to be solved; the minimum number of required stations (assuming both P and S phases are available at each station) is therefore four. If we let $G_l^{ij} \equiv \frac{\partial t_{ij}}{\partial m_l^i} \|_{\mathbf{m}_0}$ the system of double-difference equations may be written in the form

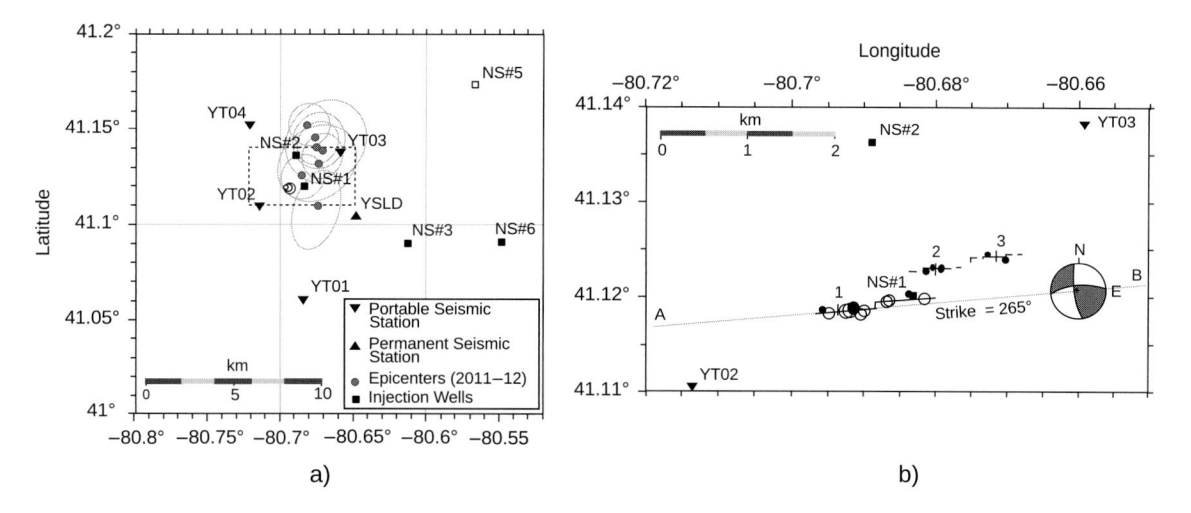

Fig. 9.9 Precise relative event locations using a double-difference relocation algorithm. (a) Locations of 12 events in the vicinity of Northstar 1 (NS#1) deep injection well. Ellipses show 95 percent confidence regions. Inverted triangles show local seismograph stations used to locate three of the events. (b) Epicentres of 12 regional events (circles) and 9 local events (black hexagons) near injection wells NS#1 and NS#2, within the area shown by the dashed rectangle in (a). Focal mechanism of the largest event in the sequence is shown by the beachball. Reproduced from Kim (2013), with permission from Wiley.

$$\mathbf{GM} = \mathbf{T}, \quad (9.11)$$

where N is the number of events, $\mathbf{M}$ is the vector of model parameters (of length $4N$) and $\mathbf{T}$ is the complete set of double-differences computed using Equation 9.10.

The double-difference algorithm determines relative event locations. As such, initial event locations are required. These are used to compute the model time required for the double-difference calculations and ideally should be obtained using a direct global-search method. Figure 9.9 shows an example of the improvement in precision of *relative* event locations that can be obtained using this approach.

9.2.3 Moment–Tensor Inversion

Source mechanisms of induced earthquakes are routinely determined, either as focal mechanisms derived from *P*-wave first-motion analysis (Eaton and Mahani, 2015), or through full moment-tensor inversions. The latter process enables the analysis of non-double-couple components, which may be useful for discriminating between natural and induced events (Zhang et al., 2016). As discussed in §3.5.1, an observed seismogram can be expressed in terms of moment-tensor components (M_{jk}) as

$$\dot{u}_i(\mathbf{x}, t) = M_{jk} \left[G_{ij,k} * \dot{s}(t) \right]. \quad (9.12)$$

This expression differs from Equation 3.44 only in the use of the time derivative, such as in the source-time function $\dot{s}(t)$, since seismometers are designed to measure ground velocity rather than displacement. In practice, however, it is common to convert the recorded signals to displacement during data preprocessing, which involves deconvolution of the instrument response for each station. Solving for **M** requires the determination of six elementary Green's functions; the problem is then formulated as direct-global search for a linear combination of fundamental Green's functions, where the optimal weights on the Green's functions determine the individual moment-tensor elements (Minson and Dreger, 2008). The objective function is commonly defined in order to minimize misfit using waveform windows for selected phases, in which observed and synthetic waveform segments are aligned by cross-correlation and a weighting scheme is used based on S/N (Steffen et al., 2012; Chen et al., 2015). This approach minimizes the effects of inaccuracies in the velocity model and/or earthquake mislocation. Nevertheless, the moment-tensor solution typically exhibits significant trade-offs with respect to hypocentre location and, although an independently determined earthquake location is often used, it is common for focal depth to be treated as a free parameter in the direct-global search scheme.

Figure 9.10 shows an example of a regional moment-tensor solution for an earthquake induced by hydraulic fracturing in western Canada (Wang et al., 2016). In this case, Green's functions were computed using a frequency-wavenumber integration method (Saikia, 1994) and the best-fitting solution was determined by maximizing the variance reduction. Two frequency bands were considered, a low-frequency band (0.05–0.1 Hz) that provided a more stable solution, and a high-frequency band (0.08–0.4 Hz) that provided greater constraints using high S/N body-wave components. The best-fitting moment tensor contains 23 percent non-DC component that is dominated by a horizontal CLVD. The focal mechanism (DC component) is characterized by near-vertical strike-slip motion that is compatible with the regional stress field (Wang et al., 2016).

9.3 Case Studies

This section provides a broad overview of recently published case studies of injection-induced in three focus regions (Figure 9.11) arising from three distinct types of fluid-injection operations. In Western Europe, induced seismicity from the development of engineered geothermal systems (EGS) has been extensively studied for operations at Soultz-sous-Forêts, France (Charléty et al., 2007) and Basel, Switzerland (Häring et al., 2008). Of the three focus regions, Western Europe is the most tectonically active, which tends to mask seismicity clusters associated with fluid injection. This region also contains the only published example of induced seismicity in Europe due to hydraulic fracturing, from a shale-gas operation near Preese-Hall in the United Kingdom (Clarke et al., 2014). The second focus region, in the US midcontinent, contains a multitude of well-documented cases of seismicity induced by saltwater disposal (SWD) including studies by Ellsworth (2013), Keranen et al. (2014), Weingarten et al. (2015) and Walsh and Zoback (2015). Finally, in western Canada a recent statistical analysis demonstrates that multistage hydraulic fracturing (HF) is the dominant cause of injection-induced seismicity (Atkinson et al., 2016). This overview of injection-induced seismicity in these regions enables comparison between fault-activation response under different injection scenarios and within different tectonic settings.

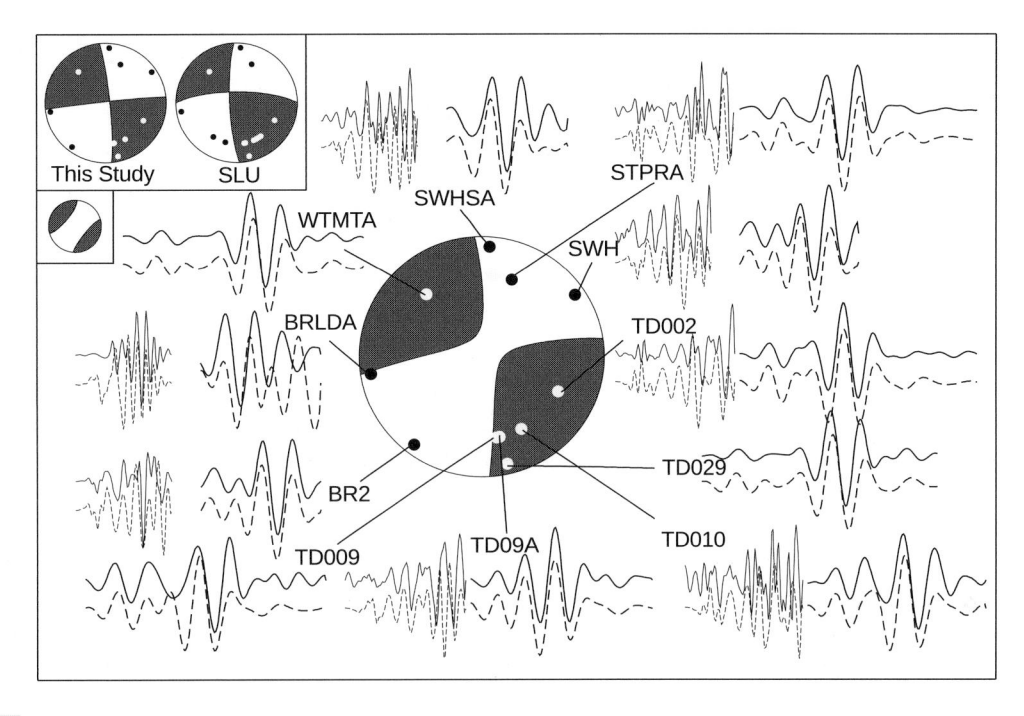

Fig. 9.10 Example of moment-tensor inversion applied to a M_W 3.9 hydraulic-fracturing (HF)-induced earthquake in western Canada that occurred on 13 June 2015. Fits are shown for observed (solid) and synthetic (dashed) waveforms in two different frequency bands: low (0.05–0.1 Hz, 11 stations) and high (0.08–0.4 Hz, 8 stations). The full moment-tensor result is shown as focal-mechanism diagram in the centre with two double-couple (DC) solutions shown in the top-left corner. The non-DC component from the full moment-tensor inversion is plotted beneath the DC solutions, for comparison. The small circles on the focal mechanisms indicate compressional (open) and dilatational (solid) P-wave first motion. SLU denotes St. Louis University. From (Wang et al., 2016), with permission from Wiley.

9.3.1 Engineered Geothermal Systems (EGS)

Engineered geothermal systems (EGS), also known as *enhanced* geothermal systems, can be defined as a heat exchanger created within deep and low permeability hot rocks using fluid-injection stimulation methods (Breede et al., 2013). The objective is to produce a hot, permeable region at depth, through which injected fluids can be efficiently circulated and brought up to the surface in order to deliver the captured heat for power conversion or other uses. The method has its roots in the hot dry rock (HDR) project conducted by Los Alamos National Laboratories (Cummings and Morris, 1979). Induced seismicity is a drawback for EGS development and has been the cause of delays and cancellation of of several projects (Majer et al., 2007).

Several of the best studied examples of induced seismicity from EGS development are located in western Europe (Evans et al., 2012). One example is the Soultz-sous-Forêts geothermal site, located within the western part of the Upper Rhine Graben in an area characterized by an extensional tectonic regime and high geothermal gradient (Charléty et al., 2007). This program commenced with drilling of a 2.0 km deep well in 1987, followed by

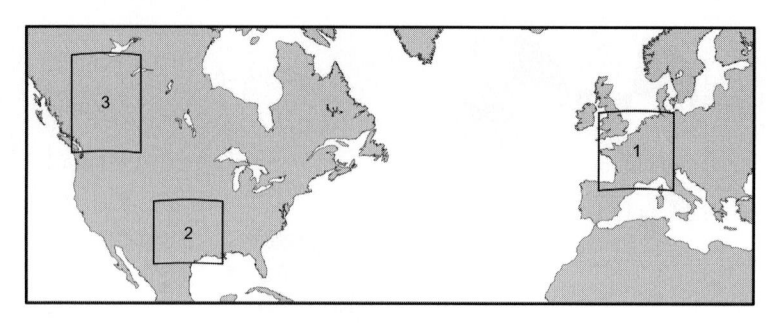

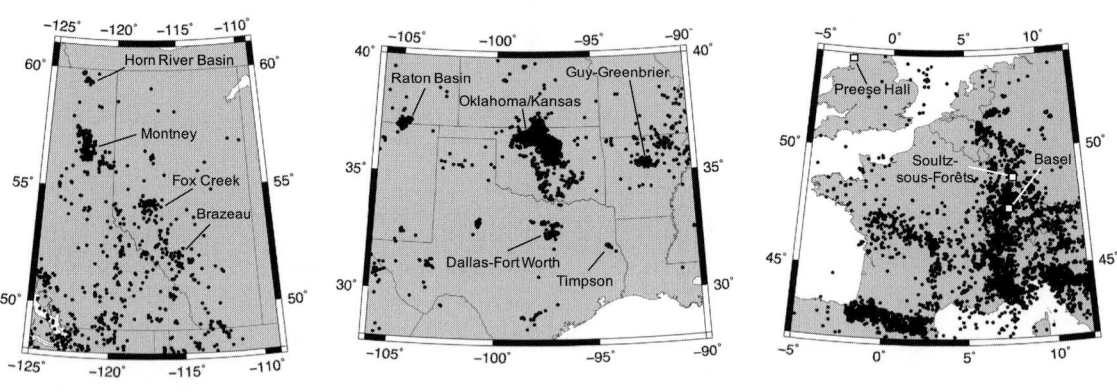

Fig. 9.11 Case histories of injection-induced seismicity. Area 1 in Western Europe includes several locations where seismicity has been induced by development of engineered geothermal systems (EGS), such as at Soultz-sous-Forêts, France (Charléty et al., 2007) and Basel, Switzerland (Häring et al., 2008). Two small earthquakes have been linked to hydraulic fracturing that occurred in 2011 near Preese Hall, United Kingdom (Clarke et al., 2014). Seismicity is shown as black dots and for area 1 it represents all $M \geq 2$ earthquakes for the period 2000–07 (inclusive) from the ISC online bulletin (International Seismological Centre, 2014). Virtually all of the seismicity here is natural, with a small fraction of EGS-induced events. Area 2 in the US midcontinent contains prominent examples of induced seismicity from saltwater disposal (Ellsworth, 2013; Weingarten et al., 2015). A large fraction of the seismicity is from central Oklahoma and southern Kansas, due to high-volume fluid disposal into a deep aquifer (Arbuckle Group) that rests unconformably on Precambrian basement. Seismicity shown in area 2 represents all $M \geq 2$ earthquakes for the period 2009–16 (inclusive) from the ANSS composite catalog (NCEDC, 2014) and includes some natural earthquakes and likely a small fraction of hydraulic-fracturing-induced events. Area 3 in western Canada contains several clusters of earthquakes that are primarily attributed to hydraulic fracturing (Atkinson et al., 2016). There are a few exceptions, including seismic activity near the Cordel Field within the Brazeau cluster that has been linked to saltwater disposal (Schultz et al., 2014). Seismicity shown in area 3 represents $M \geq 2$ events for the period 2009–16 (inclusive) from the Composite Alberta Seismicity Catalog (Fereidoni and Cui, 2015), with quarry blasts removed. All three area maps are plotted using the same scale, for ease of comparison.

subsequent drilling of four wells, three of which ultimately reached a depth of 4.5–5.0 km within fractured granite. All of the wells were subjected to *hydraulic stimulation*, with typical injection volumes in the range of 20,000–40,000 m^3 and flow rates of 2.4–4.8 m^3/min (Evans et al., 2012). In contrast to hydraulic fracturing, where injection pressures exceed the breakdown pressure, bottomhole pressures were brought only up to the level of the

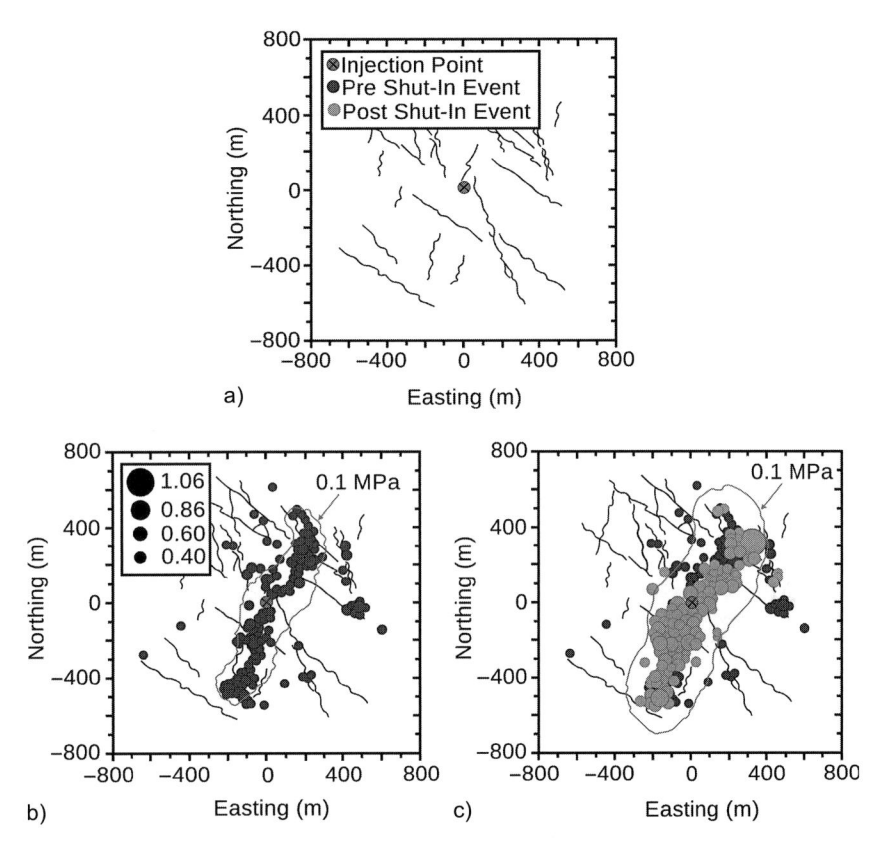

Fig. 9.12 Simulated seismicity from hydraulic stimulation, for a discrete-fracture network model. a) Naturally fractured reservoir with fluid injection point in the centre. b) Stimulated reservoir with pre-shut-in induced seismic events (dark grey dots). c) Post-shut-in events (light grey dots). The maximum horizontal stress is oriented in the north–south direction. Adapted from *Geothermics*, Vol 52, Arno Zang, et al., Analysis of induced seismicity in geothermal reservoirs – An overview, Pages 6–12, copyright 2014, with permission from Elsevier used with permission from CETI Journal.

minimum principal stress (Cornet et al., 2007). As illustrated in Figure 9.12, hydraulic stimulation is thought to operate through *mixed-mechanism* stimulation (MMS) (McClure and Horne, 2013). This is a hydro-mechanical coupled process in which pre-existing fractures can open as tensile cracks (mode I) or hydro-shears (mode II), while new fractures can form as wing cracks (mode I) from existing shear crack tips (mode II) (Yoon et al., 2014). Permeability can also be created by means of self-propping (§4.2), through the activation of shear slip on pre-existing fractures.

During development of the Soultz-sous-Forêts geothermal site, induced seismicity was monitored using both surface and downhole networks, with tens of thousands of events recorded using downhole arrays (Dorbath et al., 2009). The largest events occurred during shut-in, consistent with rate-state friction models (*cf.* Figure 9.4), including felt seismicity with a maximum magnitude of 2.9. There is no evidence that the magnitudes of induced events were influenced by changes in injection strategy (Charléty et al., 2007). Focal mechanisms reveal mixed-modes of failure, with a higher proportion of normal faults (Charléty et al., 2007). Seismicity was notably absent during some closed-loop circulation tests where injected and produced rates were balanced, but other closed-loop circulation tests generated small events when injection pressure exceeded a critical value (Evans et al., 2012). Power production commenced in 2010, and the 1.7 MW geothermal plant was inaugurated in September 2016 in connection with the European Geothermal Congress.

Development of the Deep Heat Mining project near Basel, Switzerland followed a rather different path. This project was initiated in 1996 by the Geopower Basel (GPB) consortium and was one of the first purely commercially oriented EGS projects (Giardini, 2009). Basel is an industrial centre with a population of more than 700,000. In terms of tectonic setting, it is located at the south-eastern margin of the Upper Rhine Graben, where it intersects the fold and thrust belt of the Jura Mountains of Switzerland (Deichmann and Giardini, 2009). The city has a history of earthquakes and was damaged by a magnitude 6.7 earthquake in 1356, the largest known historical earthquake in central Europe (Giardini, 2004). Nevertheless, there are economic incentives to develop geothermal sites close to a population centre, as it is more profitable to co-generate heat with electricity (Giardini, 2009).

The Basel-1 well was drilled in the outskirts of the city to a depth of 5 km, at which the temperature of the rockmass is 190° (Häring et al., 2008). Commencing on 2 December 2006, the well was stimulated with injection of 11,570 m^3 of water taken from a nearby harbour basin on the Rhine River. As shown in Figure 9.13, the injection rate increased to a maximum of 3.3 m^3/min with a wellhead pressure of 29.6 MPa. Above an injection pressure of 8 MPa and at higher flow rate, improved injectivity was noted and interpreted as initiation of MMS through coupled hydro-mechanical response of the rockmass (Häring et al., 2008). A local network of six deep borehole sensors was used to monitor seismicity, which began to increase two days after the start of the stimulation and continued to increase with rising pressure and flow rate. Based on an approved traffic light protocol (see below), the injection rate was reduced and the well was shut-in. A few hours later, a M_L 3.4 event occurred that was widely felt, with reports of high-frequency shaking lasting 1–3 seconds and a loud bang similar to an explosion (Deichmann and Giardini, 2009). Ground shaking was accompanied by minor nonstructural damage such as hairline cracks in plaster walls (Deichmann and Giardini, 2009). The Deep Heat Mining project at Basel was cancelled two years after the occurrence of M_L 3.4 induced earthquake, prompting international scientific response and new regulatory measures (Majer et al., 2012).

The Basel experience has contributed a great deal of new knowledge about EGS induced seismicity. For example, precise event relocations and focal mechanism studies suggest that induced seismicity occurred as a cascading process along a series of small en échelon faults within a pre-existing cataclastic zone (Figure 9.14). Measured focal mechanisms are characterized by either left-lateral strike faults on N–S planes or right-lateral strike-slip faults on E–W planes (Häring et al., 2008). The mainshock was characterized by dextral strike-slip displacement on a steeply dipping WNW–ESE striking fault that is favourably oriented with respect to the regional stress field (Deichmann and Giardini, 2009). Significantly, three of the four strongest events occurred several months after the injection had stopped, when bottomhole pressure at the Basel-1 well had returned to hydrostatic levels (Deichmann and Giardini, 2009).

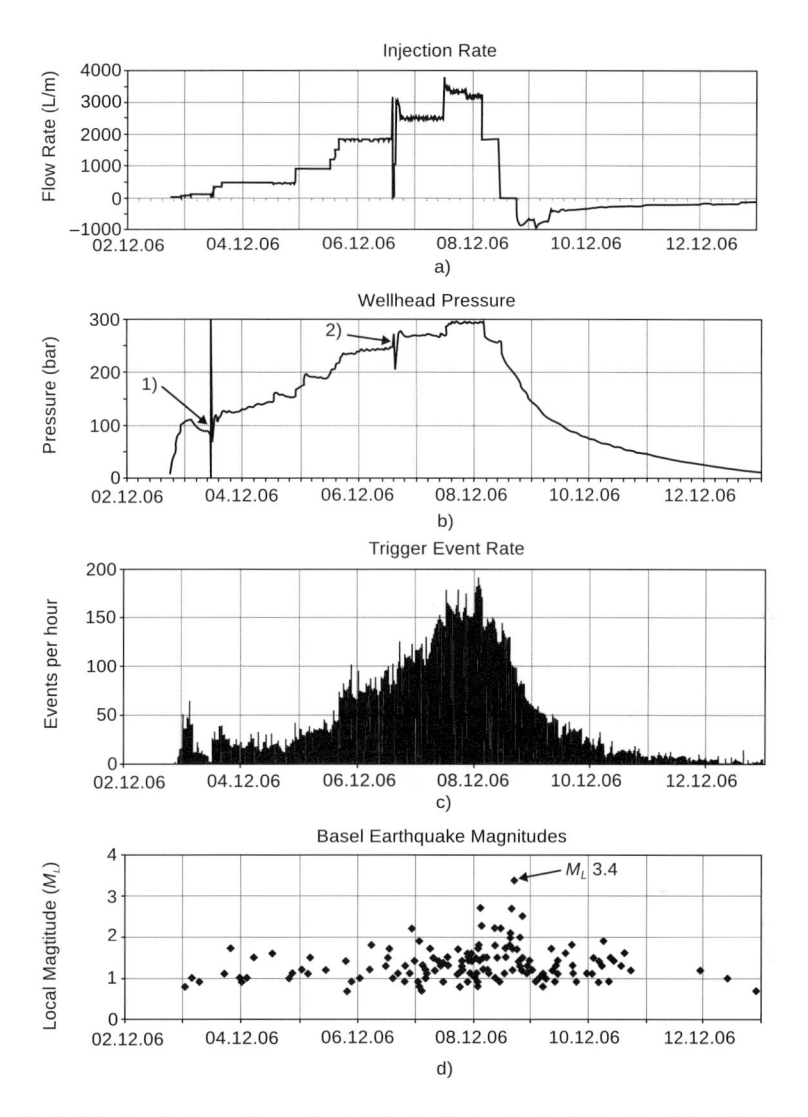

Fig. 9.13 Injection and seismicity data during and immediately following hydraulic stimulation of well Basel-1. Graphs show a) injection rates, b) wellhead pressures, c) trigger event rates and d) Basel earthquake magnitudes as determined by Swiss Seismological Survey (SED). In panel b), Transient 1 is due to a change in injection pump, and Transient 2 to the repair of a leaking wireline blowout preventer. Reproduced from *Geothermics*, Vol 37, Markus Haring, Ulrich Schanz, Florentin Lander and Ben Dyer, Characterisation of the Basel-1 enhanced geothermal system, Pages 469–495, copyright 2008, with permission from Elsevier.

In general, induced seismicity at Basel was characterized by a migration away from the borehole during the stimulation, a phenomenon known as the *Kaiser* effect. Modelling of this effect can be seen from migration of the 0.1 MPa pressure contour in Figure 9.12. The Kaiser effect is further marked by a systematic spatial heterogeneity of Gutenberg–Richter b values, with lower values (indicating a higher proportion of large events) evident at the periphery of the seismicity cloud (Bachmann et al., 2012).

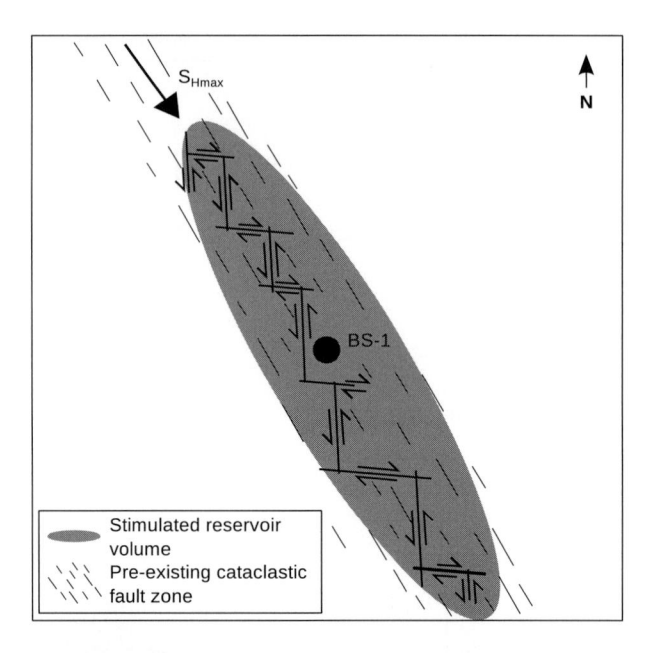

Fig. 9.14 Inferred interaction of injected fluids with pre-existing cataclastic fracture zone during hydraulic stimulation at Basel, Switzerland. Focal mechanisms suggest that slip occurred mainly along an en échelon fault system. BS-1: Basel-1 wellbore. Reproduced from *Geothermics*, Vol 37, Markus Haring, Ulrich Schanz, Florentin Lander and Ben Dyer, Characterisation of the Basel-1 enhanced geothermal system, Pages 469–495, copyright 2008, with permission from Elsevier.

Induced seismicity is well documented for other EGS projects, including St. Gallen in Switzerland (Edwards et al., 2015), the Cooper Basin in Australia (Baisch et al., 2009) and El Salvador (Kwiatek et al., 2014). There are also many documented cases of induced seismicity at non-EGS geothermal sites such as the Geysers field in California (Eberhart-Phillips and Oppenheimer, 1984). Moreover, significant prospects to leverage hydrocarbon technology for future development of EGS in sedimentary basins (Tester et al., 2007) suggest that passive seismic monitoring of induced seismicity in these areas may have growing importance.

9.3.2 Saltwater Disposal (SWD)

Statistical evidence shows that high-volume saltwater disposal (SWD) in class II injection wells[1] is nearly entirely responsible for the unprecedented increase in the rate of small-to-moderate earthquakes in the US midcontinent since 2009 (Weingarten et al., 2015). Notable injection-induced earthquakes in this region include:

- Guy-Greenbrier, Arkansas (M_W 4.7, 2011/02/27);
- Trinidad, Colorado (M_W 5.3, 2011/08/22);

[1] This is the designation given by the US Environmental Protection Agency for injection wells associated with oil and gas production (EPA, 2017).

- Prague, Oklahoma (M_W 5.6, 2011/11/05);
- Timpson, Texas (M_W 4.8, 2012/05/17);
- Fairview, Oklahoma (M_W 5.1, 2016/02/13);
- Pawnee, Oklahoma (M_W 5.8, 2016/09/03);
- Cushing, Oklahoma (M_W 5.0, 2016/11/07).

For each of these earthquakes, typical evidence for a causal connection with SWD includes proximity of hypocentres to injection wells and onset of seismic activity after the startup of injection wells (Hornbach et al., 2015). Indeed, the observed increase in seismicity rate in Oklahoma, the worst-hit state, is concentrated in areas where there have been local increases of five- to ten-fold in the rate of disposal (Walsh and Zoback, 2015). Based on correlation of seismicity with an injection-well database for the central and eastern USA, seismicity rate increase correlates primarily with injection rate rather than other factors such as cumulative injected volume, monthly wellhead pressure and proximity to basement; high-rate injection wells (i.e. those that inject at a rate greater than 50,000 m^3 per month) are much more likely to be associated with seismicity than lower-rate wells (Weingarten et al., 2015). Although SWD-induced earthquakes have been documented in other regions, such as Ohio (Kim, 2013) and western Canada (Schultz et al., 2014), the US midcontinent (Figure 9.11) is arguably the type area for this phenomenon, going back to induced earthquakes in the 1960s near the Rocky Mountain Arsenal (Healy et al., 1968).

Only a small fraction of the disposal volume derives from spent fracturing fluids (Rubinstein and Mahani, 2015); rather, most of the disposal fluids come from saline formation waters that are produced as a byproduct of oil extraction (Walsh and Zoback, 2015). What factors of unconventional oil and gas development have led to increased demand for SWD? In many conventional oil reservoirs, the use of waterflood methods is common, wherein co-produced waters are re-injected into the same formation in order to maintain production or slow its decline (Warner, 2015). For conventional reservoirs the problem of induced seismicity is thus avoided, since – at the scale of the reservoir – this practice maintains reservoir pore pressure at or below its pre-production level. In contrast, some of the major unconventional resource plays in Oklahoma, especially the Mississippi Lime play, produce massive volumes of saline water that cannot be re-injected into the same formation due to low reservoir permeability (Oklahoma Produced Water Working Group, 2017). Most of the wells in Oklahoma produce more water than oil, and reducing the volume of co-produced water means reducing oil production (Langenbruch and Zoback, 2016). Due to the high salinity of the produced formation water (40,000–300,000 ppm) the present industry practice, and most cost-effective solution, is to inject co-produced water into a permeable formation that not in hydrological contact with potable groundwater (Oklahoma Produced Water Working Group, 2017). At the scale of development of major unconventional resource plays, this is tantamount to a bulk transport of formation fluids from one stratigraphic level to a deeper level, over a significant part of the basin.

The Arbuckle Group (Fritz et al., 2012) beneath Oklahoma and Kansas, as well as the stratigraphically equivalent Ellenburger formation beneath Texas (Hornbach et al., 2015) form common target formations for SWD due to their great lateral extent and exceptionally high permeability. These units unconformably overlie crystalline rocks of Precambrian

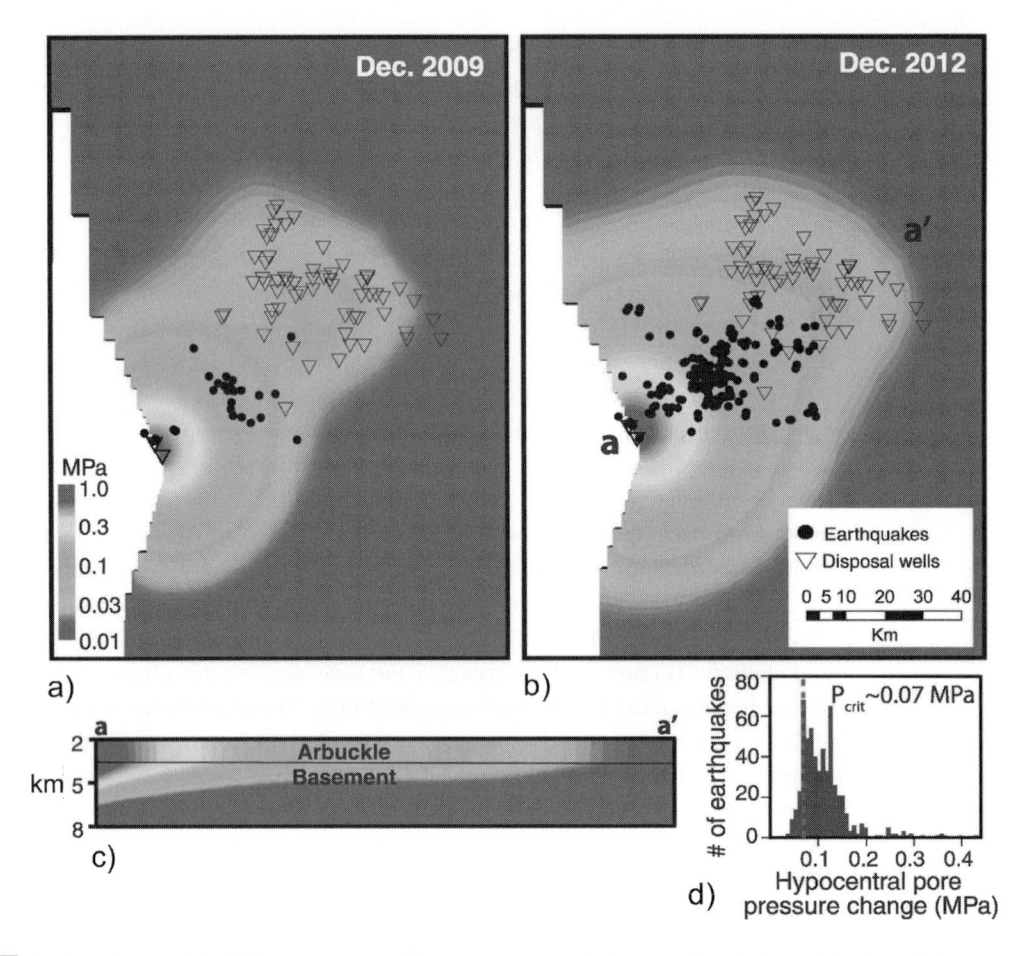

Fig. 9.15 Hydrogeologic model of diffusive expansion of pore-pressure perturbation away from high-rate injection wells in central Oklahoma. a) Modelled pressure perturbation in December 2009, showing earthquakes from 2008 to 2009. b) Modelled pressure perturbation in December 2012, showing earthquakes from 2008 to 2012. Hydraulic diffusivity is 2 m^2/s. The largest injected fluid volumes correspond with four high-rate wells, located within the area of highest pressure perturbation. c) Vertical cross section showing modelled pore pressure perturbation in 2012 along profile $a - a'$. There is a strong pore-pressure signal in the Arbuckle Group as well as the uppermost basement. d) Histogram of pore-pressure increase at all earthquake hypocentres in the catalog. A pore-pressure increase of ∼0.07 MPa is inferred to be the triggering threshold. Reproduced from Keranen et al. (2014), with permission of the American Association for the Advancement of Science. (A colour version of this figure can be found in the Plates section.)

age and appear to be in hydraulic communication with basement faults, which host the majority of earthquakes (Walsh and Zoback, 2015). Figure 9.15 shows a hydrologic model developed to simulate pore-pressure diffusion in the Arbuckle and underlying basement crust (Keranen et al., 2014). The model includes four high-rate injection wells plus 85 lower-rate injection wells, and shows that the pore-pressure perturbation is dominated by the high-rate wells. Hypocentres are concentrated near the periphery of the migrating pore-pressure front, consistent with the previously discussed Kaiser effect. A histogram showing calculated pore-pressure increase at every earthquake hypocentre (Figure 9.15d) suggests that a pore-pressure increase of ∼ 0.07 MPa is sufficient to initiate slip on faults (Keranen et al., 2014). If this model is correct, it provides a quantitative measure of crustal proximity to failure in terms of the Coulomb failure function (§2.4).

Walsh and Zoback (2016) used a recently compiled database of 26,313 basement fault segments in Oklahoma (Darold and Holland, 2015) to perform *quantitative risk assessment* (QRA) of the conditional probability of fault slip due to pore-pressure increase. The fault database does not contain information about fault dip or frictional characteristics; consequently, the analysis by Walsh and Zoback (2016) incorporates uncertainty in pore pressure, coefficient of friction, fault dip and a stress parameter ϕ_σ that represents the relative magnitudes of stress components as follows:

$$\phi_\sigma = \frac{S_2 - S_3}{S_1 - S_3}, \qquad (9.13)$$

where S_1, S_2 and S_3 are magnitudes of the three principal stresses. The results provide a basis for determining the probability of earthquake occurrence on a known fault and, in principle, could be used to assess maximum magnitude (Box 9.1). The analysis shows a relatively high probability of fault activation for the 2011 M_W 5.6 Prague earthquake a northeastern extension of a mapped fault for the 2016 M_W 5.1 Fairview earthquake sequence. The 2016 M_W 5.8 Pawnee earthquake did not occur on a mapped fault, highlighting incompleteness of the fault database.

Figure 9.16 shows a comparison of the monthly number of earthquakes of $M \geq 3$ with compiled monthly injected volume of saltwater, within an area in central and western Oklahoma (Langenbruch and Zoback, 2016). The times of occurrence for a number of significant individual earthquakes are identified on this graph, from which rate changes caused by aftershock sequences are clear. The graph shows that monthly injected volumes peaked at the end of 2014 and then started to decline, mainly in response to global economic forces. A corresponding decline is evident in the number of $M \geq 3$ earthquakes, with an apparent time lag > 6 months; note that the precise correlation between these curves is complicated by several large aftershock sequences.[2] Using these data, Langenbruch and Zoback (2016) calibrated and extended a numerical model built upon the concept of seismogenic index, introduced earlier in this chapter. They applied this approach to a large number of injection wells within a region rather than a single well, as in previous studies, in order to forecast the consequences of a regulator-imposed reduction of SWD in the Arbuckle Group to 40 percent of the 2014 rates. According to their model, the probability $P_E(M)$ to observe one or more earthquakes above a given magnitude M within one year can be expressed as

$$P_E(M) = 1 - \exp\left(-V_{Ia} 10^{\Sigma - bM}\right), \qquad (9.14)$$

where V_{Ia} is the annual injected volume above a calibrated threshold to cause induced seismicity and b is the slope parameter from the Gutenberg–Richter relationship. If the

---

[2] It is interesting to speculate whether large induced earthquakes in 2016 are analogous to transient increases during shut-in that are predicted by rate-state models (Figure 9.4) and observed following the injection at the Basel-1 well.

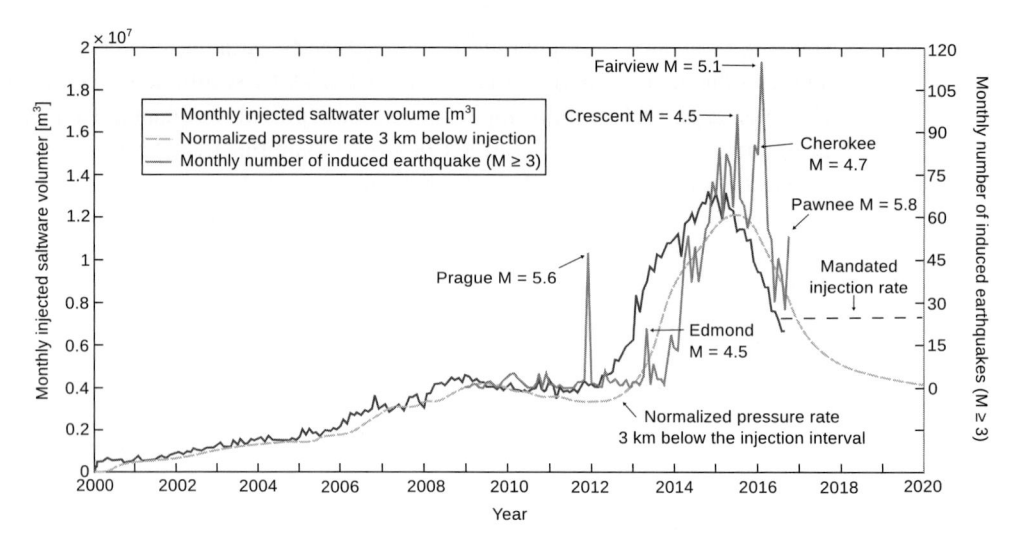

Fig. 9.16 Arbuckle monthly saltwater disposal (SWD), modelled pore pressure rate at 3 km depth and induced earthquake rate in central and western Oklahoma. Spikes in earthquake rate are from aftershock sequences following large events. Earthquake rate changes lag changes in the injection rate by several months. The dashed line shows normalized pressure rate computed for random fractures at 3 km depth below the injections, which exhibits a lag similar to the seismicity response. Reproduced from Langenbruch and Zoback (2016) with permission of the American Association for the Advancement of Science.

proposed reduction in injection volume is successful, their model predicts that overall seismicity levels will return to historic values within several years, with the caveat that the decline may take longer in areas affected by aftershock sequences. Prospects for reduced injection volumes may be enhanced by proposed alternatives to SWD for co-produced water, such as desalination for other industrial uses or pipeline transport for re-use in hydraulic fracturing operations (Oklahoma Produced Water Working Group, 2017).

9.3.3 Hydraulic Fracturing (HF) Induced Seismicity

Until recently it was widely regarded that hydraulic fracturing operations pose relatively little risk of inducing damaging earthquakes, compared with SWD and other types of causal mechanisms (National Research Council, 2013). This point of view stemmed, in large measure, from the observation that hundreds of thousands of wells had been stimulated using hydraulic fracturing with very few reports (prior to 2012) of anomalous induced earthquakes above the level of expected operationally induced microseismicity (Ellsworth, 2013). This viewpoint also reflects a notion that the net volume of fluid injected during HF well completion is relatively small in comparison to other types of injection processes (Davies et al., 2013; Rubinstein and Mahani, 2015).

One of the first reported instances of HF-induced seismicity was a M_L 2.3 event on 2011/04/01 close to the Preese Hall #1 well near Blackpool in the UK, which was undergoing HF treatment at the time of the event (Clarke et al., 2014). This small earthquake was widely reported in the media and resulted in an 18-month suspension of operations at the well. The fault that slipped during this event was unknown at the time of occurrence but has since been identified using 3-D seismic data combined with earthquake-source analysis (Clarke et al., 2014). By 2013, other reported occurrences of seismic activity correlated in time and space with HF operations included:

- 38 small earthquakes detected from 2009 to 2011 in the Horn River Basin of northwestern Canada, in the magnitude range of M_L 2.2–3.8 (BCOGC, 2012; Farahbod et al., 2015b);
- a sequence of small induced earthquakes in 2011 in south-central Oklahoma, which included 16 events in the magnitude range M_L 2.0–2.8 (Holland, 2013);
- a series of earthquakes up to M_L 3.0 in different parts of Ohio in 2013 and 2014 (Friberg et al., 2014; Skoumal et al., 2015b).

In many cases, these relatively weak events were detected using template-based methods and then correlated with reported data from HF operations.

Increasing reports of potential HF-induced seismicity in western Canada, outside of the Horn River Basin, prompted a series of further investigations. Other reported instances include:

- a sequence of 60 events in 2011–12 with magnitude up to M_L 3.0, associated with activation of a mapped normal fault due to HF operations in the Devonian-Mississippian-age Exshaw Formation (Bakken equivalent) (Schultz et al., 2015a);
- earthquake sequences starting in December 2013 within a formerly seismically quiescent region near Fox Creek, Alberta, associated with hydraulic fracturing of the Devonian Duvernay shale (Schultz et al., 2015b; Eaton and Mahani, 2015; Schultz et al., 2017);
- a sharp increase in seismicity starting in 2013 associated with unconventional hydrocarbon development within the Triassic-age Montney trend of northeastern British Columbia. This included 231 investigated seismic events in 2013 and 2014, of which 193 in the magnitude range M_L 2.5–4.4 were attributed to hydraulic fracturing (BCOGC, 2014; Mahani et al., 2017).

Atkinson et al. (2016) conducted a comprehensive study to determine whether a robust correlation exists between seismicity and hydraulic fracturing, within a 454,000 km^2 area that parallels the Rocky Mountain deformation front in the Western Canada Sedimentary Basin (WCSB). The study examined data from 12,289 hydraulically fractured (HF) wells and 1,236 disposal wells within this region, as well as earthquakes from the NRCan national catalog above the magnitude of completeness (conservatively given by $M > 3$). The correlation process used an initial screening in which event-well pairs were flagged as potential induced events if the epicentre was located within a 20 km radius of the well. In the case of hydraulic fracturing, a possible temporal correlation was deemed to exist if an event occurred within a window beginning with the commencement of hydraulic fracturing and ending three months after the completion of treatment; in the case of disposal wells, a possible temporal correlation was deemed to exist if the event occurred any time after the start of injection. All of the events that met the initial screening criteria were carefully examined to exclude spurious correlations, as outlined by Atkinson et al. (2016). Tests were also applied to ensure that the number of associations is significantly higher than expected based on random chance. In the end, the study concluded that 65 $M > 3$ earthquakes were linked to hydraulic fracturing in 39 wells (a 0.3% rate of well association) and 33 $M > 3$ earthquakes were linked to disposal wells (a 1.4% rate of well association). In a different study, Skoumal et al. (2015a) used waveform template matching to obtain remarkably similar estimates for rates of well association in Ohio (0.4% and 1.5%, respectively for HF and SWD).

Bao and Eaton (2016) carried out a detailed investigation of HF-induced seismicity in the Fox Creek area for a four-month interval in winter 2015. The study compiled injection data for all of the Duvernay HF well completions during this time interval, and compared the injection histories with a template-based seismicity catalog. At well pads within their study, the injected volumes[3] ranged from $\approx$ 30,000 m^3 to nearly 500,000 m^3. Figure 9.17 shows hypocentres that were located using the hypoDD algorithm, projected onto a cross section that passes through two horizontal wells. The largest event (M_W 3.9) nucleated near the top of crystalline basement, during shut-in two weeks after the completion of HF operations. As elaborated below, this event prompted the introduction of new traffic-light regulations for this area. The hypocentres in Figure 9.17 reveal two steeply dipping event clusters, interpreted by Bao and Eaton (2016) as two strands of a strike-slip fault system. The west strand was persistently active for several months after termination of HF operations, whereas the more distal eastern strand was active only during hydraulic fracturing. Bao and Eaton (2016) interpreted these patterns to indicate distinctive responses to activation by fluid pressurization, leading to persistent tremor, versus activation poroelastic stress changes (*cf.* Figure 9.3).

As new data comes to light, our understanding of the dynamics of fault activation by hydraulic fracturing continues to evolve. For example, Deng et al. (2016) used a numerical simulation to assess the relative contributions of pore-pressure perturbation to poroelastic stress changes for a sequence of HF-induced events in the Fox Creek area. They showed that the observed distribution of small earthquakes fits the Coulomb failure function perturbation (ΔCFS) associated with poroelastic stress, better than it fits the perturbation from pore-pressure changes. Given the relatively large inferred distance between the injection and the events (> 500 m), this result is consistent with the transient response of the east fault strand in Figure 9.17 as well as the simpler model by Segall and Lu (2015) shown in Figure 9.3. At a larger scale, Farahbod et al. (2015a) investigated the cumulative effects of multiple, near-synchronous hydraulic fracturing operations in the Horn River Basin of northeastern BC for a period of active drilling and completions from November 2006 to December 2011. They observed that larger induced events occurred only during months when a total fluid volume $> 150,000$ m^3 was injected. During these months of relatively high injection levels, industry activity was mainly confined within the $\approx$ 200 km^2 Etsho Area (BCOGC, 2012). This suggests that interactions were occurring with respect to fluids and stress changes caused by HF completions at different well pad locations, much the same as modelled interactions between different injection wells (Figure 9.15).

---

[3] Comparison of these values with injected volumes for EGS in §9.3.1 challenges a number of published statements that HF injected volumes are "small" relative to EGS. Moreover, it is noteworthy that during two months the largest 2015 Fox Creek Duvernay well pad injected a volume equivalent to ten months for a 50,000 m^3/month high-rate SWD disposal well. HF injection data are given by Bao and Eaton (2016).

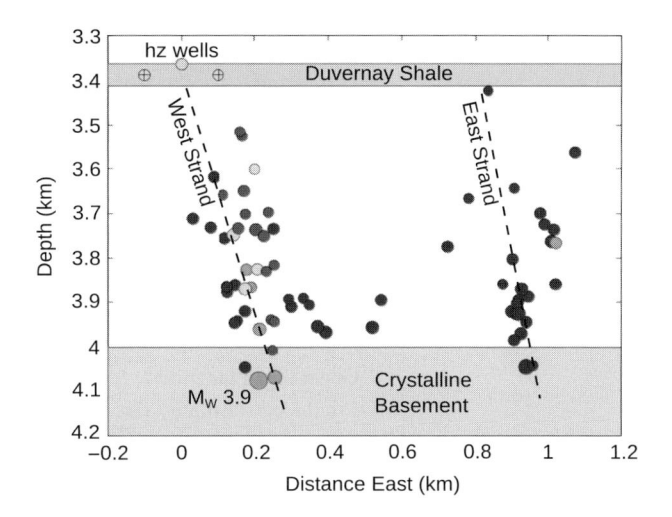

Fig. 9.17 Example of fault activation from hydraulic fracturing, presented as an east–west cross section showing hypocenters of induced events from January to March 2015. Dark blue symbols show events that occurred during hydraulic fracturing in two horizontal (hz) wells. Light blue, yellow and red symbols show post-injection events over a two-month time period. Two fault strands are evident, with contrasting temporal activation patterns. Reproduced from Bao and Eaton (2016), with permission of the American Association for the Advancement of Science. (A colour version of this figure can be found in the Plates section.)

Ghofrani and Atkinson (2016) carried out a statistical study, in which they subdivided the same region as in the study by Atkinson et al. (2016) into cells of 10 km radius. They used a hit-count density map to obtain a regionally-averaged estimate ($\approx$ 1–3%) of the prior probability of inducing seismicity within a given 314 km^2 cell. The results of this study confirm other investigations, which demonstrate that observed basin-wide patterns of induced seismicity are strongly clustered into a few small regions. This clustering invites an important question: what site-specific factors contribute to increased risk of HF-induced seismicity?

One possible explanation for this tight spatial clustering was suggested by Schultz et al. (2016), who found a significant spatial correlation between clusters of induced seismicity in the WCSB and regionally extensive margins of ancient underlying carbonate reefs. This geographic correlation points to possible basement fault controls on patterns of induced seismicity, since the reefs in question are thought to have formed atop paleobathymetric highs associated with Precambrian basement faults. Brudzinski et al. (2016) noted that, in areas where drilling in the two trends overlaps, the occurrence of HF-induced seismicity within the Utica trend contrasts with a paucity of induced seismicity due to hydraulic fracturing of the shallower Marcellus trend. The Utica formation lies within 800 m of the top of crystalline basement, suggesting that risk may scale with proximity to crystalline basement. Similarly, Eaton et al. (2016a) used pore-pressure data from two mature organic-rich shale formations in western Canada to show that, for both cases, a significant correlation exists between clusters of induced earthquakes and highly overpressured zones (i.e. areas where the vertical pore-pressure gradient is up to twice the hydrostatic gradient of 10 kPa/m). This

Table 9.1 Traffic Light Systems[1]

Jurisdiction	Amber	Red
United Kingdom	0	0.5
Ohio	0.5	1.0
Oklahoma	$M_L \geq 1.8^a$	3.5
Alberta	2.0	4.0
British Columbia	–	4.0
Colorado	2.5^b	4.5

[1] Generalized magnitude levels for HF-induced seismicity (Kao et al., 2016)
[a] Site-specific amber threshold
[b] Or any felt event

correlation implies that development of overpressure, possibly from hydrocarbon generation, may contribute locally to a critically state of stress on faults. More research is needed to undertake rigorous testing of these different hypotheses for geologic controls on the activation potential for HF-induced seismicity.

9.4 Traffic Light Systems

A *traffic light system* (TLS) is a reactive-control approach that is used to manage risks of injection-induced seismicity from HF or EGS stimulation programs. In general, a TLS has multiple discrete response thresholds that are determined used observable criteria, where each threshold invokes specific action designed to mitigate the associated risk (Kao et al., 2016; Trutnevyte and Wiemer, 2017). In practice, these systems are implemented using dedicated seismograph arrays with real-time processing, often with operational protocols for "green," "yellow" and "red" light conditions (Bommer et al., 2006). This approach has been selectively used since the 1960s, when it was applied to pump tests at Rocky Mountain Arsenal (Ellsworth, 2013). Simple TLS designs for hydraulic fracturing have been adopted by regulators in various jurisdictions, with varying threshold levels based on M_L as summarized in Table 9.1. Such design variability between jurisdictions has potential liability implications, as discussed by Eaton et al. (2016b). More sophisticated TLS approaches, which also make use of ground-motion measures in addition to public response, were introduced by Bommer et al. (2006) for geothermal operations.

In the case of HF monitoring, the most restrictive current system was established by the UK Department of Energy and Climate Change in 2013 (Kao et al., 2016). This system puts controls in place so that operators are required to assess the location of faults prior to well completion, monitor seismic activity in real time and cease operations if even minor tremors are produced. The green TLS condition, wherein the injection proceeds as planned, applies in the absence of any events above M_L 0. The detection of any induced event in the range $0 \leq M_L < 0.5$ invokes the amber condition, wherein injection proceeds with caution, likely at reduced rates, and monitoring is intensified. The red-light threshold is M_L 0.5, at which point injection is suspended immediately. In the USA, different states have implemented distinct thresholds, reflecting local conditions such as the distribution of geological faults, background seismicity and public tolerance for nuisance seismic activity. For example, in Colorado operators are required to modify operations if induced events are felt at the surface, or suspend operations in the case of an event with $M_L \geq 4.5$ (Wong et al., 2013). Ohio has established buffer zones around higher-risk areas, where an induced-seismicity mitigation plan and monitoring are required and the red-light threshold is M_L 1.0. The Oklahoma Corporation Commission has defined areas of interest to be within 10 km of any event that exceeds M_L 4.0 or swarms with two or more events exceeding M_L 3.0. TLS thresholds are established within these areas on a site-specific basis (Wong et al., 2013). In Canada, several provinces have established TLS thresholds (Kao et al., 2016). The Alberta Energy Regulator requires operators in the Kaybob-Duvernay region where Fox Creek is located to establish monitoring arrays that are capable of detecting M_L 2.0 events within 5 km of an injection well, with TLS thresholds as indicated in Table 9.1. The British Columbia Oil and Gas Commission has established permit conditions within high-risk areas that include the supplementary use of an accelerometer. Injection operations are suspended in the event of an event of M_L 4.0 or greater.

At a workshop to study traffic light protocols used in the development of low-permeability hydrocarbon resources, a number of deficiencies were noted in current TLS protocols for hydraulic fracturing (Kao et al., 2016):

1. magnitude uncertainty of induced events arising from the lack of a standardized approach for calculation of M_L;
2. lack of a link to impact/consequences of reported seismic events;
3. the need to integrate other real-time hazard measures.

Recommended actions to address these issues included incorporation of ground-motion data and standardization of methods for calculating local magnitude. Other adaptive measures were also discussed, including monitoring for a change in seismicity patterns that may be indicative of nucleation of a large event; such patterns could include a drop in b value, an increase in seismicity rate above some threshold value, or development of hypocentre lineations in an alignment that is favourable for fault slip within the regional stress regime (Maxwell et al., 2009). In addition, more general methods have been developed that apply a hazard-matrix approach (Walters et al., 2015) rather than a simple threshold scheme, in order to incorporate both geological and operational risk factors into the analysis. Further research is warranted to test these new methods and hypotheses.

TLS approaches for geothermal applications are used as pragmatic decision-making and feedback tools, based on a control principle that major ground disturbances can be avoided by timely modification of injection operations (Gaucher et al., 2015). By design, the reaction scheme is intended so that events that are not felt by the public but are detected by a fit-for-purpose monitoring system are used to take appropriate action. The first geothermal application of a TLS (Bommer et al., 2006) took into consideration vulnerability of the local building stock to determine tolerable ground-motion thresholds. A comprehensive TLS implemented for the Deep Heat Mining project near Basel included real-time

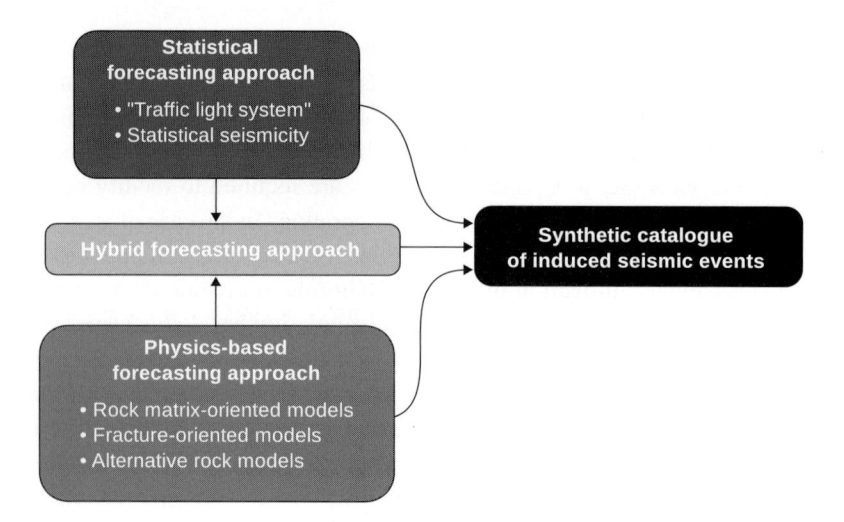

Fig. 9.18 Induced seismicity forecasting approaches for geothermal operations. From *Renewable and Sustainable Energy Reviews*, Vol 52, Emmanuel Gaucher et al., Induced seismicity in geothermal reservoirs: a review of forecasting approaches, Pages 1473–1490, copyright 2015, with permission from Elsevier.

magnitudes, peak ground motion and citizen's phone calls as warning inputs (Häring et al., 2008). Although this system ultimately did not prevent the sequence of $M_L > 3$ events from occurring, it might have prevented larger events (Ellsworth, 2013).

More generally, TLS protocols are integrated into forecasting methods for geothermal operations (Figure 9.18) using three different approaches (Gaucher et al., 2015). The first is a statistical approach that characterizes background and induced seismicity using earthquake catalog data, typically based on the ETAS model (§3.9). The second is a physics-based approach, in which thermo-hydro-mechanical processes during operations are simulated as a means to forecast induced seismicity. The physics-based approach requires information about geometrical features in the reservoir as well as initial conditions of temperature, fluid pressure and stress in order to apply various types of schemes including rock-matrix oriented and fracture-oriented models. The third is a hybrid approach, which uses physics-based simulations to model the evolution of the system up until incipient failure conditions are reached, after which a statistical approach is used. Forecast induced-seismicity models are important for calibration of TLS protocols in geothermal operations (Gaucher et al., 2015).

9.5 Probabilistic Seismic Hazard Assessment (PSHA)

Probabilistic seismic hazard analysis (PSHA) is a well-established framework to quantify ground-motion intensity for a multitude of design and engineering applications (Baker, 2008). The PSHA method has evolved considerably (McGuire, 2008) since its basic principles were first established (Cornell, 1968). Recent applications of this approach for induced seismicity hazard analysis, sometimes abbreviated as PISHA (Gaucher et al., 2015), have been developed by Atkinson et al. (2015), Petersen et al. (2016) and Petersen et al. (2017).

Following (Baker, 2008) the basic steps for PSHA at a given location can be summarized as follows.

1. Identify earthquake source regions, based on seismicity catalogs or fault models, that are capable of producing damaging ground motions.
2. Characterize the magnitude-frequency distribution using an assumed relationship for each region, such as the tapered Gutenberg–Richter formula (§3.8).
3. Determine the distribution of source-receiver distances, for example the mean hypocentral distance, based on the identified earthquake source regions.
4. Apply *ground-motion prediction equations* (GMPEs) to predict the ground-motion intensity distribution at the site for a design parameter such as *peak-ground acceleration* (PGA) or *peak ground-velocity* (PGV).
5. Use a probabilistic approach to combine information in order to determine *seismic hazard curves*, defined as the annual rate of exceedance of specified ground-motion level considering contributions from all earthquake sources.

These steps are briefly elaborated below and summarized schematically in Figure 9.19.

The first step in the procedure, determining earthquake source regions, can be accomplished using expert knowledge together with clustering analysis of the observed seismicity. Seismicity catalogs are then obtained for each source region, for which it is convenient to parameterize the magnitude-frequency distribution using the seismicity rate derived from the Gutenberg–Richter distribution. The rate parameter, a^*, quantifies the annual occurrence of positive-magnitude events and can be written as (Baker, 2008)

$$a^* = \log\left(\frac{N}{T}\right) + b M_c , \qquad (9.15)$$

where N is the number of observed events over T years, b is the Gutenberg–Richter rate parameter and M_c is the magnitude of completeness. As outlined in §3.8, M_c can be determined from the peak of the non-cumulative magnitude distribution (Wiemer and Wyss, 2000). The determination of the distance distribution is, in principle, a straightforward matter of numerical integration over the spatial probability distribution of earthquakes locations, although there are a number of important nuances arising from fault geometry and focal-depth considerations (Baker, 2008). Ground motion prediction equations (GMPEs) are parameterized using variables such as magnitude, distance, fault mechanism and focal depth and are derived from statistical regression of observed ground-motion intensity. A recent example of a GMPE developed explicitly for induced earthquakes recorded at relatively short distances is given by Atkinson (2015). Finally, assuming a *Poisson* distribution, for which each event occurs independently of the time since the last occurrence, the annual rate of exceedance (r_a) can be expressed in terms of the probability of occurrence of at least one damaging event ($P_E(M_{damage})$) as follows:

$$r_a = -\ln\left[1 - P_E(M_{damage})\right] . \qquad (9.16)$$

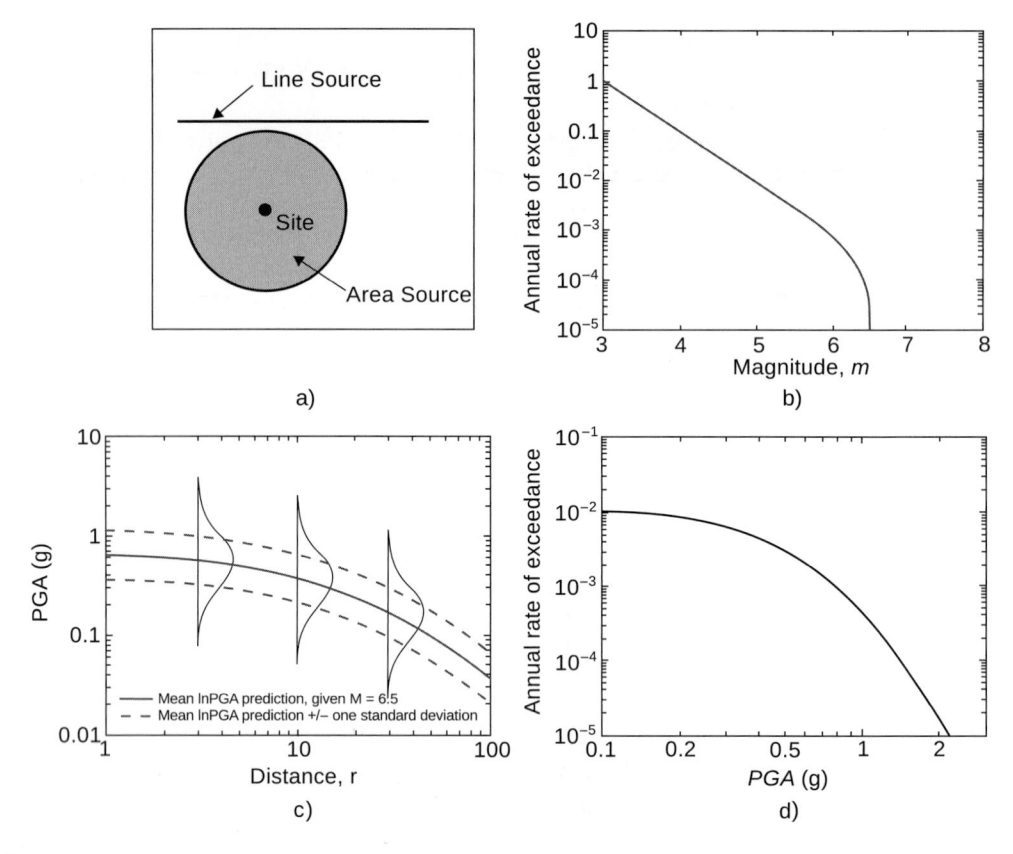

Fig. 9.19 Schematic illustration of the elements of the basic workflow for probabilistic seismic hazard analysis (PSHA). a) Identify earthquake source regions, such as a line source for a known fault or an areal source for a region of seismicity. b) Construct the magnitude-frequency distribution, such as a tapered Gutenberg–Richter distribution (§3.8) using seismicity catalogs. c) Apply ground-motion prediction equations based on a determination of the distance distribution (not shown). d) Combine information to calculate seismic hazard curves. Modified from Baker (2008), with permission.

A major challenge for application of traditional PSHA methods to induced seismicity is the non-stationary behaviour of induced earthquake sources. To address this challenge, Atkinson (2017) shows how PSHA can be combined with a traffic-light system (TLS) to produce a monitoring-and-response protocol, wherein actions are triggered if the hazard exceeds some threshold of concern. The USGS has developed a time-dependent methodology that uses seismicity rates determined from discrete time windows. Together with a logic-tree approach and estimation of area-dependent maximum magnitude that considers source regions with, or without, known induced seismicity (Petersen et al., 2016). The 2016 model has been updated and retroactively evaluated using observed seismicity during 2016 (Petersen et al., 2017).[4] Figure 9.20 compares USGS one-year seismic hazard maps

---

[4] A limitation of this approach is that it does not address hazard in the future, in areas of new development that are not currently active.

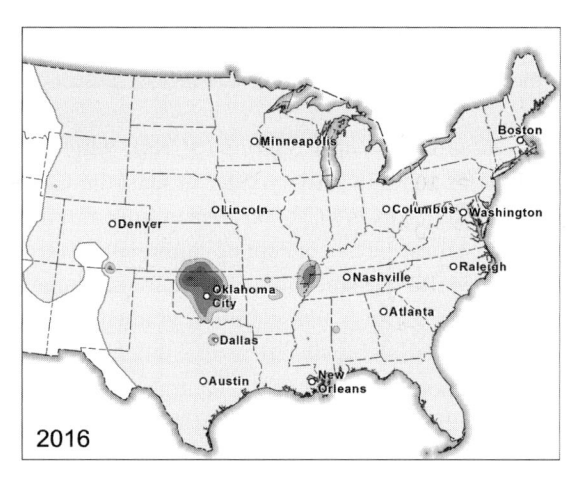

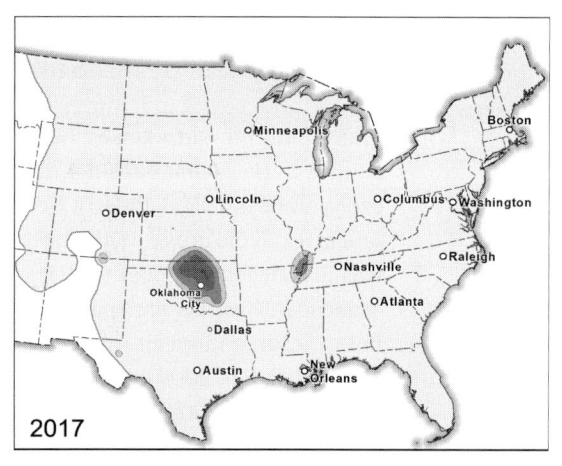

Fig. 9.20 Comparison of USGS one-year hazard forecast maps for 2016 and 2017. From Petersen et al. (2016) and Petersen et al. (2017), with permission of the USGS. (A colour version of this figure can be found in the Plates section.)

for 2016 (Petersen et al., 2016) and 2017 (Petersen et al., 2017). These maps express the chance of damage from an earthquake by averaging the exceedance probabilities for PGA at 0.12 g (1 g = 9.8 m/s^2) and for 1 Hz (1 s) spectral acceleration at 0.10 g, which are considered to represent the threshold for damaging ground shaking levels (Petersen et al., 2016). These calculations assume the US National Earthquake Hazard Reduction Program (NEHRP) site class D, or firm soil conditions (on Improved Seismic Safety Provisions and Agency, 1997). The 2016 forecast indicated high seismic hazard in all of the areas of the US midcontinent that are labelled in Figure 9.11. There was generally a high correlation between forecast high-hazard regions and actual seismicity in 2016 (Petersen et al., 2017). For example, there were 21 $M_W \geq 4$ earthquakes and 3 $M_W \geq 5$ earthquakes, including several damaging events, within the highest hazard region in Oklahoma; in addition, there were two $M_W \geq 4$ earthquakes in the Raton Basin. On the other hand, there were no $M_W \geq 2.7$ earthquakes in Texas or Arkansas during 2016, and this is reflected in reduced hazard levels in these regions for 2017.

Ultimately, it is important to go beyond hazard calculation to incorporate the concept of risk $\tilde{R}$, which can be considered as a convolution of four factors (Bommer et al., 2015)

$$\tilde{R} = \tilde{H} * \tilde{E} * \tilde{F} * \tilde{C}, \tag{9.17}$$

where $\tilde{H}$ denotes hazard (as obtained, for example, using PSHA); $\tilde{E}$ denotes exposure, which characterizes the built environment including all infrastructure elements; $\tilde{F}$ denotes the fragility, which represents the susceptibility of each element to damage or other undesirable consequences; and $\tilde{C}$ denotes consequence, which may reflect metrics such as the number of people adversely affected or the economic impact of an event.

9.6 Natural Analogs of Injection-Induced Earthquakes

Many induced earthquakes occur within *stable continental regions* (SCRs), which are defined as areas that have not experienced major tectonic activity since at least the Cretaceous (Johnston et al., 1994). Some natural earthquakes in SCRs have been linked to fluid processes in the crust, particularly *earthquake swarms*. The concept of earthquake swarms was introduced by Mogi (1963) as part of a general classification scheme for earthquake sequences. Swarms are spatio-temporal seismicity clusters that exhibit a gradual rise and fall in seismic moment release, lacking any well-defined mainshock–aftershock sequence (Ma and Eaton, 2009). Swarms constitute an important subset of the natural seismicity of SCRs (Spičák, 2000); they can occur episodically and are thought to be caused by external forcing, such as increasing fluid pressure or stress changes associated with volcanic processes (Fischer et al., 2014). Moreover, swarm seismicity has been linked to formation of hydrothermal ore deposits, where it is postulated to be a characteristic response to injection of large volumes of overpressured fluids into intrinsically low-permeability rocks (Cox, 2016).

Key differences between swarm behaviour and mainshock-dominated earthquake sequences may stem from the rate of fluid invasion into a fault zone relative to generation of co-seismic dilatancy (Cox, 2016). Natural swarms are common in areas of active volcanism, intraplate rift zones (Ibs-von Seht et al., 2008) and transform plate boundaries (Roland and McGuire, 2009), but they also occur within SCRs such as the Canadian Shield (Ma and Eaton, 2009). In West Bohemia-Vogtland, episodic earthquake swarms are associated with fluid flow and degassing of magma-derived CO_2 (Fischer et al., 2014; Alexandrakis et al., 2014). A close link between dilatancy, fluid flow and intermittent fault-slip processes has long been recognized (Sibson et al., 1975; Simpson and Richards, 1981; Segall and Rice, 1995); such *fault-valve* behaviour is characterized by a co-seismic increase in fault permeability and post-seismic fluid discharge along the rupture zone (Sibson, 1992).

Lessons learned from natural earthquake swarms may be helpful to address questions about fault-zone processes and fluid/stress systems associated with injection-induced seismicity. For example, this could provide useful perspectives on mechanisms for fault activation as well as potential risks from fluid transport to shallower depths. Conversely, studies of induced seismicity could contribute to a better understanding of natural swarm phenomena such as the episodic earthquake activity associated with fluid flow and degassing in West Bohemia-Vogtland.

9.7 Summary

This chapter starts off with a discussion of semantic distinctions that have been raised concerning the terms induced versus triggered seismicity. The term "induced" is used here to signify an association with anthropogenic activity. Pioneering studies at Rocky Mountain Arsenal (Healy et al., 1968) and Rangely (Raleigh et al., 1976) provided a foundation for our basic understanding of injection-induced seismicity and effective stress.

9.7 Summary

Various criteria have been developed to distinguish induced seismicity from natural earthquakes. Davis and Frohlich (1993) proposed a set of seven questions to assess whether an event may have been induced. Dahm et al. (2013) recommended strategies for distinguishing between induced, triggered and natural seismic events, falling into three general categories of physics-based probabilistic models, statistics-based seismicity models and source-parameter models. Zhang et al. (2016) suggested that unusually shallow focal depth provides a robust discriminator between injection-induced and natural seismicity. Skoumal et al. (2015a) found that, for a given maximum magnitude in a sequence, induced sequences contain many more events than natural sequences ("swarminess").

By analogy with petroleum systems, a fluid-systems framework is introduced, involving an injection source, pathway and critically stressed fault. This framework outlines the necessary conditions for occurrence of injection-induced seismicity. In the case of poroelastic media, the contribution of the stress effect on the Coulomb failure criterion needs to be considered in addition to the pore-pressure effect. Pore-pressure diffusion from a transient injection source is manifested by a triggering front, which marks the onset of induced seismicity, and a backfront that marks cessation.

Three distinct approaches are considered for estimation of maximum magnitude of an induced earthquake. McGarr (2014) proposed a deterministic formula that depends on the net volume of injected fluid; Shapiro et al. (2011) postulated that the rupture area for an induced earthquake is contained within the stimulated volume; and Van der Elst et al. (2016) presented statistical arguments that the maximum observed magnitude is the expected maximum value for a finite sample drawn from an unbounded Gutenberg–Richter distribution.

Template-based methods, including matched filtering and subspace detection, are becoming routine for the detection of small induced earthquakes. Recent developments include a repeating signal detector, which identifies signals of interest above a threshold S/N and then groups them using an agglomerative clustering algorithm (Skoumal et al., 2016); a Fingerprint And Similarity Thresholding (FAST) algorithm (Yoon et al., 2015) that uses feature extraction to construct waveform fingerprints; and a match-and-locate approach that uses stacked cross-correlation functions (Zhang and Wen, 2015).

Determination of earthquake hypocentre locations can be formulated as an inverse problem. Conventional linearized solutions to the inverse problem may become trapped in a local minimum. This issue can be avoided using a direct global-search method (Lomax et al., 2014). Precise relative hypocentre locations can be obtained using a widely used double-difference algorithm (Waldhauser and Ellsworth, 2000). Another tool of the trade is moment-tensor inversion, which inverts observed waveforms to obtain the best-fitting source mechanism.

Case studies of injection-induced seismicity, including earthquakes associated with enhanced geothermal systems (EGS), saltwater disposal (SWD) and hydraulic fracturing (HF), are reviewed in this chapter. Regulatory measures for managing risks of injection-induced seismicity, such as probabilistic seismic hazard assessment (PSHA) and traffic-light systems (TLS), are compared for various jurisdictions around the world. PSHA provides a quantitative framework to incorporate ground-motion intensity into engineering design and risk analysis.

Natural earthquake swarms are spatio-temporal seismicity clusters that exhibit a gradual rise and fall in seismic moment release, lacking any well-defined mainshock-aftershock sequence. Some episodic earthquake swarms may record a characteristic response to injection of large volumes of overpressured fluids into intrinsically low-permeability rocks (Cox, 2016), providing a potential natural analog for injection-induced seismicity.

9.8 Suggestions for Further Reading

- Reservoir-triggered seismicity: Talwani (1997)
- General references on fluid-induced seismicity: Nicholson and Wesson (1992), Majer et al. (2007), Guglielmi et al. (2015), Shapiro (2015) and Grigoli et al. (2017)
- PSHA methods: Reiter (1991) and McGuire (2004)

9.9 Problems

1. Calculate the seismogenic index (Equation 9.1) for cluster 1 and well pad 1 using supplementary data from Bao and Eaton (2016), assuming a b value of unity. The data are freely available in Tables S5 and S7, respectively, at:
http://science.sciencemag.org/content/suppl/2016/11/16/science.aag2583.DC1

2. Calculate the pore-pressure increase caused by a point source with a constant flux within a homogeneous poroelastic medium (Equation 9.2), using the following parameters: flux rate $q/\rho_0 = 10^{-2}$ m^3/s, dynamic viscosity $\eta = 10^{-3}$ Pa-s, permeability $k = 10^{-15}$ m^2 and hydraulic diffusivity $c = 0.1$ m^2/s. Calculate the pore pressure at distances of $r = 500$ m and 5.0 km and injection times of 30 days and 1 year. Be mindful of units.

3. Based on the seismicity and injection data for cluster 1 from Problem 1, determine and compare the maximum magnitude obtained using the method of McGarr (2014) and the method of Van der Elst et al. (2016), as described in Box 9.1.

4. Suppose that there are two seismogenic source regions at distances of 20 km and 200 km from a particular area of interest. The near region is characterized by annual Gutenberg-Richter parameters of $a = 5.2$ and $b = 1.2$, while the farther region has $a = 4.1$ and $b = 0.9$.

   a) Using the ground-motion prediction equation (GMPE) given in equation 5.11 with parameters for eastern North America from Atkinson et al. (2014), calculate the pseudoacceleration amplitude (in cm/s^2) for both distance ranges, for a period of 0.3 s and a magnitude of M4.

   b) Calculate the expected number of M4, M6 and M8 events in both regions, in a 50-year time interval.

5. **Online exercise**: West Bohemia is a region known to experience periodic natural earthquake swarms. A web exercise is provided to download and analyze data from this region.

Appendix A Glossary

Passive seismic monitoring of induced seismicity is a highly interdisciplinary field, combining aspects of seismology, continuum mechanics, fracture mechanics, geomechanics, structural geology, reservoir engineering and many other disciplines. Consequently, unfamiliar terminology can be a significant barrier to understanding. This glossary provides a concise explanation for some of the terms used in this book.

Acoustic medium: A medium in which shear waves are not considered or do not propagate, such as liquids or gases (Sheriff, 1991). Imaging methods used in exploration seismology commonly invoke an acoustic approximation and thus consider only *P*-wave propagation.

Aftershock: Earthquakes that follow the largest shock of an earthquake sequence. They are smaller than the mainshock and located within 1–2 rupture lengths distance from the mainshock. Aftershocks can continue over a period of weeks, months or years. In general, the larger the mainshock, the larger and more numerous the aftershocks, and the longer they will continue (USGS, 2017).

Anisotropy: Variation of one or more physical properties of a homogeneous material with the direction of measurement (Winterstein, 1990). A common form is transverse isotropy (TI), which is characteristic of shale, thinly bedded media and media with a single set of cracks; TI media have an infinite-fold axis of rotation (symmetry axis) and an infinite set of twofold axes of rotation in the perpendicular plane. The symmetry axis may be vertical (VTI), horizontal (HTI) or tilted (TTI). Another common form is orthorhombic, which has three mutually perpendicular symmetry axes and can arise from two sets of cracks, or a single set of vertical cracks in a layered medium (Figure 1.7). TI media are described by five independent elastic parameters, while orthorhombic media require nine independent parameters.

Anomalous seismicity: Seismicity that would not normally occur when performing an operation such as hydraulic fracturing (CAPP, 2017).

Asperity: In a literal sense, asperity means "roughness" (IRIS, 2017). An area on a fault this is stuck or locked.

b **value:** The slope of the magnitude-frequency distribution for the Gutenberg–Richter relation, which describes the relative size distribution of earthquakes. Most earthquake fault systems have a *b* value close to unity, implying that for each unit increase in magnitude there are ten times fewer earthquakes.

Backfront: A surface in space and time along which seismicity ceases due to termination of fluid injection (Shapiro, 2015).

Birefringence: Splitting of an incident wave into two waves with different polarizations; also known as shear-wave splitting (Sheriff, 1991).

Body wave: A wave propagating in the interior of an elastic medium, such as a *P* wave or *S* wave.

Borehole breakout: Elongation of an originally circular borehole cross section, due to calving of the walls in response to deviatoric stress (Bell and Gough, 1979).

Brittle deformation: Deformation from fracturing, frictional sliding or crushing of rock grains (cataclasis) (Fossen, 2016).

Calibration source: An impulsive seismic source used to calibrate a velocity model. Examples include perforation shots, which are shaped-explosive charges that are used to penetrate wellbore casing to allow egress of fracturing fluids; string shots, which are seismic detonators placed into the wellbore in order to generate an impulsive source; and sleeve-opening events, which are a multiple discrete signals generated by a sliding sleeve system used for open-hole hydraulic-fracturing completions.

Class II injection well: Defined by the US Environmental Protection Agency (EPA) as wells that inject fluids associated with oil and natural gas production. Most of the injected fluid is salt water (brine), which is brought to the surface in the process of producing oil and gas. Well types include disposal wells, enhanced recovery wells and hydrocarbon storage wells (EPA, 2015).

Clustering: Unsupervised classification of observations into groups (Jain et al., 1999). Categories include agglomerative clustering, a bottom-up approach that initially bins each event into a separate cluster and then merges the clusters until a stopping criterion is met; divisive clustering, tarts with all observations grouped into a single cluster and performs splitting until a stopping criterion is met; and monothetic clustering, which uses a single parameter such as the stage number.

Completion method: A process such as hydraulic fracturing (HF) that is used to prepare a well for production. Types of HF completions include perf and plug, which is used in a cased wellbore and requires the creation of perforations; and sliding-sleeve completion, a method that is used with packers and a sliding-sleeve assembly for HF completion in an open wellbore (Figure 4.9).

Compliance: A tensor quantity (s_{ijkl}) that relates the strain tensor to stress tensor, $\epsilon_{ij} = s_{ijkl}\sigma_{kl}$; inverse of the stress tensor.

Condition number: Ratio of the largest to the smallest eigenvalue of a matrix, used as a measure of stability of the matrix inverse.

Constitutive relation: A mathematical relationship between physical quantities that determines the response of a given material to applied forces. It is based on experimental observation or mathematical reasoning other than a fundamental conservation equation.

Covariance matrix: A symmetric matrix formed using the auto-variance and cross-variance of two time series. See Equations 6.1 and 6.2.

Creep: A process in which permanent plastic deformation occurs due to various microscale or atomic-scale mechanisms (Fossen, 2016). Dislocation creep occurs from the movement of dislocations in crystals and can lead to seismic anisotropy from lattice-preferred orientation (LPO). Diffusion creep occurs from movement of grain boundaries or vacancies in a crystal lattice.

Critically stressed: A state of incipient frictional failure, in which optimally oriented fractures or faults are approximately tangent to the failure envelope. This state is thought to be pervasive in the upper crust due to plate tectonic forces (Zoback, 2010).

Damage zone: A zone of high density of brittle deformation structures around a fault (Fossen, 2016).

Diagnostic Fracture Injection Test (DFIT): Small-volume injections that are performed to measure in-situ stress and/or permeability. An extended leakoff test (XLOT) follows a similar procedure, but includes repeated injection stages (Figure 4.3).

Diffusion: A physical mechanism that causes a quantity, such as pore pressure, heat or solute, to spread from areas of relatively high concentration to regions of relatively low concentration. For fluid injection, pore-pressure diffusion in a poroelastic medium occurs without bulk transport of the injected fluid.

Disposal well: A type of class II injection well used to inject brines or other fluids associated with the production of hydrocarbons. In the US, disposal wells represent about 20 percent of 151,000 class II injection wells (EPA, 2017).

Discrete fracture network (DFN): A mathematical model used in geomechanical simulations to represent a network of fractures.

Dispersion: Dependence of phase velocity on frequency (or wavelength) and/or wavevector. Dispersion that is related to the velocity structure can arise from wavelength-dependent velocity sensitivity and is called geometrical dispersion.

Double couple: A mathematical model for an earthquake source mechanism, consisting of two orthogonal force couples. A double-couple model fits the far-field radiated wavefield for most earthquakes. The mechanism is typically parameterized using the strike and dip of the fault plane, as well as the rake (slip vector).

Drilling-induced tensile fracture (DITF): Fractures that form in the wall of a wellbore at the azimuth of the maximum horizontal compressive stress, when the hoop stress locally goes into tension (Zoback, 2010).

Ductile deformation: Permanent deformation in response to stress, without fracturing or frictional sliding. This can occur from creep processes.

Dynamic range: The ratio of the largest reading to the minimum reading that can be recorded by and read from an instrument without change of scale (Sheriff, 1991), often expressed in decibels (dB).

Earthquake: A term used to describe both a sudden slip on a fault, as well as the resulting ground shaking and radiated seismic energy caused by the slip, or by volcanic or magmatic activity, or other sudden stress changes in the earth (USGS, 2017).

Earthquake sequence: A series of earthquakes in a given area with a well-defined mainshock–aftershock, or foreshock–mainshock–aftershock events.

Engineered (or enhanced) geothermal system (EGS): A heat exchanger created within deep and low permeability hot rocks using fluid-injection stimulation methods (Breede et al., 2013).

Enhanced recovery well: A well used to inject brine, water, steam, polymer or carbon dioxide into oil-bearing formations to recover residual hydrocarbons (EPA, 2017). In the USA approximately 80 percent of 151,000 Class II wells are enhanced recovery wells.

Epicentre: The point on the surface vertically above an earthquake's focus (USGS, 2017).

Epidemic Type Aftershock Sequence (ETAS) model: A cascading point process derived from Omori's Law that can be used to simulate the temporal patterns of earthquake sequences in a given region (Ogata et al., 1993).

Estimated stimulated volume (ESV): The estimated volume of a point cloud defined by a seismicity cluster. This is related to the stimulated reservoir volume (SRV), the volume within a reservoir inferred to be stimulated by hydraulic-fracturing completion (Mayerhofer et al., 2010).

Fabric: The spacing, arrangement, distribution, size, shape and orientation of the constituents of rocks such as minerals, grains, porosity, layering, bed boundaries, lithology contacts and fractures (Ajaya et al., 2013). A penetrative fabric is one that occurs throughout the rockmass.

Failure criterion: A mathematical model defining stress conditions under which failure may occur. Examples include Mohr–Coulomb, Hoek–Brown and Griffith failure criteria.

Fault: A discontinuity in the subsurface that has accommodated displacement between rock masses on either side of the discontinuity. Sometimes the term geologic fault is used to describe a surface where displacement occurred in the geologic past.

Faulting mechanism: Description of the rupture processes of an earthquake, including the style of faulting (i.e. normal, reverse or strike-slip) and the rupture fault plane on which it occurs (Majer et al., 2012).

Focal depth: Depth below the surface of an earthquake's focus.

Focal mechanism: A graphical representation of the faulting mechanism of an earthquake. Typically, this is represented as a lower-hemisphere projection of P-wave first motion polarity. It is also known as a beachball diagram or a fault-plane solution.

Focus: The point in the subsurface where earthquake rupture initiates.

Foreshock: A relatively smaller earthquake that precedes the largest earthquake in a sequence, which is termed the mainshock. Not all mainshocks have foreshocks (USGS, 2017).

Fractures: A sharp planar discontinuity in a medium. Categories of fractures include tensile (mode I) and shear (modes II and III). A pancake fracture is a horizontal fracture, often occurring along a bedding plane.

Green's function: The solution of a differential equation with an impulse as the exciting force (Sheriff, 1991).

Ground-motion prediction equation (GMPE): A relationship that predicts the amplitude of a specified ground-motion parameter (e.g. PGA, PGV) as a function of magnitude, distance, focal depth and site conditions (Majer et al., 2012).

Group velocity: The velocity at which the energy in a wavefield propagates.

Gutenberg–Richter relation. An empirical formula that describes the earthquake magnitude–frequency relationship. It has the form $\log_{10} N = a - bM$, where N is the number of earthquakes whose magnitude is greater than or equal to M, and a and b are constants.

Hazard: The probability that a given event will produce damage or harm. Hazard ($\tilde{H}$) is related to Risk ($\tilde{R}$) by the risk equation: $\tilde{R} = \tilde{H} * \tilde{E} * \tilde{F} * \tilde{C}$, where $\tilde{E}$ denotes exposure, $\tilde{F}$ denotes fragility and $\tilde{C}$ denotes consequence (Bommer et al., 2015).

Hooke's Law (generalized): An empirical linear relationship between stress and strain, applicable at low level of strain.

Huygens' principle: The mathematical concept that every point on a wavefront can be treated as a secondary source, and that an advancing wavefront can be constructed from the envelope of the secondary waves (Sheriff, 1991).

Hydraulic fracturing: Injecting fracturing fluids into a rock formation at a force exceeding the fracture pressure of the rock. For application to the development of low-permeability hydrocarbon resources, this process induces a network of fractures through which oil or natural gas can flow to the wellbore (CCA, 2014).

Hypocentre: The estimated location of an earthquake's focus.

Imaging condition: A criterion that is applied to extract sources from the migrated data volume.

Induced Seismicity: Seismic events that can be attributed to human activities. Examples of activities that can cause induced seismicity include geothermal development, mining, reservoir impoundment and subsurface fluid injection and withdrawal.

Instrument response: The amplitude and phase characteristics of an sensor such as a geophone or seismometer to a unit input ground motion. The instrument response needs to be corrected in order to estimate magnitude.

Intensity: The effects of earthquake ground motion on the natural or built environment. In North America, intensity is usually quantified using the Modified Mercalli Scale. Intensity is specified in Roman numerals and ranges from I (not felt except by a very few under especially favourable conditions) to XII (total damage) (USGS, 2017).

Isochron surface: A geometrical surface defined by the locus of points with constant traveltime from a source, receiver or source–receiver pair.

Joint: A fracture across which there is no discernible shear displacement.

Kaiser effect: In the case of cyclical loading, vanishing of seismicity activity for perturbation less than the maximum previously reached level as an expression of memory of the stress history of materials and rocks (Shapiro and Dinske, 2007). For an expanding cloud of injection-induced seismicity, this effect may be manifested by localization of activity near the leading edge of the cloud.

Kirsch equations: A set of equations named for Ernst Gustav Kirsch that describe the stresses induced by a circular hole in an isotropic elastic plate. The Kirsch equations equations are used to determine stresses around a wellbore with circular cross section (see Equation 4.1).

Lagrangian reference frame: A frame of reference used in seismology in which the coordinate system moves with a particle in the medium. In contrast, an Eulerian reference frame, used in fluid dynamics, is fixed in space and emphasizes the study of whatever particle happens to occupy a particular spatial location (Aki and Richards, 2002).

Magnitude: A quantitative measure of the size of an earthquake based on seismograph recordings. Several scales have been defined, but the most commonly used are (1) local magnitude (M_L), also referred to as "Richter magnitude," (2) surface-wave magnitude (M_S), (3) body-wave magnitude (m_b) and (4) moment magnitude (M_W) (USGS, 2017).

Mainshock: The largest earthquake in a sequence, sometimes preceded by one or more foreshocks, and almost always followed by many aftershocks (USGS, 2017).

Microseismic facies analysis: An approach for microseismic interpretation that is used to highlight subtle stratigraphic details, structural deformation, fracture orientation and stress compartmentalization using microseismic facies, or a body of rock with distinctive microseismic characteristics (Rafiq et al., 2016).

Microseismicity: Defined here as seismicity of magnitude less than 0.

Microtremor: A more or less continuous motion in the subsurface that is unrelated to an earthquake and that has a period of $\sim$ 1.0–9.0 s, caused by a variety of natural and anthropogenic factors.

Moment tensor: An idealized mathematical representation of the movement on a fault during an earthquake, comprised of nine generalized force couples. The tensor depends of the source strength and fault orientation (USGS, 2017).

Moveout: Relative arrival time defined by the time difference with respect to the minimum observed time.

Nodal surface: A surface across which the amplitude of the radiated wavefield passes through zero and undergoes a polarity reversal. For a double-couple source, P waves have two mutually orthogonal nodal planes.

Noise: Components of measured data that are not signal. What is considered to be noise therefore depends on what is considered to be signal. Random noise is generated by a stochastic process and has a statistically random distribution. For example, noise generated by wind is often considered to be random. White random noise has a flat spectrum, in which all frequency components have equal strength. Coherent noise, such as ground roll, exhibits waveform coherency across an array of sensors. The ratio of average noise amplitude to the amplitude of a signal of interest is called the signal-to-noise ratio (S/N).

Nucleation: Earthquake nucleation is a slip-acceleration process that occurs within a fault patch (Rubin and Ampuero, 2005).

Null hypothesis: A commonly accepted model and/or a supposition that no relationship exists between two quantities.

Nyquist: The frequency (or wavenumber) that is half of the sampling frequency for a time series. Based on the Sampling Theorem, this defines the maximum frequency in a continuous function that can be recovered from a discrete time series without aliasing.

Operationally induced seismicity: Weak (nano-, micro- and milli-) seismicity that is expected to occur during operations, such as hydraulic fracturing or development of an engineered geothermal system.

Pad: A well pad is an area at the surface that is constructed for drilling one or wells. The term pad is also used to describe the first part of a hydraulic-fracturing stage, when fracturing fluids are injected without any proppant.

Peak ground acceleration (PGA): Maximum instantaneous amplitude of the absolute value of the acceleration of the ground (Majer et al., 2012).

Peak ground velocity (PGV): Maximum instantaneous amplitude of the absolute value of the acceleration of the ground (Majer et al., 2012).

Permeability: A measure of the ease with which fluid can pass through a porous medium.

Phase velocity: The propagation velocity of a monofrequency harmonic plane. The phase velocity depends on frequency in a dispersive medium, or direction for an anisotropic medium.

Plastic deformation: Permanent deformation caused by ductile processes due to sustained stress conditions above the yield point.

Play: An area in which specific type of commercial hydrocarbon deposit occurs. A stacked play consists of several plays that are vertically stacked at one location.

Power-spectral density: Rate of change of the mean squared spectral amplitude with frequency, also known as the autospectral density function (Bendat and Piersol, 2011).

Probabilistic seismic hazard analysis (PSHA): A systematic method for estimation of the probability of ground motions that are expected to occur or be exceeded within a specified time interval (Majer et al., 2012).

Probability of exceedance: Probability at which the value of a specified parameter (e.g. PGA, PGV) is equalled or exceeded (Majer et al., 2012).

Process zone: The zone at the tip of a fracture (or rupture front) where the formation of numerous microcracks weakens the rock (Fossen, 2016).

Proppant: Granular material such as sand that is mixed with fracturing fluid in order to hold fractures open (Schlumberger, 2017).

Radiation pattern: The amplitude of waves radiated from a point source as a function of direction from the source.

Ray-centred coordinates: A local orthonormal coordinate system in which one axis is parallel to the direction of a seismic ray. The other two coordinate axes are usually defined to be coplanar and perpendicular with the sagittal plane.

Rupture area: The surface area of a fault that is affected by sudden slip during a seismic event.

Sagittal plane: The vertical plane that passes through the source and receiver points.

Self-organized criticality: A property of a dynamical system that spontaneously evolves to a critical state, with little or no dependence on the initial conditions. An example of this behaviour is a growing pile of granular material, whose slope evolves toward a material-dependent angle of repose (Bak et al., 1988).

Seismic moment: A measure of the size of an earthquake based on the product of the rupture area, the average amount of slip, and the force that was required to overcome fault friction. Seismic moment can also be calculated from the amplitude spectra of seismic waves (USGS, 2017).

Seismicity: Earthquakes or other seismic activity within a given area.

Seismogenic: Capable of producing an earthquake. The seismogenic index is a measure of the induced-seismicity susceptibility to fluid injection.

Signal: Components of measured data that carries information of interest. The ratio of average noise amplitude to the amplitude of signal is called the signal-to-noise ratio (S/N).

Slickwater: A fracturing fluid that is predominantly composed of fresh water, with surfactants and friction reducers used as additives in order to reduce surface tension and pumping resistance.

Sommerfeld radiation conditions: A boundary value in which the contribution from sources at infinite distances is zero.

Stress: A tensor that defines the force per unit area that acts on a planar surface inside a medium. There are various categories of stress, including normal stress, which acts perpendicular to the surface and shear stress, which acts tangentially to the surface. For any stress tensor, there exists a frame of reference in which it can be represented as a diagonal tensor (matrix). The diagonal elements are called the principal stresses. The mean stress is the arithmetic average of the principal stresses. Lithostatic stress is a stress tensor in which all of the principal stresses are identical and equal to the mean stress. Hydrostatic stress is similar, except that the principal stresses are due to the weight of an overlying column of water. The deviatoric stress tensor is the difference between the full stress tensor and the lithostatic stress. Effective stress is determined by subtracting the product of the pore pressure and Biot's coefficient from the diagonalized (principal) stress tensor. Tectonic stresses arise from geologic processes such as movement of the tectonic plates.

Stress drop: The co-seismic reduction in shear stress acting on a fault (i.e. the difference between the shear stress on the fault before an earthquake to the shear stress after an earthquake).

Stratabound fracture network: A fracture network in which fracture height is controlled or influenced by mechanical bed thickness.

Stiffness: A tensor quantity (c_{ijkl}) that relates the stress tensor to strain tensor, $\sigma_{ij} = c_{ijkl}\epsilon_{kl}$; inverse of the compliance tensor.

Surface wave: A seismic wave that propagates along the surface of a medium.

Swarm: A spatiotemporal cluster of earthquakes that lacks a well-defined mainshock–aftershock sequence or foreshock–mainshock–aftershock sequence. Swarminess, given by the ratio of the number of events in a sequence above a specified magnitude to the maximum magnitude in the sequence, can be used to quantify this characteristic (Skoumal et al., 2016).

Tensor: A generalization of vectors that provides a multidimensional representation of a physical quantity. Tensors are invariant under transformation of coordinate system. The order (also called rank) of a tensor is the number of required indices. The Einstein summation convention means that repeated indices within products of tensors implies summation.

Triaxial test: A rock-mechanics compressional testing procedure in which the axial stress acting on a cylindrical sample differs from the perpendicular confining stress (Zoback, 2010). A true triaxial test uses three distinct orthogonal normal stress values.

Triggered seismic event: A seismic event that is the result of failure along a pre-existing zone of weakness, for example a fault that is already critically stressed and is pushed to failure by a stress perturbation from natural or anthropogenic activities (Majer et al., 2012). In earthquake seismology, natural triggering has been recognized due to both static stress changes (i.e. long-term changes to the stress field caused by an earthquake) and dynamic stress changes (e.g. transient stress changes from a propagating seismic wave).

Triggering front: A surface that expands away from an injection point and marks the initiation of seismicity.

Uncertainty: A measure of the accuracy and precision with which a measured value is known. Uncertainty that derives from random errors is called aleotoric uncertainty, from the Greek *aleo*, which means rolling of a dice (Der Kiureghian and Ditlevsen, 2009). Another type of uncertainty is called epistemic uncertainty, from the Greek *episteme*, which means knowledge.

Unconventional resource: Oil and gas resources whose porosity, permeability, fluid trapping mechanism or other characteristics differ from conventional hydrocarbon reservoirs (CCA, 2014).

Viscoelastic medium: A medium that exhibits an instantaneous deformation in response to stress, followed by a gradual continuous deformation process.

Viscosity: A measure of the resistance of a fluid to flow.

Voigt notation: A compact notation to represent Generalized Hooke's law (see Equation 1.11).

Appendix B Signal-Processing Essentials

A basic understanding of signal processing and Fourier analysis is essential for many aspects of passive seismic monitoring. Concepts of waveforms, spectra, deconvolution and cross-correlation are basic tools that are second nature to practicing seismologists. The following material is intended for general reference; for additional background, theorems or proofs, the interested reader is referred to any standard textbook, such as classic books by Oppenheim and Schafer (1975) and Bracewell (1986).

B.1 Fourier Transform

The Fourier transform of an integrable function $u(t)$ is given by

$$U(f) = \int_{-\infty}^{\infty} u(t) e^{-i2\pi t f} dt, \tag{B.1}$$

and the inverse Fourier transform is

$$u(t) = \int_{-\infty}^{\infty} U(f) e^{i2\pi t f} df, \tag{B.2}$$

where f denotes frequency, if t represents time. In this case, $u(t)$ is a *time series* and $U(f)$ is called its *spectrum*. In general, Fourier analysis is not limited to time series and can be applied to any type of signal (e.g. a topographic profile). It is also easily generalizable to multiple dimensions. Alert readers are cautioned that Equations B.1 and B.2 are sometimes normalized in a different manner (Bracewell, 1986).

The functions $u(t)$ and $U(f)$ are called a Fourier transform pair; this relationship is written symbolically as

$$u(t) \longleftrightarrow U(f). \tag{B.3}$$

In general, $U(f)$ is always a complex-valued function, regardless of whether $u(t)$ is real or complex. Applying Euler's relation, the Fourier spectrum can be expressed as

$$U(f) = |U(f)| e^{i\phi(s)}, \tag{B.4}$$

where $|U(f)|$ is the *amplitude* spectrum and

$$\phi(s) = \tan^{-1} \frac{Im\{U(f)\}}{Re\{U(f)\}} \tag{B.5}$$

is the *phase* spectrum. If $u(t) \in \Re$ then $U(f)$ has the property

$$U(f) = U^*(-f), \tag{B.6}$$

where the * superscript denotes complex conjugate. In other words, if $U(f)$ is the Fourier transform of a real-valued function, the real component of $U(f)$ has *even* symmetry

$$Re\{U(f)\} = Re\{U(-f)\}, \tag{B.7}$$

while the imaginary part has *odd* symmetry,

$$Im\{U(f)\} = -Im\{U(-f)\}. \tag{B.8}$$

A function with this complex symmetry property is called *hermitian*. The Fourier transform has the following additional properties (Sheriff, 1991).

Linearity and superposition apply in both domains. Consequently:

$$a(t) + b(t) \longleftrightarrow A(f) + B(f). \tag{B.9}$$

Compressing a time function expands its frequency spectrum and reduces its amplitude by the same factor:

$$u(kt) \longleftrightarrow \frac{1}{k} U\left(\frac{f}{k}\right). \tag{B.10}$$

The derivative of a function is related to its Fourier transform by:

$$u'(t) \longleftrightarrow i 2\pi f\, U(f). \tag{B.11}$$

There are also a number of special functions whose Fourier transforms are important in seismology.

The Dirac δ function:

$$\delta(t) \longleftrightarrow \Delta(s) = 1, \quad -\infty < s < \infty. \tag{B.12}$$

The Dirac δ function has the following *sifting* property:

$$\int_{-\infty}^{\infty} \delta(t-a) u(t) dt = u(a). \tag{B.13}$$

Next, we consider a *symmetric boxcar function of width τ*:

$$\Pi(t) \longleftrightarrow \tau\, \text{sinc}(f\tau) = \frac{1}{\pi f} \sin(\pi f \tau), \tag{B.14}$$

where the boxcar function $\Pi(t)$ is defined as:

$$\Pi(t) = \begin{cases} 1, & |t| \leq \tau/2 \\ 0, & |t| > \tau/2 \end{cases}. \tag{B.15}$$

The Fourier-domain expression on the right side of Equation B.14 introduces an important function, called the *sinc* function (also known as the interpolating function), which is defined as

$$\text{sinc}\, x \equiv \frac{\sin \pi x}{\pi x}. \tag{B.16}$$

We continue with a *triangle function with base* 2τ:

$$\Lambda(t) \longleftrightarrow \tau \operatorname{sinc}^2(s\tau) = \frac{1}{\pi^2 s^2 \tau} \sin^2(\pi s\tau), \qquad (B.17)$$

where $\Lambda(t)$ is defined as

$$\Lambda(t) = \begin{cases} 1 - |t|/\tau, & |t| \leq \tau \\ 0, & |t| > \tau \end{cases}. \qquad (B.18)$$

Next, the Fourier transform of a function with Gaussian form is also a function with a Gaussian form:

$$g(t) = e^{-\pi t^2} \longleftrightarrow e^{-\pi s^2} = G(s). \qquad (B.19)$$

The above expressions can be written using a standard Gaussian formula by applying the Fourier scaling rule in Equation B.10. Finally, the Fourier transform of a *comb* function with sample interval τ, also known as the sampling function, *is also a comb function*:

$$\operatorname{comb}(t,\tau) \equiv \sum_{n=-\infty}^{\infty} \tilde{\delta}(t - n\tau) \longleftrightarrow \sum_{n=-\infty}^{\infty} \tilde{\delta}(s - n/\tau), \qquad (B.20)$$

where $\tilde{\delta}$ denotes a δ function that is scaled to have unit amplitude. In the time domain the comb function is a series of equally spaced unit impulse functions separated by the time interval τ, such that:

$$\operatorname{comb}(t)\, u(t) = \sum_{n=-\infty}^{\infty} u(n)\, \tilde{\delta}(t - n). \qquad (B.21)$$

In general, the reciprocal Fourier relationships for the above formulae can be expressed by switching the arguments t and f.

B.1.1 Discrete Fourier Transform

While the continuous forms of the above formula are useful for theoretical derivations and to obtain insights into their properties, in practice time series that are stored on any digital device contain *discrete* data samples. For a time series u_k, $k = 1, 2, 3, \ldots, N$, the discrete Fourier transform is defined as (Bracewell, 1986)

$$U = \frac{1}{N} \sum_{k=1}^{N} u_k e^{-i2\pi(k/N)\tau}, \qquad (B.22)$$

where U is uniformly sampled in the frequency domain and the quantity k/N is analogous to frequency f in the continuous transform. The original time series may be recovered by applying the discrete inverse Fourier transform,

$$u = \sum_{k=1}^{N} U_k e^{i2\pi(k/N)\tau}. \qquad (B.23)$$

The Fast Fourier Transform (FFT) algorithm was developed by Cooley and Tukey (1965) and is now used almost universally for digital computations. The efficiency improvement is quantified by considering the number of operations; by using the FFT algorithm, the number of calculations scales as $O(N \log N)$, whereas a direct implementation of Equations B.22 and B.23 leads to an algorithm that scales as $O(N^2)$.

B.1.2 Sampling Theorem and Nyquist Frequency

Digital sampling of a continuous signal can be represented as multiplication by a comb function, as in Equation B.21. The *sampling theorem* states that a *bandlimited* continuous function that is sampled at a regular sampling interval τ can be fully reconstructed from the discrete samples if there are two or more samples per cycle of the highest frequency present. The remarkable implication of this theorem is that, subject to the latter condition, no information is lost by reducing a continuous function to a set of discrete samples.

The concept of a bandlimited signal means that the spectrum of the signal has *compact support*; in other words, outside of a finite frequency band, the amplitude of the spectrum is identically zero. For sampling at a given rate τ, the highest frequency that can be recovered is called the *Nyquist* frequency, which is given by

$$f_N = \frac{1}{2\tau}. \tag{B.24}$$

If the continuous signal contains frequencies above f_N, frequency ambiguity arises, known as *aliasing*. In practice, continuous signals from a sensor such as a seismometer or a geophone are preconditioned, prior to digital sampling, by applying an analog electronic filter called an *anti-alias* filter. This ensures that no frequency ambiguity due to aliasing is present in the acquired digital data stream.

B.2 Convolution

In mathematical terms, a linear filter can be expressed as a *convolution*; in continuous form, application of the filter $b(t)$ to an input signal $a(t)$ is written as

$$a(t) * b(t) = \int_{-\infty}^{\infty} a(t') b(t - t') \, dt', \tag{B.25}$$

where the symbol * (no superscript) denotes convolution. This can be expressed in discrete form as

$$a * b = \sum_{k=1}^{N} a_k b_{N-k}. \tag{B.26}$$

The *convolution theorem* states that the Fourier transform of the convolution of two functions is given by the product of the two spectra:

$$a(t) * b(t) \longleftrightarrow A(f)B(f). \tag{B.27}$$

The concept of convolution is important to many aspects of seismology. For example, a seismogram is often represented as the convolution of a source time function with a Green's function (see Equation 3.29), which represents the *path effects* – that is, the filtering effect of the Earth along the path from the source to the receiver. This concept can be expressed as

$$u(t) = s(t) * p(t), \tag{B.28}$$

where $u(t)$ denotes the seismogram, $s(t)$ denotes the source time function and $p(t)$ captures the filtering effects along the path from the source to the receiver. There are several important properties of filters that are associated with physical systems. For example, such filters are *causal*, which simply means that

$$p(t) = 0, \ t < 0. \tag{B.29}$$

A more complete representation of a seismogram includes the instrument response, $r(t)$ (Wielandt, 2002):

$$u(t) = s(t) * p(t) * r(t). \tag{B.30}$$

Let us suppose that the discrete time samples for an instrument response function with N points are:

$$r_k = [r_0, r_1, r_2, \ldots, r_N]. \tag{B.31}$$

The instrument response can be represented as a *transfer function*, which may be written as

$$r(z) = r_0 + r_1 z + r_2 z^2 + \ldots + r_N z^N, \tag{B.32}$$

where z is a symbolic representation of a unit time-delay function, $e^{i2\pi f \tau}$. This polynomial representation of a digital filter is also knows as a *z-transform*. Taking this approach one step further, the polynomial expression above can be factored and rewritten as follows:

$$r(z) = (z - a)(z - b)(z - c) \ldots (z - n), \tag{B.33}$$

such that the entire filter is represented as a convolution of two-term wavelets (doublets). The parameters a, b, c, etc. are called the *zeros* of the filter; if the amplitude of each of the zeros is less than unity, the filter is called *minimum phase*. In effect, this property means that, for a given amplitude spectrum, the filter time series is maximally front-loaded. A more complete representation of a filter transfer function also incorporates *poles*. A filter that is represented by a transfer function with N zeros (a_k) and M poles (b_k) can be expressed as

$$R(z) = A \frac{(z - a_1)(z - a_2) \ldots (z - a_N)}{(z - p_1)(z - p_2) \ldots (z - p_M)}, \tag{B.34}$$

where A is a scalar factor that represents the *sensitivity* of the instrument. It is very common for a seismometer's instrument response to be expressed in this way.

The assumption that source wavelets are minimum phase is central to *deconvolution*, a key data processing element in exploration seismology. Deconvolution (with or without the minimum phase assumption) is also widely used in global seismology. The process of deconvolution can be expressed in the Fourier domain as

$$b^{-1}(t) * a(t) \longleftrightarrow \frac{A(f)}{B(f)}. \tag{B.35}$$

Since the spectrum $B(f)$ may contain very small (or even zero) spectral components, a simple heuristic stabilization approach is to approximate the deconvolution operator by adding a small constant value ϵ (known as *pre-whitening*):

$$\frac{A(f)}{B(f) + \epsilon}. \tag{B.36}$$

B.3 Cross-Correlation and Autocorrelation

Cross-correlation is a measure of the similarity of two signals. In continuous form, it is written as

$$\phi_{ab}(t) = a(t) \star b(t) \equiv \int_{-\infty}^{\infty} a(t') b(t' - t) dt'. \tag{B.37}$$

Comparison of this expression with the definition of convolution in Equation B.25 shows that correlation may be written as

$$\phi_{ab}(t) = a(t) * b(-t). \tag{B.38}$$

The *autocorrelation* of a signal is simply

$$\phi_{aa}(t) = a(t) \star a(t). \tag{B.39}$$

The autocorrelation function is always symmetrical with respect to $t = 0$. High-amplitude positive peaks at nonzero lag are indicative of repeating signal, such as repeating earthquakes. In the case of cross-correlation between two signals a and b in which one is a time-lagged version of the other, the time lag can be measured using the lag time of the positive peak in $\phi_{ab}(t)$.

B.4 Time–Frequency Analysis

The Fourier transform of a signal resolves the response at different frequencies, but offers no time resolution. Thus, if a change in frequency response occurs at a particular time, the Fourier transform provides no direct measure of when that change occurred. Many schemes have been developed that provide some measure of resolution in both time and frequency, as recently reviewed by Tary et al. (2014b). These schemes enable simultaneous time-frequency analysis of a signal, which can provide a very powerful approach. Here, only one type of time-frequency analysis is considered, known as the *short-time* Fourier transform (STFT). The output of this method is called a *spectragram*. The continuous STFT of a signal $u(t)$ can be written as:

$$U_S(\tau,f) = \int_{-\infty}^{\infty} u(t)w(t-\tau)e^{-i2\pi tf} dt, \qquad (B.40)$$

where $w(t)$ is a windowing function that is symmetrical around $t = 0$, often a Gaussian function. Tradeoffs between time and frequency resolution can be adjusted by varying the width of the Gaussian window function.

B.5 Complex Trace Analysis and Analytical Signal

Given a real signal $u(t)$, the *analytical signal* is a complex-valued function that is given by

$$\tilde{u}(t) \equiv u(t) + i\mathcal{H}(u(t)), \qquad (B.41)$$

where $\mathcal{H}$ denotes the *Hilbert* transform, which is given by

$$\mathcal{H}(u(t)) \equiv \frac{-1}{\pi t} * u(t). \qquad (B.42)$$

In practice the Hilbert transform can be accomplished by applying a $\pi/2$ constant phase shift to a signal, which renders an originally symmetrical signal to be asymmetric. The analytical signal can be used to construct a number of useful functions, including:

Amplitude envelope: A positive function that forms the envelope for a signal $u(t)$, given in terms of the analytic signal by

$$A(t) = |\tilde{u}(t)|. \qquad (B.43)$$

Instantaneous phase: By expressing the analytic signal cin the form $\tilde{u}(t) = A(t)e^{i\gamma(t)}$, the instantaneous phase is given by

$$\gamma(t) = \frac{\text{Re}(\tilde{u}(t))}{\text{Im}(\tilde{u}(t))}. \qquad (B.44)$$

Instantaneous frequency: This provides a measure of the frequency of a signal at a particular time and is given by $\dot{\gamma}(t)$.

Collectively, the use of the analytic signal to compute the amplitude envelope, instantaneous phase and frequency is called *complex trace analysis*.

Appendix C Data Formats

C.1 Microseismic Data Formats

Various data formats are used for digital storage of microseismic data. A legacy data format that is still widely used in the oil and gas industry is called SEG Y (Barry et al., 1975). This format was introduced in 1975 by the Committee on Technical Standards of the Society of Exploration Geophysicists (SEG) for exchange of *demultiplexed* data and was revised in 2002 (Norris and Faichney, 2002). The term demultiplexed means that data samples are stored as sequential time series, in contrast to *multiplexed* data in which data samples are stored sequentially by channel number. The designation "Y" was chosen as this format specification replaced the previous standard, known as the SEG "Ex" format.

The SEG Y format was developed for use with magnetic tapes and contains three distinct types of data blocks. The first two blocks at the start of a standard SEG Y file represent the reel-identification header. This header starts with a block contains a 3200-byte free-form textual description of the data. This typically uses American Standard Code for Information Interchange (ASCII) format, but it was originally specified to use the Extended Binary Coded Decimal Interchange Code (EBCDIC) format developed by IBM. This was defined to contain 40 images of 80-character punch cards, where each card image should contain the character "C" in the first column. The second 400-byte binary block begins at byte 3201 of the reel-identification header and contains parameters that affect the whole file and that are crucial for processing the data. Some of the fields are considered mandatory, others are optional, while others are left undefined to allow for flexibility. Examples of defined reel-identification header binary fields include:

- Sample interval in microseconds (bytes 3217–3218);
- Number of samples per trace (bytes 3221–3222);
- Data sample format code (bytes 3225–3226);
- Number of 3200-byte extended textual file header records following the binary header (bytes 3505–3506).

These parameters are stored as integer values and the first three example parameters listed above are considered mandatory.

The data sample format code includes various options for floating point and integer format, such as Institute of Electrical and Electronics Engineers (IEEE) floating-point (format code = 5). The binary values may be stored in *big-endian* format, in which case the most significant byte is stored first, or *little-endian* format which has the opposite byte order.[1] Which of the two binary formats is native to a specific processing environment depends on the operation system.

According to [rev 1] of the SEG Y format (Norris and Faichney, 2002), if bytes 3505–3506 are nonzero then a series of extended textual headers is presented. These are organized into stanzas and may be used, for example, to describe aspects of a dataset that are relevant for microseismic acquisition and processing (Norris and Faichney, 2002).

The reel-identification header is followed by data traces. Each data trace begins with a 240-byte binary trace header. Examples of defined trace-header fields include:

- Receiver elevation (bytes 41–44);
- Datum elevation at receiver (bytes 53–56);
- Receiver coordinate – *x* (bytes 73–76);
- Receiver coordinate – *y* (bytes 85–88).

Each trace header is followed by N data samples, where each sample is stored in the format specified in the reel-identification header and N is the number of samples per trace. The SEG Y format is common for exchange of microseismic data, but is not well suited to continuous data recordings due to trace-length limitations. For standard SEG Y format, the maximum number of samples in a single trace is $2^{15} = 32{,}768$. This limitation is imposed by the mandatory header field which uses a 2-byte integer to store the number of samples. The specification of this header field can be traced by to the original SEG Y format in the 1970s that did not envision the use of this format for continuous data. As a consequence of this limitation, microseismic data delivered in SEG Y format contains a multitude of individual SEG Y files, each corresponding to a short time window. This large set of files require concatenation prior to processing the data.

Another format that is commonly used for exchange of microseismic data is the SEG-2 format (Pullan, 1990). This format was developed by the SEG's Engineering and Groundwater Geophysics Committee and is recommended for raw or processed data in a small computer environment. It is similar to SEG Y inasmuch as it uses a set of ordered blocks, a File Descriptor Block followed by one or more Trace Descriptor Blocks and one or more Data Blocks. Greater flexibility is achieved with the use of pointers, which indicate the locations of blocks with respect to the beginning of a file. It is thus a free-format standard that allows for flexible storage of ancillary data. A variant of the SEG-2 format, called SEG2-M format, was prepared by a special SEG committee in 2010. This format is based on SEG-2 format, with recommendations for modifications and/or deletion of some of the SEG-2 tags.

Another legacy format, known as SEG-D, is used for multiplexed data that are stored as a byte stream. It is similarly organized into general headers, channel set headers, etc. (Levin et al., 2007). Prior to processing, files that are stored in SEG-D format need to be demultiplexed (i.e. transposed into a time series).

---

[1] The etymology of this designation comes from Gulliver's travels; in Lilliput, soft-boiled eggs are cracked open at the *little end*, whereas in Blefuscu they are opened at the *big end*.

C.2 Induced Seismicity Data Formats

Induced seismicity data are commonly distributed using standardize earthquake data formats. These include SEED and SAC formats.

The Standard for the Exchange of Earthquake Data (SEED) format (Ahern and Dost, 2012) is an international standard format used for exchange of digital seismological time series data and related metadata. It is designed for digital data measured at one point in space with a uniform sampling rate, but it can be used to store data from many different stations in a single file. It was first officially released as version 2.0 in 1980 and uses a self-defining archive format with compression. Full SEED volumes include metadata such as instrument response files, whereas mini-SEED contains only the time-series data. There are widely used tools such as *rdseed* that can read and/or convert SEED files into other formats.

The Seismic Analysis Code (SAC) format is a time-series data format that is designed for the general-purpose interactive software of the same name (Helffrich et al., 2013). SAC is widely used by earthquake seismologists, and makes use of its own unique data format specifications. The SAC software package was originally developed by Lawrence Livermore National Laboratory and is currently being developed by the Incorporated Research Institutes in Seismology (IRIS).

References

Abercrombie, R. E. 1995. Earthquake source scaling relationships from 1 to 5 ML using seismograms recorded at 2.5-km depth. *Journal of Geophysical Research: Solid Earth*, **100**(B12), 24015–24036.

AbuAisha, M., Eaton, D. W., Priest, J., and Wong, R. 2017. Hydro-mechanically coupled FDEM framework to investigate near-wellbore hydraulic fracturing in homogeneous and fractured rock formations. *Journal of Petroleum Science and Engineering*, **154**, 100–113.

Adachi, J., Siebrits, E., Peirce, A., and Desroches, J. 2007. Computer simulation of hydraulic fractures. *International Journal of Rock Mechanics and Mining Sciences*, **44**(5), 739–757.

Ahern, T., and Dost, B. 2012. *SEED Reference Manual: Standard for the Exchange of Earthquake Data*. Tech. rept. International Federation of Digital Seismograph Networks.

Ajaya, B., Aso, I. I., Terry, I. J., Walker, K., Wutherirch, K., Caplan, J., Gerdom, D., Clark, B. D., Ganguly, U., Li, X., Xu, Y., Yang, H., Liu, H., Luo, Y., and Waters, G. 2013. Stimulation design for unconventional resources. *Oilfield Review*, **25**(2), 34–46.

Akaike, H. 1998. Information theory and an extension of the maximum likelihood principle. Pages 199–213 of: *Selected Papers of Hirotugu Akaike*. Springer.

Aki, K. 1965. Maximum likelihood estimate of b in the formula $\log N = a - bM$ and its confidence limits. *Bulletin of the Earthquake Research Institute*, **43**, 237–239.

Aki, K., and Richards, P. G. 2002. *Quantitative Seismology*. Vol. I. University Science Books.

Akram, J. 2014. *Downhole Microseismic Monitoring: Processing, Algorithms and Error Analysis*. Ph.D. thesis, University of Calgary.

Akram, J., and Eaton, D. W. 2016a. Refinement of arrival-time picks using a cross-correlation based workflow. *Journal of Applied Geophysics*, **135**, 55–66.

Akram, J., and Eaton, D. W. 2016b. A review and appraisal of arrival-time picking methods for downhole microseismic data. *Geophysics*, **81**(2), KS71–KS91.

Albright, J. N., and Pearson, C. F. 1982. Acoustic emissions as a tool for hydraulic fracture location: Experience at the Fenton Hill Hot Dry Rock site. *Society of Petroleum Engineers Journal*, **22**(04), 523–530.

Alexandrakis, C., Calò, M., Bouchaala, F., and Vavryčuk, V. 2014. Velocity structure and the role of fluids in the West Bohemia Seismic Zone. *Solid Earth*, **5**(2), 863.

Allen, D. T., Torres, V. M., Thomas, J., Sullivan, D. W., Harrison, M., Hendler, A., Herndon, S. C., Kolb, C. E., Fraser, M. P., Hill, A. D., Lamb, B. K., Miskimins, J., Sawyer,

R. F., and Seinfeld, J. H. 2013. Measurements of methane emissions at natural gas production sites in the United States. *Proceedings of the National Academy of Sciences*, **110**(44), 17768–17773.

Allmann, B. P., and P. M. Shearer 2009. Global variations of stress drop for moderate to large earthquakes, *J. Geophys. Res.*, 114, B01310, doi:10.1029/2008JB005821.

Amadei, B., and Stephansson, O. 1997. *Rock Stress and its Measurement*. Chapman & Hall.

Anderson, E. M. 1951. *The Dynamics of Faulting and Dyke Formation with Applications to Britain*. Hafner Pub. Co.

Anderson, T. L. 2005. *Fracture Mechanics: Fundamentals and Applications*. 3rd edn. CRC Press.

API. 2014. *API Recommended Practice 13C*. Tech. rept. American Petroleum Institute.

Artman, B. 2006. Imaging passive seismic data. *Geophysics*, **71**(4), SI177–SI187.

Artman, B., Podladtchikov, I., and Witten, B. 2010. Source location using time-reverse imaging. *Geophysical Prospecting*, **58**(5), 861–873.

Atkinson, G. M. 2015. Ground-motion prediction equation for small-to-moderate events at short hypocentral distances, with application to induced-seismicity hazards. *Bulletin of the Seismological Society of America*, **105**(2A), 981–992.

Atkinson, G. M. 2017. Strategies to prevent damage to critical infrastructure due to induced seismicity. *FACETS*, **2**(1), 374–394.

Atkinson, G. M., and Assatourians, K. 2017. Are ground-motion models derived from natural events applicable to the estimation of expected motions for induced earthquakes? *Seismological Research Letters*, **88**(2A), 430–441.

Atkinson, G. M., Kaka, S. I., Eaton, D., Bent, A., Peci, V., and Halchuk, S. 2008. A very close look at a moderate earthquake near Sudbury, Ontario. *Seismological Research Letters*, **79**(1), 119–131.

Atkinson, G. M., Greig, D. W., and Yenier, E. 2014. Estimation of moment magnitude (M) for small events (M < 4) on local networks. *Seismological Research Letters*, **85**(5), 1116–1124.

Atkinson, G. M., Ghofrani, H., and Assatourians, K. 2015. Impact of induced seismicity on the evaluation of seismic hazard: some preliminary considerations. *Seismological Research Letters*, **86**(3), 1009–1021.

Atkinson, G. M., Eaton, D. W., Ghofrani, H., Walker, D., Cheadle, B., Schultz, R., Shcherbakov, R., Tiampo, K., Gu, Y. J., Harrington, R. M., Liu, Y., Van der Baan, M., and Kao, H. 2016. Hydraulic fracturing and seismicity in the Western Canada Sedimentary Basin. *Seismological Research Letters*, **87**(3), 631–647.

Avseth, P., Mukerji, T., and Mavko, G. 2010. *Quantitative Seismic Interpretation: Applying Rock Physics Tools to Reduce Interpretation Risk*. Cambridge University Press.

Aydin, A. 2000. Fractures, faults, and hydrocarbon entrapment, migration and flow. *Marine and Petroleum Geology*, **17**(7), 797–814.

Babcock, E. A. 1978. Measurement of subsurface fractures from dipmeter logs. *AAPG Bulletin*, **62**(7), 1111–1126.

Bachmann, C. E., Wiemer, S., Goertz-Allmann, B. P., and Woessner, J. 2012. Influence of pore-pressure on the event-size distribution of induced earthquakes. *Geophysical Research Letters*, **39**(9).

Backus, G. E. 1962. Long-wave elastic anisotropy produced by horizontal layering. *Journal of Geophysical Research*, **67**(11), 4427–4440.

Bagaini, C. 2005. Performance of time-delay estimators. *Geophysics*, **70**(4), V109–V120.

Baig, A., and Urbancic, T. 2010. Microseismic moment tensors: a path to understanding frac growth. *The Leading Edge*, **29**(3), 320–324.

Baisch, S., Vörös, R., Weidler, R., and Wyborn, D. 2009. Investigation of fault mechanisms during geothermal reservoir stimulation experiments in the Cooper Basin, Australia. *Bulletin of the Seismological Society of America*, **99**(1), 148–158.

Bak, P., and Tang, C. 1989. Earthquakes as a self-organized critical phenomenon. *Journal of Geophysical Research: Solid Earth*, **94**(B11), 15635–15637.

Bak, P., Tang, C., and Wiesenfeld, K. 1988. Self-organized criticality. *Physical Review A*, **38**(1), 364.

Baker, J. W. 2008. *Probabilistic Seismic Hazard Analysis*. Jack W. Baker.

Bakun, W. H., and Joyner, W. B. 1984. The ML scale in central California. *Bulletin of the Seismological Society of America*, **74**(5), 1827–1843.

Bao, X., and Eaton, D. W. 2016. Fault activation by hydraulic fracturing in western Canada. *Science*, **354**(6318), 1406–1409.

Baranova, V., Mustaqeem, A., and Bell, S. 1999. A model for induced seismicity caused by hydrocarbon production in the Western Canada Sedimentary Basin. *Canadian Journal of Earth Sciences*, **36**(1), 47–64.

Barati, R. and Liang, J.-T. 2014. A review of fracturing fluid systems used for hydraulic fracturing of oil and gas wells. *J. Appl. Polym. Sci.*, 131, 40735, doi: 10.1002/app.40735.

Barber, C. B., Dobkin, D. P., and Huhdanpaa, H. 1996. The quickhull algorithm for convex hulls. *ACM Transactions on Mathematical Software (TOMS)*, **22**(4), 469–483.

Barenblatt, G. I. 1962. The mathematical theory of equilibrium cracks in brittle fracture. *Advances in Applied Mechanics*, **7**, 55–129.

Barrett, S. A., and Beroza, G. C. 2014. An empirical approach to subspace detection. *Seismological Research Letters*, **85**(3), 594–600.

Barry, K. M., Cavers, D. A., and Kneale, C. W. 1975. Recommended standards for digital tape formats. *Geophysics*, **40**(2), 344–352.

Barth, A., Reinecker, J., and Heidbach, O. 2016. Guidelines for the analysis of earthquake focal mechanism solutions. Pages 15–26 of: *WSM Scientific Technical Report*, vol. WSM STR 16-01. World Stress Map Project.

Barton, C. A., Zoback, M. D., and Burns, K. L. 1988. In-situ stress orientation and magnitude at the Fenton Geothermal Site, New Mexico, determined from wellbore breakouts. *Geophysical Research Letters*, **15**(5), 467–470.

BCOGC. 2012. *Investigation of Observed Seismicity in the Horn River Basin*. Tech. rept. BC Oil and Gas Commission.

BCOGC. 2014. *Investigation of Observed Seismicity in the Montney Trend*. Tech. rept. BC Oil and Gas Commission.

Beeler, N. M. 2001. Stress drop with constant, scale independent seismic efficiency and overshoot. *Geophysical Research Letters*, **28**(17), 3353–3356.

Belayouni, N., Gesret, A., Daniel, G., and Noble, M. 2015. Microseismic event location using the first and reflected arrivals. *Geophysics*, **80**(6), WC133–WC143.

Bell, J. S., and Babcock, E. A. 1986. The stress regime of the Western Canadian Basin and implications for hydrocarbon production. *Bulletin of Canadian Petroleum Geology*, **34**(3), 364–378.

Bell, J. S., and Bachu, S. 2003. In situ stress magnitude and orientation estimates for Cretaceous coal-bearing strata beneath the plains area of central and southern Alberta. *Bulletin of Canadian Petroleum Geology*, **51**(1), 1–28.

Bell, J. S., and Gough, D. I. 1979. Northeast–southwest compressive stress in Alberta evidence from oil wells. *Earth and Planetary Science Letters*, **45**(2), 475–482.

Bell, M., Kraaijevanger, H., and Maisons, C. 2000. Integrated downhole monitoring of hydraulically fractured production wells. In: *SPE European Petroleum Conference*. Society of Petroleum Engineers.

Ben-Zion, Y., and Sammis, C. G. 2003. Characterization of fault zones. Pages 677–715 of: *Seismic Motion, Lithospheric Structures, Earthquake and Volcanic Sources: The Keiiti Aki Volume*, 1 edn. Pageoph Topical Volumes. Birkhauser Basel.

Bendat, J. S., and Piersol, A. G. 2011. *Random Data: Analysis and Measurement Procedures*. Vol. 729. John Wiley & Sons.

Benz, H. M., McMahon, N. D., Aster, R. C., McNamara, D. E., and Harris, D. B. 2015. Hundreds of earthquakes per day: the 2014 Guthrie, Oklahoma, earthquake sequence. *Seismological Research Letters*, **86**(5), 1318–1325.

Beresnev, I. A. 2001. What we can and cannot learn about earthquake sources from the spectra of seismic waves. *Bulletin of the Seismological Society of America*, **91**(2), 397–400.

Beresnev, I. A. 2003. Uncertainties in finite-fault slip inversions: to what extent to believe? (a critical review). *Bulletin of the Seismological Society of America*, **93**(6), 2445–2458.

Bethmann, F., Deichmann, N., and Mai, P. M. 2011. Scaling relations of local magnitude versus moment magnitude for sequences of similar earthquakes in Switzerland. *Bulletin of the Seismological Society of America*, **101**(2), 515–534.

Bilek, S. L., and Lay, T. 1999. Rigidity variations with depth along interplate megathrust faults in subduction zones. *Nature*, **400**(6743), 443–446.

Biot, M. A. 1962a. General theory of 3-dimensional consolidation. *Journal of Applied Physics*, **12**, 155–164.

Biot, M. A. 1962b. Mechanics of deformation and acoustic propagation in porous media. *Journal of Applied Physics*, **33**(4), 1482–1498.

Biryukov, A. 2016. *Design Optimization for a Local Seismograph Network: Application to the Liard Basin*. Tech. rept. University of Calgary.

Boatwright, J. 1980. A spectral theory for circular seismic sources; simple estimates of source dimension, dynamic stress drop, and radiated seismic energy. *Bulletin of the Seismological Society of America*, **70**(1), 1–27.

Boese, C. M., Wotherspoon, L., Alvarez, M., and Malin, P. 2015. Analysis of anthropogenic and natural noise from multilevel borehole seismometers in an urban environment, Auckland, New Zealand. *Bulletin of the Seismological Society of America*, **105**(1), 285–299.

Bohnhoff, M., Dresen, G., Wellsworth, W. L., and Ito, H. 2010. Passive seismic monitoring of natural and induced earthquakes: Case studies, future directions and socio-economic relevance. Pages 261–285 of: Cloetingh, S., and Jegendank, J. (eds.), *New Frontiers in Integrated Solid Earth Sciences*. International Year of Planet Earth. Netherlands: Springer.

Bommer, J. J., Oates, S., Cepeda, J. M., Lindholm, C., Bird, J., Torres, R., Marroquín, G., and Rivas, J. 2006. Control of hazard due to seismicity induced by a hot fractured rock geothermal project. *Engineering Geology*, **83**(4), 287–306.

Bommer, J. J., Crowley, H., and Pinho, R. 2015. A risk-mitigation approach to the management of induced seismicity. *Journal of Seismology*, **19**(2), 623–646.

Boness, N. L., and Zoback, M.D. 2006. A multiscale study of the mechanisms controlling shear velocity anisotropy in the San Andreas Fault Observatory at Depth. *Geophysics*, **71**(5), F131–F146.

Bonnet, E., Bour, O., Odling, N. E., Davy, P., Main, I., Cowie, P., and Berkowitz, B. 2001. Scaling of fracture systems in geological media. *Reviews of Geophysics*, **39**(3), 347–383.

Boore, D. M., and Boatwright, J. 1984. Average body-wave radiation coefficients. *Bulletin of the Seismological Society of America*, **74**(5), 1615–1621.

Boroumand, N., and Eaton, D. W. 2012. Comparing energy calculations-hydraulic fracturing and microseismic monitoring. In: *74th EAGE Conference and Exhibition incorporating EUROPEC 2012*.

Boroumand, N., and Eaton, D. W. 2015. Energy-based hydraulic fracture numerical simulation: Parameter selection and model validation using microseismicity. *Geophysics*, **80**(5), W33–W44.

Bott, M. H. P. 1959. The mechanics of oblique slip faulting. *Geological Magazine*, **96**(02), 109–117.

Boyd, O. S. 2006. An efficient Matlab script to calculate heterogeneous anisotropically elastic wave propagation in three dimensions. *Computers & Geosciences*, **32**(2), 259–264.

Bracewell, R. N. 1986. *The Fourier Transform and its Applications*. New York: McGraw-Hill.

Breede, K., Dzebisashvili, K., Liu, X., and Falcone, G. 2013. A systematic review of enhanced (or engineered) geothermal systems: past, present and future. *Geothermal Energy*, **1**(1), 4.

Brown, E. T. 1970. Strength of models of rock with intermittent joints. *Journal of Soil Mechanics & Foundations Div*, **96**(SM6), 1935–1949.

Brown, J. E., Thrasher, R. S., and Behrmann, L. A. 2000. Fracturing Operations. Pages 11-1 – 11-33 of: Economides, M. J., and Nolte, K. G. (eds.), *Reservoir Stimulation*, 3rd edn. John Wiley & Sons.

Brown, S. R., and Bruhn, R. L. 1998. Fluid permeability of deformable fracture networks. *Journal of Geophysical Research: Solid Earth*, **103**(B2), 2489–2500.

Brudzinski, M. R., Skoumal, R., and Currie, B. S. 2016. Proximity of wastewater disposal and hydraulic fracturing to crystalline basement affects the likelihood of induced seismicity in the Central and Eastern United States. In: *AGU 2016 Fall Meeting*. American Geophysical Union.

Brune, J. N. 1970. Tectonic stress and the spectra of seismic shear waves from earthquakes. *Journal of Geophysical Research*, **75**(26), 4997–5009.

Brune, J. N. 1971. Correction. *Journal of Geophysical Research*, **76**, 5002.

Building Seismic Safety Council. 2003. *NEHRP Recommended Provisions for Seismic Regulations for New Buildings and Other Structures*. Tech. rept. FEMA-450. Federal Emergency Management Agency.

Burridge, R., and Knopoff, L. 1964. Body force equivalents for seismic dislocations. *Bulletin of the Seismological Society of America*, **54**(6A), 1875–1888.

Byerlee, J. 1978. Friction of rocks. *Pure and Applied Geophysics*, **116**(4), 615–626.

Caffagni, E., Eaton, D., Van der Baan, M., and Jones, J. P. 2014. Regional seismicity: a potential pitfall for identification of long-period long-duration events. *Geophysics*, **80**(1), A1–A5.

Caffagni, E., Eaton, D. W., Jones, J. P., and Van der Baan, M. 2016. Detection and analysis of microseismic events using a Matched Filtering Algorithm (MFA). *Geophysical Journal International*, **206**(1), 644–658.

Cai, M., Kaiser, P. K., and Martin, C. D. 1998. A tensile model for the interpretation of microseismic events near underground openings. *Seismicity Caused by Mines, Fluid Injections, Reservoirs, and Oil Extraction*, 67–92.

CAPP. 2017. *CAPP Hydraulic Fracturing Operating Practice: Anomalous induced seismicity: assessment, monitoring, mitigation and response*. www.capp.ca/~/media/capp/customer-portal/publications/217532.pdf Accessed: 2017/07/27.

Carter, J. A., and Frazer, L. N. 1984. Accommodating lateral velocity changes in Kirchhoff migration by means of Fermat's principle. *Geophysics*, **49**(1), 46–53.

Carter, J. A., Barstow, N., Pomeroy, P. W., Chael, E. P., and Leahy, P. J. 1991. High-frequency seismic noise as a function of depth. *Bulletin of the Seismological Society of America*, **81**(4), 1101–1114.

Cary, P. W., and Eaton, D. W. 1993. A simple method for resolving large converted-wave (P-SV) statics. *Geophysics*, **58**(3), 429–433.

Castellanos, F., and Van der Baan, M. 2013. Microseismic event locations using the double-difference algorithm. *CSEG Recorder*, **38**(3), 26–37.

CCA. 2014. *Environmental Impacts of Shale Gas Extraction in Canada*. Tech. rept. Council of Canadian Academies.

Cesca, S., Rohr, A., and Dahm, T. 2013. Discrimination of induced seismicity by full moment tensor inversion and decomposition. *Journal of Seismology*, **17**(1), 147–163.

Chaisri, S., and Krebes, E. S. 2000. Exact and approximate formulas for P-SV reflection and transmission coefficients for a nonwelded contact interface. *Journal of Geophysical Research: Solid Earth*, **105**(B12), 28045–28054.

Chambers, K., Kendall, J-M., Brandsberg-Dahl, S., and Rueda, J. 2010. Testing the ability of surface arrays to monitor microseismic activity. *Geophysical Prospecting*, **58**(5), 821–830.

Chambers, K., Dando, B.D.E., Jones, G.A., Velasco, R., and Wilson, S. 2014. Moment tensor migration imaging. *Geophysical Prospecting*, **62**(4), 879–896.

Chapman, M. 2003. Frequency-dependent anisotropy due to meso-scale fractures in the presence of equant porosity. *Geophysical Prospecting*, **51**(5), 369–379.

Charléty, J., Cuenot, N., Dorbath, L., Dorbath, C., Haessler, H., and Frogneux, M. 2007. Large earthquakes during hydraulic stimulations at the geothermal site of Soultz-sous-Forêts. *International Journal of Rock Mechanics and Mining Sciences*, **44**(8), 1091–1105.

Chen, W., Ni, S., Kanamori, H., Wei, S., Jia, Z., and Zhu, L. 2015. CAPjoint, a computer software package for joint inversion of moderate earthquake source parameters with local and teleseismic waveforms. *Seismological Research Letters*, **86**(2A), 432–441.

Chester, F. M., Evans, J. P., and Biegel, R. L. 1993. Internal structure and weakening mechanisms of the San Andreas fault. *Journal of Geophysical Research: Solid Earth*, **98**(B1), 771–786.

Cieslik, K., and Artman, B. 2016. Signal to noise analysis of densely sampled microseismic data. In: *2016 Convention, CSPG CSEG CWLS, Expanded Abstracts*.

Cipolla, C., and Wallace, J. 2014. Stimulated reservoir volume: a misapplied concept? In: *SPE Hydraulic Fracturing Technology Conference*. Society of Petroleum Engineers.

Cipolla, C. L., Maxwell, S. C., Mack, M. G., and Downie, R. C. 2011. A practical guide to interpreting microseismic measurements. In: *SPE North American Unconventional Gas Conference and Exhibition*. The Woodlands, Texas: Society of Petroleum Engineers.

Cipolla, C. L., Maxwell, S. C., and Mack, M. G. 2012. Engineering guide to the application of microseismic interpretations. In: *SPE Hydraulic Fracturing Technology Conference*. Society of Petroleum Engineers.

Claerbout, J. F. 1985. *Imaging the Earth's Interior*. Oxford: Blackwell Scientific Publications.

Clarke, H., Eisner, L., Styles, P., and Turner, P. 2014. Felt seismicity associated with shale gas hydraulic fracturing: the first documented example in Europe. *Geophysical Research Letters*, **41**(23), 8308–8314.

Clarkson, C. R., and Williams-Kovacs, J. D. 2013a. Modeling two-phase flowback of multifractured horizontal wells completed in shale. *SPE Journal*, **18**(04), 795–812.

Clarkson, C. R., and Williams-Kovacs, J. D. 2013b. A new method for modeling multi-phase flowback of multi-fractured horizontal tight oil wells to determine hydraulic fracture properties. In: *SPE Annual Technical Conference and Exhibition*. New Orleans, Louisiana: Society of Petroleum Engineers.

Clarkson, C. R., Qanbari, F., and Williams-Kovacs, J. D. 2014. Innovative use of rate-transient analysis methods to obtain hydraulic-fracture properties for low-permeability reservoirs exhibiting multiphase flow. *The Leading Edge*, **33**(10), 1108–1122.

Close, D., Cho, D., Horn, F., and Edmundson, H. 2009. The sound of sonic: a historical perspective and introduction to acoustic logging. *CSEG Recorder*, **34**(5), 34–43.

Constien, V. G., Hawkins, G. W., Prud'homme, R. K., and Navarret, R. 2000. Performance of Fracturing Materials. Pages 8–1 – 8–26 of: Economides, M. J., and Nolte, K. G. (eds.), *Reservoir Stimulation*. John Wiley & Sons.

References

Cooley, J. W., and Tukey, J. W. 1965. An algorithm for the machine calculation of complex Fourier series. *Mathematics of Computation*, **19**(90), 297–301.

Cornell, C. A. 1968. Engineering seismic risk analysis. *Bulletin of the Seismological Society of America*, **58**(5), 1583–1606.

Cornet, F. H., Bérard, T., and Bourouis, S. 2007. How close to failure is a granite rock mass at a 5km depth? *International Journal of Rock Mechanics and Mining Sciences*, **44**(1), 47–66.

Courtney, E. C. 2000. *The Mechanical Behavior of Materials.* Waveland Press.

Cox, S. F. 2016. Injection-driven swarm seismicity and permeability enhancement: implications for the dynamics of hydrothermal ore systems in high fluid-flux, overpressured faulting regimes. *Economic Geology*, **111**(3), 559–587.

Crampin, S., Chesnokov, E. M., and Hipkin, R. G. 1984. Seismic anisotropy – the state of the art: II. *Geophysical Journal International*, **76**(1), 1–16.

Cummings, R. G., and Morris, G. E. 1979. *Economic Modelling of Electricity Production from Hot Dry Rock Geothermal Reservoirs: Methodology and Analyses.* Tech. rept. EPRI-EA-630. United States Department of Energy.

Dahm, T., Becker, D., Bischoff, M., Cesca, S., Dost, B., Fritschen, R., Hainzl, S., Klose, C. D., Kühn, D., Lasocki, S., Meier, T., Ohrnberger, M., Rivalta, E., Wegler, U., and Husen, S. 2013. Recommendation for the discrimination of human-related and natural seismicity. *Journal of Seismology*, **17**(1), 197–202.

Daley, T. M., Freifeld, B. M., Ajo-Franklin, J., Dou, S., Pevzner, R., Shulakova, V., Kashikar, S., Miller, D. E., Goetz, J., Henninges, J., and Lueth, S. 2013. Field testing of fiber-optic distributed acoustic sensing (DAS) for subsurface seismic monitoring. *The Leading Edge*, **32**(6), 699–706.

Daneshy, A. A. 1978. Numerical solution of sand transport in hydraulic fracturing. *Journal of Petroleum Technology*, **30**(1), 132–140.

Daniels, J. L., Waters, G. A., Le Calvez, J. H., Bentley, D., and Lassek, J. T. 2007. Contacting more of the Barnett Shale through an integration of real-time microseismic monitoring, petrophysics, and hydraulic fracture design. In: *SPE Annual Technical Conference and Exhibition.* Society of Petroleum Engineers.

Dankbaar, J. W. M. 1985. Separation of P- and S-waves. *Geophysical Prospecting*, **33**(7), 970–986.

Darbyshire, F. A., Eaton, D. W., and Bastow, I. D. 2013. Seismic imaging of the lithosphere beneath Hudson Bay: episodic growth of the Laurentian mantle keel. *Earth and Planetary Science Letters*, **373**, 179–193.

Darold, A. P., and Holland, A. A. 2015. *Preliminary Oklahoma optimal fault orientations.* Tech. rept. Open File Report OF4. Oklahoma Geological Survey.

Das, I., and Zoback, M. D. 2013. Long-period, long-duration seismic events during hydraulic stimulation of shale and tight-gas reservoirs – Part 1: Waveform characteristics. *Geophysics*, **78**(6), KS97–KS108.

Davies, D., Kelly, E. J., and Filson, J. R. 1971. Vespa process for analysis of seismic signals. *Nature*, **232**, 8–13.

Davies, R., Foulger, G., Bindley, A., and Styles, P. 2013. Induced seismicity and hydraulic fracturing for the recovery of hydrocarbons. *Marine and Petroleum Geology*, **45**, 171–185.

Davis, S. D., and Frohlich, C. 1993. Did (or will) fluid injection cause earthquakes? – criteria for a rational assessment. *Seismological Research Letters*, **64**(3–4), 207–224.

De Meersman, K., Van der Baan, M., and Kendall, J.-M. 2006. Signal extraction and automated polarization analysis of multicomponent array data. *Bulletin of the Seismological Society of America*, **96**(6), 2415–2430.

De Meersman, K., Kendall, J.-M., and Van der Baan, M. 2009. The 1998 Valhall microseismic data set: an integrated study of relocated sources, seismic multiplets, and S-wave splitting. *Geophysics*, **74**(5), B183–B195.

Deichmann, N., and Giardini, D. 2009. Earthquakes induced by the stimulation of an enhanced geothermal system below Basel (Switzerland). *Seismological Research Letters*, **80**(5), 784–798.

Deng, K., Liu, Y., and Harrington, R. M. 2016. Poroelastic stress triggering of the December 2013 Crooked Lake, Alberta, induced seismicity sequence. *Geophysical Research Letters*, **43**(16), 8482–8491.

Denlinger, R. P., and Bufe, C. G. 1982. Reservoir conditions related to induced seismicity at the Geysers steam reservoir, northern California. *Bulletin of the Seismological Society of America*, **72**(4), 1317–1327.

Der Kiureghian, A., and Ditlevsen, O. 2009. Aleatory or epistemic? Does it matter? *Structural Safety*, **31**(2), 105–112.

Detring, J., and Williams-Stroud, S. C. 2012. Using microseismicity to understand subsurface fracture systems and increase the effectiveness of completions: Eagle Ford formation, Texas. In: *SPE Canadian Unconventional Resources Conference*. Society of Petroleum Engineers.

Dettmer, J., Benavente, R., Cummins, P. R., and Sambridge, M. 2014. Trans-dimensional finite-fault inversion. *Geophysical Journal International*, **199**(2), 735–751.

Di Bona, M. 2016. A local magnitude scale for crustal earthquakes in Italy. *Bulletin of the Seismological Society of America*, **106**(1), 242–258.

Dieterich, J. 1994. A constitutive law for rate of earthquake production and its application to earthquake clustering. *Journal of Geophysical Research: Solid Earth*, **99**(B2), 2601–2618.

Dieterich, J. H. 1972. Time-dependent friction in rocks. *Journal of Geophysical Research*, **77**(20), 3690–3697.

Dieterich, J. H. 1978. Time-dependent friction and the mechanics of stick-slip. *Pure and Applied Geophysics*, **116**(4-5), 790–806.

Dinske, C., and Shapiro, S. A. 2013. Seismotectonic state of reservoirs inferred from magnitude distributions of fluid-induced seismicity. *Journal of Seismology*, **17**(1), 13–25.

Dohmen, T., Zhang, J., Barker, L., and Blangy, J. P. 2017. Microseismic magnitudes and b-values for delineating hydraulic fracturing and depletion. *SPE Journal*, **SPE 186096**.

References

Dorbath, L., Cuenot, N., Genter, A., and Frogneux, M. 2009. Seismic response of the fractured and faulted granite of Soultz-sous-Forêts (France) to 5 km deep massive water injections. *Geophysical Journal International*, **177**(2), 653–675.

Duhault, J. L. J. 2012. Cardium microseismic west central Alberta: a case history. *CSEG Recorder*, **37**(8), 48–57.

Duncan, P. M. 2005. Is there a future for passive seismic? *First Break*, **23**(6), 111–115.

Duncan, P. M., and Eisner, L. 2010. Reservoir characterization using surface microseismic monitoring. *Geophysics*, **75**(5), 139–146.

Dusseault, M., and McLennan, J. 2011. Massive multistage hydraulic fracturing: where are we. In: *45th US Rock Mechanics/Geomechanics Symposium, San Francisco*.

Dutta, N. C., and Odé, H. 1979. Attenuation and dispersion of compressional waves in fluid-filled porous rocks with partial gas saturation (White model) – Part II: Results. *Geophysics*, **44**(11), 1789–1805.

Dziewonski, A. M., Chou, T.-A., and Woodhouse, J. H. 1981. Determination of earthquake source parameters from waveform data for studies of global and regional seismicity. *Journal of Geophysical Research: Solid Earth*, **86**(B4), 2825–2852.

Earle, P. S., and Shearer, P. M. 1994. Characterization of global seismograms using an automatic-picking algorithm. *Bulletin of the Seismological Society of America*, **84**(2), 366–376.

Eaton, D. W. 1989. The free surface effect: implications for amplitude-versus-offset inversion. *Canadian Journal of Exploration Geophysics*, **25**, 97–103.

Eaton, D. W. 2014. Alberta Telemetered Seismograph Network (ATSN): Real-time monitoring of seismicity in northern Alberta. *CSEG Recorder*, **39**(9), 30–33.

Eaton, D. W. 2016. Injection-induced seismicity: an academic perspective. *Canadian Energy Technology and Innovation Journal*, **2**(4), 34–41.

Eaton, D. W., and Caffagni, E. 2015. Enhanced downhole microseismic processing using matched filtering analysis (MFA). *First Break*, **33**(7), 49–55.

Eaton, D. W., and Forouhideh, F. 2011. Solid angles and the impact of receiver-array geometry on microseismic moment-tensor inversion. *Geophysics*, **76**(6), WC77–WC85.

Eaton, D. W., and Maghsoudi, S. 2015. 2b... or not 2b? Interpreting magnitude distributions from microseismic catalogs. *First Break*, **33**(10), 79–86.

Eaton, D. W., and Mahani, A. B. 2015. Focal mechanisms of some inferred induced earthquakes in Alberta, Canada. *Seismological Research Letters*, **86**(4), 1078–1085.

Eaton, D. W., Adams, J., Asudeh, I., Atkinson, G. M., Bostock, J. F., Cassidy, J. F., Ferguson, I. J., Samson, C., Snyder, D. B., Timapo, K. F., and Unsworth, M. J. 2005, 169–176. Investigating Canada's lithosphere and earthquake hazards with portable arrays. *EOS Transactions of the American Geophysical Union*, **86**(17).

Eaton, D. W., Akram, J., St-Onge, A., and Forouhideh, F. 2011. Determining microseismic event locations by semblance-weighted stacking. In: *Proceedings of the CSPG CSEG CWLS Convention*.

Eaton, D. W., Davidsen, J., Pedersen, P. K., and Boroumand, N. 2014a. Breakdown of the Gutenberg–Richter relation for microearthquakes induced by hydraulic fracturing: influence of stratabound fractures. *Geophysical Prospecting*, **62**(4), 806–818.

Eaton, D. W., Rafiq, A., Pedersen, P., and Van der Baan, M. 2014b. Microseismic expression of natural fracture activation in a tight sand reservoir. Pages 19–22 of: *Proceedings of the 1st International Conference on Discrete Fracture Network Engineering*.

Eaton, D. W., Caffagni, E., Van der Baan, M., and Matthews, L. 2014c. Passive seismic monitoring and integrated geomechanical analysis of a tight-sand reservoir during hydraulic-fracture treatment, flowback and production. Pages 1537–1545 of: *Unconventional Resources Technology Conference (URTEC)*. Society of Exploration Geophysicists, American Association of Petroleum Geologists, Society of Petroleum Engineers.

Eaton, D. W., Van der Baan, M., Birkelo, B., and Tary, J.-B. 2014d. Scaling relations and spectral characteristics of tensile microseisms: Evidence for opening/closing cracks during hydraulic fracturing. *Geophysical Journal International*, **196**(3), 1844–1857.

Eaton, D. W., Cheadle, B., and Fox, A. 2016a. A causal link between overpressured hydrocarbon source rocks and seismicity induced by hydraulic fracturing. In: *SSA 2016 Annual Meeting*. Seismological Society of America.

Eaton, D. W., Van der Baan, M., and Ingelson, A. 2016b. Terminology for fluid-injection induced seismicity in oil and gas operations. *CSEG Recorder*, **41**(4), 24–28.

Eaton, J. P. 1992. Determination of amplitude and duration magnitudes and site residuals from short-period seismographs in Northern California. *Bulletin of the Seismological Society of America*, **82**(2), 533–579.

Eberhart-Phillips, D., and Oppenheimer, D. H. 1984. Induced seismicity in The Geysers geothermal area, California. *Journal of Geophysical Research: Solid Earth*, **89**(B2), 1191–1207.

Economides, M. J. and Nolte, K. G. 2000. *Reservoir Stimulation*. 3rd edn. Vol. 18. Wiley New York.

Edwards, B., Kraft, T., Cauzzi, C., Kästli, P., and Wiemer, S. 2015. Seismic monitoring and analysis of deep geothermal projects in St Gallen and Basel, Switzerland. *Geophysical Journal International*, **201**(2), 1020–1037.

Ehlig-Economides, C. A., and Economides, M. J. 2000. Formation Characterization: Well and Reservoir Testing. Pages 2–1 – 2–25 of: Economides, M. J. (ed.), *Reservoir Stimulation*, 3rd edn. John Wiley & Sons.

EIA. 2015. *World Shale Resource Assessments*. Tech. rept. Energy Information Agency.

Eisner, L., Abbott, D., Barker, W. B., Lakings, J., and Thornton, M. P. 2008. Noise suppression for detection and location of microseismic events using a matched filter. Pages 1431–1435 of: *SEG Technical Program Expanded Abstracts 2008*. Society of Exploration Geophysicists.

Eisner, L., Fischer, T., and Rutledge, J. T. 2009a. Determination of S-wave slowness from a linear array of borehole receivers. *Geophysical Journal International*, **176**(1), 31–39.

Eisner, L., Duncan, P. M., Heigl, W. M., and Keller, W. R. 2009b. Uncertainties in passive seismic monitoring. *The Leading Edge*, **28**(6), 648–655.

Eisner, L., Hulsey, B. J., Duncan, P., Jurick, D., Werner, H., and Keller, W. 2010. Comparison of surface and borehole locations of induced seismicity. *Geophysical Prospecting*, **58**(5), 809–820.

Eisner, L., Thornton, M., and Griffin, J. 2011a. Challenges for microseismic monitoring. Pages 1519–1523 of: *SEG Technical Program Expanded Abstracts 2011*. Society of Exploration Geophysicists.

Eisner, L., De La Pena, A., Wessels, S., Barker, W., and Heigl, W. 2011b. Why surface monitoring of microseismic events works. In: *Third EAGE Passive Seismic Workshop-Actively Passive 2011*.

Ejofodomi, E. A., Yates, M., Downie, R., Itibrout, T., and Catoi, O. A. 2010. Improving well completion via real-time microseismic monitoring: a west Texas case study. In: *Tight Gas Completions Conference*. Society of Petroleum Engineers.

Ekström, G., Nettles, M., and Dziewoński, A. M. 2012. The global CMT project 2004–2010: Centroid-moment tensors for 13,017 earthquakes. *Physics of the Earth and Planetary Interiors*, **200–201**, 1–9.

El-Isa, Z. H., and Eaton, D. W. 2014. Spatiotemporal variations in the b-value of earthquake magnitude–frequency distributions: Classification and causes. *Tectonophysics*, **615**, 1–11.

Ellsworth, W. L. 2013. Injection-induced earthquakes. *Science*, **341**(6142), 1225942.

EPA. 2004. *Evaluation of Impacts to Underground Sources of Drinking Water by Hydraulic Fracturing of Coalbed Methane Reservoirs*. Tech. rept. EPA 816-R-04-003. United States Environmental Protection Agency.

EPA. 2017. *Class II Oil and Gas Related Injection Wells*. Tech. rept. US Environmental Protection Agency, www.epa.gov/uic/class-ii-oil-and-gas-related-injection-wells.

Esmersoy, C., and Miller, D. 1989. Backprojection versus backpropagation in multidimensional linearized inversion. *Geophysics*, **54**(7), 921–926.

Esmersoy, C., Koster, K., Williams, M., Boyd, A., and Kane, M. 1994. Dipole shear anisotropy logging. Pages 1139–1142 of: *SEG Technical Program Expanded Abstracts 1994*. Society of Exploration Geophysicists.

Evans, K. F., Zappone, A., Kraft, T., Deichmann, N., and Moia, F. 2012. A survey of the induced seismic responses to fluid injection in geothermal and CO^2 reservoirs in Europe. *Geothermics*, **41**, 30–54.

Ewy, R. T. 1999. Wellbore-stability predictions by use of a modified Lade criterion. *SPE Drilling & Completion*, **14**(02), 85–91.

Farahbod, A. M., Kao, H., Cassidy, J. F., and Walker, D. 2015a. How did hydraulic-fracturing operations in the Horn River Basin change seismicity patterns in northeastern British Columbia, Canada? *The Leading Edge*, **34**(6), 658–663.

Farahbod, A. M., Kao, H., Walker, D. M., and Cassidy, J. F. 2015b. Investigation of regional seismicity before and after hydraulic fracturing in the Horn River Basin, northeast British Columbia. *Canadian Journal of Earth Sciences*, **52**(2), 112–122.

Fereidoni, A., and Cui, L. 2015. *Composite Alberta Seismicity Catalog*, www.inducedseismicity.ca/catalogues.

Feroz, A., and Van der Baan, M. 2013. Uncertainties in microseismic event locations for horizontal, vertical, and deviated boreholes. Pages 592–596 of: *SEG Technical Program Expanded Abstracts 2013*. Society of Exploration Geophysicists.

Fink, M. 1999. Time-reversed acoustics. *Scientific American*, **281**(5), 91–97.

Firdaouss, M., Guermond, J.-L., and Le Quéré, P. 1997. Nonlinear corrections to Darcy's law at low Reynolds numbers. *Journal of Fluid Mechanics*, **343**, 331–350.

Fischer, T., and A. Guest 2011. Shear and tensile earthquakes caused by fluid injection, *Geophys. Res. Lett.*, 38, L05307, doi:10.1029/2010GL045447.

Fischer, T., Horálek, J., Hrubcová, P., Vavryčuk, V., Bräuer, K., and Kämpf, H. 2014. Intra-continental earthquake swarms in West-Bohemia and Vogtland: a review. *Tectonophysics*, **611**, 1–27.

Fisher, M. K., and Warpinski, N. R. 2012. Hydraulic-fracture-height growth: Real data. *SPE Production & Operations*, **27**(1), 8–19.

Fisher, M. K., Wright, C. A., Davidson, B. M., Goodwin, A. K., Fielder, E. O., Buckler, W. S., and Steinsberger, N. P. 2002. Integrating fracture mapping technologies to optimize stimulations in the Barnett Shale. In: *SPE Annual Technical Conference and Exhibition*. San Antonio, Texas: Society of Petroleum Engineers.

Flumerfelt, R. 2015. Appraisal and development of the Midland Basin Wolfcamp Shale. *Houston Geological Society Bulletin*, **57**(7), 9–11.

Fossen, H. 2016. *Structural Geology*. 2nd edn. Cambridge University Press.

Fossen, H., Schultz, R. A., Shipton, Z. K., and Mair, K. 2007. Deformation bands in sandstone: a review. *Journal of the Geological Society*, **164**(4), 755–769.

Foulger, G. R., Julian, B. R., Hill, D. P., Pitt, A. M., Malin, P. E., and Shalev, E. 2004. Non-double-couple microearthquakes at Long Valley caldera, California, provide evidence for hydraulic fracturing. *Journal of Volcanology and Geothermal Research*, **132**(1), 45–71.

Fowler, C. M. R. 2004. *The Solid Earth: An Introduction to Global Geophysics*. 2nd edn. Cambridge University Press.

Freed, A. F. 2005. Earthquake triggering by static, dynamic, and postseismic stress transfer. *Annu. Rev. Earth Planet. Sci.*, **33**, 335–367.

Friberg, P. A., Besana-Ostman, G. M., and Dricker, I. 2014. Characterization of an earthquake sequence triggered by hydraulic fracturing in Harrison County, Ohio. *Seismological Research Letters*, **85**(6), 1295–1307.

Fritz, R. D., Medlock, P., Kuykendall, M. J., and Wilson, J. L. 2012. The geology of the Arbuckle Group in the midcontinent: sequence stratigraphy, reservoir development, and the potential for hydrocarbon exploration. Pages 203–273 of: Derby, J. R., Fritz, R. D., Longacre, S. A., Morgan, W. A., and Sternbach, C. A. (eds), *The Great American Carbonate Bank: The Geology and Economic Resources of the Cambrian–Ordovician Sauk megasequence of Laurentia*. AAPG Memoir, vol. 98. AAPG.

Gadde, P. B., Liu, Y., Norman, J., Bonnecaze, R., and Sharma, M. M. 2004. Modeling proppant settling in water-fracs. In: *SPE Annual Technical Conference and Exhibition*. Society of Petroleum Engineers.

Gajewski, D., and Tessmer, E. 2005. Reverse modelling for seismic event characterization. *Geophysical Journal International*, **163**(1), 276–284.

Garagash, D. I., and L. N. Germanovich 2012. Nucleation and arrest of dynamic slip on a pressurized fault, *J. Geophys. Res.*, 117, B10310, doi:10.1029/2012JB009209.

Garbin, H. D., and Knopoff, L. 1975. Elastic moduli of a medium with liquid-filled cracks. *Quarterly of Applied Mathematics*, **33**(3), 301–303.

Gassman, F. 1951. Uber die elastisitat poroser medien. *Naturforschenden Gesellschaft Vierteljahrschrift, Zurich*, **96**(1), 1–23.

Gaucher, E., Schoenball, M., Heidbach, O., Zang, A., Fokker, P., van Wees, J., and Kohl, T. 2015. Induced seismicity in geothermal reservoirs: a review of forecasting approaches. *Renewable and Sustainable Energy Reviews*, **52**, 1473–1490.

Geertsma, J., and De Klerk, F. 1969. A rapid method of predicting width and extent of hydraulically induced fractures. *Journal of Petroleum Technology*, **21**(12), 1571–1581.

Geiger, L. 1912. Probability method for the determination of earthquake epicenters from the arrival time only. *Bulletin of St. Louis University*, **8**(1), 56–71.

Gephart, J. W., and Forsyth, D. W. 1984. An improved method for determining the regional stress tensor using earthquake focal mechanism data: application to the San Fernando earthquake sequence. *Journal of Geophysical Research: Solid Earth*, **89**(B11), 9305–9320.

Ghofrani, H., and Atkinson, G. M. 2016. A preliminary statistical model for hydraulic fracture-induced seismicity in the Western Canada Sedimentary Basin. *Geophysical Research Letters*, **43**(19), 10, 164–10, 172.

Giardini, D. 2004. *Seismic hazard assessment of Switzerland, 2004*. Swiss Seismological Service: ETH.

Giardini, D. 2009. Geothermal quake risks must be faced. *Nature*, **462**(7275), 848–849.

Gibowicz, S. J., and Kijko, A. 1994. *An Introduction to Mining Seismology*. Vol. 55. Academic Press.

Gilbert, F. 1971. Excitation of the normal modes of the Earth by earthquake sources. *Geophysical Journal International*, **22**(2), 223–226.

Glover, K., Bozarth, T., Cui, A., and Wust, R. 2015. Lithological controls on mechanical anisotropy in shales to predict in situ stress magnitudes and potential for shearing of laminations during fracturing. In: *SPE/CSUR Unconventional Resources Conference*. Society of Petroleum Engineers.

Goertz-Allmann, B. P., Kühn, D., Oye, V., Bohloli, B., and Aker, E. 2014. Combining microseismic and geomechanical observations to interpret storage integrity at the In Salah CCS site. *Geophysical Journal International*, **198**(1), 447–461.

Grassberger, P., and Procaccia, I. 1983. Measuring the strangeness of strange attractors. *Physica D: Nonlinear Phenomena*, **9**(1–2), 189–208.

Grechka, V. 2010. Data-acquisition design for microseismic monitoring. *The Leading Edge*, **29**(3), 278–282.

Griffith, A. A. 1921. The phenomena of rupture and flow in solids. *Philosophical Transactions of the Royal Society of London, Series A*, **221**, 163–198.

Grigoli, F., Cesca, S., Priolo, E., Rinaldi, A. P., Clinton, J. F., Stabile, T. A., Dost, B., Fernandez, M. G., Wiemer, S., and Dahm, T. 2017. Current challenges in monitoring, discrimination, and management of induced seismicity related to underground industrial activities: a European perspective. *Reviews of Geophysics*.

Grob, M., and Van der Baan, M. 2011. Inferring in-situ stress changes by statistical analysis of microseismic event characteristics. *The Leading Edge*, **30**(11), 1296–1301.

Gruner, J. W. 1932. The crystal structure of kaolinite. *Zeitschrift für Kristallographie-Crystalline Materials*, **83**(1–6), 75–88.

Guglielmi, Y., Cappa, F., Avouac, J.-P., Henry, P., and Elsworth, D. 2015. Seismicity triggered by fluid injection–induced aseismic slip. *Science*, **348**(6240), 1224–1226.

Gulrajani, S. N., and Nolte, K. G. 2000. Fracture Evalution Using Pressure Diagnostics. Pages 9–1 – 9–63 of: Economides, M. J., and Nolte, K. G. (eds.), *Reservoir Stimulation*, 3rd edn. John Wiley & Sons.

Gutenberg, B. 1945a. Amplitudes of P, PP, and S and magnitude of shallow earthquakes. *Bulletin of the Seismological Society of America*, **35**(2), 57–69.

Gutenberg, B. 1945b. Amplitudes of surface waves and magnitudes of shallow earthquakes. *Bulletin of the Seismological Society of America*, **35**(1), 3–12.

Gutenberg, B., and Richter, C. F. 1944. Frequency of earthquakes in California. *Bulletin of the Seismological Society of America*, **34**(4), 185–188.

Hakala, M., Hudson, J. A., and Christiansson, R. 2003. Quality control of overcoring stress measurement data. *International Journal of Rock Mechanics and Mining Sciences*, **40**(7), 1141–1159.

Halleck, P. M. 2000. Appendix: Understanding perforator penetration and flow performance. Pages A11–1 – A11–12 of: Economides, M. J., and Nolte, K. G. (eds.), *Reservoir Stimulation*. John Wiley & Sons.

Hanks, T. C., and Kanamori, H. 1979. A moment magnitude scale. *Journal of Geophysical Research*, **84**(B5), 2348–2350.

Hansen, S. M., and Schmandt, B. S. 2015. Automated detection and location of microseismicity at Mount St. Helens with a large-N geophone array. *Geophysical Research Letters*, **42**(18), 7390–7397.

Hardebeck, J. L., and Hauksson, E. 2001. Stress orientations obtained from earthquake focal mechanisms: what are appropriate uncertainty estimates? *Bulletin of the Seismological Society of America*, **91**(2), 250–262.

Häring, M. O., Schanz, U., Ladner, F., and Dyer, B. C. 2008. Characterisation of the Basel 1 enhanced geothermal system. *Geothermics*, **37**(5), 469–495.

Harris, D. B. 2006. *Subspace Detectors: Theory*. US Department of Energy.

Hashin, Z., and Shtrikman, S. 1962. A variational approach to the theory of the elastic behaviour of polycrystals. *Journal of the Mechanics and Physics of Solids*, **10**(4), 343–352.

Hashin, Z., and Shtrikman, S. 1963. A variational approach to the theory of the elastic behaviour of multiphase materials. *Journal of the Mechanics and Physics of Solids*, **11**(2), 127–140.

Haskell, N. A. 1964. Total energy and energy spectral density of elastic wave radiation from propagating faults. *Bulletin of the Seismological Society of America*, **54**(6A), 1811–1841.

Hayes, B. J. R., Christopher, J. E., Rosenthal, L., Los, G., McKercher, B., Minken, D., Tremblay, Y. M., Fennell, J., and Smith, D. G. 1994. Cretaceous Mannville Group of the western Canada sedimentary basin. Pages 317–334 of: *Geological Atlas of the Western Canada sedimentary basin*, vol. 4. Canadian Society of Petroleum Geologists and Alberta Research Council.

Healy, J. H., Rubey, W. W., Griggs, D. T., and Raleigh, C. B. 1968. The Denver earthquakes. *Science*, **161**(3848), 1301–1310.

Heidbach, O., Tingay, T., Barth, A., Reinecker, J., Kurfeß, D., and Müller, B. 2010. Global crustal stress pattern based on the World Stress Map database release 2008. *Tectonophysics*, **482**(1–4), 3–15.

Helffrich, G., Wookey, J., and Bastow, I. 2013. *The Seismic Analysis Code: A Primer and User's Guide*. Cambridge University Press.

Helmstetter, A., D. Sornette, and J.-R. Grasso 2003. Mainshocks are aftershocks of conditional foreshocks: How do foreshock statistical properties emerge from aftershock laws, *J. Geophys. Res.*, 108, 2046, doi:10.1029/2002JB001991, B1.

Herrmann, R. B., Park, S.-K., and Wang, C.-Y. 1981. The Denver earthquakes of 1967–1968. *Bulletin of the Seismological Society of America*, **71**(3), 731–745.

Hill, R. 1963. Elastic properties of reinforced solids: some theoretical principles. *Journal of the Mechanics and Physics of Solids*, **11**(5), 357–372.

Hirata, T. 1989. A correlation between the b value and the fractal dimension of earthquakes. *Journal of Geophysical Research: Solid Earth*, **94**(B6), 7507–7514.

Hoek, E., and Brown, E. T. 1997. Practical estimates of rock mass strength. *International Journal of Rock Mechanics and Mining Sciences*, **34**(8), 1165–1186.

Hoek, E., Stagg, K. G., and Zienkiewicz, O. C. 1968. Brittle fracture of rock. Pages 99–124 of: *Rock Mechanics in Engineering Practice*. Wiley Series in Numerical Methods in Engineering Series. Wiley.

Hoek, E., Carranza-Torres, C., and Corkum, B. 2002. Hoek–Brown failure criterion – 2002 edition. *Proceedings of NARMS-TAC Conference*, **1**, 267–273.

Holland, A. A. 2013. Earthquakes triggered by hydraulic fracturing in south-central Oklahoma. *Bulletin of the Seismological Society of America*, **103**(3), 1784–1792.

Hornbach, M. J., DeShon, H. R., Ellsworth, W. L., Stump, B. W., Hayward, C., Frohlich, C., Oldham, H. R., Olson, J. E., Magnani, M. B., Brokaw, C., and Luetgert, J. H. 2015. Causal factors for seismicity near Azle, Texas. *Nature Communications*, **6**, 6728.

Hough, S. E. 2014. Shaking from injection-induced earthquakes in the central and eastern United States. *Bulletin of the Seismological Society of America*, **104**(5), 2619–2626.

Hudson, J. A. 1981. Wave speeds and attenuation of elastic waves in material containing cracks. *Geophysical Journal International*, **64**(1), 133–150.

Hudson, J. A., Pearce, R. G., and Rogers, R. M. 1989. Source type plot for inversion of the moment tensor. *Journal of Geophysical Research: Solid Earth*, **94**(B1), 765–774.

Hudson, J. A., Liu, E., and Crampin, S. 1996. The mechanical properties of materials with interconnected cracks and pores. *Geophysical Journal International*, **124**(1), 105–112.

Husen, S., Kissling, E., and von Deschwanden, A. 2013. Induced seismicity during the construction of the Gotthard Base Tunnel, Switzerland: hypocenter locations and source dimensions. *Journal of Seismology*, **17**(1), 63–81.

Hutton, L. K., and Boore, D. M. 1987. The ML scale in southern California. *Bulletin of the Seismological Society of America*, **77**(6), 2074–2094.

Ibs-von Seht, M., Plenefisch, T., and Klinge, K. 2008. Earthquake swarms in continental rifts—a comparison of selected cases in America, Africa and Europe. *Tectonophysics*, **452**(1), 66–77.

Igonin, N., and Eaton, D. 2017. A comparison of surface and near-surface acquisition techniques for induced seismicity and microseismic monitoring. In: *79th EAGE Conference and Exhibition*.

Inamdar, A. A., Ogundare, T. M., Malpani, R., Atwood, W. K., Brook, K., Erwemi, A. M., and Purcell, D. 2010. Evaluation of stimulation techniques using microseismic mapping in the Eagle Ford Shale. In: *Tight Gas Completions Conference*. San Antonio, Texas: Society of Petroleum Engineers.

Ingate, S. F., Husebye, E.S., and Christoffersson, A. 1985. Regional arrays and optimum data processing schemes. *Bulletin of the Seismological Society of America*, **75**(4), 1155–1177.

International Seismological Centre. 2014. *On-line Bulletin*, www.isc.ac.uk.

IRIS. 2017. *Background Page to Accompany the Animations on the Website: IRIS Animations*. www.iris.edu/hq/inclass/downloads/optional/261. Accessed: 2017/07/27.

Irving, J. D., Knoll, M. D., and Knight, R. J. 2007. Improving crosshole radar velocity tomograms: a new approach to incorporating high-angle traveltime data. *Geophysics*, **72**(4), J31–J41.

Irwin, G. R. 1948. Fracture dynamics. *Fracturing of Metals*, **152**.

Ishimoto, M., and Iida, K. 1939. Observations of earthquakes registered with the microseismograph constructed recently. *Bulletin of the Earthquake Research Institute*, **17**(443–478), 391.

on Improved Seismic Safety Provisions, BSSC Program, and Agency, United States. Federal Emergency Management. 1997. *NEHRP Recommended Provisions for Seismic Regulations for New Buildings and Other Structures: Provisions*. Vol. 302. FEMA.

Jaeger, J. C., Cook, N. G. W., and Zimmerman, R. 2009. *Fundamentals of Rock Mechanics*. 4th edn. Wiley-Blackwell.

Jain, A. K., Murty, M. N., and Flynn, P. J. 1999. Data clustering: a review. *ACM Computing Surveys (CSUR)*, **31**(3), 264–323.

Johnson, D. H., and Dudgeon, D. E. 1992. *Array Signal Processing: Concepts and Techniques*. Simon & Schuster.

Johnston, A. C., Coppersmith, K. J., Kanter, L. R., and Cornell, C. A. 1994. *The Earthquakes of Stable Continental Regions*. Tech. rept. TR-102261. Electric Power Research Insitute (EPRI).

Jones, A. G., Evans, R. L., and Eaton, D. W. 2009. Velocity–conductivity relationships for mantle mineral assemblages in Archean cratonic lithosphere based on a review of laboratory data and Hashin–Shtrikman extremal bounds. *Lithos*, **109**(1–2), 131–143.

Jones, G. A., Raymer, D., Chambers, K., and Kendall, J.-M. 2010. Improved microseismic event location by inclusion of a priori dip particle motion: a case study from Ekofisk. *Geophysical Prospecting*, **58**(5), 727–737.

Jones, G. A., Kendall, J.-M., Bastow, I. D., and Raymer, D. G. 2014. Locating microseismic events using borehole data. *Geophysical Prospecting*, **62**(1), 34–49.

Jones, L. M., and Molnar, P. 1979. Some characteristics of foreshocks and their possible relationship to earthquake prediction and premonitory slip on faults. *Journal of Geophysical Research: Solid Earth*, **84**(B7), 3596–3608.

Jost, M. L., and Herrmann, R. B. 1989. A student's guide to and review of moment tensors. *Seismological Research Letters*, **60**(2), 37–57.

Julian, B. R., Miller, A. D., and Foulger, G. R. 1998. Non-double-couple earthquakes 1. Theory. *Reviews of Geophysics*, **36**(4), 525–549.

Jurkevics, A. 1988. Polarization analysis of three-component array data. *Bulletin of the Seismological Society of America*, **78**(5), 1725–1743.

Kagan, Y. Y. 2002. Seismic moment distribution revisited: I. Statistical results. *Geophysical Journal International*, **148**(3), 520–541.

Kagan, Y. Y. 2010. Earthquake size distribution: power-law with exponent? *Tectonophysics*, **490**(1–2), 103–114.

Kalahara, K. W. 1996. Estimation of in-situ stress profiles from well-logs. In: *SPWLA 37th Annual Logging Symposium*. New Orleans, Louisiana: Society of Petrophysicists and Well-log Analysts.

Kanamori, H. 1977. The energy release in great earthquakes. *Journal of Geophysical Research*, **82**(20), 2981–2987.

Kanamori, H. 1983. Magnitude scale and quantification of earthquakes. *Tectonophysics*, **93**(3–4), 185–199.

Kanamori, H., and Anderson, D. L. 1975. Theoretical basis of some empirical relations in seismology. *Bulletin of the Seismological Society of America*, **65**(5), 1073–1095.

Kanamori, H., and Brodsky, E. E. 2004. The physics of earthquakes. *Reports on Progress in Physics*, **67**(8), 1429–1496.

Kao, H., and Shan, S.-J. 2004. The source-scanning algorithm: mapping the distribution of seismic sources in time and space. *Geophysical Journal International*, **157**(2), 589–594.

Kao, H., Eaton, D. W., Atkinson, G. M., Maxwell, S., and Mahani, A. B. 2016. *Technical Meeting on the Traffic Light Protocols (TLP) for Induced Seismicity: Summary and Recommendations*. Open File Report 8075. Geological Survey of Canada.

Kent, A. H., Eaton, D. W., and Maxwell, S. C. 2017. Microseismic response and geomechanical principles of short interval re-injection (SIR) treatments. In: *Unconventional Resources Technology Conference (URTEC)*.

Keranen, K. M., Weingarten, M., Abers, G. A., Bekins, B. A., and Ge, S. 2014. Sharp increase in central Oklahoma seismicity since 2008 induced by massive wastewater injection. *Science*, **345**(6195), 448–451.

Kern, L. R., Perkins, T. K., and Wyant, R. E. 1959. The mechanics of sand movement in fracturing. *Journal of Petroleum Technology*, **11**(7), 55–57.

Kim, W.-Y. 2013. Induced seismicity associated with fluid injection into a deep well in Youngstown, Ohio. *Journal of Geophysical Research: Solid Earth*, **118**(7), 3506–3518.

King, G. C. P., Stein, R. S., and Lin, J. 1994. Static stress changes and the triggering of earthquakes. *Bulletin of the Seismological Society of America*, **84**(3), 935–953.

King, G. E. 2012. Hydraulic fracturing 101: what every representative, environmentalist, regulator, reporter, investor, university researcher, neighbor and engineer should know about estimating frac risk and improving frac performance in unconventional gas and oil wells. In: *SPE Hydraulic Fracturing Technology Conference*. The Woodlands, Texas: Society of Petroleum Engineers.

King Hubbert, M. 1956. Darcy's law and the field equations of the flow of underground fluids. *AIME Petroleum Transactions*, **207**, 222–239.

King Hubbert, M., and Rubey, W. W. 1959. Role of fluid pressure in mechanics of overthrust faulting I. Mechanics of fluid-filled porous solids and its application to overthrust faulting. *Geological Society of America Bulletin*, **70**(2), 115–166.

Knapp, R. W., and Steeples, D. W. 1986. High-resolution common-depth-point seismic reflection profiling: Instrumentation. *Geophysics*, **51**(2), 276–282.

Knopoff, L., and Randall, M. J. 1970. The compensated linear-vector dipole: a possible mechanism for deep earthquakes. *Journal of Geophysical Research*, **75**(26), 4957–4963.

Kohli, A. H., and Zoback, M. D. 2013. Frictional properties of shale reservoir rocks. *Journal of Geophysical Research: Solid Earth*, **118**(9), 5109–5125.

Kratz, M., Hill, A., and Wessels, S. 2012. Identifying fault activation in unconventional reservoirs in real time using microseismic monitoring. In: *SPE/EAGE European Unconventional Resources Conference & Exhibition – From Potential to Production*.

Kratz, M., Teran, O., and Thornton, M. 2015. Use of automatic moment tensor inversion in real time microseismic imaging. Pages 1544–1549 of: *Unconventional Resources Technology Conference*. Society of Exploration Geophysicists, American Association of Petroleum Geologists, Society of Petroleum Engineers.

Krey, T., and Helbig, K. 1956. A theorem concerning anisotropy of stratified media and its significance for reflection seismics. *Geophysical Prospecting*, **4**(3), 294–302.

Kumar, D., and Ahmed, I. 2011. Seismic Noise. Pages 1157–1161 of: *Encyclopedia of Solid Earth Geophysics*. Springer.

Kwiatek, G., Bulut, F., Bohnhoff, M., and Dresen, G. 2014. High-resolution analysis of seismicity induced at Berlín geothermal field, El Salvador. *Geothermics*, **52**, 98–111.

Lakings, J. D., Duncan, P. M., Neale, C., and Theiner, T. 2006. Surface based microseismic monitoring of a hydraulic fracture well stimulation in the Barnett shale. Pages 605–608 of: *SEG Technical Program Expanded Abstracts 2006*. Society of Exploration Geophysicists.

Lamontagne, M., Lavoie, D., Ma, S., Burke, K. B. S., and Bastow, I. 2015. Monitoring the earthquake activity in an area with shale gas potential in southeastern New Brunswick, Canada. *Seismological Research Letters*, **86**(4), 1068–1077.

Langenbruch, C., and Zoback, M. D. 2016. How will induced seismicity in Oklahoma respond to decreased saltwater injection rates? *Science Advances*, **2**(11), e1601542.

Lavrov, A. 2016. Dynamics of stresses and fractures in reservoir and cap rock under production and injection. *Energy Procedia*, **86**, 381–390.

Lay, T., and Wallace, T. C. 1995. *Modern Global Seismology*. Vol. 58. Academic Press.

Lee, W. H. K., Jennings, P., Kisslinger, C., and Kanamori, H. 2002. *International Handbook of Earthquake & Engineering Seismology*. Academic Press.

Leeman, E. R. 1968. The determination of the complete state of stress in rock in a single borehole—laboratory and underground measurements. Pages 31–38 of: *International Journal of Rock Mechanics and Mining Sciences & Geomechanics Abstracts*, vol. 5. Elsevier.

Leonard, M. 2010. Earthquake fault scaling: self-consistent relating of rupture length, width, average displacement, and moment release. *Bulletin of the Seismological Society of America*, **100**(5A), 1971–1988.

Leonard, M. 2014. Self-consistent earthquake fault-scaling relations: update and extension to stable continental strike-slip faults. *Bulletin of the Seismological Society of America*, 2953–2965.

Levin, F. K. 1979. Seismic velocities in transversely isotropic media. *Geophysics*, **44**(5), 918–936.

Levin, S. A., Lewis, J., Hagelund, R., and Barrs, B. D. 2007. SEG-D for the next generation. *The Leading Edge*, **26**(7), 854–855.

Li, F., Rich, J., Marfurt, K. J., and Zhou, H. 2014. Automatic event detection on noisy microseismograms. Pages 2363–2367 of: *SEG Technical Program Expanded Abstracts 2014*. Society of Exploration Geophysicists.

Li, L. X., and Wang, T. J. 2005. A unified approach to predict overall properties of composite materials. *Materials Characterization*, **54**(1), 49–62.

Liang, C., Thornton, M. P., Morton, P., Hulsey, B. J., Hill, A., and Rawlins, P. 2009. Improving signal-to-noise ratio of passsive seismic data with an adaptive FK filter. Pages 1703–1707 of: *SEG Technical Program Expanded Abstracts 2009*. Society of Exploration Geophysicists.

Lindsay, R., and Van Koughnet, R. 2001. Sequential backus averaging: upscaling well logs to seismic wavelengths. *The Leading Edge*, **20**(2), 188–191.

Lomax, A., Michelini, A., and Curtis, A. 2014. Earthquake location, direct, global-search methods. Pages 1–33 of: *Encyclopedia of Complexity and Systems Science*. Springer.

Louis, L., Baud, P., and Wong, T.-F. 2007. Characterization of pore-space heterogeneity in sandstone by X-ray computed tomography. London: *Geological Society, Special Publications*, **284**(1), 127–146.

Lund, B., and Slunga, R. 1999. Stress tensor inversion using detailed microearthquake information and stability constraints: application to Ölfus in southwest Iceland. *Journal of Geophysical Research: Solid Earth*, **104**(B7), 14947–14964.

Luo, Y., Marhoon, M., Al Dossary, S., and Alfaraj, M. 2002. Edge-preserving smoothing and applications. *The Leading Edge*, **21**(2), 136–158.

Ma, S., and Eaton, D. W. 2009. Anatomy of a small earthquake swarm in southern Ontario, Canada. *Seismological Research Letters*, **80**(2), 214–223.

Ma, S., and Eaton, D. W. 2011. Combining double-difference relocation with regional depth-phase modelling to improve hypocentre accuracy. *Geophysical Journal International*, **185**(2), 871–889.

Mack, M. G., and Warpinski, N. R. 2000. Mechanics of Hydraulic Fracturing. Pages 6–1 – 4–49. of: Economides, M. J., and Nolte, K. G. (eds.), *Reservoir Stimulation*. John Wiley & Sons.

Macosko, C. W. 1994. *Rheology: Principles, Measurements, and Applications*. New York: VCH.

Madariaga, R. 1976. Dynamics of an expanding circular fault. *Bulletin of the Seismological Society of America*, **66**(3), 639–666.

Madariaga, R. 2007. Seismic Source Theory. Pages 59–82 of: Kanamori, H. (ed.), *Earthquake Seismology*. Treatise on Geophysics, no. 4. Elsevier.

Maghsoudi, S., Eaton, D. W., and Davidsen, J. 2016. Nontrivial clustering of microseismicity induced by hydraulic fracturing. *Geophysical Research Letters*, **43**(20), 10672–10679.

Mahani, A. B., Kao, H., Walker, D., Johnson, J., and Salas, C. 2016. *Regional Monitoring of Induced Seismicity in Northeastern British Columbia*. Tech. rept. Report 2016-1. Geoscience BC.

Mahani, A. B., Schultz, R., Kao, H., Walker, D., Johnson, J., and Salas, C. 2017. Fluid injection and seismic activity in the Northern Montney Play, British Columbia, Canada, with special reference to the 17 August 2015 Mw 4.6 induced earthquake. *Bulletin of the Seismological Society of America*, **107**(2), 542–552.

Majer, E., Nelson, J., Robertson-Tait, A., Savy, J., and Wong, I. 2012. *Protocol for addressing induced seismicity associated with enhanced geothermal systems*. Tech. rept. DOE/EE-0662. US Department of Energy.

Majer, E. L., and McEvilly, T. V. 1979. Seismological investigations at The Geysers geothermal field. *Geophysics*, **44**(2), 246–269.

Majer, E. L., Baria, R., Stark, M., Oates, S., Bommer, J., Smith, W., and Asanuma, H. 2007. Induced seismicity associated with enhanced geothermal systems. *Geothermics*, **36**(3), 185–222.

Marinos, V., Marinos, P., and Hoek, E. 2005. The geological strength index: applications and limitations. *Bulletin of Engineering Geology and the Environment*, **64**(1), 55–65.

Martakis, N., Kapotas, S., and Tselentis, G. 2006. Integrated passive seismic acquisition and methodology. Case Studies. *Geophysical Prospecting*, **54**(6), 829–847.

Martin, A. R., Cramer, D. D., Nunez, O., and Roberts, N. R. 2012. A method to perform multiple diagnostic fracture injection tests simultaneously in a single wellbore. In: *SPE Hydraulic Fracturing Technology Conference*. The Woodlands, Texas: Society of Petroleum Engineers.

Massé, R. P. 1981. Review of seismic source models for underground nuclear explosions. *Bulletin of the Seismological Society of America*, **71**(4), 1249–1268.

Masters, J. A. 1979. Deep basin gas trap, western Canada. *AAPG bulletin*, **63**(2), 152–181.

Maurer, H., Curtis, A., and Boerner, D. E. 2010. Recent advances in optimized geophysical survey design. *Geophysics*, **75**(5), 75A177–75A194.

Maxwell, S. C. 2009. Microseismic location uncertainty. *CSEG Recorder*, **34**(4), 41–46.

Maxwell, S. C. 2010. Microseismic: Growth born from success. *The Leading Edge*, **29**(3), 338–343.

Maxwell, S. C. 2014. *Microseismic Imaging of Hydraulic Fracturing: Improved Engineering of Unconventional Shale Reservoirs*. Distinguished Instructor Series. Society of Exploration Geophysicists.

Maxwell, S. C., and Cipolla, C. L. 2011. What does microseismicity tell us about hydraulic fracturing? In: *SPE Annual Technical Conference and Exhibition*. Society of Petroleum Engineers.

Maxwell, S. C., and Le Calvez, J. H. 2010. Horizontal vs. vertical borehole-based microseismic monitoring: which is better? In: *SPE Unconventional Gas Conference*. Society of Petroleum Engineers.

Maxwell, S. C., and Parker, R. 2012. Microseismic monitoring of ball drops during hydraulic fracturing using sliding sleeves. *CSEG Recorder*, **37**(8), 23–30.

Maxwell, S. C., and Urbancic, T. I. 2001. The role of passive microseismic monitoring in the instrumented oil field. *The Leading Edge*, **20**(6), 636–639.

Maxwell, S. C., Urbancic, T. I., Steinsberger, N., Zinno, R., et al. 2002. Microseismic imaging of hydraulic fracture complexity in the Barnett shale. In: *SPE Annual Technical Conference and Exhibition*. Society of Petroleum Engineers.

Maxwell, S. C., Jones, M., Parker, R., Miong, S., Leaney, S., Dorval, D., D'Amico, D., Logel, J., Anderson, E., and Hammermaster, K. 2009. Fault activation during hydraulic fracturing. Pages 1552–1556 of: *SEG Technical Program Expanded Abstracts 2009*. Society of Exploration Geophysicists.

Maxwell, S. C., Rutledge, J., Jones, R., and Fehler, M. 2010. Petroleum reservoir characterization using downhole microseismic monitoring. *Geophysics*, **75**(5), 75A129–75A137.

Maxwell, S. C., Raymer, D., Williams, M., and Primiero, P. 2012. Tracking microseismic signals from the reservoir to surface. *The Leading Edge*, **31**, 1301–1308.

Maxwell, S. C., Mack, M., Zhang, F., Chorney, D., Goodfellow, S. D., and Grob, M. 2015. Differentiating wet and dry microseismic events induced during hydraulic fracturing. Pages 1513–1524 of: *Unconventional Resources Technology Conference*. Society of Exploration Geophysicists, American Association of Petroleum Geologists, Society of Petroleum Engineers.

Mayerhofer, M. J., Lolon, E., Warpinski, N. R., Cipolla, C. L., Walser, D. W., and Rightmire, C. L. 2010. What is stimulated reservoir volume? *SPE Production & Operations*, **25**(01), 89–98.

McClain, W. C. 1971. *Seismic mapping of hydraulic fractures*. Tech. rept. ORNL-TM-3502. Oak Ridge National Laboratory.

McClure, M., and Horne, R. 2013. Is pure shear stimulation always the mechanism of stimulation in EGS. Pages 11–13 of: *Proceedings, Thirtyeight Workshop on Geothermal Reservoir Engineering*.

McFadden, P. L., Drummond, B. J., and Kravis, S. 1986. The Nth-root stack: theory, applications, and examples. *Geophysics*, **51**(10), 1879–1892.

McGarr, A. 1976. Seismic moments and volume changes. *Journal of Geophysical Research*, **81**(8), 1487–1494.

McGarr, A. 2014. Maximum magnitude earthquakes induced by fluid injection. *Journal of Geophysical Research: Solid Earth*, **119**(2), 1008–1019.

McGarr, A., Simpson, D., and Seeber, L. 2002. Case histories of induced and triggered seismicity. Pages 647–661 of: Lee, W. H., Kanamori, H., Jennings, P. C., and Kisslinger, C. (eds), *International Handbook of Earthquake & Engineering Seismology, Part A*. Academic Press.

McGillivray, P. 2005. Microseismic and time-lapse seismic monitoring of a heavy oil extraction process at Peace River, Canada. *CSEG Recorder*, **30**(1), 5–9.

McGuire, R. K. 2004. *Analysis of Seismic Hazard and Risk*. Oakland, California: Earthquake Engineering Research Center.

McGuire, R. K. 2008. Probabilistic seismic hazard analysis: early history. *Earthquake Engineering & Structural Dynamics*, **37**(3), 329–338.

McKenna, J. P. 2013. Magnitude-based calibrated discrete fracture network methodology. In: *SPE Annual Technical Conference and Exhibition*. Society of Petroleum Engineers.

McKenna, J. P. 2014. Where did the proppant go? In: *Unconventional Resources Technology Conference (URTEC)*.

McMechan, G. A. 1982. Determination of source parameters by wavefield extrapolation. *Geophysical Journal International*, **71**(3), 613–628.

Michael, A. J. 1984. Determination of stress from slip data: faults and folds. *Journal of Geophysical Research: Solid Earth*, **89**(B13), 11517–11526.

Miller, A. D., Julian, B. R., and Foulger, G. R. 1998. Three-dimensional seismic structure and moment tensors of non-double-couple earthquakes at the Hengill–Grensdalur volcanic complex, Iceland. *Geophysical Journal International*, **133**(2), 309–325.

Minson, S. E., and Dreger, D. S. 2008. Stable inversions for complete moment tensors. *Geophysical Journal International*, **174**(2), 585–592.

Mitchum, R. M., Vail, P. R., and Thompson, S. 1977. Seismic stratigraphy and global changes of sea level: Part 2. The depositional sequence as a basic unit for stratigraphic analysis: Section 2. Application of seismic reflection configuration to stratigraphic interpretation. Pages 53–62 of: *Seismic Stratigraphy–Applications to Hydrocarbon Exploration*, vol. Memoir 26. AAPG.

Mogi, K. 1963. Some discussions on aftershocks, foreshocks and earthquake swarms: the fracture of a semi-infinite body caused by an inner stress origin and its relation to the earthquake phenomena. *Bulletin of the Earthquake Research Institute*, **41**, 615–658.

Molenaar, M. M., Hill, D., Webster, P., Fidan, E., and Birch, W. 2012. First downhole application of distributed acoustic sensing for hydraulic-fracturing monitoring and diagnostics. *SPE Drilling & Completion*, **27**(01), 32–38.

Montgomery, C. T., and Smith, M. B. 2010. Hydraulic fracturing: history of an enduring technology. *Journal of Petroleum Technology*, **62**(12), 26–40.

Moradi, S. 2016. *Time-Lapse Numerical Modeling for a Carbon Capture and Storage (CCS) Project in Alberta, Using a Poroelastic Velocity-Stress Staggered-Grid Finite-Difference Method*. Ph.D. thesis, University of Calgary.

Moriya, H., Niitsuma, H., and Baria, R. 2003. Multiplet-clustering analysis reveals structural details within the seismic cloud at the Soultz geothermal field, France. *Bulletin of the Seismological Society of America*, **93**(4), 1606–1620.

Muhuri, S. K., Dewers, T. A., Scott, T. E., and Reches, Z. 2003. Interseismic fault strengthening and earthquake-slip instability: friction or cohesion? *Geology*, **31**(10), 881–884.

Munjiza, A., Owen, D. R. J., and Bicanic, N. 1995. A combined finite-discrete element method in transient dynamics of fracturing solids. *Engineering Computations*, **12**(2), 145–174.

Musgrave, M. J. P. 2003. *Crystal Acoustics: Introduction to the Study of Elastic Waves and Vibrations in Crystals*. Acoustical Society of America.

References

Nagel, N., Sheibani, F., Lee, B., Agharazi, A., and Zhang, F. 2014. Fully-coupled numerical evaluations of multiwell completion schemes: the critical role of in-situ pressure changes and well configuration. In: *SPE Hydraulic Fracturing Technology Conference*. The Woodlands, Texas: Society of Petroleum Engineers.

Nagel, N. B., Garcia, X., Sanchez, M. A., and Lee, B. 2012. Understanding SRV: a numerical investigation of wet vs. dry microseismicity during hydraulic fracturing. In: *SPE Annual Technical Conference and Exhibition*. Society of Petroleum Engineers.

National Energy Board. 2016. *The Unconventional Gas Resources of Mississippian-Devonian Shales in the Liard Basin of British Columbia, the Northwest Territories and Yukon*. Tech. rept. MR-14. National Energy Board.

National Research Council. 2013. *Induced Seismicity Potential in Energy Technologies*. National Academies Press.

NCEDC. 2014. *Northern California Earthquake Data Center. UC Berkeley Seismological Laboratory. Dataset. doi:10.7932/NCEDC*.

Neidell, N. S., and Taner, M. T. 1971. Semblance and other coherency measures for multichannel data. *Geophysics*, **36**(3), 482–497.

Nettles, M., and Ekström, G. 1998. Faulting mechanism of anomalous earthquakes near Bardarbunga Volcano, Iceland. *Journal of Geophysical Research: Solid Earth*, **103**(B8), 17973–17983.

Newman, M. E. J. 2005. Power laws, Pareto distributions and Zipf's law. *Contemporary Physics*, **46**(5), 323–351.

Nguyen, D. H., and Cramer, D. D. 2013. Diagnostic fracture injection testing tactics in unconventional reservoirs. In: *SPE Hydraulic Fracturing Technology Conference*. The Woodlands, Texas: Society of Petroleum Engineers.

Nicholson, C., and Wesson, R. L. 1992. Triggered earthquakes and deep well activities. *Pure and Applied Geophysics*, **139**(3–4), 561–578.

Nordgren, R. P. 1972. Propagation of a vertical hydraulic fracture. *Society of Petroleum Engineers Journal*, **12**(04), 306–314.

Norris, M. W., and Faichney, A. K. 2002. *SEG Y rev 1 Data Exchange format*. Tech. rept. Society of Exploration Geophysicists.

Odling, N. E., Gillespie, P., Bourgine, B., Castaing, C., Chiles, J.-P., Christensen, n.p., Fillion, E., Genter, A., Olsen, C., Thrane, L., Trice, R., Aarseth, E., Walsh, J.J., and Watterson, J. 1999. Variations in fracture system geometry and their implications for fluid flow in fractured hydrocarbon reservoirs. *Petroleum Geoscience*, **5**(4), 373–384.

Ogata, Y., Matsu'ura, R. S., and Katsura, K. 1993. Fast likelihood computation of epidemic type aftershock-sequence model. *Geophysical Research Letters*, **20**(19), 2143–2146.

Okada, Y. 1992. Internal deformation due to shear and tensile faults in a half-space. *Bulletin of the Seismological Society of America*, **82**(2), 1018–1040.

Oklahoma Produced Water Working Group. 2017. *Oklahoma Water for 2060: Produced Water Reuse and Recycling*. Tech. rept. Oklahoma Water Resources Board.

Ong, O. N., Schmitt, D. R., Kofman, R. S., and Haug, K. 2016. Static and dynamic pressure sensitivity anisotropy of a calcareous shale. *Geophysical Prospecting*, **64**(4), 875–897.

Oppenheim, A. V., and Schafer, R. W. 1975. *Digital Signal Processing*. Prentice-Hall.

Oprsal, I., and Eisner, L. 2014. Cross-correlation—an objective tool to indicate induced seismicity. *Geophysical Journal International*, **196**(3), 1536–1543.

Oye, V., and Roth, M. 2003. Automated seismic event location for hydrocarbon reservoirs. *Computers & Geosciences*, **29**(7), 851–863.

Pandolfi, D., Rebel-Schissele, E., Chambefort, M., and Bardainne, T. 2013. New design and advanced processing for frac jobs monitoring. In: *4th EAGE Passive Seismic Workshop*.

Pao, Y.-H., and Varatharajulu, V. 1976. Huygens' principle, radiation conditions, and integral formulas for the scattering of elastic waves. *Journal of the Acoustical Society of America*, **59**, 1361–1371.

Pap, A. 1983. *Source and receiver arrays*. Tech. rept. Amoco Research.

Park, C. B., Miller, R. D., and Xia, J. 1999. Multichannel analysis of surface waves. *Geophysics*, **64**(3), 800–808.

Parkes, G., and Hegna, S. 2011. A marine seismic acquisition system that provides a full "ghost-free" solution. Pages 37–41 of: *SEG Technical Program Expanded Abstracts 2011*. Society of Exploration Geophysicists.

Parotidis, M., Shapiro, S. A., and Rothert, E. 2004. Back front of seismicity induced after termination of borehole fluid injection. *Geophysical Research Letters*, **31**(2).

Parry, R. H. G. 2004. *Mohr Circles, Stress Paths and Geotechnics*. 2nd edn. London: Spon Press.

Passarelli, L., Maccaferri, F., Rivalta, E., Dahm, T., and Boku, E. A. 2013. A probabilistic approach for the classification of earthquakes as "triggered" or "not triggered." *Journal of Seismology*, **17**(1), 165–187.

Pawlak, A., Eaton, D. W., Bastow, I. D., Kendall, J., Helffrich, G., Wookey, J., and Snyder, D. 2011. Crustal structure beneath Hudson Bay from ambient-noise tomography: Implications for basin formation. *Geophysical Journal International*, **184**(1), 65–82.

Perkins, T. K., and Kern, L. R. 1961. Widths of hydraulic fractures. *Journal of Petroleum Technology*, **13**(09), 937–949.

Pesicek, J. D., Child, D., Artman, B., and Cieślik, K. 2014. Picking versus stacking in a modern microearthquake location: comparison of results from a surface passive seismic monitoring array in Oklahoma. *Geophysics*, **79**(6), KS61–KS68.

Peters, D. C., and Crosson, R. S. 1972. Application of prediction analysis to hypocenter determination using a local array. *Bulletin of the Seismological Society of America*, **62**(3), 775–788.

Petersen, M. D., Mueller, C. S., Moschetti, M. P., Hoover, S. M., Llenos, A. L., Ellsworth, W. L., Michael, A. J., Rubinstein, J. L., McGarr, A. F., and Rukstales, K. S. 2016. *2016 one-year seismic hazard forecast for the Central and Eastern United States from induced and natural earthquakes*. USGS Numbered Series 2016-1035. Reston, Virginia: US Geological Survey. IP-073237.

Petersen, M. D., Mueller, C. S., Moschetti, M. P., Hoover, S. M., Shumway, A. M., McNamara, D. E., Williams, R. A., Llenos, A. L., Ellsworth, W. L., Michael, A. J., Rubinstein, J. L., McGarr, A. F., and Rukstales, K. S. 2017. 2017 one-year seismic-hazard forecast for the Central and Eastern United States from induced and natural earthquakes. *Seismological Research Letters*, **88**(3), 772–783.

Peterson, J. 1993. *Observations and Modeling of Seismic Background Noise*. Tech. rept. OFR 93-322. USGS.

Peyret, O., Drew, J., Mack, M., Brook, K., Cipolla, C., and Maxwell, S. C. 2012. Subsurface to surface microseismic monitoring for hydraulic fracturing. In: *SPE Paper 159670*.

Pike, K. A. 2014. *Microseismic Data Processing, Modeling and Interpretation in the Presence of Coals: A Falher Member Case Study*. Ph.D. thesis, University of Calgary.

Pinder, G. F., and Gray, W. G. 2008. *Essentials of Multiphase Flow in Porous Media*. John Wiley & Sons.

Postma, G. W. 1955. Wave propagation in a stratified medium. *Geophysics*, **20**(4), 780–806.

Potocki, D. J. 2012. Understanding induced fracture complexity in different geological settings using DFIT net fracture pressure. In: *SPE Canadian Unconventional Resources Conference*. Calgary, Alberta: Society of Petroleum Engineers.

Power, D. V., Schuster, C. L., Hay, R., and Twombly, J. 1976. Detection of hydraulic fracture orientation and dimensions in cased wells. *Journal of Petroleum Technology*, **28**(09), 1–116.

Press, W. H., Teukolsky, S. A., Vetterling, W. T., and Flannery, B. P. 2007. *Numerical Recipes: The Art of Scientific Computing*. New York: Cambridge University Press.

Pullan, S. E. 1990. Recommended standard for seismic(/radar) data files in the personal computer environment. *Geophysics*, **55**(9), 1260–1271.

Putnis, A. 1992. *An Introduction to Mineral Sciences*. Cambridge University Press.

Rabinowitz, N., and Steinberg, D. M. 1990. Optimal configuration of a seismographic network: a statistical approach. *Bulletin of the Seismological Society of America*, **80**(1), 187–196.

Rafiq, A., Eaton, D. W., McDougall, A., and Pedersen, P. K. 2016. Reservoir characterization using microseismic facies analysis integrated with surface seismic attributes. *Interpretation*, **4**(2), T167–T181.

Raleigh, C. B., Healy, J. H., and Bredehoeft, J. D. 1976. An experiment in earthquake control at Rangely, Colorado. *Science*, **191**(4233), 1230–1237.

Ray, B., Lewis, C., Martysevich, V., Shetty, D. A., Walters, H. G., Bai, J., and Ma, J. 2017. An investigation into proppant dynamics in hydraulic fracturing. In: *SPE Hydraulic Fracturing Technology Conference and Exhibition*. Society of Petroleum Engineers.

Reinecker, J., Stephansson, O., and Zang, A. 2016. Guidelines for the analysis of overcoring data. Pages 43–48 of: *WSM Scientific Technical Report*. World Stress Map Project.

Reiter, L. 1991. *Earthquake Hazard Analysis: Issues and Insights*. Columbia University Press.

Reynolds, M. M., Thomson, S., Quirk, D. J., Dannish, M. B., Peyman, F., and Hung, A. 2012. A direct comparison of hydraulic fracture geometry and well performance between cemented liner and openhole packer completed horizontal wells in a tight gas reservoir. In: *SPE Hydraulic Fracturing Technology Conference*. The Woodlands, Texas: Society of Petroleum Engineers.

Rich, J., and Ammerman, M. 2010. Unconventional geophysics for unconventional plays. In: *SPE Unconventional Gas Conference*. Pittsburgh, Pennsylvania: Society of Petroleum Engineers.

Richter, C. F. 1935. An instrumental earthquake magnitude scale. *Bulletin of the Seismological Society of America*, **25**(1), 1–32.

Rio, P., Mukerji, T., Mavko, G., and Marion, D. 1996. Velocity dispersion and upscaling in a laboratory-simulated VSP. *Geophysics*, **61**(2), 584–593.

Robein, E., Cerda, F., Drapeau, D., Maurel, L., Gaucher, E., and Auger, E. 2009. Multi-network microseismic monitoring of fracturing jobs–Neuquen TGR application. In: *71st EAGE Conference and Exhibition incorporating SPE EUROPEC 2009*.

Roberts, A. 2001. Curvature attributes and their application to 3D interpreted horizons. *First Break*, **19**(2), 85–100.

Roberts, R. G., Christoffersson, A., and Cassidy, F. 1989. Real-time event detection, phase identification and source location estimation using single station three-component seismic data. *Geophysical Journal International*, **97**(3), 471–480.

Roche, V., and Van der Baan, M. 2015. The role of lithological layering and pore pressure on fluid-induced microseismicity. *Journal of Geophysical Research: Solid Earth*, **120**(2), 923–943.

Rodinov, Y., Parker, R., Jones, M., Chen, Z., Maxwell, S., and Matthews, L. 2012. Optimization of stimulation strategies using real-time microseismic monitoring in Horn River Basin: GeoConvention 2012, CSEG. In: *Geoconvention 2012, Expanded Abstracts*.

Roland, E., and McGuire, J. J. 2009. Earthquake swarms on transform faults. *Geophysical Journal International*, **178**(3), 1677–1690.

Ross, D. J. K., and Bustin, R. M. 2008. Characterizing the shale gas resource potential of Devonian–Mississippian strata in the Western Canada sedimentary basin: application of an integrated formation evaluation. *AAPG Bulletin*, **92**(1), 87–125.

Rost, S., and Thomas, C. 2002. Array seismology: methods and applications. *Reviews of Geophysics*, **40**(3), 2–1–2–27.

Roux, P.-F., Kostadinovic, J., Bardainne, T., Rebel, E., Chmiel, M., Van Parys, M., Macault, R., and Pignot, L. 2014. Increasing the accuracy of microseismic monitoring using surface patch arrays and a novel processing approach. *First Break*, **32**(7), 95–101.

Rubin, A. M., and Ampuero, J.-P. 2005. Earthquake nucleation on (aging) rate and state faults. *Journal of Geophysical Research: Solid Earth*, **110**(B11).

Rubinstein, J. L., and Mahani, A. B. 2015. Myths and facts on wastewater injection, hydraulic fracturing, enhanced oil recovery, and induced seismicity. *Seismological Research Letters*.

Rudnicki, J. W. 1986. Fluid mass sources and point forces in linear elastic diffusive solids. *Mechanics of Materials*, **5**(4), 383–393.

Rüger, A. 1997. P-wave reflection coefficients for transversely isotropic models with vertical and horizontal axis of symmetry. *Geophysics*, **62**(3), 713–722.

Rutledge, J. T., and Phillips, W. S. 2003. Hydraulic stimulation of natural fractures as revealed by induced microearthquakes, Carthage Cotton Valley gas field, east Texas. *Geophysics*, **68**(2), 441–452.

Rutledge, J. T., Downie, R. C., Maxwell, S. C., and Drew, J. E. 2013. Geomechanics of hydraulic fracturing inferred from composite radiation patterns of microseismicity. In: *SPE Annual Technical Conference and Exhibition*. Society of Petroleum Engineers.

Rutqvist, J., Rinaldi, A. P., Cappa, F., and Moridis, G. J. 2013. Modeling of fault reactivation and induced seismicity during hydraulic fracturing of shale-gas reservoirs. *Journal of Petroleum Science and Engineering*, **107**, 31–44.

Saari, J. 1991. Automated phase picker and source location algorithm for local distances using a single three-component seismic station. *Tectonophysics*, **189**(1–4), 307–315.

Saikia, C. K. 1994. Modified frequency-wavenumber algorithm for regional seismograms using Filon's quadrature: modelling of Lg waves in eastern North America. *Geophysical Journal International*, **118**(1), 142–158.

Sato, H., Ono, K., Johnston, C. T., and Yamagishi, A. 2005. First-principles studies on the elastic constants of a 1:1 layered kaolinite mineral. *American Mineralogist*, **90**(11–12), 1824–1826.

Sattari, A. 2017. *Finite-Element Modelling of Fault Slip*. Ph.D. thesis, University of Calgary.

Sayers, C. M., and Kachanov, M. 1995. Microcrack-induced elastic wave anisotropy of brittle rocks. *Journal of Geophysical Research: Solid Earth*, **100**(B3), 4149–4156.

Schimmel, M., and Paulssen, H. 1997. Noise reduction and detection of weak, coherent signals through phase-weighted stacks. *Geophysical Journal International*, **130**(2), 497–505.

Schisselé, E., and Meunier, J. 2009. Detection of micro-seismic events using a surface receiver network. In: *EAGE Workshop on Passive Seismic*.

Schlumberger. 2017. *Schlumberger Oilfield Glossary*. www.glossary.oilfield.slb.com Accessed: 2017/07/27.

Schmitt, D. R., and Zoback, M. D. 1993. Infiltration effects in the tensile rupture of thin walled cylinders of glass and granite: implications for the hydraulic fracturing breakdown equation. Pages 289–303 of: *International Journal of Rock Mechanics and Mining Sciences & Geomechanics Abstracts*, vol. 30. Elsevier.

Schmitt, D. R., Currie, C. A., and Zhang, L. 2012. Crustal stress determination from boreholes and rock cores: fundamental principles. *Tectonophysics*, **580**, 1–26.

Schneider, W. A. 1978. Integral formulation for migration in two and three dimensions. *Geophysics*, **43**, 49–76.

Schoenberg, M., and Helbig, K. 1997. Orthorhombic media: modeling elastic wave behavior in a vertically fractured earth. *Geophysics*, **62**(6), 1954–1974.

Schoenberg, M., and Sayers, C. M. 1995. Seismic anisotropy of fractured rock. *Geophysics*, **60**(1), 204–211.

Scholz, C., Molnar, P., and Johnson, T. 1972. Detailed studies of frictional sliding of granite and implications for the earthquake mechanism. *Journal of Geophysical Research*, **77**(32), 6392–6406.

Scholz, C. H. 1998. Earthquakes and friction laws. *Nature*, **391**(6662), 37–42.

Scholz, C. H. 2002. *The Mechanics of Earthquakes and Faulting*. 2nd edn. Cambridge University Press.

Schön, J. H. 2015. *Physical Properties of Rocks: Fundamentals and Principles of Petrophysics*. 2nd edn. Developments in Petroleum Science, Vol. 65. Elsevier.

Schorlemmer, D., and Woessner, J. 2008. Probability of detecting an earthquake. *Bulletin of the Seismological Society of America*, **98**(5), 2103–2117.

Schorlemmer, D., Wiemer, S., and Wyss, M. 2005. Variations in earthquake-size distribution across different stress regimes. *Nature*, **437**(7058), 539–542.

Schubarth, S., and Milton-Tayler, D. 2004. Investigating how proppant packs change under stress. In: *SPE Annual Technical Conference and Exhibition*. Society of Petroleum Engineers.

Schultz, R., and Stern, V. 2015. The Regional Alberta Observatory for Earthquake Studies Network (RAVEN). *CSEG Recorder*, **40**(8), 34–37.

Schultz, R., Stern, V., and Gu, Y. J. 2014. An investigation of seismicity clustered near the Cordel Field, west central Alberta, and its relation to a nearby disposal well. *Journal of Geophysical Research: Solid Earth*, **119**(4), 3410–3423.

Schultz, R., Mei, S., Paná, D., Stern, V., Gu, Y. J., Kim, A., and Eaton, D. W. 2015a. The Cardston earthquake swarm and hydraulic fracturing of the Exshaw Formation (Alberta Bakken play). *Bulletin of the Seismological Society of America*, **105**(6), 2871–2884.

Schultz, R., Stern, V., Novakovic, M., Atkinson, G. M., and Gu, Y. J. 2015b. Hydraulic fracturing and the Crooked Lake sequences: insights gleaned from regional seismic networks. *Geophysical Research Letters*, **42**(8), 2750–2758.

Schultz, R., Corlett, H., Haug, K., Kocon, K., MacCormack, K., Stern, V., and Shipman, T. 2016. Linking fossil reefs with earthquakes: geologic insight to where induced seismicity occurs in Alberta. *Geophysical Research Letters*, **43**, 2534–2542.

Schultz, R., Wang, R., Gu, Y. J., Haug, K., and Atkinson, G. M. 2017. A seismological overview of the induced earthquakes in the Duvernay play near Fox Creek, Alberta. *Journal of Geophysical Research: Solid Earth*, **122**(1), 492–505.

Seber, G. A. F. 2009. *Multivariate Observations*. Vol. 252. John Wiley & Sons.

Segall, P. 1989. Earthquakes triggered by fluid extraction. *Geology*, **17**(10), 942–946.

Segall, P., and Lu, S. 2015. Injection-induced seismicity: poroelastic and earthquake nucleation effects. *Journal of Geophysical Research: Solid Earth*, **120**(7), 5082–5103.

Segall, P., and Rice, J. R. 1995. Dilatancy, compaction, and slip instability of a fluid-infiltrated fault. *Journal of Geophysical Research: Solid Earth*, **100**(B11), 22155–22171.

Settari, A. 1985. A new general model of fluid loss in hydraulic fracturing. *Society of Petroleum Engineers Journal*, **25**(04), 491–501.

Shapiro, S. A. 2015. *Fluid-Induced Seismicity*. Cambridge University Press.

Shapiro, S. A., and Dinske, C. 2007. Violation of the Kaiser effect by hydraulic-fracturing-related microseismicity. *Journal of Geophysics and Engineering*, **4**(4), 378.

Shapiro, S. A., and Dinske, C. 2009. Fluid-induced seismicity: pressure diffusion and hydraulic fracturing. *Geophysical Prospecting*, **57**(2), 301–310.

Shapiro, S. A., Patzig, R., Rothert, E., and Rindschwentner, J. 2003. Triggering of seismicity by pore-pressure perturbations: permeability-related signatures of the phenomenon. *Pure and Applied Geophysics*, **160**(5–6), 1051–1066.

Shapiro, S. A., Dinske, C., Langenbruch, C., and Wenzel, F. 2010. Seismogenic index and magnitude probability of earthquakes induced during reservoir fluid stimulations. *The Leading Edge*, **29**(3), 304–309.

Shapiro, S. A., Krüger, O. S., Dinske, C., and Langenbruch, C. 2011. Magnitudes of induced earthquakes and geometric scales of fluid-stimulated rock volumes. *Geophysics*, **76**(6), WC55–WC63.

Shcherbakov, R., D. L. Turcotte, and J. B. Rundle 2004. A generalized Omori's law for earthquake aftershock decay, *Geophys. Res. Lett.*, 31, L11613, doi:10.1029/2004GL019808.

Shearer, P. M. 2009. *Introduction to Seismology*. 2 edn. Cambridge University Press.

Shemeta, J., and Anderson, P. 2010. It's a matter of size: magnitude and moment estimates for microseismic data. *The Leading Edge*, **29**(3), 296–302.

Sheng, P. 1990. Effective-medium theory of sedimentary rocks. *Physical Review B*, **41**(7), 4507–4512.

Sheriff, R.E. 1991. *Encyclopedic Dictionary of Exploration Geophysics*. 3rd edn. Society of Exploration Geophysicists.

Shimazaki, K., and Nakata, T. 1980. Time-predictable recurrence model for large earthquakes. *Geophysical Research Letters*, **7**(4), 279–282.

Shimizu, H., Ueki, S., and Koyama, J. 1987. A tensile–shear crack model for the mechanism of volcanic earthquakes. *Tectonophysics*, **144**(1–3), 287–300.

Sibson, R. H. 1992. Implications of fault-valve behaviour for rupture nucleation and recurrence. *Tectonophysics*, **211**(1–4), 283–293.

Sibson, R. H., Moore, J., and Rankin, A. H. 1975. Seismic pumping—a hydrothermal fluid transport mechanism. *Journal of the Geological Society*, **131**(6), 653–659.

Silver, P. G., and Chan, W. W. 1991. Shear wave splitting and subcontinental mantle deformation. *Journal of Geophysical Research: Solid Earth*, **96**(B10), 16429–16454.

Silver, P. G., and Jordan, T. H. 1982. Optimal estimation of scalar seismic moment. *Geophysical Journal International*, **70**(3), 755–787.

Simmons, G., and Wang, H. 1971. *Single crystal elastic constants and calculated aggregate properties*. 2nd edn. M.I.T. Press.

Simpson, D. W. 1986. Triggered earthquakes. *Annual Review of Earth and Planetary Sciences*, **14**(1), 21–42.

Simpson, D. W., and Richards, P. G. (eds). 1981. *Fluid Flow Accompanying Faulting: Field Evidence and Models*. AGU Maurice Ewing Series.

Simpson, D. W., Leith, W. S., and Scholz, C. H. 1988. Two types of reservoir-induced seismicity. *Bulletin of the Seismological Society of America*, **78**(6), 2025–2040.

Sjöberg, J., Christiansson, R., and Hudson, J. A. 2003. ISRM suggested methods for rock stress estimation—Part 2: overcoring methods. *International Journal of Rock Mechanics and Mining Sciences*, **40**(7–8), 999–1010.

Skoumal, R. J., Brudzinski, M. R., and Currie, B. S. 2015a. Distinguishing induced seismicity from natural seismicity in Ohio: demonstrating the utility of waveform template matching. *Journal of Geophysical Research: Solid Earth*, **120**(9), 6284–6296.

Skoumal, R. J., Brudzinski, M. R., and Currie, B. S. 2015b. Earthquakes induced by hydraulic fracturing in Poland Township, Ohio. *Bulletin of the Seismological Society of America*, **105**(1), 189–197.

Skoumal, R. J., Brudzinski, M. R., and Currie, B. S. 2016. An efficient repeating signal detector to investigate earthquake swarms. *Journal of Geophysical Research: Solid Earth*, **121**(8), 5880–5897.

Smith, M. B., and Montgomery, C. T. 2015. *Hydraulic Fracturing*. CRC Press.

Smith, M. B., and Shlyapobersky, J. W. 2000. Basics of hydraulic fracturing. Pages 5–1 – 5–28 of: Economides, M. J., and Nolte, K. G. (eds.), *Reservoir Stimulation*, vol. 18. Chichester: John Wiley & Sons Ltd.

Smith, M. L., and Dahlen, F. A. 1973. The azimuthal dependence of Love and Rayleigh wave propagation in a slightly anisotropic medium. *Journal of Geophysical Research*, **78**(17), 3321–3333.

Smith, R. J., Alinsangan, N. S., and Talebi, S. 2002. Microseismic response of well casing failures at a thermal heavy oil operation. In: *SPE/ISRM Rock Mechanics Conference*. Society of Petroleum Engineers.

Smith, T. M., Sondergeld, C. H., and Rai, C. S. 2003. Gassmann fluid substitutions: a tutorial. *Geophysics*, **68**(2), 430–440.

Sneddon, I. N., and Elliot, H. A. 1946. The opening of a Griffith crack under internal pressure. *Quarterly of Applied Mathematics*, **4**(3), 262–267.

Snelling, P., and Taylor, N. 2013. Optimization of a shallow microseismic array design for hydraulic fracture monitoring a horn river basin case study. *CSEG Recorder*, **38**(3), 22–25.

Snelling, P. E., de Groot, M., and Hwang, K. 2013. Characterizing hydraulic fracture behaviour in the Horn River Basin with microseismic data. Pages 4502–4507 of: *SEG Technical Program Expanded Abstracts 2013*. Society of Exploration Geophysicists.

Soltanzadeh, M., Fox, A., Rahim, N., Davies, G., and Hume, D. 2015. Application of mechanical and mineralogical rock properties to identify fracture fabrics in the Devonian Duvernay formation in Alberta. Pages 1668–1681 of: *Unconventional Resources Technology Conference, San Antonio, Texas, 20–22 July 2015*. Society of Exploration Geophysicists, American Association of Petroleum Geologists, Society of Petroleum Engineers.

Song, F., Warpinski, N. R., Toksöz, M. N., and Kuleli, H. S. 2014. Full-waveform based microseismic event detection and signal enhancement: an application of the subspace approach. *Geophysical Prospecting*, **62**(6), 1406–1431.

Speight, J. G. 2016. *Handbook of Hydraulic Fracturing*. John Wiley & Sons.

Spičák, A. 2000. Earthquake swarms and accompanying phenomena in intraplate regions: a review. *Studia Geophysica et Geodaetica*, **44**(2), 89–106.

St-Onge, A. 2011. Akaike information criterion applied to detecting first arrival times on microseismic data. Pages 1658–1662 of: *SEG Technical Program Expanded Abstracts 2011*. Society of Exploration Geophysicists.

St-Onge, A., and Eaton, D. W. 2011. Noise examples from two microseismic datasets. *CSEG Recorder*, **36**(10), 46–49.

Stanek, F., and Eisner, L. 2013. New model explaining inverted source mechanisms of microseismic events induced by hydraulic fracturing. Pages 2201–2205 of: *SEG Technical Program Expanded Abstracts 2013*. Society of Exploration Geophysicists.

Steffen, R., Eaton, D. W., and Wu, P. 2012. Moment tensors, state of stress and their relation to post-glacial rebound in northeastern Canada. *Geophysical Journal International*, **189**(3), 1741–1752.

Steffen, R., Wu, P., Steffen, H., and Eaton, D. W. 2014. On the implementation of faults in finite-element glacial isostatic adjustment models. *Computers & Geosciences*, **62**, 150–159.

Stein, S., and Wysession, M. 2009. *An Introduction to Seismology, Earthquakes, and Earth Structure*. John Wiley & Sons.

Stesky, R. M., Brace, W. F., Riley, D. K., and Robin, P.-Y. F. 1974. Friction in faulted rock at high temperature and pressure. *Tectonophysics*, **23**(1–2), 177–203.

Stork, A. L., Verdon, J. P., and Kendall, J.-M. 2014. The robustness of seismic moment and magnitudes estimated using spectral analysis. *Geophysical Prospecting*, **62**(4), 862–878.

Suckale, J. 2010. Moderate-to-large seismicity induced by hydrocarbon production. *The Leading Edge*, **29**(3), 310–319.

Surjaatmadja, J. B., Bezanson, J., Lindsay, S. D., Ventosilla, P. A., and Rispler, K. A. 2008. New hydra-jet tool demonstrates improved life for perforating and fracturing applications. In: *SPE/ICoTA Coiled Tubing and Well Intervention Conference and Exhibition*. Society of Petroleum Engineers.

Talwani, P. 1997. On the nature of reservoir-induced seismicity. *Pure and Applied Geophysics*, **150**(3–4), 473–492.

Tan, Y., and Engelder, T. 2016. Further testing of the bedding-plane-slip model for hydraulic-fracture opening using moment-tensor inversions. *Geophysics*, **81**(5), KS159–KS168.

Taner, M. T. 2001. Seismic attributes. *CSEG Recorder*, **26**(7), 48–56.

Tapley, W. C., and Tull, J. E. 1992. *SAC-Seismic Analysis Code: Users Manual*. Tech. rept. Lawrence Livermore National Laboratory.

Tary, J. B., Baan, M., and Eaton, D. W. 2014a. Interpretation of resonance frequencies recorded during hydraulic fracturing treatments. *Journal of Geophysical Research: Solid Earth*, **119**(2), 1295–1315.

Tary, J. B., Herrera, R. H., Han, J., and Van der Baan, M. 2014b. Spectral estimation—What is new? What is next? *Reviews of Geophysics*, **52**(4), 723–749.

Taylor, S. R., Arrowsmith, S. J., and Anderson, D. N. 2010. Detection of short time transients from spectrograms using scan statistics. *Bulletin of the Seismological Society of America*, **100**(5A), 1940–1951.

Telesca, L. 2010. Analysis of the cross-correlation between seismicity and water level in the Koyna area of India. *Bulletin of the Seismological Society of America*, **100**(5A), 2317–2321.

Tester, J. W., Anderson, B. J., Batchelor, A. S., Blackwell, D. D., DiPippo, R., Drake, E. M., Garnish, J., Livesay, B., Moore, M. C., Nichols, K., Petty, S., Toksoz, M. N., Veath, R. W., Roy, B., Augustine, C., Murphy, E., Negraru, P., and Richards, M. 2007. Impact of enhanced geothermal systems on US energy supply in the twenty-first century.

Philosophical Transactions of the Royal Society of London A: Mathematical, Physical and Engineering Sciences, **365**(1853), 1057–1094.

Thiercelin, M. C., and Roegiers, J.-C. 2000. Formation characterization: rock mechanics. Pages 3–1 – 3–35 of: Economides, M. J., and Nolte, K. G. (eds.), *Reservoir Stimulation*, 3rd edn. John Wiley & Sons.

Thomsen, L. 1986. Weak elastic anisotropy. *Geophysics*, **51**(10), 1954–1966.

Thornton, M. 2012. Resolution and location uncertainties in surface microseismic monitoring. In: *Geoconvention 2012*.

Thornton, M., and Mueller, M. 2013. Uncertainty in surface microseismic monitoring. In: *Geoconvention 2013*.

Tian, X., Zhang, W., and Zhang, J. 2017. Cross double-difference inversion for simultaneous velocity model update and microseismic event location. *Geophysical Prospecting*, **doi:10.1111/1365-2478.12556**.

Townend, J., and Zoback, M. D. 2000. How faulting keeps the crust strong. *Geology*, **28**(5), 399–402.

Trifunac, M. D. 1974. A three-dimensional dislocation model for the San Fernando, California, earthquake of February 9, 1971. *Bulletin of the Seismological Society of America*, **64**(1), 149–172.

Trojanowski, J., and Eisner, L. 2017. Comparison of migration-based location and detection methods for microseismic events. *Geophysical Prospecting*, **65**(1), 47–63.

Trutnevyte, E., and Wiemer, S. 2017. Tailor-made risk governance for induced seismicity of geothermal energy projects: an application to Switzerland. *Geothermics*, **65**, 295–312.

Tsvankin, I. 1997. Anisotropic parameters and P-wave velocity for orthorhombic media. *Geophysics*, **62**(4), 1292–1309.

Tsvankin, I., and Thomsen, L. 1994. Nonhyperbolic reflection moveout in anisotropic media. *Geophysics*, **59**(8), 1290–1304.

Unwin, A. T., and Hammond, P. S. 1995. Computer simulations of proppant transport in a hydraulic fracture. In: *SPE Western Regional Meeting*. Society of Petroleum Engineers.

Urbancic, T., and Wuestefeld, A. 2013. Black box recording of passive seismicity: pitfalls of not understanding your acquisition instrumentation and its limitations. In: *Geoconvention 2013*.

USGS. 2017. *Earthquake Glossary – USGS Earthquake Hazards Program.* https://earthquake.usgs.gov/learn/glossary Accessed: 2017/07/27.

Utsu, T. 1961. A statistical study on the occurrence of aftershocks. *Geophysical Magazine*, **30**(4), 521–605.

Valcke, S. L. A., Casey, M., Lloyd, G. E., Kendall, J.-M., and Fisher, Q. J. 2006. Lattice preferred orientation and seismic anisotropy in sedimentary rocks. *Geophysical Journal International*, **166**(2), 652–666.

Van der Baan, M., Eaton, D. W., and Dusseault, M. 2013. Microseismic monitoring developments in hydraulic fracture stimulation. Pages 439–466 of: *ISRM International Conference for Effective and Sustainable Hydraulic Fracturing*. International Society for Rock Mechanics.

Van der Baan, M., Eaton, D. W., and Preisig, G. 2016. Stick-split mechanism for anthropogenic fluid-induced tensile rock failure. *Geology*, **44**(7), 503–506.

Van der Elst, N. J., Savage, H. M., Keranen, K. M., and Abers, G. A. 2013. Enhanced remote earthquake triggering at fluid-injection sites in the midwestern United States. *Science*, **341**(6142), 164–167.

Van der Elst, N. J., Page, M. T., Weiser, D. A., Goebel, T. H. W., and Hosseini, S. M. 2016. Induced earthquake magnitudes are as large as (statistically) expected. *Journal of Geophysical Research: Solid Earth*, **121**(6), 4575–4590.

Van Domelen, M. S. 2017. A practical guide to modern diversion technology. In: *SPE Oklahoma City Oil and Gas Symposium*. Society of Petroleum Engineers.

Van Thienen-Visser, K., and Breunese, J. N. 2015. Induced seismicity of the Groningen gas field: history and recent developments. *The Leading Edge*, **34**(6), 664–671.

Van Trees, H. L. 1968. *Detection, Estimation, and Modulation Theory*. John Wiley & Sons.

Vasudevan, K., Eaton, D. W., and Davidsen, J. 2010. Intraplate seismicity in Canada: a graph theoretic approach to data analysis and interpretation. *Nonlinear Processes in Geophysics*, **17**(5), 513.

Vavryčuk, V. 2005. Focal mechanisms in anisotropic media. *Geophysical Journal International*, **161**(2), 334–346.

Vavryčuk, V. 2006. Calculation of the slowness vector from the ray vector in anisotropic media. Pages 883–896 of: *Proceedings of the Royal Society of London A: Mathematical, Physical and Engineering Sciences*, vol. 462. The Royal Society.

Vavryčuk, V. 2007. On the retrieval of moment tensors from borehole data. *Geophysical Prospecting*, **55**(3), 381–391.

Vavryčuk, V. 2011. Tensile earthquakes: Theory, modeling, and inversion, *J. Geophys. Res.*, 116, B12320, doi:10.1029/2011JB008770.

Vavryčuk, V. 2014. Iterative joint inversion for stress and fault orientations from focal mechanisms. *Geophysical Journal International*, **199**(1), 69–77.

Vavryčuk, V. 2015. Moment tensor decompositions revisited. *Journal of Seismology*, **19**(1), 231–252.

Vengosh, A., Jackson, R. B., Warner, N., Darrah, T. H., and Kondash, A. 2014. A critical review of the risks to water resources from unconventional shale gas development and hydraulic fracturing in the United States. *Environmental Science & Technology*, **48**(15), 8334–8348.

Verdon, J. P., and Wüstefeld, A. 2013. Measurement of the normal/tangential fracture compliance ratio (ZN/ZT) during hydraulic fracture stimulation using S-wave splitting data. *Geophysical Prospecting*, **61**(s1), 461–475.

Virieux, J. 1986. P-SV wave propagation in heterogeneous media: velocity-stress finite-difference method. *Geophysics*, **51**(4), 889–901.

Virues, C., Hendrick, J., and Kashikar, S. 2016. Development of limited discrete fracture network using surface microseismic event detection testing in Canadian Horn River Basin. Pages 1346–1361 of: *Unconventional Resources Technology Conference, San Antonio, Texas, 1–3 August 2016*. Society of Exploration Geophysicists, American Association of Petroleum Geologists, Society of Petroleum Engineers.

Vlček, J., Fischer, T., and Vilhelm, J. 2016. Back-projection stacking of P- and S-waves to determine location and focal mechanism of microseismic events recorded by a surface array. *Geophysical Prospecting*, **64**(6), 1428–1440.

Waldhauser, F. 2001. *HypoDD-A Program to Compute Double-Difference Hypocenter Locations*. Tech. rept. Open File Report 01-113. USGS.

Waldhauser, F., and Ellsworth, W. L. 2000. A double-difference earthquake location algorithm: method and application to the northern Hayward fault, California. *Bulletin of the Seismological Society of America*, **90**(6), 1353–1368.

Walker, R. N. 1997. Cotton Valley hydraulic fracture imaging project. In: *SPE Annual Technical Conference and Exhibition*. Society of Petroleum Engineers.

Wallace, R. E. 1951. Geometry of shearing stress and relation to faulting. *The Journal of Geology*, **59**(2), 118–130.

Walsh, F. R., and Zoback, M. D. 2015. Oklahoma's recent earthquakes and saltwater disposal. *Science Advances*, **1**(5), e1500195.

Walsh, F. R., and Zoback, M. D. 2016. Probabilistic assessment of potential fault slip related to injection-induced earthquakes: Application to north-central Oklahoma, USA. *Geology*, **44**(12), 991–994.

Walter, W. R., and Brune, J. N. 1993. Spectra of seismic radiation from a tensile crack. *Journal of Geophysical Research: Solid Earth*, **98**(B3), 4449–4459.

Walters, R. J., Zoback, M. D., Baker, J. W., and Beroza, G. C. 2015. Characterizing and responding to seismic risk associated with earthquakes potentially triggered by fluid disposal and hydraulic fracturing. *Seismological Research Letters*, **86**(4), 1110–1118.

Wang, H. F. 2017. *Theory of Linear Poroelasticity with Applications to Geomechanics and Hydrogeology*. Princeton University Press.

Wang, M., and Pan, N. 2008. Predictions of effective physical properties of complex multiphase materials. *Materials Science and Engineering: R: Reports*, **63**(1), 1–30.

Wang, R., Gu, Y. J., Schultz, R., Kim, A., and Atkinson, G. M. 2016. Source analysis of a potential hydraulic-fracturing-induced earthquake near Fox Creek, Alberta. *Geophysical Research Letters*, **43**, 564–573.

Wapenaar, K., and Fokkema, J. 2006. Green's function representations for seismic interferometry. *Geophysics*, **71**(4), SI33–SI46.

Warner, H. R. 2015. *The Reservoir Engineering Aspects of Waterflooding*. Society of Petroleum Engineers.

Warpinski, N. 2009. Microseismic monitoring: inside and out. *Journal of Petroleum Technology*, **61**(11), 80–85.

Warpinski, N., Kramm, R. C., Heinze, J. R., and Waltman, C. K. 2005a. Comparison of single- and dual-array microseismic mapping techniques in the Barnett Shale. In: *SPE Annual Technical Conference and Exhibition*. Society of Petroleum Engineers.

Warpinski, N. R., Sullivan, R. B., Uhl, J., Waltman, C., and Machovoie, S. 2005b. Improved microseismic fracture mapping using perforation timing measurements for velocity calibration. *SPE Journal*, **10**(01), 14–23.

Warpinski, N. R. 1989. Elastic and viscoelastic calculations of stresses in sedimentary basins. *SPE Formation Evaluation*, **4**(04), 522–530.

Warpinski, N. R., and Wolhart, S. 2016. A validation assessment of microseismic monitoring. In: *SPE Hydraulic Fracturing Technology Conference*. Society of Petroleum Engineers.

Warpinski, N. R., Branagan, P. T., Peterson, R. E., and Wolhart, S. L. 1998. An interpretation of M-site hydraulic fracture diagnostic results. In: *SPE Rocky Mountain Regional/Low-Permeability Reservoirs Symposium*. Society of Petroleum Engineers.

Warpinski, N. R., Mayerhofer, M. J., Vincent, M. C., Cipolla, C. L., and Lolon, E. P. 2009. Stimulating unconventional reservoirs: maximizing network growth while optimizing fracture conductivity. *Journal of Canadian Petroleum Technology*, **48**(10), 39–51.

Warpinski, N. R., Du, J., and Zimmer, U. 2012. Measurements of hydraulic-fracture-induced seismicity in gas shales. *SPE Production & Operations*, **27**(SPE-151597-PA), 240–252.

Warpinski, N. R., Mayerhofer, M. J., Davis, E. J., and Holley, E. H. 2014. Integrating fracture diagnostics for improved microseismic interpretation and stimulation modeling. Pages 1518–1536 of: *Unconventional Resources Technology Conference (URTEC)*.

Watt, J. P., Davies, G. F., and O'Connell, R. J. 1976. The elastic properties of composite materials. *Reviews of Geophysics*, **14**(4), 541–563.

Webb, S. C. 2002. Seismic noise on land and on the sea floor. Chap. 19 of: *International Handbook of Earthquake & Engineering Seismology, Part A*, vol. 81A. International Association of Seismology and Physics of the Earth's Interior.

Weingarten, M., Ge, S., Godt, J. W., Bekins, B. A., and Rubinstein, J. L. 2015. High-rate injection is associated with the increase in US mid-continent seismicity. *Science*, **348**(6241), 1336–1340.

Welzl, E. 1991. Smallest enclosing disks (balls and ellipsoids). *New Results and New Trends in Computer Science*, 359–370.

Weng, X., Kresse, O., Cohen, C.-E., Wu, R., and Gu, H. 2011. Modeling of hydraulic-fracture-network propagation in a naturally fractured formation. *SPE Production & Operations*, **26**(4), 368–380.

Westfall, P. H. 2014. Kurtosis as peakedness, 1905–2014. RIP. *The American Statistician*, **68**(3), 191–195.

White, A. J., Traugott, M. O., and Swarbrick, R. E. 2002. The use of leak-off tests as means of predicting minimum in-situ stress. *Petroleum Geoscience*, **8**(2), 189–193.

Wielandt, E. 2002. Seismometry. Chap. 18 of: *International Handbook of Earthquake & Engineering Seismology, Part A*, vol. 81A. International Association of Seismology and Physics of the Earth's Interior.

Wiemer, S., and Wyss, M. 2000. Minimum magnitude of completeness in earthquake catalogs: examples from Alaska, the western United States, and Japan. *Bulletin of the Seismological Society of America*, **90**(4), 859–869.

Winterstein, D. F. 1990. Velocity anisotropy terminology for geophysicists. *Geophysics*, **55**(8), 1070–1088.

Woessner, J., and Wiemer, S. 2005. Assessing the quality of earthquake catalogues: estimating the magnitude of completeness and its uncertainty. *Bulletin of the Seismological Society of America*, **95**(2), 684–698.

Wong, I., Nemser, E., Bott, J., and Dober, M. 2013. *White Paper: Induced Seismicity and Traffic Light Systems as Related to Hydraulic Fracturing in Ohio*. Tech. rept. Seismic Hazards Group.

Wyss, M., and Molnar, P. 1972. Efficiency, stress drop, apparent stress, effective stress, and frictional stress of Denver, Colorado, earthquakes. *Journal of Geophysical Research*, **77**(8), 1433–1438.

Xia, J., Miller, R. D., and Park, C. B. 1999. Estimation of near-surface shear-wave velocity by inversion of Rayleigh waves. *Geophysics*, **64**(3), 691–700.

Xue, J., Gu, H., and Cai, C. 2017. Model-based amplitude versus offset and azimuth inversion for estimating fracture parameters and fluid content. *Geophysics*, **82**(2), M1–M17.

Yenier, E. 2017. A local magnitude relation for earthquakes in the western Canada Sedimentary Basin. *Bulletin of the Seismological Society of America*, 1421–1431.

Yilmaz, Ö. 2001. *Seismic Data Analysis: Processing, Inversion, and Interpretation of Seismic Data*. Society of Exploration Geophysicists.

Yoon, C. E., O'Reilly, O., Bergen, K. J., and Beroza, G. C. 2015. Earthquake detection through computationally efficient similarity search. *Science Advances*, **1**(11), e1501057.

Yoon, J. S., Zang, A., and Stephansson, O. 2014. Numerical investigation on optimized stimulation of intact and naturally fractured deep geothermal reservoirs using hydro-mechanical coupled discrete particles joints model. *Geothermics*, **52**, 165–184.

Young, G. B., and Braile, L. W. 1976. A computer program for the application of Zoeppritz's amplitude equations and Knott's energy equations. *Bulletin of the Seismological Society of America*, **66**(6), 1881–1885.

Zecevic, M., Daniel, G., and Jurick, D. 2016. On the nature of long-period long-duration seismic events detected during hydraulic fracturing. *Geophysics*, **81**(3), KS113–KS121.

Zhang, G., Qu, C., Shan, X., Song, X., Zhang, G., Wang, C., Hu, J.-C., and Wang, R. 2011a. Slip distribution of the 2008 Wenchuan Ms 7.9 earthquake by joint inversion from GPS and InSAR measurements: a resolution test study. *Geophysical Journal International*, **186**(1), 207–220.

Zhang, Y., Eisner, L., Barker, W., Mueller, M., and Smith, K. 2011b. Consistent imaging of hydraulic fracture treatments from permanent arrays using a calibrated velocity model. In: *3rd EAGE Passive Seismic Workshop*.

Zhang, H., Thurber, C., and Rowe, C. 2003. Automatic P-wave arrival detection and picking with multiscale wavelet analysis for single-component recordings. *Bulletin of the Seismological Society of America*, **93**(5), 1904–1912.

Zhang, H., Sarkar, S., Toksöz, M. N., Kuleli, H. S., and Al-Kindy, F. 2009. Passive seismic tomography using induced seismicity at a petroleum field in Oman. *Geophysics*, **74**(6), WCB57–WCB69.

Zhang, H., Eaton, D. W., Li, G., Liu, Y., and Harrington, R. M. 2016. Discriminating induced seismicity from natural earthquakes using moment tensors and source spectra. *Journal of Geophysical Research: Solid Earth*, **121**(2), 972–993.

Zhang, M., and Wen, L. 2015. An effective method for small event detection: match and locate (M&L). *Geophysical Journal International*, **200**(3), 1523–1537.

Zhebel, O., and Eisner, L. 2014. Simultaneous microseismic event localization and source mechanism determination. *Geophysics*, **80**(1), KS1–KS9.

Zimmer, U. 2011. Microseismic design studies. *Geophysics*, **76**(6), WC17–WC25.

Zoback, M. D. 2010. *Reservoir Geomechanics*. Cambridge University Press.

Zoback, M. D., Mastin, L., and Barton, C. 1986. In-situ stress measurements in deep boreholes using hydraulic fracturing, wellbore breakouts, and stonely wave polarization. In: *ISRM International Symposium*. Stockholm: International Society for Rock Mechanics.

Zoback, M. D., Apel, R., Baumgärtner, J., Brudy, M., Emmermann, R., Engeser, B., Fuchs, K., Kessels, W., Rischmüller, H., Rummel, F., and Vernik, L. 1993. Upper-crustal strength inferred from stress measurements to 6 km depth in the KTB borehole. *Nature*, **365**, 633–635.

Zoback, M. D., Barton, C. A., Brudy, M., Castillo, D. A., Finkbeiner, T., Grollimund, B. R., Moos, D. B., Peska, P., Ward, C. D., and Wiprut, D. J. 2003. Determination of stress orientation and magnitude in deep wells. *International Journal of Rock Mechanics and Mining Sciences*, **40**(7–8), 1049–1076.

Zoback, M. L. 1992. First- and second-order patterns of stress in the lithosphere: The World Stress Map Project. *Journal of Geophysical Research: Solid Earth*, **97**(B8), 11703–11728.

Index

acausal pulse, 71
accelerometer, 147, 252
    force-balance, 147
accuracy, 182, 202
acoustic impedance, 65
acoustic wave speed in water, 166
aftershock, 85, 90
Akaike Information Criterion (AIC), 168
aleatoric uncertainty, 202, 212
aliasing, 139, 295
alternating tensor, 56
amplitude envelope, 298
analytical signal, 162, 298
Anderson fault classification, 36, 76, 102
anelastic, 29
anelastic media, 69
anisotropy, 11
    weak, 68
ansatz, 53
arrival-time picking, 165
asperity, 40, 102
attribute, 209
    validation, 210
attributes
    cluster, 216
    global, 216
autocorrelation, 170, 297
automatic gain control (AGC), 175
autoregressive process, 168

backfront, 248
background model, 143
bandlimited signal, 295
Bayesian inversion, 255
beachball diagram; see also focal mechanism, 78
beamforming, 174, 194
bi-wing fracture, 115
Biot coefficient, 22, 99
birefringence, 66
Boatwright source spectrum, 93
body waves, 53
Bond transformation, 9
borehole breakout, 98
bottom-hole pressure, 100
boxcar function, 293

Brune source spectrum, 93
bulk modulus, K, 10

calibration sources, 144
causal filter, 296
centroid moment tensor, 73
checkerboard tests, 151
Christoffel matrix, 54
class II injection well, 264
clustering
    agglomerative, 213
    divisive, 213
clustering analysis, 213
co-seismic rupture, 43
coda, 193
coefficient of Earth pressure at rest, 105
coefficient of internal friction, 37
coefficient of static friction, 40
coherency scanning, 199
cohesion, 37
comb function, 294
comminution, 31
compact support, 295
compensated linear-vector dipole (CLVD), 74
complex trace analysis, 298
compliance, 19
constitutive relations, 3
convex hull, 219
convolution, 85, 295
convolution theorem, 295
corner frequency, 86
Coulomb criterion, 36
creep, 46
critical angle, 63
critical stress state, 40
critically damped geophone, 146
critically stressed fault, 246
cross-correlation, 170
    pilot trace, 170
crossed-dipole sonic logging, 97
crystal-plastic deformation, 47
cumulative effects, 270

damage zone, 31
damped harmonic oscillator, 70

Index

dashpot, 47
data format
    SAC, 301
    SEED, 301
    SEG Y, 299
    SEG2-M, 300
DC-shift, 162
decibel (dB), 139
deconvolution, 296
deep-downhole array, 134
deformation
    brittle, 29
    ductile, 29
demultiplexed, 299
detection limit, 138
deviated well, 161
    survey, 162
diagnostic fracture-injection test, DFIT, 100
Dieterich–Ruina constitutive relationship, 42
diffraction, 68
diffraction stack, 199
diffusion equation, 23
dilatation, 10
Dirac δ function, 61, 293
direction cosine, 61
discrete fracture network (DFN), 117
dislocation creep, 47
dispersion, 55
    geometrical, 59
dispersion relation
    Love wave, 59
distributed acoustic sensing (DAS), 147
divergence operator, 22
divergence theorem, 52
diversion, 234
double-couple source, 75
double-couple sources, 227
drag folds, 46
drained fractures, 20
drilling-induced tensile fracture, DITF, 99
ductile deformation, 46
dyadic, 5

earthquake
    nucleation, 42
    scaling relations, 85
earthquake
    cycle, 43
earthquake swarms, 278
effective medium, 16
elastic medium, 3
elastic modulus, 8
elementary moment tensor, 75
engineered geothermal system (EGS), 259
envelope, 60
envelope stacking, 199
epicentral distance, Δ, 83

epicentre, 78
Epidemic Type Aftershock Sequence (ETAS), 91, 244
epistemic uncertainty, 212
equal distance time (EDT) method, 181
equations of motion, 52
estimated stimulated volume (ESV), 209, 218
Euler's identity, 53
even/odd complex symmetry, 293
event catalogue, 160
event detection, 165
extended-leakoff tests, XLOT, 101

fabric element, 30
far-field terms, 61
fast Fourier transform (FFT), 294
fault, 30
    dip, $\tilde{\delta}$, 76
    rake, $\tilde{\lambda}$, 76
    strike, $\tilde{\phi}$, 76
    active, 30
    auxiliary plane, 78
    damage zone, 30
    oblique slip, 77
    quiescent, 30
    seismogenic, 30
    thrust, 77
    velocity strengthening, 43
    velocity weakening, 43
fault zone, 30
fault-propagation folds, 46
fault-valve behaviour, 278
filter
    bandpass, 196
    poles, 296
    zeros, 296
fines migration, 109
finite-difference method, 230
finite-discrete element method (FDEM), 120
flexural waves, 97
flow regimes, 121, 124
flowback, 121, 229
fluid leakoff, 102
fluid viscosity, 20
focal depth, 78
focal sphere, 78
focal-mechanism diagram; see also beachball, 78
focal-mechanism inversion, 96
focus, 78
footwall, 76
foreshocks, 91
formation-breakdown pressure, F_B, 101
Fourier transform, 292
    short-time (STFT), 297
Fréchet-derivative matrix, 151
fracture, 29
    closed, 30
    dilatant, 29

Index

incipient, 117
network, 29
open, 30
self-propping, 109
sets, 29
toughness, 32
fracture closure pressure, FCP, 101
fracture modes, 30
fracture-propagation pressure, P_F, 101
free-surface boundary condition, 57
free-surface effect, 192
friction, 40
fundamental mode, 59

gain, 196
generalized-strain model, 105
Geological Strength Index (GSI), 39
geometrical spreading, 62
geophone, 146
  low frequency, 252
geophone polarity
  tap test, 165
Gibbs phenomenon, 166
gradient operator, 56
Green's function, 61
greenhouse gas (GHG) emissions, 107
Griffith failure criterion, 37
ground roll, 58
ground-motion intensity
  peak-ground acceleration (PGA), 275
  peak-ground velocity (PGV), 275
ground-motion prediction equation (GMPE), 153, 275
group velocity, v_g, 60
Gutenberg-Richter relationship, 88

hanging wall, 76
Hashin–Shtrikman extremal bounds, 17
Helmholtz potentials, 56
hermitian, 293
heterogeneity, 16
Hilbert transform, 298
hodogram analysis, 162
Hoek–Brown criterion, 38
Hooke's Law, 8
  generalized, 3
hoop stress, 98
  infiltration, 99
horizontal well
  elements, 107
Hudson diagram, 81
Huygens' principle, 197
hydraulic diffusivity tensor, 23
hydraulic fracture stages, 107
hydraulic fracturing, 95, 106
hydraulic stimulation, 260
hydraulic-fracturing breakdown equation, 102
hydrostatic pressure, 10

hypocentre, 78

ill-posed problem, 96
imaging condition, 200, 201
induced seismicity, 241
  activation potential, 272
  anomalous, 242
  case studies, 258
  fluid injection, 242
  hydraulic fracture, 268
  liability implications, 272
  monitoring, 250
  operatonal, 242
initial tensile strength, T_0, 102
injection energy, 125
instantaneous phase, 195, 298
instantaneous shut-in pressure, ISIP, 101
instrument response
  sensitivity, 296
instrument-response correction, 195
intermediate reference frame, 162
internal friction angle, 37
isochron surface, 198
isotropic material, 9
isotropic source (ISO), 74

joints, 29

Kaiser effect, 263, 266
Kelvin–Christoffel equations, 54
KGD fracture model, 117
kinetic-energy density, 62
Kirchhoff migration, 190
Kirsch equations, 97
Kronecker delta, 9
kurtosis, 167

Lagrangian reference frame, 7
Lamé parameters, 9
landing zone, 107
lattice-preferred orientation (LPO), 12, 283
leakoff coefficient (C_L), 116
linear array, 139
linear elastic fracture mechanics (LEFM), 31
lithosphere, 96
lithostatic stress, 14
Love wave, 58
  fundamental mode, 59
  higher-order modes, 59
lower-hemisphere project, 78
lumped-parameter models, 118

magnitude, 82
  body-wave, m_b, 83
  deterministic upper limit, 250
  local magnitude, M_L, 83
  maximum plausible, 250

saturation, 251
seismicity rate, 250
surface-wave, M_S, 83
magnitude of completness, 89
magnitude-distance scatter plots, 137
mainshock, 85
massive multi-stage hydraulic fracturing, MMHF, 106
matched filtering (MF), 174
matched-filtering, 214
mechanical Earth model, 125
microseismic facies analysis (MFA), 210
microseismic monitoring, 129
microseismic noise peak, 149
microseismic-depletion delineation (MDD), 235
microseismic-facies analysis, 224
minimum phase filter, 296
minimum-phase filter, 166
modified Lade criterion, 39
Mohr circle, 7
Mohr diagram, 6, 34
    2-D, 7
    3-D, 7
Mohr envelope, 34
Mohr–Coulomb failure criterion, 37
moment tensor, 72, 227
    principal strain axes, 74
most-positive curvature, 225
moveout, 166, 175, 194
Multichannel Analysis of Surface Waves (MASW), 58

near-field term, 61
net pressure, P_{net}, 101
Newtonian fluid, 21
nodal axes, 76
nodal plane, 76, 173
noise
    coherent, 130
    cultural, 148
    pink, 149
    random, 130
    white, 149
non-double-couple sources, 227
Nyquist frequency, 295

Omori's Law, 91
Omori's Law, generalized, 94
open-hole completion, 112
origin time, 254
overburden, 193
overcoring stress analysis, 102
overdetermined inverse problem, 255
overtones, 59

packer, 112
pancake fractures, 115
passive-seismic monitoring, 129
path effects, 295

penny-shaped cracks, 20
perf and plug completion, 144
periodic thin layer (PTL), 18
permeability, 20, 22
phase, 53
phase velocity, 60
phase-weighted stack (PWS), 195
physics-based probabilistic models, 244
PKN fracture model, 116
plane wave solution, 53
plane-strain approximation, 115
plastic, 46
Poisson distribution, 275
Poisson solid, 26, 57
Poisson's ratio, ν, 10
polarization analysis, 162
polarization ellipsoid, 164
polarization vector, 53
pore-pressure diffusion, 23
poroelastic medium, 21
potency tensor, 81
power law, 89
power-spectral density (PSD), 149
precision, 202
principal stresses, 5
probability-density function (PDF), 255
process zone, 31
proppant
    bridging, 121
pseudo-3-D models, 118

quality factor, Q, 70
quantitative risk assessment (QRA), 267

radiation patterns, 73
rate-state friction, 42
ray parameter, 63
refraction, 63
repeating signal detector, 253
reservoir conductivity, 102
residual pressure, P_R, 101
residual tensile strength, T_R, 101
resolution, 202
Reuss estimate, 16
reverse-time imaging (RTI), 190
Ricker wavelet, 199
right-hand rule, 76
rise time, 85
risk equation, 277
rock fabric, 9
root-mean-squared (RMS), 149
rupture time constant, 85

sagittal plane, 55
saltwater disposal, 264
sampling theorem, 295
scattering, 63

screen out, 111, 234
seismic hazard curve, 275
seismic moment (M_0), 75, 83
seismic source, 72
seismogenic index (Σ), 246, 267
seismometer, 146
    broadband, 251
self-organized criticality (SOC), 221
self-similarity, 88
semblance, 195
sensitivity kernel, 58
shape-preferred orientation (SPO), 12
shear-tensile source model, 38, 80
shear-wave singularity (in anisotropic media), 68
shear-wave splitting, 21, 55, 66
short-interval refracturing (SIR), 113
short-time average/long-time average (STA/LTA) method, 166
shut in, 101
sifting property, 293
signal
    fidelity, 134
sinc function, 86, 293
single-phase events, 173
single-station method, 254
singular-value decomposition (SVD), 151, 178
slickwater, 109
slip surface, 31
slowness law, 42
slowness vector, 53
Snell's Law, 63, 193
Sommerfeld radiation condition, 198
sonic log, 183
    dipole, 183
source-scanning algorithm (SSA), 190
source-type diagrams, 94
spectragram, 297
spectral balancing, 196
spectrum, 292
    amplitude, 292
    phase, 293
spherical divergence, 63
STA/LTA method, 252
stable-continental region (SCR), 88
stable-continental regions (SCRs), 141, 278
stacked cross-correlation function (SCCF), 175
stacking, 139
statics corrections, 196
stick-split failure mechanism, 232
stiffness matrix, isotropic, 27
stiffness tensor, 8
stimulated reservoir volume (SRV), 209, 216
STK/LTK method, 168
strain, 7
strain component
    normal, 8
    shear, 8
strain hardening, 47
strain weakening, 40, 47
strain-energy density, 62
stratabound fracture network, 30
stress
    effective, 34, 242
    normal, 5
    shear, 5
    uniaxial, 10
stress drop, 43, 87
stress polygon, 41
stress tensor, 4
stress-intensity factor, K_I, 32
strike-slip fault
    left lateral, 77
    right lateral, 77
string shots, 145
supercritical fluid, 27
surface waves, 53, 56
surface-ghost reflection, 232
swarm, 215

tensile strength, 38
tensile stress, 10
tensor, 4, 5
    order, 5
    summation convention, 5
time series, 292
tortuosity, 112
traction, 3
traffic light system (TLS), 272
transfer function, 146, 296
transverse isotropy
    horizontal (HTI), 68
    tilted (TTI), 68
    vertical (VTI), 68
transverse isotropy (TI), 14
    horizontal (HTI), 14
    vertical (VTI), 14
triaxial test, 34
trigger, 168
triggered seismicity, 242
triggering front, 235, 247
triplication, 68
tube waves, 165

ultracataclasite, 30
uncertainty, 202
unconfined compressive strength (UCS), 34, 35
uniaxial elastic strain (UES) model, 125
uniaxial-elastic-strain model, 105
unit cell, 11
Universal Coordinated Time (UTC), 213
UTM coordinates, 160

velocity

group, 53
phase, 53
velocity analysis, 196
velocity model, 143
vertical stress (S_V), 36
vespa process, 194
vespagram, 194
viscoelastic, 46
viscosity, 21
Voigt estimate, 16
Voigt notation, 8
Voigt–Reuss–Hill estimate, 17

Wallace–Bott, 96
waterflood, 265
wave equation, 56
wave potential
    scalar, 56
    vector, 56

wave surface, 66
waveform covariance matrix, 164
wavefront, 53
wavenumber, 139
wavevector, 53
welded-contact boundary condition, 53
welded-contact boundary conditions, 64
well completion, 107
well pad, 113
wireline depth, 162

yield point, 47
Young's modulus, E, 10

z-transform, 296
zipper frac, 114
Zoeppritz equations, 63
Zoeppritz graphs, 94